ENGINEERING METROLOGY AND MEASUREMENTS

N.V. RAGHAVENDRA
Professor & Head
Department of Mechanical Engineering
The National Institute of Engineering
Mysore

L. KRISHNAMURTHY
Professor
Department of Mechanical Engineering
The National Institute of Engineering
Mysore

OXFORD
UNIVERSITY PRESS

OXFORD
UNIVERSITY PRESS

Oxford University Press is a department of the University of Oxford.
It furthers the University's objective of excellence in research, scholarship,
and education by publishing worldwide. Oxford is a registered trade mark of
Oxford University Press in the UK and in certain other countries.

Published in India by
Oxford University Press
22 Workspace, 2nd Floor, 1/22 Asaf Ali Road, New Delhi 110 002

First published in 2013
13th impression 2024

ISBN-13: 978-0-19-808549-2
ISBN-10: 0-19-808549-4

Typeset in Times
by Trinity Designers & Typesetters, Chennai
Printed in India by Nutech Print Services India

For product information and current price, please visit www.india.oup.com

Dedicated to our revered guru and mentor,
Dr T.R. Seetharam

Preface

The origin of metrology can be traced to the Industrial Revolution, which began in Western Europe and the United States in the beginning of the 19th century. This period saw a transition from manual to mechanized production and the setting up of factories to manufacture iron and textiles. There was a paradigm shift from artisan-oriented production methods to mass production. An artisan produced an article the same way a storage shelf is built in a closet—by trial and error till the parts fit. Mass production called for division of labour and precise definition of production tasks. Tasks became specialized, requiring skilled people who worked on only a portion of the job, but completed it quickly and efficiently. The workers' wages were determined by a 'piece-rate' system. They were only paid for the good parts; thus it became necessary to define what a good part was. This led to the design of inspection gauges and the need for an inspector who could decide whether a part was good or not. In 1913, Henry Ford, an American idustrialist, perfected the assembly line system to produce cars. In order to ensure quality as well as high production rates, new methods of inspection and quality control were initiated, which perhaps formed the basis of modern metrology.

Engineering metrology deals with the applications of measurement science in manufacturing processes. It provides a means of assessing the suitability of measuring instruments, their calibration, and the quality control of manufactured components. A product that is not manufactured according to metrological specifications will have to incur heavy costs to comply with the specifications later. Any compromise in quality creates rapid negative sentiments in the market and the cost of recovering the original market position would be quite high. Today, metrological error has a far greater impact on cost than in the past. Hence, an organization should strive towards a *zero-defect* regime in order to survive in a highly competitive market. Ensuring this aspect of manufacturing is the responsibility of a quality control engineer, who must be completely familiar with the basics of measurement, standards and systems of measurement, tolerances, measuring instruments, and their limitations.

The science of mechanical measurements has its roots in physics. It is an independent domain of knowledge dealing with the measurement of various physical quantities such as pressure, temperature, force, and flow.

ABOUT THE BOOK

Engineering Metrology and Measurements is a core subject for mechanical, production, and allied disciplines in all the major universities in India. Although there are a few good books available on metrology, the coverage of topics on mechanical measurements is either scanty or superficial, necessitating students to refer to different books on mechanical measurements. This book provides a comprehensive coverage of both metrology and mechanical measurements.

Divided into three parts, the first part of the book comprising Chapters 1–11, begins with a comprehensive outline of the field of engineering metrology and its importance in mechanical engineering design and manufacturing. The basic concepts of engineering standards, limits, fits, and tolerances, for ensuring interchangeability of machine components are then discussed.

This is followed by a discussion on metrology of linear and angular measurements. Later in the book, comparators, followed by the metrology of gears, screw threads, and surface finish metrology are discussed. The chapter on miscellaneous metrology talks about laser-based instrumentation and coordinate measuring machines. The last chapter in this section features inspection methods and quality control.

The second part of the book comprising Chapters 12–16 focuses on mechanical measurements. The coverage is restricted to measurement techniques and systems that are complementary to engineering metrology. The topics covered are the basics of transducers and the measurement of force, torque, strain, temperature, and pressure.

The third part of the book comprising Chapter 17 details nanometrology techniques and instrumentation. Nanotechnology has opened a new world of scientific research and applications. India has also joined the bandwagon and today, we see a phenomenal investment in the research and development of this discipline, both in the government and private sectors. There is abundant scope for pursuing higher studies both in India and abroad. We hope this section on nanometrology will further stimulate the curiosity of the students and motivate them to take up higher studies in this new and interesting field.

The book is designed to meet the needs of undergraduate students of mechanical engineering and allied disciplines. The contents of this book have been chosen after careful perusal of the syllabi of the undergraduate (B.E./B. Tech) and diploma programmes in India. The topics are explained lucidly and are supported by self-explanatory illustrations and sketches. The following are a few key features of the book.

KEY FEATURES

- Covers both metrology and mechanical measurements in one volume
- Offers guidelines for the proper use and maintenance of important instruments, such as vernier callipers, autocollimators, slip gauges, and pyrometers
- Provides simple solved examples, numerical exercises in all relevant chapters, theoretical review questions, and multiple-choice questions with answers at the end of every chapter
- Introduces the principles of nanometrology, a topic that has emerged from the popular discipline of nanotechnology, in an exclusive chapter, highlighting its applications in the production processes
- Includes an appendix containing 20 laboratory experiments with comprehensive procedures, observation templates, and model characteristics, with select experiments presenting photographs of the actual instruments to gain a visual understanding of the equipment used

ONLINE RESOURCES

To aid the faculty and students using this book, the companion website of this book http://oupinheonline.com/book/raghavendra-engineering-metrology-measurements/9780198085492 provides the following resources:

For instructors
- A solutions manual for the numerical exercises given in the book

• A complete chapter-wise PowerPoint presentation to aid classroom teaching

For students
• Multiple-choice questions to test students' understanding of the subject

CONTENTS AND COVERAGE

The book is divided into three parts: Engineering Metrology (Chapters 1–11), Mechanical Measurements (Chapters 12–16), and Nano Impact on Metrology (Chapter 17). A chapter-wise scheme of the book is presented here.

Chapter 1 deals with the basic principles of engineering metrology. It gives an overview of the subject along with its importance. It also talks about general measurement, methods of measurement, errors associated with any measurement, and the types of errors.

Chapter 2 sets the standards of measurement. These standards acts as a reference point for the dimensional measurements.

Chapter 3 presents the limits, fits, and tolerances in design and manufacturing. An understanding of these concepts helps in the interchangeability of manufactured components.

Chapter 4 discusses linear measurements that form one of the most important constituents of metrology. The chapter throws light on surface plates and V-blocks, over which the measurand is inspected. It discusses the scaled, vernier, and micrometer instruments in detail. The chapter ends with a detailed explanation of slip gauges.

Chapter 5 elaborates on angular measurements. The fact that not all measurands can be measured by linear methods stresses the significance of this topic. This chapter deals with devices such as protractors, sine bars, angle gauges, spirit levels, and other optical instruments used for angular measurements.

Chapter 6 aids in the comprehension of comparators. In several instances, a measurement may be carried out on the basis of a comparison with the existing standards of measurements. This chapter discusses the instruments that work on this common principle.

Chapter 7 explains optical measurements and interferometry. Optical measurement provides a simple, accurate, and reliable means of carrying out inspection and measurements in the industry. This chapter gives insights into some of the important instruments and techniques that are widely used. Interferometers, which use laser as a source, are also discussed in detail.

Chapter 8 focuses on the metrological inspection of gears and screw threads. Gears are the main elements in a transmission system. Misalignment and gear runout will result in vibrations, chatter, noise, and loss of power. Therefore, one cannot understate the importance of precise measurement and inspection techniques for gears. Similarly, the geometric aspects of screw threads are quite complex and hence, thread gauging is an integral part of a unified thread gauging system.

Chapter 9 analyses the metrology of surface finish. Two apparently flat contacting surfaces are assumed to be in perfect contact throughout the area of contact. However, in reality, there are peaks and valleys between surface contacts. Since mechanical engineering is primarily concerned with machines and moving parts that are designed to precisely fit with each other, surface metrology has become an important topic in engineering metrology.

Chapter 10 comprises miscellaneous metrology, which details certain measurement principles and techniques that cannot be classified under any of the aforementioned dimensional measurements. Coordinate measuring machines (CMM), machine tool test alignment, automated inspection, and machine vision form the core of this chapter.

Chapter 11 lays emphasis on inspection and quality control. Inspection is the scientific examination of work parts to ensure adherence to dimensional accuracy, surface texture, and other related attributes. This chapter encompasses the basic functions of inspection and statistical quality control—*total quality management* (TQM) and *six sigma*—the customer-centric approaches towards achieving high quality of products, processes, and delivery.

Chapter 12 helps in understanding mechanical measurements. Mechanical measurements are (physical) quantity measurements unlike the dimensional measurements discussed in Chapters 1–11.

Chapter 13 explains the principle and working of transducers. Transducers are generally defined as devices that transform physical quantities in the form of input signals into corresponding electrical output signals. Since many of the measurement principles learnt in earlier chapters require a transducer to transmit the obtained signal into an electrical form, the study of transducers is inevitable.

Chapter 14 elucidates the physical quantities of measurement: force, torque, and strain.

Chapter 15 illustrates the concept of temperature measurements—the principles involved in temperature measurement and devices such as resistance temperature detector (RTD), thermocouple, liquid in glass thermometer, bimetallic strip thermometers, and pyrometers.

Chapter 16 defines yet another important physical quantity, pressure. It helps us in getting acquainted with instruments such as manometers, elastic transducers, and vacuum and high pressure measurement systems.

Chapter 17 helps us appreciate the applications of nanotechnology in metrology. It explains the basic principles of nanotechnology and its application in the manufacturing of nanoscale elements that are made to perfection.

Appendix A introduces the universal measuring machine.

Appendix B simplifies the theory of flow measurement. Although a broader subset of mechanical measurements, flow measurement is an independent field of study. Students are introduced to this field in a typical course on fluid mechanics. Here we have tried to present only the basics of flow measurement with a synopsis of measurement devices such as the orifice meter, venturi meter, pitot tube, and rotameter.

Appendix C comprises 20 laboratory experiments with photographs of some of the equipment used in measurement. The appendix also provides a step-by-step procedure to conduct the experiments and an observation of results.

Appendix D presents the control chart associated with statistical quality control. These values help understand certain problems discussed in Chapter 11.

ACKNOWLEDGEMENTS

We attribute our efforts for completing this book to Dr T.R. Seetharam and Dr G.L. Shekar, who have inspired us and shaped our careers. Dr. Seetharam, Professor (retired) in Mechanical Engineering and former Principal, National Institute of Engineering (NIE), Mysore, is an embodiment of scholarship and simplicity. He has motivated thousands of students, who are now in noteworthy positions in organizations all over the world. He mentored us during our formative years at the NIE and instilled in us the spirit to strive for excellence. Dr G.L. Shekar, the present Principal of NIE has been a friend, philosopher, and guide. He is a bundle of unlimited energy and has initiated a large number of research and industry-related projects at the NIE. We are happy to be associated with many of these projects, which have broadened our horizon of knowledge and provided a highly stimulating work environment.

We thank our college management, colleagues, and students, who encouraged us to work on this book. Special thanks to our esteemed colleagues, Dr B.K. Sridhara, Dr T.N. Shridhar, and Dr M.V. Achutha, for their valuable suggestions and continuous encouragement. We acknowledge the contributions of our former colleagues, Mr Manil Raj and Mr N.S. Prasad, in the preparation of the laboratory experiments provided as an appendix in the book. Special thanks to Mr K. Chandrashekar, Scientist B, Centre for Nanotechnology, NIE, for sourcing a large number of e-books on nanotechnology. Ms Pooja K., Software Engineer, Delphi Automotive Systems Pvt. Ltd, Bangalore, provided useful inputs for key chapters in Part II of the book and we thank her for the same.

We are extremely grateful to our families, who graciously accepted our inability to attend to family chores during the course of writing this book, and especially for their extended warmth and encouragement. Without their support, we would not have been able to venture into writing this book.

Last, but not the least, we express our heartfelt thanks to the editorial team at the Oxford University Press, who guided us through this project.

We eagerly look forward to your feedback. You can reach us by e-mail at raghu62.nie@gmail.com and kitty_nie@yahoo.co.in

N.V. Raghavendra
L. Krishnamurthy

Features of the Book

CHAPTER 1 Basic Principles of Engineering Metrology

CHAPTER 17 Nanometrology

Metrology and Mechanical Measurements
One of the very few textbooks that lay equal emphasis on both metrology and mechanical measurements.

Illustrative Schematics
The figures have been drawn in such a way so as to aid in the quick understanding of the concepts described in the ensuing text.

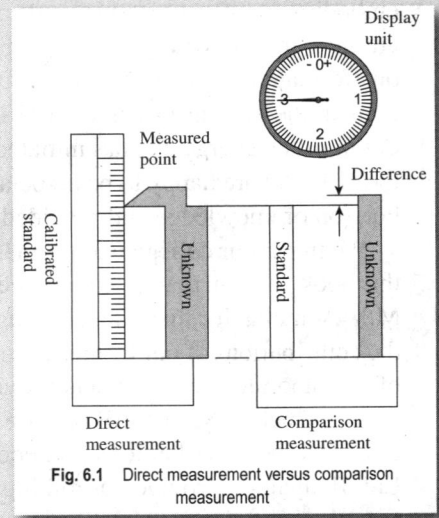

Fig. 6.1 Direct measurement versus comparison measurement

Example 2.2 A calibrated metre end bar, which in the calibration of two bars X and Y, eac with the metre bar, the sum of L_X and L_Y are compared, it is observed that X is 0.00 X and Y.

A QUICK OVERVIEW

• Mass production, an idea of the last industrial revolution, has become very popular and resent manufacturing

MULTIPLE-CHOICE QUESTIONS

1. The modern metre is
   (a) the length of the path travelled by light in vacuum during a time interval of 1/29,97,92,45

REVIEW QUESTIONS

1. Explain the role of standards of measurements in modern industry.
   volution of standards.

PROBLEMS

1. It is required to calibrate four length bars A, B, C, and D, each having a b

Answers to Multiple-choice Questions

1. (a)	2. (b)	3. (d)	4. (c)	5.
9. (d)	10. (a)	11. (c)	12. (b)	1

Pedagogy
Provides numerical examples, exercises, end-chapter review questions, and multiple-choice questions with answers.

Instrument Guidelines

Provides step-by-step guidelines for the use and maintenance of select instruments.

A bevel protractor is a precision angle-measuring instrument. measurement, one should follow these guidelines:
1. The instrument should be thoroughly cleaned before use. It is not re pressed air for cleaning, as it can drive particles into the instrument
2. It is important to understand that the universal bevel protractor does the angle on the work part. It measures the angle between its own between the base plate and the adjustable blade. Therefore, one sl intimate contact between the protractor and the features of the part.
3. An easy method to determine if the blade is in contact with the wol behind it and adjust the blade so that no light leaks between the two
4. It should always be ensured that the instrument is in a plane parallel In the absence of this condition, the angle measured will be errone
5. The accuracy of measurement also depends on the surface quality

Chapter on Nanometrology

With the field of nanotechnology having an impact on almost every discipline of science and technology, this is the only Indian textbook to discuss this topic.

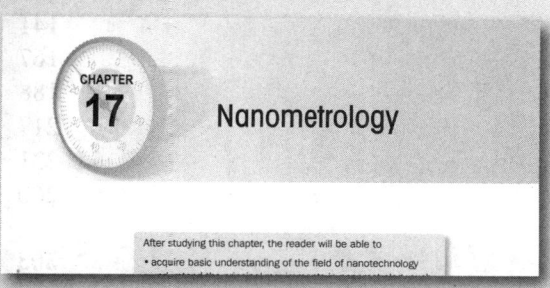

CHAPTER
17
Nanometrology

After studying this chapter, the reader will be able to
• acquire basic understanding of the field of nanotechnology

Laboratory Experiments

Appendix C contains 20 experiments usually demonstrated in a typical laboratory course on the subject. Select experiments also feature photographs of the devices used for experimentation. It contains procedures, observation templates, and model characteristics of all relevant experiments.

C.4 STUDY OF BEVEL PROTRACTORS

C.4.1 Aim
To determine the angle of a given specimen

C.4.2 Apparatus
The following items are used for measurement:
1. Bevel protractor
2. Specimen

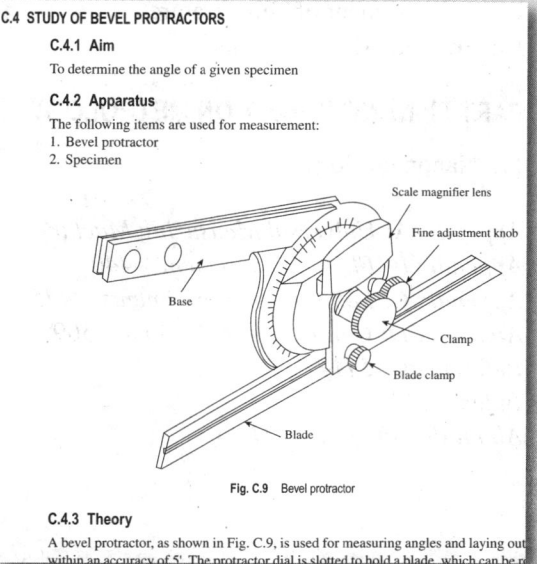

Fig. C.9 Bevel protractor

C.4.3 Theory
A bevel protractor, as shown in Fig. C.9, is used for measuring angles and laying out within an accuracy of 5'. The protractor dial is slotted to hold a blade, which can be r

Brief Contents

Detailed Contents

PART III: NANO IMPACT ON METROLOGY   411

PART I

Engineering Metrology

- Basic Principles of Engineering Metrology
- Standards of Measurement
- Limits, Fits, and Tolerances
- Linear Measurement
- Angular Measurement
- Comparators
- Optical Measurement and Interferometry
- Metrology of Gears and Screw Threads
- Metrology of Surface Finish
- Miscellaneous Metrology
- Inspection and Quality Control

Basic Principles of Engineering Metrology

After studying this chapter, the reader will be able to

- understand the importance of metrology
- appreciate the significance of inspection
- appreciate the concepts of accuracy and precision
- explain the objectives of metrology and measurements
- understand the general measurement concepts
- elucidate the different sources and types of errors
- compare the different types of measurements

1.1 INTRODUCTION

The importance of metrology as a scientific discipline gained momentum during the industrial revolution. Continuing technological advancement further necessitated refinement in this segment. Metrology is practised almost every day, often unknowingly, in our day-to-day tasks. Measurement is closely associated with all the activities pertaining to scientific, industrial, commercial, and human aspects. Its role is ever increasing and encompasses different fields such as communications, energy, medical sciences, food sciences, environment, trade, transportation, and military applications. Metrology concerns itself with the study of measurements. It is of utmost importance to measure different types of parameters or physical variables and quantify each of them with a specific unit. Thus, measurement is an act of assigning an accurate and precise value to a physical variable. The physical variable then gets transformed into a measured variable. Meaningful measurements require common measurement standards and must be performed using them. The common methods of measurement are based on the development of international specification standards. These provide appropriate definitions of parameters and protocols that enable standard measurements to be made and also establish a common basis for comparing measured values. In addition, metrology is also concerned with the reproduction, conservation, and transfer of units of measurements and their standards. Measurements provide a basis for judgements about process information, quality assurance, and process control.

Design is one of the major aspects of all branches of engineering. A product/system comprising several elements has to be properly designed to perform the required (desired) function. In order to test whether functioning of the elements constituting the product/system meets the design expectation, and to finally assess the functioning of the whole system, measurements are inevitable. Another associated aspect is to provide proper operation and maintenance of such a product/system. Measurement is a significant source for acquiring very important and necessary data about both these aspects of engineering, without which the function or analysis cannot be performed properly.

Hence, measurements are required for assessing the performance of a product/system, performing analysis to ascertain the response to a specific input function, studying some fundamental principle or law of nature, etc. Measurements contribute to a great extent to the design of a product or process to be operated with maximum efficiency at minimum cost and with desired maintainability and reliability.

Metrology helps extract high-quality information regarding the completion of products, working condition, and status of processes in an operational and industrial environment. A high product quality along with effectiveness and productivity is a must, in order to survive economically in this competitive global market. The task of attaining workpiece accuracy in modern industrial production techniques has gained much significance through constantly increasing demands on the quality of the parts produced. In order to achieve high product quality, metrology has to be firmly integrated into the production activity. Hence, metrology forms an inseparable key element in the process of manufacturing. This needs focus on the additional expense caused throughout the whole manufacturing process, due to worldwide competition. The quality of the products influences various production attributes such as continuity, production volume and costs, productivity, reliability, and efficiency of these products with respect to their application or their consumption in a diverse manner. Thus, it is desirable to use the resources in an optimal manner and strive to achieve cost reduction in manufacturing.

1.2 METROLOGY

Metrology literally means science of measurements. In practical applications, it is the enforcement, verification, and validation of predefined standards. Although metrology, for engineering purposes, is constrained to measurements of length, angles, and other quantities that are expressed in linear and angular terms, in a broader sense, it is also concerned with industrial inspection and its various techniques. Metrology also deals with establishing the units of measurements and their reproduction in the form of standards, ascertaining the uniformity of measurements, developing methods of measurement, analysing the accuracy of methods of measurement, establishing uncertainty of measurement, and investigating the causes of measuring errors and subsequently eliminating them.

The word metrology is derived from the Greek word 'metrologia', which means measure. Metrology has existed in some form or other since ancient times. In the earliest forms of metrology, standards used were either arbitrary or subjective, which were set up by regional or local authorities, often based on practical measures like the length of an arm.

It is pertinent to mention here the classic statement made by Lord Kelvin (1824–1907), an eminent scientist, highlighting the importance of metrology: 'When you can measure what you are speaking about and express it in numbers, you know something about it; but when you cannot measure it, when you cannot express it in numbers, your knowledge of it is of a meagre and unsatisfactory kind. It may be the beginning of knowledge, but you have scarcely in your thought advanced to the stage of science.'

Another scientist Galileo (1564–1642) has clearly formulated the comprehensive goal of metrology with the following statement: 'Measure everything that is measurable and make measurable what is not so.'

Metrology is an indispensable part of the modern day infrastructure. In fact, it plays an important role in our lives, either directly or indirectly, in various ways. In this competitive world, economic success of most of the manufacturing industries critically depends on the quality and reliability of the products manufactured—requirements in which measurement plays a key role. It has become increasingly essential to conform to the written standards and specifications and mutual recognition of measurements and tests, to trade in national and international markets. This can be achieved by the proper application of measurement methods that enhance the quality of products and the productive power of plants.

Metrology not only deals with the establishment, reproduction, protection, maintenance, and transfer or conversion of units of measurements and their standards, but is also concerned with the correctness of measurement. In addition to encompassing different industrial sectors, it also plays a vital role in establishing standards in different fields that affect human beings, such as health sciences, safety, and environment. Hence, one of the major functions of metrology is to establish international standards for measurements used by all the countries in the world in both science and industry.

Modern manufacturing technology is based on precise reliable dimensional measurements. The term 'legal metrology' applies to any application of metrology that is subjected to national laws or regulations. There will be mandatory and legal bindings on the units and methods of measurements and measuring instruments. The scope of legal metrology may vary considerably from one country to another. The main objective is to maintain uniformity of measurement in a particular country. Legal metrology ensures the conservation of national standards and guarantees their accuracy in comparison with the international standards, thereby imparting proper accuracy to the secondary standards of the country. Some of the applications of legal metrology are industrial measurement, commercial transactions, and public health and human safety aspects.

A group of techniques employed for measuring small variations that are of a continuous nature is termed as 'dynamic metrology'. These techniques find application in recording continuous measurements over a surface and have obvious advantages over individual measurements of a distinctive character.

The metrology in which part measurement is substituted by process measurement is known as 'deterministic metrology'. An example of deterministic metrology is a new technique known as 3D error compensation by computer numerical control (CNC) systems and expert systems, leading to fully adaptive control. This technology is adopted in high-precision manufacturing machinery and control systems to accomplish micro and nanotechnology accuracies.

1.3 NEED FOR INSPECTION

Industrial inspection has acquired significance in recent times and has a systematic and scientific approach. Prior to the industrial revolution, craftsmen used to assemble the different parts by hand and, in the process, consumed a lot of time. They were entirely responsible for the quality of their products. Inspection was an integral function of production. Since the industrial revolution, many new manufacturing techniques have been developed to facilitate mass production of components.

In modern manufacturing techniques, a product has to be disintegrated into different components. Manufacture of each of these components is then treated as an independent process.

F.W. Taylor, who has been acknowledged as the father of scientific management of manufacturing industry, created the modern philosophy of production and also the philosophy of production metrology and inspection. He decomposed a job into multiple tasks, thereby isolating the tasks involved in inspection from the production tasks. This culminated in the creation of a separate quality assurance department in manufacturing industries, which is assigned the task of inspection and quality control.

Inspection is defined as a procedure in which a part or product characteristic, such as a dimension, is examined to determine whether it conforms to the design specification. Basically, inspection is carried out to isolate and evaluate a specific design or quality attribute of a component or product. Industrial inspection assumed importance because of mass production, which involved interchangeability of parts. The various components that come from different locations or industries are then assembled at another place. This necessitates that parts must be so assembled that satisfactory mating of any pair chosen at random is possible. In order to achieve this, dimensions of the components must be well within the permissible limits to obtain the required assemblies with a predetermined fit. Measurement is an integral part of inspection. Many inspection methods rely on measurement techniques, that is, measuring the actual dimension of a part, while others employ the gauging method. The gauging method does not provide any information about the actual value of the characteristic but is faster when compared to the measurement technique. It determines only whether a particular dimension of interest is well within the permissible limits or not. If the part is found to be within the permissible limits, it is accepted; otherwise it is rejected. The gauging method determines the dimensional accuracy of a feature, without making any reference to its actual size, which saves time. In inspection, the part either passes or fails. Thus, industrial inspection has become a very important aspect of quality control.

Inspection essentially encompasses the following:

1. Ascertain that the part, material, or component conforms to the established or desired standard.
2. Accomplish interchangeability of manufacture.
3. Sustain customer goodwill by ensuring that no defective product reaches the customers.
4. Provide the means of finding out inadequacies in manufacture. The results of inspection are recorded and reported to the manufacturing department for further action to ensure production of acceptable parts and reduction in scrap.
5. Purchase good-quality raw materials, tools, and equipment that govern the quality of the finished products.
6. Coordinate the functions of quality control, production, purchasing, and other departments of the organizations.
7. Take the decision to perform rework on defective parts, that is, to assess the possibility of making some of these parts acceptable after minor repairs.
8. Promote the spirit of competition, which leads to the manufacture of quality products in bulk by eliminating bottlenecks and adopting better production techniques.

1.4 ACCURACY AND PRECISION

We know that accuracy of measurement is very important for manufacturing a quality product. *Accuracy* is the degree of agreement of the measured dimension with its true magnitude. It can also be defined as the maximum amount by which the result differs from the true value or as the nearness of the measured value to its true value, often expressed as a percentage. True value may be defined as the mean of the infinite number of measured values when the average deviation due to the various contributing factors tends to zero. In practice, realization of the true value is not possible due to uncertainties of the measuring process and hence cannot be determined experimentally. Positive and negative deviations from the true value are not equal and will not cancel each other. One would never know whether the quantity being measured is the true value of the quantity or not.

Precision is the degree of repetitiveness of the measuring process. It is the degree of agreement of the repeated measurements of a quantity made by using the same method, under similar conditions. In other words, precision is the repeatability of the measuring process. The ability of the measuring instrument to repeat the same results during the act of measurements for the same quantity is known as *repeatability*. Repeatability is random in nature and, by itself, does not assure accuracy, though it is a desirable characteristic. Precision refers to the consistent reproducibility of a measurement. Reproducibility is normally specified in terms of a scale reading over a given period of time. If an instrument is not precise, it would give different results for the same dimension for repeated readings. In most measurements, precision assumes more significance than accuracy. It is important to note that the scale used for the measurement must be appropriate and conform to an internationally accepted standard.

It is essential to know the difference between precision and accuracy. Accuracy gives information regarding how far the measured value is with respect to the true value, whereas precision indicates quality of measurement, without giving any assurance that the measurement is correct. These concepts are directly related to random and systematic measurement errors.

Figure 1.1 also clearly depicts the difference between precision and accuracy, wherein several measurements are made on a component using different types of instruments and the results plotted.

It can clearly be seen from Fig. 1.1 that precision is not a single measurement but is associated with a process or set of measurements. Normally, in any set of measurements performed by the same instrument on the same component, individual measurements are distributed around the mean value and precision is the agreement of these values with each other. The difference

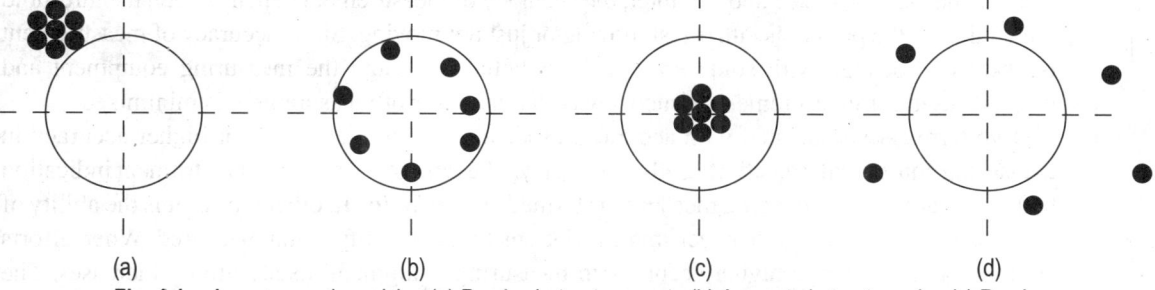

| (a) | (b) | (c) | (d) |

Fig. 1.1 Accuracy and precision (a) Precise but not accurate (b) Accurate but not precise (c) Precise and accurate (d) Not precise and not accurate

between the true value and the mean value of the set of readings on the same component is termed as an *error*. Error can also be defined as the difference between the indicated value and the true value of the quantity measured.

$$E = V_m - V_t$$

where E is the error, V_m the measured value, and V_t the true value.

The value of E is also known as the absolute error. For example, when the weight being measured is of the order of 1 kg, an error of ± 2 g can be neglected, but the same error of ± 2 g becomes very significant while measuring a weight of 10 g. Thus, it can be mentioned here that for the same value of error, its distribution becomes significant when the quantity being measured is small. Hence, % error is sometimes known as relative error. Relative error is expressed as the ratio of the error to the true value of the quantity to be measured. Accuracy of an instrument can also be expressed as % error. If an instrument measures V_m instead of V_t, then,

$$\% \text{ error} = \frac{\text{Error}}{\text{True value}} \times 100$$

Or $\qquad \% \text{ error} = \dfrac{V_m - V_t}{V_t} \times 100$

Accuracy of an instrument is always assessed in terms of error. The instrument is more accurate if the magnitude of error is low. It is essential to evaluate the magnitude of error by other means as the true value of the quantity being measured is seldom known, because of the uncertainty associated with the measuring process. In order to estimate the uncertainty of the measuring process, one needs to consider the systematic and constant errors along with other factors that contribute to the uncertainty due to scatter of results about the mean. Consequently, when precision is an important criterion, mating components are manufactured in a single plant and measurements are obtained with the same standards and internal measuring precision, to accomplish interchangeability of manufacture. If mating components are manufactured at different plants and assembled elsewhere, the accuracy of the measurement of two plants with true standard value becomes significant.

In order to maintain the quality of manufactured components, accuracy of measurement is an important characteristic. Therefore, it becomes essential to know the different factors that affect accuracy. Sense factor affects accuracy of measurement, be it the sense of feel or sight. In instruments having a scale and a pointer, the accuracy of measurement depends upon the threshold effect, that is, the pointer is either just moving or just not moving. Since accuracy of measurement is always associated with some error, it is essential to design the measuring equipment and methods used for measurement in such a way that the error of measurement is minimized.

Two terms are associated with accuracy, especially when one strives for higher accuracy in measuring equipment: sensitivity and consistency. The ratio of the change of instrument indication to the change of quantity being measured is termed as *sensitivity*. In other words, it is the ability of the measuring equipment to detect small variations in the quantity being measured. When efforts are made to incorporate higher accuracy in measuring equipment, its sensitivity increases. The permitted degree of sensitivity determines the accuracy of the instrument. An instrument cannot be more accurate than the permitted degree of sensitivity. It is very pertinent to mention here that unnecessary use of a more sensitive instrument for measurement than required is a disadvantage. When successive readings of the measured quantity obtained from the measuring instrument are same all the time, the equipment is said to be *consistent*. A highly accurate instrument possesses both sensitivity and consistency. A highly sensitive instrument need not be consistent, and the degree of consistency determines the accuracy of the instrument. An instrument that is both consistent and sensitive need not be accurate, because its scale may have been calibrated with a wrong standard. Errors of measurement will be constant in such instruments, which can be taken care of by calibration. It is also important to note that as the magnification increases, the range of measurement decreases and, at the same time, sensitivity increases. Temperature variations affect an instrument and more skill is required to handle it. *Range* is defined as the difference between the lower and higher values that an instrument is able to measure. If an instrument has a scale reading of 0.01–100 mm, then the range of the instrument is 0.01–100 mm, that is, the difference between the maximum and the minimum value.

1.4.1 Accuracy and Cost

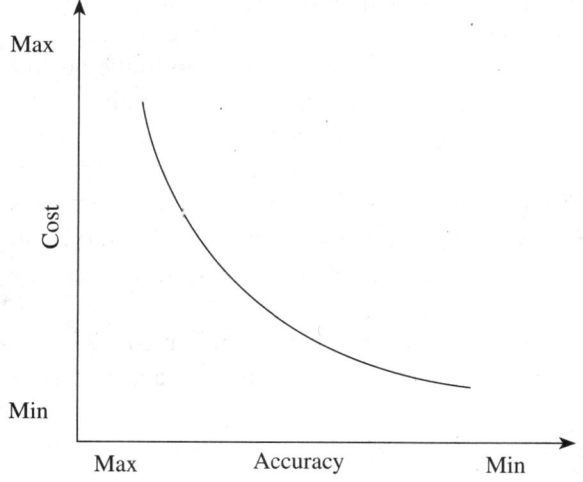

Fig. 1.2 Relationship of accuracy with cost

It can be observed from Fig. 1.2 that as the requirement of accuracy increases, the cost increases exponentially. If the tolerance of a component is to be measured, then the accuracy requirement will normally be 10% of the tolerance values. Demanding high accuracy unless it is absolutely required is not viable, as it increases the cost of the measuring equipment and hence the inspection cost. In addition, it makes the measuring equipment unreliable, because, as discussed in Section 1.4, higher accuracy increases sensitivity. Therefore, in practice, while designing the measuring equipment, the desired/required accuracy to cost considerations depends on the quality and reliability of the component/product and inspection cost.

1.5 OBJECTIVES OF METROLOGY AND MEASUREMENTS

From the preceding discussions, we know that accuracy of measurement is very important for the production of a quality product, and hence it is imperative to mention here that the basic objective of any measurement system is to provide the required accuracy at minimum cost. In addition, metrology is an integral part of modern engineering industry consisting of various departments, namely design, manufacturing, assembly, research and development, and engineering departments. The objectives of metrology and measurements include the following:
1. To ascertain that the newly developed components are comprehensively evaluated and designed within the process, and that facilities possessing measuring capabilities are available in the plant

2. To ensure uniformity of measurements
3. To carry out process capability studies to achieve better component tolerances
4. To assess the adequacy of measuring instrument capabilities to carry out their respective measurements
5. To ensure cost-effective inspection and optimal use of available facilities
6. To adopt quality control techniques to minimize scrap rate and rework
7. To establish inspection procedures from the design stage itself, so that the measuring methods are standardized
8. To calibrate measuring instruments regularly in order to maintain accuracy in measurement
9. To resolve the measurement problems that might arise in the shop floor
10. To design gauges and special fixtures required to carry out inspection
11. To investigate and eliminate different sources of measuring errors

1.6 GENERAL MEASUREMENT CONCEPTS

We know that the primary objective of measurement in industrial inspection is to determine the quality of the component manufactured. Different quality requirements, such as permissible tolerance limits, form, surface finish, size, and flatness, have to be considered to check the conformity of the component to the quality specifications. In order to realize this, quantitative information of a physical object or process has to be acquired by comparison with a reference. The three basic elements of measurements (schematically shown in Fig. 1.3), which are of significance, are the following:

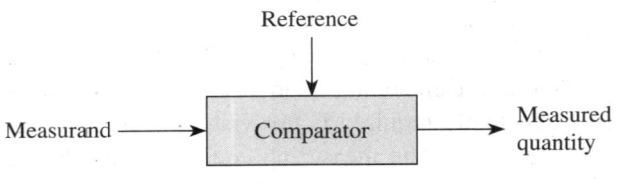

Fig. 1.3 Elements of measurement

1. *Measurand*, a physical quantity such as length, weight, and angle to be measured
2. *Comparator*, to compare the measurand (physical quantity) with a known standard (reference) for evaluation
3. *Reference*, the physical quantity or property to which quantitative comparisons are to be made, which is internationally accepted

All these three elements would be considered to explain the direct measurement using a calibrated fixed reference. In order to determine the length (a physical quantity called measurand) of the component, measurement is carried out by comparing it with a steel scale (a known standard).

1.6.1 Calibration of Measuring Instruments

It is essential that the equipment/instrument used to measure a given physical quantity is validated. The process of validation of the measurements to ascertain whether the given physical quantity conforms to the original/national standard of measurement is known as *traceability* of the standard. One of the principal objectives of metrology and measurements is to analyse the uncertainty of individual measurements, the efforts made to validate each measurement with a given equipment/instrument, and the data obtained from it. It is essential that traceability (which is often performed by a calibration laboratory having conformity with a proven quality system with such standards) should disseminate to the consumers. Calibration is a means of achieving traceability. One of the essential aspects of metrology is that the results of measurements obtained should be meaningful. To accomplish this, calibration of any measuring system/instrument is very essential. Calibration is the procedure used to establish a relationship between the values of the quantities indicated by the measuring instrument and the corresponding values realized by standards under specified conditions. It refers to the process of establishing the characteristic relationship between the values of the physical quantity applied to the instrument and the corresponding positions of the index, or creating a chart of quantities being measured versus readings of the instrument. If the instrument has an arbitrary scale, the indication has to be multiplied by a factor to obtain the nominal value of the quantity measured, which is referred to as *scale factor*. If the values of the variable involved remain constant (not time dependent) while calibrating a given instrument, this type of calibration is known as *static calibration*, whereas if the value is time dependent or time-based information is required, it is called *dynamic calibration*. The relationship between an input of known dynamic behaviour and the measurement system output is determined by dynamic calibration.

The main objective of all calibration activities is to ensure that the measuring instrument will function to realize its accuracy objectives. General calibration requirements of the measuring systems are as follows: (a) accepting calibration of the new system, (b) ensuring traceability of standards for the unit of measurement under consideration, and (c) carrying out calibration of measurement periodically, depending on the usage or when it is used after storage.

Calibration is achieved by comparing the measuring instrument with the following: (a) a primary standard, (b) a known source of input, and (c) a secondary standard that possesses a higher accuracy than the instrument to be calibrated. During calibration, the dimensions and tolerances of the gauge or accuracy of the measuring instrument is checked by comparing it with a standard instrument or gauge of known accuracy. If deviations are detected, suitable adjustments are made in the instrument to ensure an acceptable level of accuracy. The limiting factor of the calibration process is repeatability, because it is the only characteristic error that cannot be calibrated out of the measuring system and hence the overall measurement accuracy is curtailed. Thus, repeatability could also be termed as the minimum uncertainty that exists between a measurand and a standard. Conditions that exist during calibration of the instrument should be similar to the conditions under which actual measurements are made. The standard that is used for calibration purpose should normally be one order of magnitude more accurate than the instrument to be calibrated. When it is intended to achieve greater accuracy, it becomes imperative to know all the sources of errors so that they can be evaluated and controlled.

1.7 ERRORS IN MEASUREMENTS

While performing physical measurements, it is important to note that the measurements obtained are not completely accurate, as they are associated with uncertainty. Thus, in order to analyse the measurement data, we need to understand the nature of errors associated with the measurements.

Therefore, it is imperative to investigate the causes or sources of these errors in measurement systems and find out ways for their subsequent elimination. Two broad categories of errors in measurement have been identified: systematic and random errors.

1.7.1 Systematic or Controllable Errors

A systematic error is a type of error that deviates by a fixed amount from the true value of measurement. These types of errors are controllable in both their magnitude and their direction, and can be assessed and minimized if efforts are made to analyse them. In order to assess them, it is important to know all the sources of such errors, and if their algebraic sum is significant with respect to the manufacturing tolerance, necessary allowance should be provided to the measured size of the workpiece. Examples of such errors include measurement of length using a metre scale, measurement of current with inaccurately calibrated ammeters, etc. When the systematic errors obtained are minimum, the measurement is said to be extremely accurate. It is difficult to identify systematic errors, and statistical analysis cannot be performed. In addition, systematic errors cannot be eliminated by taking a large number of readings and then averaging them out. These errors are reproducible inaccuracies that are consistently in the same direction. Minimization of systematic errors increases the accuracy of measurement. The following are the reasons for their occurrence:
1. Calibration errors
2. Ambient conditions
3. Deformation of workpiece
4. Avoidable errors

Calibration Errors

A small amount of variation from the nominal value will be present in the actual length standards, as in slip gauges and engraved scales. Inertia of the instrument and its hysteresis effects do not allow the instrument to translate with true fidelity. Hysteresis is defined as the difference between the indications of the measuring instrument when the value of the quantity is measured in both the ascending and descending orders. These variations have positive significance for higher-order accuracy achievement. Calibration curves are used to minimize such variations. Inadequate amplification of the instrument also affects the accuracy.

Ambient Conditions

It is essential to maintain the ambient conditions at internationally accepted values of standard temperature (20 °C) and pressure (760 mmHg) conditions. A small difference of 10 mmHg can cause errors in the measured size of the component. The most significant ambient condition affecting the accuracy of measurement is temperature. An increase in temperature of 1 °C results in an increase in the length of C25 steel by 0.3 μm, and this is substantial when precision measurement is required. In order to obtain error-free results, a correction factor for temperature has to be provided. Therefore, in case of measurements using strain gauges, temperature compensation is provided to obtain accurate results. Relative humidity, thermal gradients, vibrations, and CO_2 content of the air affect the refractive index of the atmosphere. Thermal expansion occurs due to heat radiation from different sources such as lights, sunlight, and body temperature of operators.

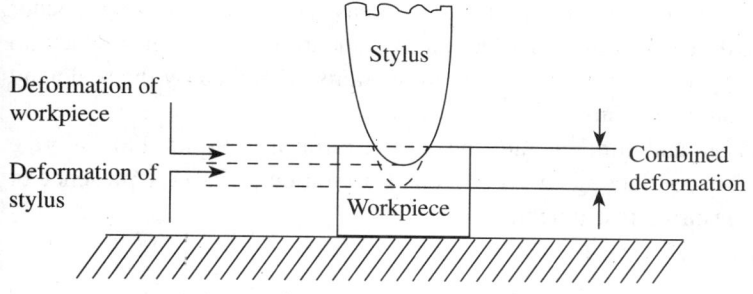

Fig. 1.4   Elastic deformation due to stylus pressure

Deformation of Workpiece

Any elastic body, when subjected to a load, undergoes elastic deformation. The stylus pressure applied during measurement affects the accuracy of measurement. Due to a definite stylus pressure, elastic deformation of the workpiece and deflection of the workpiece shape may occur, as shown in Fig. 1.4. The magnitude of deformation depends on the applied load, area of contact, and mechanical properties of the material of the given workpiece. Therefore, during comparative measurement, one has to ensure that the applied measuring loads are same.

Avoidable Errors

These include the following:

Datum errors   Datum error is the difference between the true value of the quantity being measured and the indicated value, with due regard to the sign of each. When the instrument is used under specified conditions and a physical quantity is presented to it for the purpose of verifying the setting, the indication error is referred to as the datum error.

Reading errors   These errors occur due to the mistakes committed by the observer while noting down the values of the quantity being measured. Digital readout devices, which are increasingly being used for display purposes, eliminate or minimize most of the reading errors usually made by the observer.

Errors due to parallax effect   Parallax errors occur when the sight is not perpendicular to the instrument scale or the observer reads the instrument from an angle. Instruments having a scale and a pointer are normally associated with this type of error. The presence of a mirror behind the pointer or indicator virtually eliminates the occurrence of this type of error.

Effect of misalignment These occur due to the inherent inaccuracies present in the measuring instruments. These errors may also be due to improper use, handling, or selection of the instrument. Wear on the micrometer anvils or anvil faces not being perpendicular to the axis results in misalignment, leading to inaccurate measurements. If the alignment is not proper, sometimes sine and cosine errors also contribute to the inaccuracies of the measurement.

Zero errors   When no measurement is being carried out, the reading on the scale of the instrument should be zero. A zero error is defined as that value when the initial value of a physical quantity indicated by the measuring instrument is a non-zero value when it should have actually been zero. For example, a voltmeter might read 1 V even when it is not under any electromagnetic influence. This voltmeter indicates 1 V more than the true value for all subsequent measurements made. This error is constant for all the values measured using the same instrument. A *constant error* affects all measurements in a measuring process by the same amount or by an amount proportional to the magnitude of the quantity being measured. For example, in a planimeter, which is used to measure irregular areas, a constant error might occur because of an error in the scale used in the construction of standard or, sometimes, when an incorrect conversion factor is used in conversion between the units embodied by the scale and those in which the results of the measurements are expressed.

Therefore, in order to find out and eliminate any systematic error, it is required to calibrate the measuring instrument before conducting an experiment. Calibration reveals the presence of any systematic error in the measuring instrument.

1.7.2 Random Errors

Random errors provide a measure of random deviations when measurements of a physical quantity are carried out repeatedly. When a series of repeated measurements are made on a component under similar conditions, the values or results of measurements vary. Specific causes for these variations cannot be determined, since these variations are unpredictable and uncontrollable by the experimenter and are random in nature. They are of variable magnitude and may be either positive or negative. When these repeated measurements are plotted, they follow a normal or Gaussian distribution. Random errors can be statistically evaluated, and their mean value and standard deviation can be determined. These errors scatter around a mean value. If n measurements are made using an instrument, denoted by v_1, v_2, v_3, ..., v_n, then arithmetic mean is given as

$$\bar{v} = \frac{v_1 + v_2 + v_3 \ldots \ldots \ldots v_n}{n}$$

and standard deviation σ is given by the following equation:

$$\sigma = \pm \sqrt{\sum \frac{\left(v - \bar{v}\right)^2}{n}}$$

Standard deviation is a measure of dispersion of a set of readings. It can be determined by taking the root mean square deviation of the readings from their observed numbers, which is given by the following equation:

$$\sigma = \pm \sqrt{\sum \frac{\left(v_1 - \bar{v}\right)^2 + \left(v_2 - \bar{v}\right)^2 + \ldots + \left(v_n - \bar{v}\right)^2}{n}}$$

Random errors can be minimized by calculating the average of a large number of observations. Since precision is closely associated with the repeatability of the measuring process, a precise instrument will have very few random errors and better repeatability. Hence, random errors limit the precision of the instrument. The following are the likely sources of random errors:

1. Presence of transient fluctuations in friction in the measuring instrument
2. Play in the linkages of the measuring instruments
3. Error in operator's judgement in reading the fractional part of engraved scale divisions
4. Operator's inability to note the readings because of fluctuations during measurement
5. Positional errors associated with the measured object and standard, arising due to small variations in setting

Figure 1.5 clearly depicts the relationship between systematic and random errors with respect to the measured value. The measure of a system's accuracy is altered by both systematic and random errors. Table 1.1 gives the differences between systematic and random errors.

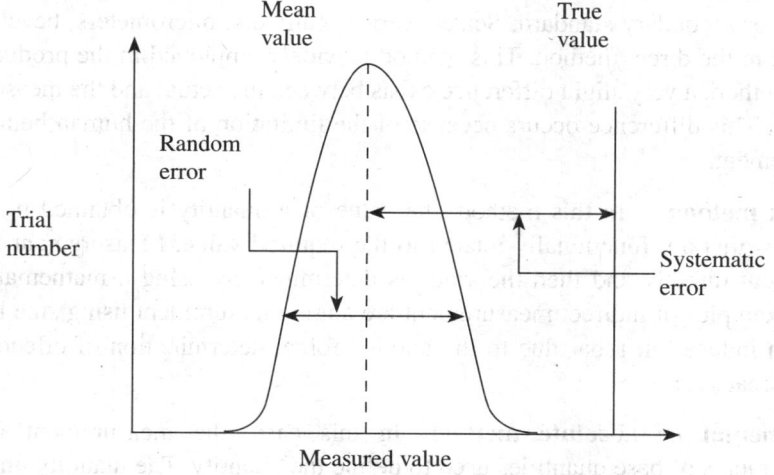

Fig. 1.5 Relationship between systematic and random errors with measured value

Table 1.1 Differences between systematic and random errors

Systematic error	Random error
Not easy to detect	Easy to detect
Cannot be eliminated by repeated measurements	Can be minimized by repeated measurements
Can be assessed easily	Statistical analysis required
Minimization of systematic errors increases the accuracy of measurement	Minimization of random errors increases repeatability and hence precision of the measurement
Calibration helps reduce systematic errors	Calibration has no effect on random errors
Characterization not necessary	Characterized by mean, standard deviation, and variance
Reproducible inaccuracies that are consistently in the same direction	Random in nature and can be both positive and negative

1.8 METHODS OF MEASUREMENT

When precision measurements are made to determine the values of a physical variable, different methods of measurements are employed. Measurements are performed to determine the magnitude of the value and the unit of the quantity under consideration. For instance, the length of a rod is 3 m, where the number, 3, indicates the magnitude and the unit of measurement is metre. The choice of the method of measurement depends on the required accuracy and the amount of permissible error. Irrespective of the method used, the primary objective is to minimize the uncertainty associated with measurement. The common methods employed for making measurements are as follows:

Direct method In this method, the quantity to be measured is directly compared with the primary or secondary standard. Scales, vernier callipers, micrometers, bevel protractors, etc., are used in the direct method. This method is widely employed in the production field. In the direct method, a very slight difference exists between the actual and the measured values of the quantity. This difference occurs because of the limitation of the human being performing the measurement.

Indirect method In this method, the value of a quantity is obtained by measuring other quantities that are functionally related to the required value. Measurement of the quantity is carried out directly and then the value is determined by using a mathematical relationship. Some examples of indirect measurement are angle measurement using sine bar, measurement of strain induced in a bar due to the applied force, determination of effective diameter of a screw thread, etc.

Fundamental or absolute method In this case, the measurement is based on the measurements of base quantities used to define the quantity. The quantity under consideration is directly measured, and is then linked with the definition of that quantity.

Comparative method In this method, as the name suggests, the quantity to be measured is compared with the known value of the same quantity or any other quantity practically related to it. The quantity is compared with the master gauge and only the deviations from the master gauge are recorded after comparison. The most common examples are comparators, dial indicators, etc.

Transposition method This method involves making the measurement by direct comparison, wherein the quantity to be measured (V) is initially balanced by a known value (X) of the same quantity; next, X is replaced by the quantity to be measured and balanced again by another known value (Y). If the quantity to be measured is equal to both X and Y, then it is equal to

$$V = \sqrt{XY}$$

An example of this method is the determination of mass by balancing methods and known weights.

Coincidence method This is a differential method of measurement wherein a very minute difference between the quantity to be measured and the reference is determined by careful observation of the coincidence of certain lines and signals. Measurements on vernier calliper and micrometer are examples of this method.

Deflection method This method involves the indication of the value of the quantity to be measured directly by deflection of a pointer on a calibrated scale. Pressure measurement is an example of this method.

Complementary method The value of the quantity to be measured is combined with a known value of the same quantity. The combination is so adjusted that the sum of these two values is equal to the predetermined comparison value. An example of this method is determination of the volume of a solid by liquid displacement.

Null measurement method In this method, the difference between the value of the quantity to be measured and the known value of the same quantity with which comparison is to be made is brought to zero.

Substitution method It is a direct comparison method. This method involves the replacement of the value of the quantity to be measured with a known value of the same quantity, so selected that the effects produced in the indicating device by these two values are the same. The Borda method of determining mass is an example of this method.

Contact method In this method, the surface to be measured is touched by the sensor or measuring tip of the instrument. Care needs to be taken to provide constant contact pressure in order to avoid errors due to excess constant pressure. Examples of this method include measurements using micrometer, vernier calliper, and dial indicator.

Contactless method As the name indicates, there is no direct contact with the surface to be measured. Examples of this method include the use of optical instruments, tool maker's microscope, and profile projector.

Composite method The actual contour of a component to be checked is compared with its maximum and minimum tolerance limits. Cumulative errors of the interconnected elements of the component, which are controlled through a combined tolerance, can be checked by this method. This method is very reliable to ensure interchangeability and is usually effected through the use of composite GO gauges. The use of a GO screw plug gauge to check the thread of a nut is an example of this method.

A QUICK OVERVIEW

- The importance and necessity of metrology have greatly increased with the industrial revolution, which emphasizes the importance of metrology in industries.
- Inspection is defined as a procedure in which a part or product characteristic, for example a dimension, is examined to determine whether it conforms to the design specification. Basically, inspection is carried out to isolate and evaluate a specific design or to measure the quality attribute of a component or product.
- We know that accuracy of measurement is very important in order to manufacture a quality product. *Accuracy* is the degree of agreement of

the measured dimension with its true magnitude. It can also be defined as the maximum amount by which the result differs from true value.
- *Precision* is the degree of repetitiveness of the measuring process. It is the degree of agreement of the repeated measurements of a quantity made by using the same method, under similar conditions.
- The ability of the measuring instrument to repeat the same results during the act of measurements for the same quantity is known as *repeatability*.
- The difference between the true value and the mean value of the set of readings on the same component is termed as an *error*. Error can also

be defined as the difference between the indicated value and the true value of the quantity measured.

- The ratio of the change of instrument indication to the change of quantity being measured is termed as *sensitivity*. When successive readings of the measured quantity obtained from the measuring instrument are the same all the time, the equipment is said to be *consistent*.
- *Range* is defined as the difference between the lower and higher values that an instrument is able to measure.
- The process of validation of the measurements to ascertain whether the given physical quantity conforms to the original/national standard of measurement is known as *traceability* of the standard.

- Calibration is the procedure used to establish a relationship between the values of the quantities indicated by the measuring instrument and the corresponding values realized by standards under specified conditions.
- Systematic error is a type of error that deviates by a fixed amount from the true value of measurement. These types of errors are controllable in both their magnitude and direction.
- Random errors provide a measure of random deviations when measurements of a physical quantity are carried out repeatedly. These errors can be statistically evaluated, and their mean value and standard deviation can be determined.

MULTIPLE-CHOICE QUESTIONS

1. When a set of readings of a measurement has a wide range, it indicates
   (a) high precision   (c) low precision
   (b) high accuracy   (d) low accuracy

2. The difference between the lower and higher values that an instrument is able to measure is called
   (a) accuracy   (c) range
   (b) sensitivity   (d) error

3. When determining the uncertainty for a particular measurement device, the common uncertainty factors that should be included are
   (a) errors in the measurement technique and method
   (b) random variability of the measurement process
   (c) technician's error
   (d) all of these

4. The aim of calibration is to
   (a) meet customer requirement
   (b) detect deterioration of accuracy
   (c) comply with ISO 9000 standard requirements
   (d) practise measurement procedures

5. Which of the following defines parallax error?
   (a) Same as observational error
   (b) Apparent shift of an object when the position of the observer is altered

   (c) Error caused by the distance between the scale and the measured feature
   (d) Mean of the values of measurements when the object is observed from the right and from the left

6. The best way to eliminate parallax error is to
   (a) use a magnifying glass while taking measurements
   (b) use a mirror behind the readout pointer or indicator
   (c) centre the scale along the measured feature
   (d) take consistent measurements from one side only

7. Which of the following errors is eliminated or minimized by zero setting adjustment on a dial indicator?
   (a) Parallax error   (c) Alignment error
   (b) Inherent error   (d) Computational error

8. The best standard characteristic when calibrating an instrument for production inspection is
   (a) ten times the accuracy
   (b) new and seldom used since calibration
   (c) permanently assigned to this particular instrument
   (d) using a standard similar in size, shape, and material, with part being measured

9. Interpretation of repeated measurement res-

ults on the same feature is considered the instrument's
  (a) accuracy   (c) range
  (b) precision   (d) sensitivity
10. When a steel rule is used, which of the following is the source of measurement uncertainty?
  (a) Inherent instrument error, temperature error, and manipulative error
  (b) Attachment error, manipulative error, and temperature error
  (c) Attachment error, bias, and inherent instrument error
  (d) Inherent instrument error, manipulative error, and observational error
11. Accuracy is defined as
  (a) a measure of how often an experimental value can be repeated
  (b) the closeness of a measured value to the real value
  (c) the number of significant figures used in a measurement
  (d) none of these

12. Conformity of a physical quantity to the national standard of measurement is known as
  (a) calibration   (c) traceability
  (b) sensitivity   (d) repeatability
13. Systematic errors are
  (a) controllable errors
  (b) random errors
  (c) uncontrollable errors
  (d) none of these
14. When a series of repeated measurements that are made on a component under similar conditions are plotted, it follows
  (a) log-normal distribution
  (b) Weibull distribution
  (c) binomial distribution
  (d) Gaussian distribution
15. Random errors can be assessed
  (a) experimentally
  (b) by performing sensitivity analysis
  (c) statistically
  (d) empirically

REVIEW QUESTIONS

1. Define metrology. Explain the significance of metrology.
2. Differentiate between sensitivity and consistency.
3. Define measurand and comparator.
4. Briefly explain legal and deterministic metrology.
5. Distinguish between direct and indirect measurements. Give two examples of each.
6. Describe any five methods of measurement.
7. Define the following:
  (a) Range of measurement
  (b) Sensitivity
  (c) Consistency
  (d) Repeatability

  (e) Calibration
  (f) Traceability
8. Differentiate between accuracy and precision.
9. What are the possible sources of errors in measurements? Briefly explain them.
10. Discuss the need for inspection.
11. Explain the important elements of measurements.
12. Differentiate between systematic and random errors.
13. Discuss the relationship between accuracy and cost.
14. List the objectives of metrology.
15. Discuss the different reasons for the occurrence of systematic errors.

Answers to Multiple-choice Questions

1. (a)	2. (c)	3. (d)	4. (b)	5. (b)	6. (b)	7. (c)	8. (a)
9. (b)	10. (d)	11. (b)	12. (c)	13. (a)	14. (d)	15. (c)	

2

Standards of Measurement

After studying this chapter, the reader will be able to

- understand the importance of and need for standards
- describe the evolution of standards and the role of the National Physical Laboratory
- elucidate the different material standards, their contribution, and disadvantages, and appreciate the significance of wavelength standards
- compare the characteristics of line and end standards
- explain the transfer from line standard to end standard
- calibrate end bars

2.1 INTRODUCTION

Human beings have always been innovative and have exploited the natural resources of the earth to manufacture products and devices that satisfy their basic needs and desires. They have always experimented with the form, size, and performance of the products they have invented. During the medieval period, the measurement process underwent an evolution and people accepted the process in specific trades, but no common standards were set. Generally, these measurement standards were region-dependent, and as trade and commerce grew, the need for standardization was also realized. In fact, it would be impossible to imagine today's modern world without a good system of standards of measurement.

Mass production, an idea generated during the last industrial revolution, has become very popular and synonymous with the present manufacturing industry, and a necessity for manufacturing identical parts. Today, almost all manufacturing units practise the principle of interchangeability of manufacture. In order to accomplish complete interchangeability of manufacture in industries, it is essential to a have a measurement system that can adequately define the features of the components/products to the accuracy required.

2.2 STANDARDS AND THEIR ROLES

In order to make measurements a meaningful exercise, some sort of comparison with a known quantity is very essential. It is necessary to define a unit value of any physical quantity under consideration such that it will be accepted internationally. It is not sufficient to only define these unit values of physical quantities; these should also be measureable. A standard is defined as the fundamental value of any known physical quantity, as established by national and international organizations of authority, which can be reproduced. Fundamental units of physical quantities such as length, mass, time, and temperature form the basis for establishing a measurement system.

In the present world of globalization, it is impossible to perform trade in national and international arenas without standards. In fact, a good system of standards is essential for fair international trade and commerce; it also helps accomplish complete interchangeability of manufacture. Adhering to globally accepted standards allows the manufacturer to convince the customers about the quality of the product.

Standards play a vital role for manufacturers across the world in achieving consistency, accuracy, precision, and repeatability in measurements and in supporting the system that enables the manufacturers to make such measurements.

2.3 EVOLUTION OF STANDARDS

The need for accurate measurements has been recognized by human beings since early times, which is evident from the history of standards. Length standards were one of the earliest standards set by human beings. A brief look at history reveals the following interesting facts. The Egyptian cubit was the earliest recorded length standard—the length of the Pharaoh's forearm plus the width of his palm. The royal cubit was the first master standard made out of black granite, which was used in Egyptian pyramid construction. The actual foot length of the Greek monarch was defined as a foot. King Henry I decreed the length from the tip of nose to the end of the middle finger, when the arm is fully stretched as one yard.

One of the important prerequisites for progress in science is to gain mastery over the science of measurement. Any progress made in manufacturing and other business sectors in the international arena necessitates that activities at high scientific and technical levels be accomplished by progress in metrology. Further, automation in manufacturing industries demands a very high level of accuracy, precision, and reliability. It is pertinent to mention here that human beings' knowledge of nature and the universe, their adaptability to the purpose, and their ability to measure precisely, provides a basis for the science of metrology.

The metric system, which was accepted by France in 1795, coexisted with medieval units until 1840, when it was declared as the exclusive system of weights and measures. In 1798, Eli Whitney introduced the concept of manufacturing interchangeable components for assembling guns. This led to the development of standardization of manufacturing activities to achieve interchangeability. Based on a 4-year investigation, John Quincy Adams in 1821 submitted a report on the metric system and the modernization of our measurement system to the United States Congress. Highlighting the importance of measurement in his report, he stated, 'Weights and measures may be ranked among the necessaries of life to every individual of human society. They enter into the economical arrangements and daily concerns of every family. They are necessary to every occupation of human industry; to the distribution and security of every species of property; to every transaction of trade and commerce; to the labors of the husbandman; to the ingenuity of the artificer; to the studies of the philosopher; to the researches of the antiquarian; to the navigation of the mariner, and the marches of the soldier; to all the exchanges of peace, and all the operations of war. The knowledge of them, as in established use, is among the first elements of education, and is often learned by those who learn nothing else, not even to read and write. This knowledge is riveted in the memory by the habitual application of it to the employments of men throughout life'.

By 1860, in order to keep pace with scientific inventions, there arose a need for better metric standards. In 1855, the imperial standard yard was developed in England, which was quite accurate. In 1872, the first international prototype metre was developed in France. The International Metric Convention, which was held in France in 1875, universally accepted the metric system, and provisions were also made to set up the International Bureau of Weights and Measures (BIPM) in Paris, which was signed by 17 countries. This convention also agreed to precisely define the metric standards for mass and length and established permanent mechanisms to propose and implement further improvements in the metric system. In 1866, the USA passed an act of Congress to employ the metric system of weights and measures in all contracts, dealings, and court proceedings. In the USA, since 1893, the internationally accepted metric standards have served as the basic measurement standards. Around 35 countries, including continental Europe and most of South America, officially adopted the metric system in 1900.

The internationally adopted standards were required to extend support to the rapid increase in trade between industrialized countries. This resulted in the establishment of international organizations for standardization such as the International Electrotechnical Commission (IEC) in 1906 and the International Organization for Standardization (ISO) in 1947.

In October 1960, at the 11th General Conference on Weights and Measures held in Paris, the original metric standards were redefined in accordance with the 20th-century standards of measurement and a new revised and simplified international system of units, namely the SI units was devised. SI stands for *systeme international d'unites* (international system of units). The seven basic units established under the SI unit system are given in Table 2.1. The 11th General Conference recommended a new standard of length, known as wavelength standard, according to which, metre is defined as 16,50,763.73 × wavelengths of the red–orange radiation of krypton 86 atom in vacuum.

In the 17th General Conference of Weights and Measures held on 20 October 1983, the modern metre was defined as the length of the path travelled by light in vacuum during a time interval of 1/29,97,92,458 of a second.

Table 2.1 Basic units of SI system

Quantity	Unit	Symbol
Length	Metre	m
Mass	Kilogram	kg
Time	Second	s
Thermodynamic temperature	Kelvin	K
Amount of substance	Mole	mol
Electric current	Ampere	A
Luminous intensity	Candela	Cd

In 1999, a mutual recognition arrangement (MRA) was signed by the Committee of Weights and Measures to cater to the growing need for an open, transparent, and comprehensive method to provide users with reliable quantitative information on the comparability of national metrology services. It also provides the technical basis for wider agreements negotiated for international trade, commerce, and regulatory affairs.

In 1999, directors of the National Metrology Institutes (NMIs) signed an MRA for national measurement standards and for calibration and measurement certificates issued by NMIs.

2.4 NATIONAL PHYSICAL LABORATORY

The National Physical Laboratory (NPL) was established in UK in 1900. It is a public institution for standardizing and verifying instruments, testing materials, and determining physical constants. NPL India (NPLI) was established in 1947 in New Delhi under the Council of Scientific and Industrial Research (CSIR). It also has to comply with the statutory obligation of realizing, establishing, maintaining, reproducing, and updating the national standards of measurement and calibration facilities for different parameters.

The main purpose of establishing NPLI is to reinforce and carry out research and development activities in the areas of physical sciences and key physics-based technologies. NPLI is also responsible for maintaining national standards of measurements and ensuring that they conform to international standards. It is established to support industries and national and private agencies in their research and development activities by carrying out calibration and testing, precision measurements, and development of processes and devices. It also ascertains that the national standards of measurements are traceable to the international standards. NPLI also shoulders the responsibility of assisting in research and development activities in the fields of material development, radio and atmospheric sciences, superconductivity and cryogenics, etc.

The major exercise of NPLI is to compare at regular intervals, the national standards with the corresponding standards maintained by the NMIs of other countries in consultation with the International Committee of Weights and Measures and the member nations of the Asia Pacific Metrology Programme. This exercise is essential to establish equivalence of national standards of measurement at NPL with those at other NMIs so that the calibration certificates issued by NPL would have global acceptability.

2.5 MATERIAL STANDARD

Two standard systems for linear measurement that have been accepted and adopted worldwide are English and metric (yard and metre) systems. Most countries have realized the importance and advantages of the metric system and accepted metre as the fundamental unit of linear measurement.

Scientists across the world have always been in the pursuit of a suitable unit for length, and consistent efforts have been made to keep the unit of length constant irrespective of the environmental conditions. The problem with material standards used earlier was that the materials used for defining the standards could change their size with temperature and other conditions. In order to keep the fundamental unit unchanged, great care and attention had to be exercised to maintain the same conditions. The natural and invariable unit for length was finalized as the primary standard when they found that wavelength of monochromatic light was not affected by environmental conditions. They were able to express easily the previously defined yard and metre in terms of the wavelength of light.

Yard or metre is defined as the distance between two scribed lines on a bar of metal maintained under certain conditions of temperature and support. These are legal line standards and are governed by the Act of Parliament for their use.

2.5.1 Yard

The imperial standard yard is a bronze bar 1 sq. inch in cross-section and 38 inches in length, having a composition of 82% Cu, 13% tin, and 5% Zn. The bar contains holes of ½-inch diameter × ½-inch depth. It has two round recesses, each located one inch away from either end and extends up to the central plane of the bar. A highly polished gold plug having a diameter of 1/10 of an inch comprises three transversely engraved lines and two longitudinal lines that are inserted into each of these holes such that the lines lie in the neutral plane. The top surface of the plug lies on the neutral axis. Yard is then defined as the distance between the two central transverse lines of the plug maintained at a temperature of 62 °F. Yard, which was legalized in 1853, remained a legal standard until it was replaced by the wavelength standard in 1960. One of the advantages of maintaining the gold plug lines at neutral axis is that this axis remains unaffected due to bending of the beam. Another advantage is that the gold plug is protected from getting accidentally damaged. Three orthographic views of the imperial standard yard are shown in Fig. 2.1. It is important to note that an error occurs in the neutral axis because of the support provided at the ends. This error can be minimized by placing the supports in such a way that the slope at the ends is zero and the flat end faces of the bar are mutually parallel to each other.

Airy points can be defined as the points at which a horizontal rod is optionally supported to prevent it from bending. These points are used to support a length standard in such a way as to minimize the error due to bending. Sir George Biddell Airy (1801–92) showed that the distance d between the supports can be determined by the formula $\dfrac{1}{\sqrt{n^2 - 1}} \times L$, where n is the number of supports and L, the length of the bar. When it is supported by two points, $n = 2$. Substituting this value in the preceding formula, the distance between the supports is obtained as $0.577L$. This means that the supports should be at an equal distance from each end to a position where they are $0.577L$ apart. Normally, airy points are marked for length bars greater than 150 mm.

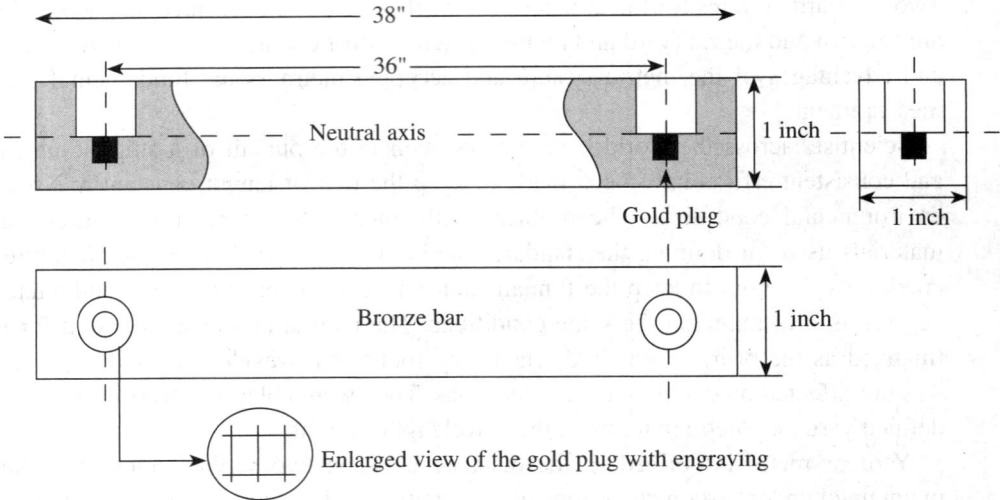

38"

36"

Neutral axis

Gold plug

1 inch

1 inch

Bronze bar

1 inch

Enlarged view of the gold plug with engraving

Fig. 2.1 Imperial standard yard

The distance between two supports for international yard and international prototype metre is marked as 29.94 inches and 58.9 mm, respectively.

2.5.2 Metre

This standard is also known as international prototype metre, which was established in 1875. It is defined as the distance between the centre positions of the two lines engraved on the highly polished surface of a 102 cm bar of pure platinum–iridium alloy (90% platinum and 10% iridium) maintained at 0 °C under normal atmospheric pressure and having the cross-section of a web, as shown in Fig. 2.2. The top surface of the web contains graduations coinciding with the neutral axis of the section. The web-shaped section offers two major advantages. Since the section is uniform and has graduations on the neutral axis, it allows the whole surface to be graduated. This type of cross-section provides greater rigidity for the amount of metal involved and is economical even though an expensive metal is used for its construction. The bar is inoxidizable and can have a good polish, which is required for obtaining good-quality lines. It is supported by two rollers having at least 1 cm diameter, which are symmetrically located in the same horizontal plane at a distance of 751 mm from each other such that there is minimum deflection.

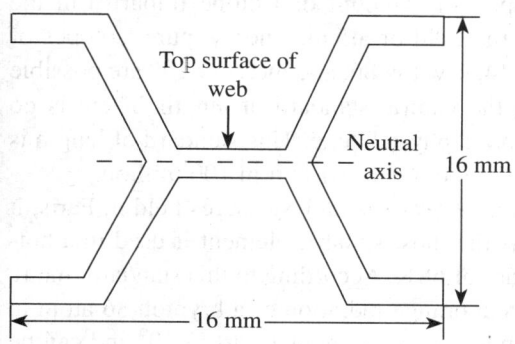

Fig. 2.2   International prototype metre

2.5.3 Disadvantages of Material Standards

The following disadvantages are associated with material standards:
1. Material standards are affected by changes in environmental conditions such as temperature, pressure, humidity, and ageing, resulting in variations in length.
2. Preservation of these standards is difficult because they must have appropriate security to prevent their damage or destruction.
3. Replicas of material standards are not available for use at other places.
4. They cannot be easily reproduced.
5. Comparison and verification of the sizes of gauges pose considerable difficulty.
6. While changing to the metric system, a conversion factor is necessary.

These disadvantages prompted scientists to look for a standard that remains unaffected by the variations in environmental conditions. Consequently, the 11th General Conference of Weights and Measures, which was held in Paris in 1960, recommended a new standard of length, known as wavelength standard, measured in terms of wavelengths of the red–orange radiation of the krypton 86 isotope gas.

2.6 WAVELENGTH STANDARD

It is very clear from the methods discussed earlier that comparison and verification of the sizes of the gauges pose considerable difficulty. This difficulty arises because the working standard used as a reference is derived from a physical standard and successive comparisons are required to establish the size of a working standard using the process discussed earlier, leading to errors that are unacceptable. By using wavelengths of a monochromatic light as a natural and invariable unit of length, the dependency of the working standard on the physical standard can be eliminated. The definition of a standard of length relative to the metre can easily be expressed in terms of the wavelengths of light.

The use of the interference phenomenon of light waves to provide a working standard may thus be accepted as ultimate for all practical purposes. However, there were some objections to the use of the light wavelength standard because of the impossibility of producing pure monochromatic light, as wavelength depends upon the amount of isotope impurity in the elements. However, with rapid advancements in the field of atomic energy, pure isotopes of natural elements have been produced. Cadmium 114, krypton 86, and mercury 198 are possible sources of radiation of wavelengths suitable for the natural standard of length. There is no need to preserve the wavelength standard as it is not a physical one. This standard of length is reproducible, and the error of reproduction can be of the order of 1 part in 100 million.

Finally, in 1960, at the 11th General Conference of Weights and Measures held in Paris, it was recommended and decided that krypton 86 is the most suitable element if used in a hot-cathode discharge lamp maintained at a temperature of 68 K. According to this standard, metre is defined as 1,650,763.73 × wavelengths of the red–orange radiation of a krypton 86 atom in vacuum. This standard can be reproduced with an accuracy of about 1 part in 10^9 and can be accessible to any laboratory.

2.6.1 Modern Metre

The modern metre was defined in the 17th General Conference of Weights and Measures held on 20 October 1983. According to this, the metre is the length of the path travelled by light in vacuum during a time interval of 1/299,792,458 of a second. This standard is technologically more accurate and feasible when compared to the red–orange radiation of a krypton 86 atom and can be realized in practice through the use of an iodine-stabilized helium–neon laser. The reproducibility of the modern metre is found to be 3 parts in 10^{11}, which could be compared to measuring the earth's mean circumference to an accuracy of about 1 mm.

2.7 SUBDIVISIONS OF STANDARDS

The imperial standard yard and metre defined in Sections 2.5.1 and 2.5.2 are master standards that cannot be used for daily measurement purposes. In order to facilitate measurement at different locations depending upon the relative importance of standard, they are subdivided into the following four groups:

Primary standards For defining the unit precisely, there shall be one and only one material standard. Primary standards are preserved carefully and maintained under standard atmospheric conditions so that they do not change their values. This has no direct application to a measuring problem encountered in engineering. These are used only for comparing with secondary standards. International yard and international metre are examples of standard units of length.

Secondary standards These are derived from primary standards and resemble them very closely with respect to design, material, and length. Any error existing in these bars is recorded by comparison with primary standards after long intervals. These are kept at different locations under strict supervision and are used for comparison with tertiary standards (only when it is absolutely essential). These safeguard against the loss or destruction of primary standards.

Tertiary standards Primary and secondary standards are the ultimate controls for standards; these are used only for reference purposes and that too at rare intervals. Tertiary standards are reference standards employed by NPL and are used as the first standards for reference in laboratories and workshops. These standards are replicas of secondary standards and are usually used as references for working standards.

Working standards These are used more frequently in workshops and laboratories. When compared to the other three standards, the materials used to make these standards are of a lower grade and cost. These are derived from fundamental standards and suffer from loss of instrumental accuracy due to subsequent comparison at each level in the hierarchical chain. Working standards include both line and end standards.

Accuracy is one of the most important factors to be maintained and should always be traceable to a single source, usually the national standards of the country. National laboratories of most of the developed countries are in close contact with the BIPM. This is essential because ultimately all these measurements are compared with the standards developed and maintained by the bureaus of standards throughout the world. Hence, there is an assurance that items manufactured to identical dimensions in different countries will be compatible with each other, which helps in maintaining a healthy trade. Figure 2.3 shows the hierarchical classification of standards and Table 2.2 shows a classification based on purpose.

The accuracy of a particular standard depends on subsequent comparisons of standards in the hierarchical ladder. While assessing the accuracy of a standard it is also important to consider factors such as the care taken during comparison, the

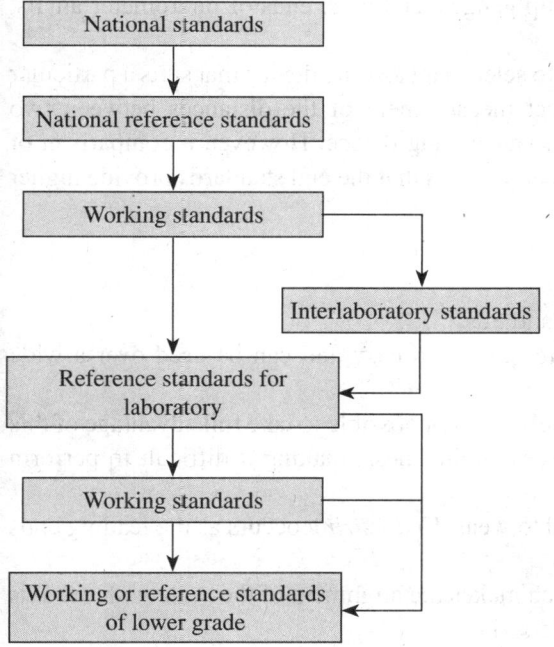

Fig. 2.3 Hierarchical classification of standards

Table 2.2 Classification of standards based on purpose

Standard	Purpose
Reference	Reference
Calibration	Calibration of inspection and working standards
Inspection	Used by inspectors
Working standards	Used by operators

procedure followed for comparison when the recent comparison was made, and the stability of the particular standard. It is imperative that there is a degradation of accuracy in the standards that follow the national standards in the hierarchical classification of standards, as shown in Fig. 2.3.

Comparisons with national standards are seldom performed, as frequent comparisons may degrade their accuracy. For frequent comparisons of the lower-order standards, national reference standards are generally used. For calibration purposes, normally working standards are employed. These working standards stand third in the hierarchical classification and are sometimes called national working standards. Interlaboratory standards and reference standards for laboratories, which are of good quality, are derived from working standards.

2.8 LINE AND END MEASUREMENTS

We all know that sometimes distances have to be measured between two lines or two surfaces or between a line and a surface. When the distance between two engraved lines is used to measure the length, it is called line standard or line measurement. The most common examples are yard and metre. The rule with divisions marked with lines is widely used.

When the distance between two flat parallel surfaces is considered a measure of length, it is known as end standard or end measurement. The end faces of the end standards are hardened to reduce wear and lapped flat and parallel to a very high degree of accuracy. The end standards are extensively used for precision measurement in workshops and laboratories. The most common examples are measurements using slip gauges, end bars, ends of micrometer anvils, vernier callipers, etc.

For an accurate measurement, it is required to select a measuring device that suits a particular measuring situation. For example, for a direct measurement of the distances between two edges, a rule is not suitable because it is a line-measuring device. However, a comparison of characteristics of the line and end standards clearly shows that the end standards provide higher accuracy than line standards.

2.8.1 Characteristics of Line Standards

The following are the characteristics of line standards:
1. Measurements carried out using a scale are quick and easy and can be used over a wide range.
2. Even though scales can be engraved accurately, it is not possible to take full advantage of this accuracy. The engraved lines themselves possess thickness, making it difficult to perform measurements with high accuracy.
3. The markings on the scale are not subjected to wear. *Undersizing* occurs as the leading ends are subjected to wear.
4. A scale does not have a built-in datum, which makes the alignment of the scale with the axis of measurement difficult. This leads to undersizing.
5. Scales are subjected to parallax effect, thereby contributing to both positive and negative reading errors.
6. A magnifying lens or microscope is required for close tolerance length measurement.

2.8.2 Characteristics of End Standards

End standards comprise a set of standard blocks or bars using which the required length is created. The characteristics of these standards are as follows:

1. These standards are highly accurate and ideal for making close tolerance measurement.
2. They measure only one dimension at a time, thereby consuming more time.
3. The measuring faces of end standards are subjected to wear.
4. They possess a built-in datum because their measuring faces are flat and parallel and can be positively located on a datum surface.
5. Groups of blocks/slip gauges are wrung together to create the required size; faulty wringing leads to inaccurate results.
6. End standards are not subjected to parallax errors, as their use depends on the feel of the operator.
7. Dimensional tolerance as close as 0.0005 mm can be obtained.

The end and line standards are initially calibrated at $20 \pm \frac{1}{2}\,°C$. Temperature changes influence the accuracy of these standards. Care should be taken in the manufacturing of end and line standards to ensure that change of shape with time is minimum or negligible. Table 2.3 gives a complete comparison of line and end standards.

Table 2.3   Comparison of line and end standards

Characteristics	Line standard	End standard
Principle of measurement	Distance between two engraved lines is used as a measure of length	Distance between two flat and parallel surfaces is used as a measure of length
Accuracy of measurement	Limited accuracy of ±0.2 mm; magnifying lens or microscope is required for high accuracy	High accuracy of measurement; close tolerances upto ±0.0005 mm can be obtained
Ease and time of measurement	Measurements made using a scale are quick and easy	Measurements made depend on the skill of the operator and are time consuming
Wear	Markings on the scale are not subjected to wear. Wear may occur on leading ends, which results in undersizing	Measuring surfaces are subjected to wear
Alignment	Alignment with the axis of measurement is not easy, as they do not contain a built-in datum	Alignment with the axis of measurement is easy, as they possess a built-in datum
Manufacture	Manufacturing process is simple	Manufacturing process is complex
Cost	Cost is low	Cost is high
Parallax effect	Subjected to parallax effect	No parallax error; their use depends on the feel of the operator
Wringing	Does not exist	Slip gauges are wrung together to build the required size
Examples	Scale (yard and metre)	Slip gauges, end bars, ends of micrometer anvils, and vernier callipers

2.8.3 Transfer from Line Standard to End Standard

We know that primary standards are basically line standards and that end standards are practical workshop standards. Line standards are highly inconvenient for general measurement purposes and are usually used to calibrate end standards, provided that the length of primary line standard is accurately known. There is a probability of the existence of a very small error in the primary standard, which may not be of serious concern. It is important to accurately determine the error in the primary standard so that the lengths of the other line standards can be precisely evaluated when they are compared with it.

From the aforementioned discussions it is clear that when measurements are made using end standards, the distance is measured between the working faces of the measuring instrument, which are flat and mutually parallel. A composite line standard is used to transfer a line standard to an end standard.

Figure 2.4 shows a primary line standard of a basic length of 1 m whose length is accurately known. A line standard having a basic length of more than 1 m is shown in Fig. 2.5. This line standard consists of a central length bar that has a basic length of 950 mm. Two end blocks of 50 mm each are wrung on either end of the central bar. Each end block contains an engraved line at the centre.

The composite line standard whose length is to be determined is compared with the primary line standard, and length L is obtained as using the following formula:

$$L = L_1 + b + c$$

The four different ways in which the two end blocks can be arranged using all possible combinations and then compared with the primary line standard are

$$L = L_1 + b + c$$
$$L = L_1 + b + d$$
$$L = L_1 + a + c$$
$$L = L_1 + a + d$$

Summation of these four measurements gives

$$4L = 4L_1 + 2a + 2b + 2c + 2d$$

$$= 4L_1 + 2(a + b) + 2(c + d)$$

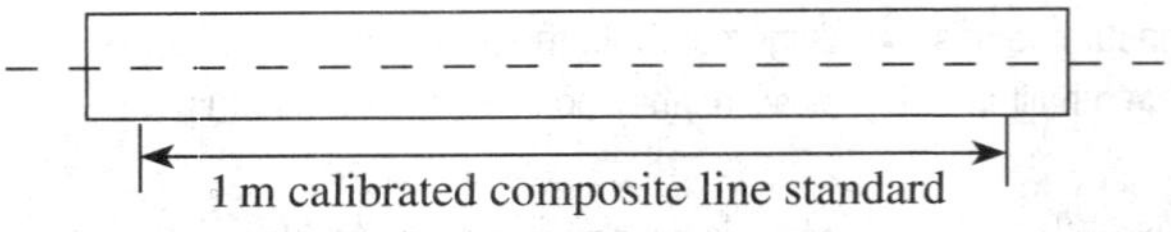

Now, the combination of blocks $(a + b)$ and $(c + d)$ are unlikely to be of the same length. The two are therefore compared; let the difference between them be x, as shown in Fig. 2.6.

$$(c + d) = (a + b) + x$$

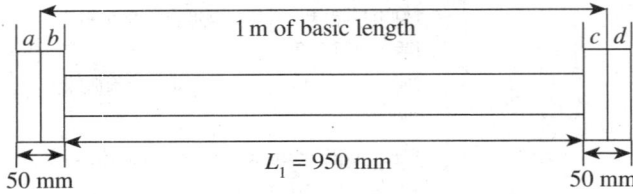

Fig. 2.4 Basic length of primary line standard

Fig. 2.5 Basic length of primary line standard with end blocks

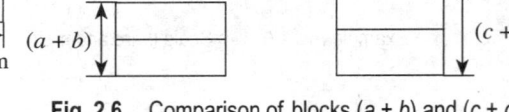

Fig. 2.6 Comparison of blocks $(a + b)$ and $(c + d)$

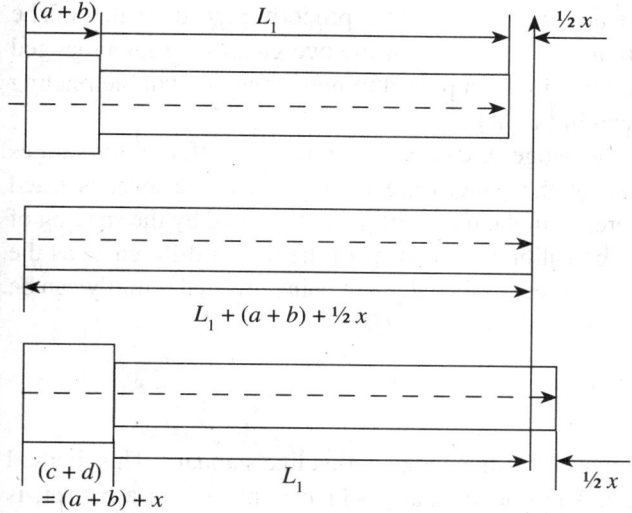

Fig. 2.7 End standard of known length as obtained from the standard

Substituting the value of $(c + d)$,

$$4L = 4L_1 + 2(a + b) + 2[(a + b) + x)]$$
$$4L = 4L_1 + 2(a + b) + 2(a + b) + 2x$$
$$4L = 4L_1 + 4(a + b) + 2x$$

Dividing by 4, we get

$$L = L_1 + (a + b) + \tfrac{1}{2}x$$

An end standard of known length can now be obtained consisting of either $L_1 + (a + b)$ or $L_1 + (c + d)$, as shown in Fig. 2.7. The length of $L_1 + (a + b)$ is $L_1 + (a + b) + \tfrac{1}{2}x$ less $\tfrac{1}{2}x$, where $(a + b)$ is shorter of the two end blocks. The length of $L_1 + (c + d)$ is $L_1 + (a + b) + \tfrac{1}{2}x$ plus $\tfrac{1}{2}x$, where $(c + d)$ is longer of the two end blocks. The calibrated composite end bar can be used to calibrate a solid end standard of the same basic length.

2.9 BROOKES LEVEL COMPARATOR

The Brookes level comparator (Fig. 2.8) is used to calibrate standards by comparing with a master standard. End standards can be manufactured very accurately using a Brookes level comparator. A.J.C. Brookes devised this simple method in 1920 and hence the name. The Brookes level comparator has a very accurate spirit level. In order to achieve an accurate comparison, the spirit level is supported on balls so that it makes only a point contact with the gauges.

The table on which the gauges are placed for comparison are first levelled properly using the spirit level. The two gauges (the master standard gauge and the standard gauge) that are

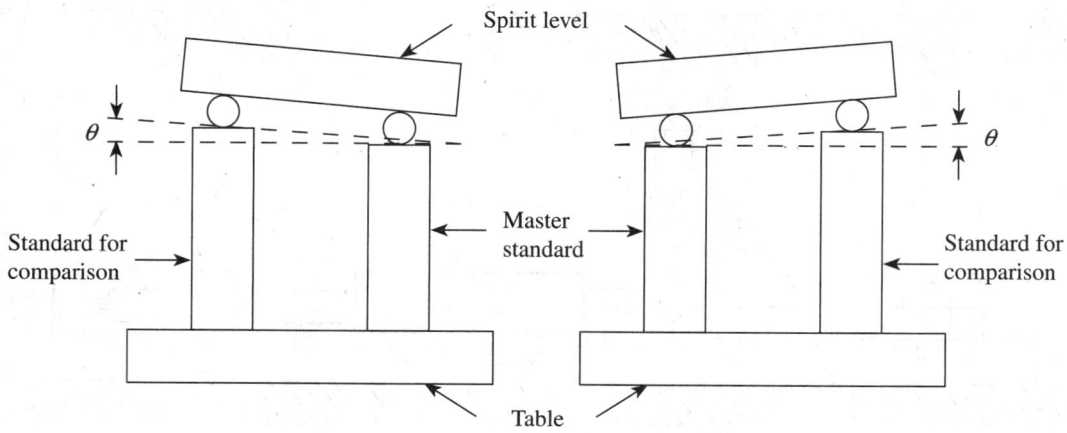

Fig. 2.8 Calibration using a Brookes level comparator

to be compared are wrung on the table and the spirit level is properly placed on them. The bubble reading is recorded at this position. The positions of the two gauges are interchanged by rotating the table by 180°. The spirit level is again placed to note down the bubble reading at this position. The arrangement is shown in Fig. 2.8.

The two readings will be the same if the gauges are of equal length and different for gauges of unequal lengths. When the positions of the gauges are interchanged, the level is tilted through an angle equal to twice the difference in the height of gauges divided by the spacing of level supports. The bubble readings can be calibrated in terms of the height difference, as the distance between the two balls is fixed. The effect of the table not being levelled initially can be eliminated because of the advantage of turning the table by 180°.

2.10 DISPLACEMENT METHOD

The displacement method is used to compare an edge gauge with a line standard. This method is schematically represented in Fig. 2.9. The line standard, which is placed on a carrier, is positioned such that line A is under the cross-wires of a fixed microscope, as seen in Fig. 2.9(a). The spindle of the micrometer is rotated until it comes in contact with the projection on the carrier and then the micrometer reading is recorded. The carrier is moved again to position line B under the cross-wires of the microscope. At this stage, the end gauge is inserted as shown in Fig. 2.9(b) and the micrometer reading is recorded again. Then the sum of the length of the line standard and the difference between the micrometer readings will be equal to the length of the end gauge.

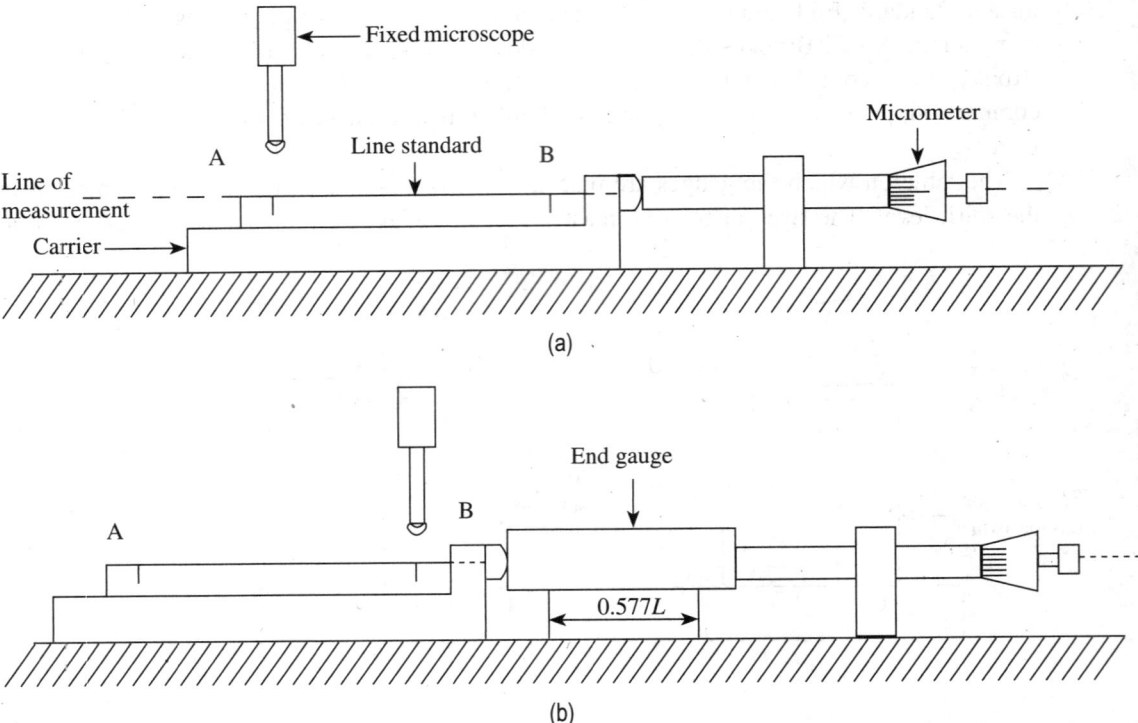

Fig. 2.9 Displacement method (a) With line standard only (b) With line and end standards combined

2.11 CALIBRATION OF END BARS

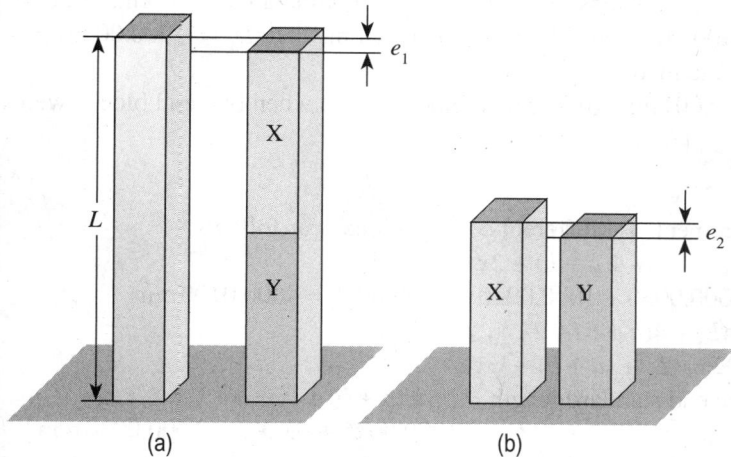

(a)   (b)

Fig. 2.10 Calibration of end bars (a) Comparison of metre bar and end bars wrung together (b) Comparison of individual end bars

In order to calibrate two bars having a basic length of 500 mm with the help of a one piece metre bar, the following procedure is adopted.

The metre bar to be calibrated is wrung to a surface plate. The two 500 mm bars to be calibrated are wrung together to form a bar that has a basic length of 1 m, which in turn is wrung to the surface plate beside the metre bar, as shown in Fig. 2.10(a). The difference in height e_1 is obtained. The two 500 mm bars are then compared to determine the difference in length, as shown in Fig. 2.10(b).

Let L_X and L_Y be the lengths of the two 500 mm bars. Let e_1 be the difference in height between the calibrated metre bar and the combined lengths of X and Y. Let the difference between the lengths of X and Y be e_2. Let L be the actual length of the metre bar.

Then the first measurement gives a length of $L \pm e_1 = L_X + L_Y$, depending on whether the combined length of L_X and L_Y is longer or shorter than L.

The second measurement yields a length of $L_X \pm e_2 = L_Y$, again depending on whether X is longer or shorter than Y.

Then substituting the value of L_Y from the second measurement in the first measurement, we get

$$L \pm e_1 = L_X + L_X \pm e_2 = 2L_X \pm e_2$$

or

$$2L_X = L \pm e_1 \pm e_2$$

Therefore, $L_X = (L \pm e_1 \pm e_2)/2$ and $L_Y = L_X \pm e_2$

For calibrating three, four, or any other number of length standards of the same basic size, the same procedure can be followed. One of the bars is used as a reference while comparing the individual bars and the difference in length of the other bar is obtained relative to this bar.

2.12 NUMERICAL EXAMPLES

Example 2.1   It is required to obtain a metre standard from a calibrated line standard using a composite line standard. The actual length of the calibrated line standard is 1000.015 mm. The composite line standard comprises a length bar having a basic length of 950 mm and two end blocks, $(a + b)$ and $(c + d)$, each having a basic length of 50 mm. Each end block contains an engraved line at the centre.

Four different measurements were obtained when comparisons were made between the calibrated line standard and the composite bar using all combinations of end blocks: L_1 = 1000.0035 mm, L_2 = 1000.0030 mm, L_3 = 1000.0020 mm, and L_4 = 1000.0015 mm. Determine the actual length of the metre bar.

Block $(a + b)$ was found to be 0.001 mm greater than block $(c + d)$ when two end blocks were compared with each other.

Solution

The sum of all the four measurements for different combinations is as follows:
$$4L = L_1 + L_2 + L_3 + L_4 = 4L_1 + 4(a + b) + 2x$$
However, $4L = 1000.0035 + 1000.0030 + 1000.0020 + 1000.0015 = 4000.0100$ mm
Therefore, $4L = 4000.0100 = 4L_1 + 4(a + b) + 2x$
Dividing by 4, we get $1000.0025 = L_1 + (a + b) + \frac{1}{2}x$
When block $(a + b)$ is used, the end standard length $= L_1 + (a + b) + \frac{1}{2}x$
$$= 1000.0025 + 0.0005 = 1000.0030 \text{ mm}$$
When block $(c + d)$ is used, the end standard length $= L_1 + (a + b) - \frac{1}{2}x$
$$= 1000.0025 - 0.0005 = 1000.0020 \text{ mm}$$

Example 2.2   A calibrated metre end bar, which has an actual length of 1000.0005 mm, is to be used in the calibration of two bars X and Y, each having a basic length of 500 mm. When compared with the metre bar, the sum of L_X and L_Y is found to be shorter by 0.0003 mm. When X and Y are compared, it is observed that X is 0.0004 mm longer than Y. Determine the actual length of X and Y.

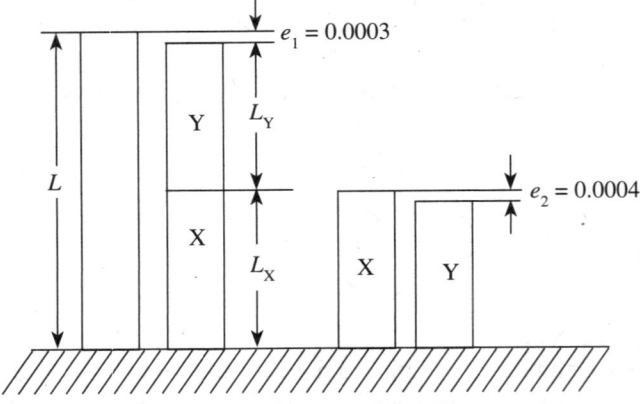

Fig. 2.11   Arrangement of end bars

Solution

The arrangement of gauges is as shown in Fig. 2.11.
From first principles, we have
$$L - e_1 = L_X + L_Y$$
However, $L_X = L_Y + e_2$
Therefore, we get $L - e_1 = 2L_Y + e_2$
Or, $L_Y = (L - e_1 - e_2)/2$
Substituting the values for L, e_1, and e_2, we get
$$L_Y = (1000.0005 - 0.0003 - 0.0004)/2$$
$$L_Y = 999.9998/2 \text{ mm}$$
i.e., $L_Y = 499.9999$ mm

We have $L_X = L_Y + e_2$
$$L_X = 499.9999 + 0.0004 = 500.0003 \text{ mm}$$

Example 2.3   Three 200 mm gauges to be calibrated are measured on a level comparator by wringing them together and then comparing them with a 600 mm gauge. The 600 mm gauge has an actual length of 600.0025 mm, and the three gauges together have a combined length of 600.0035 mm. When the three gauges are intercompared, it is found that gauge A is longer than gauge B by 0.0020 mm but shorter than gauge C by 0.001 mm. Determine the length of each gauge.

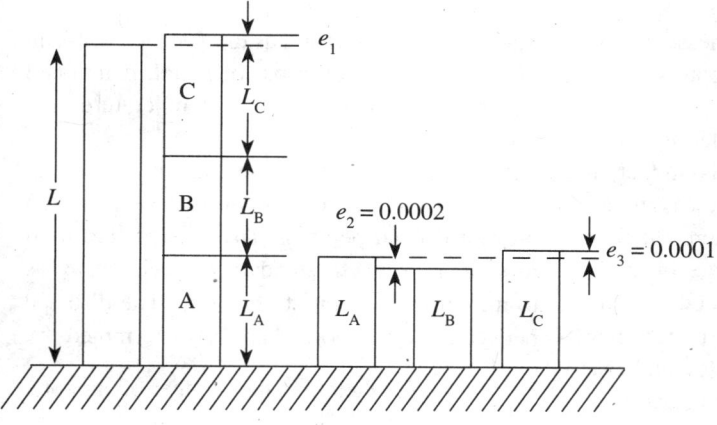

Fig. 2.12 Arrangement of end bars

Solution

Figure 2.12 shows the arrangement of gauges.

Let L_A, L_B, and L_C be the lengths of gauges A, B, and C, respectively. The combined length of L_A, L_B, and L_C is greater than L by $e_1 = 600.0035 - 600.0025 = 0.001$ mm.

In addition, we have, $e_2 = 0.0020$ mm and $e_3 = 0.0010$ mm.

From Fig. 2.12, we have

$$L = L_A + L_B + L_C - e_1$$
$$L_A = L_B + e_2$$
$$L_C = L_A + e_3$$
$$L_C = L_B + e_2 + e_3$$

Thus,

$$L = L_B + e_2 + L_B + L_B + e_2 + e_3 - e_1$$

Simplifying,

$$L = 3L_B + 2e_2 + e_3 - e_1$$
$$\text{or, } 3L_B = L - 2e_2 + e_3 - e_1$$
$$L_B = (L - 2e_2 - e_3 + e_1)/3$$

Substituting the values, we get

$$L_B = 600.0025 - 2 \times 0.0020 - 0.0010 + 0.001$$
$$L_B = 599.9985/3$$
$$L_B = 199.9995 \text{ mm}$$
$$L_A = L_B + e_2$$
$$L_A = 199.9995 + 0.0020 = 200.0015 \text{ mm}$$
$$L_C = L_A + e_3$$
$$L_C = 200.0015 + 0.0010 = 200.0025 \text{ mm}$$

A QUICK OVERVIEW

- Mass production, an idea of the last industrial revolution, has become very popular and synonymous with the present manufacturing industry, and is a necessity for manufacturing identical parts. Today, almost all manufacturing units practise the principle of interchangeability of manufacture. In order to accomplish complete interchangeability of manufacture in industries, it is essential to a have a measurement system that can adequately define the features of the components/products to the accuracy required and standards of sufficient accuracy to aid the measuring system.

- Metrology, which literally means science of measurements, not only deals with the establishment, reproduction, protection, mainten-ance, and transfer or conversion of units of measurements and their standards, but is also concerned with the correctness of measurement. One of the major functions of metrology is to establish international standards for measurement used by all countries in the world in both science and industry.

- It is impossible to perform trade in national and international arenas without standards. In fact, a good system of standards is very much essential for fair international trade and commerce; it also helps accomplish complete interchangeability of manufacture. Establishment of a good system of standards allows the manufacturer to define the required attributes to the accuracy and the standards having adequate accuracy to act as a support mechanism.

- Standards play a vital role for manufacturers across the world in achieving consistency, accuracy, precision, and repeatability in measurements and in supporting the system that enables the manufacturers to make such measurements. In fact, any progress made in manufacturing and other business sectors at the international arena necessitates that activities at high scientific and technical levels have to be accomplished by progress in metrology. Further automation in manufacturing industries demands a very high level of accuracy, precision, and reliability.

- The National Physical Laboratory (NPL) established in UK in 1900 is a public institution for standardizing and verifying instruments, testing materials, and determining physical constants. NPLI (India), which was established in 1947 at New Delhi under the CSIR, has to comply with the statutory obligation of realizing, establishing, maintaining, reproducing, and updating the national standards of measurement and calibration facilities for different parameters.

- Yard or metre is defined as the distance between two scribed lines on a bar of metal maintained under certain conditions of temperature and support.

- Airy points can be defined as the points at which a horizontal rod is optionally supported to prevent it from bending. These points are used to support a length standard in such a way as to minimize error due to bending. The distance between the supports can be determined by $\dfrac{1}{\sqrt{n^2-1}} \times L$, where n is the number of supports and L the length of the bar. When it is supported by two points, $n = 2$; the supports should be at an equal distance from each end to a position where they are $0.577L$ apart.

- The modern metre is defined as the length of the path travelled by light in vacuum during a time interval of 1/299,792,458 of a second. The reproducibility of the modern metre is found to be 3 parts in 10^{11}, which could be compared to measuring the earth's mean circumference to an accuracy of about 1 mm.

- When the distance between two engraved lines is used to measure the length, it is called line standard or line measurement. The most common examples are yard and metre.

- When the distance between two flat parallel surfaces is considered as a measure of length, it is known as end standard or end measurement. The most common examples are measurements using slip gauges, end bars, ends of micrometer anvils, vernier callipers, etc.

- Subdivision of line standards can be carried out using Brookes level comparator.

MULTIPLE-CHOICE QUESTIONS

1. The modern metre is
   - (a) the length of the path travelled by light in vacuum during a time interval of 1/29,97,92,458 of a second
   - (b) 16,50,763.73 × wavelengths of the red–orange radiation of a krypton 86 atom in vacuum
   - (c) the length of the path travelled by light in vacuum during a time interval of 1/399,792,458 of a second
   - (d) 1,660,793.73 × wavelengths of the red–orange radiation of a krypton 86 atom in vacuum

2. Reproducibility of the modern metre is of the order of
   (a) 1 part in 100 million
   (b) 3 parts in 10^{11}
   (c) 3 parts in 10^9
   (d) 1 part in 1000 million

3. When a measurement is made between two flat parallel surfaces, it is called
   (a) line measurement
   (b) direct measurement
   (c) standard measurement
   (d) end measurement

4. A public institution for standardizing and verifying instruments, testing materials, and determining physical constants is called
   (a) BPL
   (b) IPL
   (c) NPL
   (d) NMI

5. Subdivision of end standards is carried out using
   (a) Crook's level comparator
   (b) Brookes level comparator
   (c) Johansson Mikrokator
   (d) Sigma electronic comparator

6. When airy points support a length standard at two points, they will be apart by a distance of
   (a) $0.577L$
   (b) $0.575L$
   (c) $0.757L$
   (d) $0.775L$

7. Which one of following is true?
   (a) Line standard does not have parallax error.
   (b) End standard does not have parallax error.
   (c) Both line and end standards have parallax error.
   (d) Both line and end standards do not have parallax error.

8. Which of the following formulae can be used to determine the distance d between the supports?

   (a) $\dfrac{1}{\sqrt{n^2-1}} \times L^2$

   (b) $\dfrac{1}{\sqrt{n^2-1}} \times L$

   (c) $\dfrac{1}{\sqrt{n^2-1}}$

   (d) $\dfrac{L}{\sqrt{n-1}}$

9. Alignment with the axis of measurement is easy in end standards because they possess
   (a) parallax effect   (c) airy points
   (b) high accuracy   (d) a built-in datum

10. A line standard is transferred to an end standard by using
    (a) a composite line standard
    (b) a built-in datum
    (c) workshop standards
    (d) airy points

11. Both line and end standards are initially calibrated at
    (a) $18 \pm 1\,°C$ (c) $20 \pm \frac{1}{2}\,°C$
    (b) $18 \pm \frac{1}{2}\,°C$ (d) $20 \pm 1\,°C$

12. In a line standard, distance is measured between
    (a) two flat parallel surfaces
    (b) two engraved lines
    (c) two points
    (d) two inclined surfaces

13. Wringing of slip gauges is used in
    (a) line measurement
    (b) primary standards
    (c) both line and end measurements
    (d) end measurement

14. Comparison of the characteristics of line and end standards clearly shows that the accuracy
    (a) in line standard is greater than in end standard
    (b) in end standard is greater than in line standard
    (c) of both are equal
    (d) cannot be determined by the comparison of characteristics only

15. In the hierarchical classification of standards, the accuracy in the standards
    (a) is degraded
    (b) is improved
    (c) does not change
    (d) is not related to hierarchical classifications

REVIEW QUESTIONS

1. Explain the role of standards of measurements in modern industry.
2. Write a brief note on the evolution of standards.
3. What are material standards? List their disadvantages.
4. Describe the following with neat sketches:
   (a) Imperial standard yard
   (b) International prototype of metre
5. What are airy points? State the condition to achieve it.
6. Explain why length bars should be supported correctly in the horizontal position.
7. What is the current definition of metre? What is its reproducibility?
8. Write a note on wavelength standards.
9. List the advantages of wavelength standards.
10. With the help of a block diagram, explain the hierarchical classification of standards.
11. Explain the contribution of NPL to metrology.
12. List the objectives of NPL.
13. How are standards subdivided?
14. Distinguish between primary, secondary, tertiary, and working standards.
15. What do you understand by line and end standards? Give examples.
16. Discuss the characteristics of line and end standards.
17. With a neat sketch, explain how Brookes level comparator is used to subdivide end standards.
18. Differentiate between line and end standards.
19. Describe the procedure to transfer from line standard to end standard.
20. With an example, explain how end standards are derived from line standards.

PROBLEMS

1. It is required to calibrate four length bars A, B, C, and D, each having a basic length of 250 mm. A calibrated length bar of 1000 mm is to be used for this purpose. The 1000 mm bar has an actual length of 999.9991 mm. It is also observed that
$L_B = L_A + 0.0001$ mm
$L_C = L_A + 0.0005$ mm
$L_D = L_A + 0.0001$ mm
$L_A + L_B + L_C + L_D = L + 0.0003$ mm
Determine L_A, L_B, L_C, and L_D.

2. A calibrated end bar having an actual length of 500.0005 mm is to be used to calibrate two end bars A and B, each having a basic length of 250 mm. On comparison, the combined length $L_A + L_B$ is found to be shorter than the 500 mm end bar by 0.0003 mm. When the two end bars A and B are intercompared with each other, A is found to be 0.0006 mm longer than B. Determine L_A and L_B.

3. Three 200 mm end bars (P, Q, and R) are measured by first wringing them together and comparing with a 600 mm bar. They are then intercompared. The 600 mm bar has a known error of 40 μm and the combined length of the three end bars is found to be 64 μm less than the 600 mm bar. It is also observed that bar P is 18 μm longer than bar Q and 23 μm longer than bar S. Determine the lengths of the three end bars.

Limits, Fits, and Tolerances

After studying this chapter, the reader will be able to

- understand the importance of manufacturing components to specified sizes
- elucidate the different approaches of interchangeable and selective assembly
- appreciate the significance of different types of limits, fits, and tolerances in design and manufacturing fields, which are required for efficient and effective performance of components/products
- utilize the principle of limit gauging and its importance in inspection in industries
- design simple GO and NOT gauges used in workshops/ inspection

3.1 INTRODUCTION

Although any two things found in nature are seldom identical, they may be quite similar. This is also true in the case of manufacturing of different components for engineering applications. No two parts can be produced with identical measurements by any manufacturing process. A manufacturing process essentially comprises five m's—man, machine, materials, money, and management.

Variations in any of the first three elements induce a change in the manufacturing process. All the three elements are subjected to natural and characteristic variations. In any production process, regardless of how well it is designed or how carefully it is maintained, a certain amount of natural variability will always exist. These natural variations are random in nature and are the cumulative effect of many small, essentially uncontrollable causes. When these natural variations in a process are relatively small, we usually consider this to be an acceptable level of process performance.

Usually, variability arises from improperly adjusted machines, operator error, tool wear, and/or defective raw materials. Such characteristic variability is generally large when compared to the natural variability. This variability, which is not a part of random or chance cause pattern, is referred to as 'assignable causes'. Characteristic variations can be attributed to assignable causes that can easily be identified and controlled. However, this has to be achieved economically, which brings in the fourth element. Characteristic variability causes variations in the size of components. If the process can be kept under control, that is, all the assignable and controllable causes of variations have been eliminated or controlled, the size variations will be well within the prescribed limits. These variations can be modified through operator or management action.

Production processes must perform consistently to meet the production and design requirements. In order to achieve this, it is essential to keep the process under control. Thus, when the process is under control, distribution of most of the measured values will be around the mean value in a more or less symmetrical way, when plotted on a chart. It is therefore impossible to produce a part to an exact size or basic size and some variations, known as tolerances, need to be allowed. Some variability in dimension within certain limits must be tolerated during manufacture, however precise the process may be. The permissible level of tolerance depends on the functional requirements, which cannot be compromised.

No component can be manufactured precisely to a given dimension; it can only be made to lie between two limits, upper (maximum) and lower (minimum). The designer has to suggest these tolerance limits, which are acceptable for each of the dimensions used to define shape and form, and ensure satisfactory operation in service. When the tolerance allowed is sufficiently greater than the process variation, no difficulty arises. The difference between the upper and lower limits is termed *permissive tolerance.*

For example, a shaft has to be manufactured to a diameter of 40 ± 0.02 mm. This means that the shaft, which has a basic size of 40 mm, will be acceptable if its diameter lies anywhere between the limits of sizes, that is, an upper limit of 40.02 mm and a lower limit of 39.98 mm. Then permissive tolerance is equal to 40.02 − 39.98 = 0.04. *Basic* or *nominal size* is defined as the size based on which the dimensional deviations are given.

In any industry, a manufactured product consists of many components. These components when assembled should have a proper fit, in order for the product to function properly and have an extended life. Fit depends on the correct size relationships between the two mating parts. Consider the example of rotation of a shaft in a hole. Enough clearance must be provided between the shaft and the hole to allow an oil film to be maintained for lubrication purpose. If the clearance is too small, excessive force would be required to rotate the shaft. On the other hand, if the clearance is too wide, there would be vibrations and rapid wear resulting in ultimate failure. Therefore, the desired clearance to meet the requirements has to be provided. Similarly, to hold the shaft tightly in the hole, there must be enough interference between the two so that forces of elastic compression grip them tightly and do not allow any relative movement between them.

An ideal condition would be to specify a definite size to the hole and vary the shaft size for a proper fit or vice versa. Unfortunately, in practice, particularly in mass production, it is not possible to manufacture a part to the exact size due to the inherent inaccuracy of manufacturing methods. Even if a part is manufactured to the exact size by chance, it is not possible to measure it accurately and economically during machining. In addition, attempts to manufacture to the exact size can increase the production cost.

Dimensional variations, although extremely small, do exist because of the inevitable inaccuracies in tooling, machining, raw material, and operators. If efforts are made to identify and reduce or eliminate common causes of variation, that is, if the process is kept under control, then the resultant frequency distribution of dimensions produced will have a normal

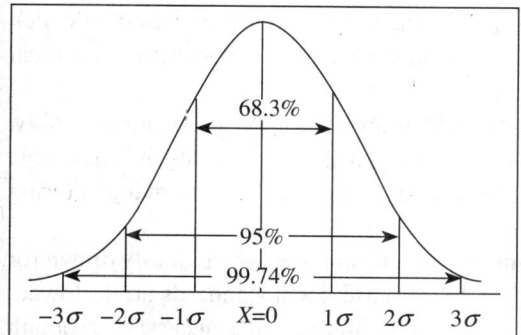

Fig. 3.1 Normal or Gaussian frequency distribution

or Gaussian distribution, that is, 99.74% parts will be well within $\pm 3\sigma$ limits of the mean value, as shown in Fig. 3.1. Thus, it is possible to express the uncertainty in measurement as a multiple of standard deviation. This value is determined by the probability that a measurement will fall outside the stated limits.

3.2 PRINCIPLE OF INTERCHANGEABILITY

For manufacturing a large number of components, it is not economical to produce both the mating parts (components) using the same operator. Further, such parts need to be manufactured within minimum possible time without compromising on quality. To enable the manufacture of identical parts, mass production, an idea of the last industrial revolution that has become very popular and synonymous with the present manufacturing industry, becomes inevitable.

Modern production techniques require that a complete product be broken into various component parts so that the production of each part becomes an independent process, leading to specialization. The various components are manufactured in one or more batches by different persons on different machines at different locations and are then assembled at one place.

To achieve this, it is essential that the parts are manufactured in bulk to the desired accuracy and, at the same time, adhere to the limits of accuracy specified. Manufacture of components under such conditions is called *interchangeable manufacture*.

When interchangeable manufacture is adopted, any one component selected at random should assemble with any other arbitrarily chosen mating component. In order to assemble with a predetermined fit, the dimensions of the components must be confined within the permissible tolerance limits. By *interchangeable assembly*, we mean that identical components, manufactured by different operators, using different machine tools and under different environmental conditions, can be assembled and replaced without any further modification during the assembly stage and without affecting the functioning of the component when assembled. Production on an interchangeable basis results in an increased productivity with a corresponding reduction in manufacturing cost. Modern manufacturing techniques that complement mass production of identical parts facilitating interchangeability of components have been developed. When components are produced in bulk, unless they are interchangeable, the purpose of mass production is not fulfilled.

For example, consider the assembly of a shaft and a part with a hole. The two mating parts are produced in bulk, say 1000 each. By interchangeable assembly any shaft chosen randomly should assemble with any part with a hole selected at random, providing the desired fit.

Another major advantage of interchangeability is the ease with which replacement of defective or worn-out parts is carried out, resulting in reduced maintenance cost. In addition, the operator, by performing the same limited number of operations, becomes a specialist in that work. By achieving specialization in labour, there will be a considerable reduction in manufacturing and assembly time and enhancement in quality. Interchangeable manufacture increases productivity and reduces production and time costs.

In order to achieve interchangeability, certain standards need to be followed, based on which interchangeability can be categorized into two types—universal interchangeability and local interchangeability.

When the parts that are manufactured at different locations are randomly chosen for assembly, it is known as universal interchangeability. To achieve universal interchangeability, it is desirable that common standards be followed by all and the standards used at various manufacturing locations be traceable to international standards.

When the parts that are manufactured at the same manufacturing unit are randomly drawn for assembly, it is referred to as local interchangeability. In this case, local standards are followed, which in turn should be traceable to international standards, as this becomes necessary to obtain the spares from any other source.

3.2.1 Selective Assembly Approach

Today's consumers desire products that are of good quality and, at the same time, reliable and available at attractive prices. Further, in order to achieve interchangeability, it is not economical to manufacture parts to a high degree of accuracy. It is equally important to produce the part economically and, at the same time, maintain the quality of the product for trouble-free operation. Sometimes, for instance, if a part of minimum limit is assembled with a mating part of maximum limit, the fit obtained may not fully satisfy the functional requirements of the assembly. The reason may be attributed to the issues of accuracy and uniformity that may not be satisfied by the certainty of the fits given under a fully interchangeable system. It should be realized that, in practice, complete interchangeability is not always feasible; instead, selective assembly approach can be employed. Attaining complete interchangeability in these cases involves some extra cost in inspection and material handling, as selective assembly approach is employed wherein the parts are manufactured to wider tolerances. In selectively assembly, despite being manufactured to rather wide tolerances, the parts fit and function as if they were precisely manufactured in a precision laboratory to very close tolerances.

The issue of clearances and tolerances when manufacturing on an interchangeable basis is rather different from that when manufacturing on the basis of selective assembly. In interchangeability of manufacture, minimum clearance should be as small as possible as the assembling of the parts and their proper operating performance under allowable service conditions. Maximum clearance should be as great as the functioning of the mechanisms permits. The difference between maximum clearance and minimum clearance establishes the sum of the tolerances on companion parts. To manufacture the parts economically on interchangeable basis, this allowable difference must be smaller than the normal permissible manufacturing conditions. In such situations, selective assembly may be employed. This method enables economical manufacture of components as per the established tolerances. In selective assembly, the manufactured components are classified into groups according to their sizes. Automatic gauging is employed for this purpose. Both the mating parts are segregated according to their sizes, and only matched groups of mating parts are assembled. This ensures complete protection and elimination of defective assemblies, and the matching costs are reduced because the parts are produced with wider tolerances.

Selective assembly finds application in aerospace and automobile industries. A very pertinent and practical example is the manufacture and assembly of ball and bearing units, as the tolerances desired in such industries are very narrow and impossible to achieve economically by any sophisticated machine tools. Balls are segregated into different groups depending on their size to enable the assembly of any bearing with balls of uniform size. In a broader sense, a combination of both interchangeable and selective assemblies exists in modern-day manufacturing industries, which help to manufacture quality products.

3.3 TOLERANCES

To satisfy the ever-increasing demand for accuracy, the parts have to be produced with less dimensional variation. Hence, the labour and machinery required to manufacture a part has become more expensive. It is essential for the manufacturer to have an in-depth knowledge of the tolerances to manufacture parts economically but, at the same time, adhere to quality and reliability aspects. In fact, precision is engineered selectively in a product depending on the functional requirements and its application. To achieve an increased compatibility between mating parts to enable interchangeable assembly, the manufacturer needs to practise good tolerance principles. Therefore, it is necessary to discuss some important principles of tolerances that are usually employed for manufacturing products.

We know that it is not possible to precisely manufacture components to a given dimension because of the inherent inaccuracies of the manufacturing processes. The components are manufactured in accordance with the permissive tolerance limits, as suggested by the designer, to facilitate interchangeable manufacture. The permissible limits of variations in dimensions have to be specified by the designer in a logical manner, giving due consideration to the functional requirements. The choice of the tolerances is also governed by other factors such as manufacturing process, cost, and standardization.

Tolerance can be defined as the magnitude of permissible variation of a dimension or other measured value or control criterion from the specified value. It can also be defined as the total variation permitted in the size of a dimension, and is the algebraic difference between the upper and lower acceptable dimensions. It is an absolute value.

The basic purpose of providing tolerances is to permit dimensional variations in the manufacture of components, adhering to the performance criterion as established by the specification and design. If high performance is the sole criterion, then functional requirements dictate the specification of tolerance limits; otherwise, the choice of setting tolerance, to a limited extent, may be influenced and determined by factors such as methods of tooling and available manufacturing equipment. The industry follows certain approved accuracy standards, such as ANSI (American National Standards Institute) and ASME (American Society of Mechanical Engineers), to manufacture different parts.

3.3.1 Computer-aided Modelling

Nowadays, computers are widely being employed in the design and manufacture of parts. Most leading design tools such as AEROCADD, AUTOCAD, and Solid Works, which are currently being used in industries, are equipped with tolerance features. The algorithms and programming codes that are in existence today are aimed at enhancing the accuracy with minimum material wastage. These programs have the capability of allotting tolerance ranges for different miniature parts of complex mechanical systems.

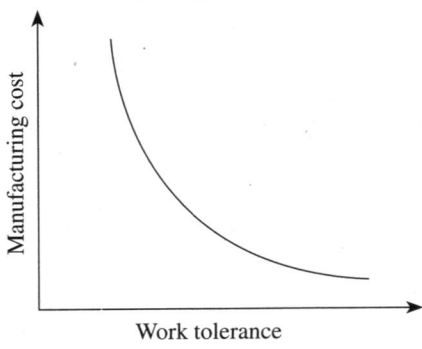

Fig. 3.2 Relationship between work tolerance and manufacturing cost

3.3.2 Manufacturing Cost and Work Tolerance

It is very pertinent to relate the production of components within the specified tolerance zone to its associated manufacturing cost. As the permissive tolerance goes on decreasing, the manufacturing cost incurred to achieve it goes on increasing exponentially. When the permissive tolerance limits are relaxed without degrading the functional requirements, the manufacturing cost decreases. This is clearly illustrated in Fig. 3.2.

Further, in order to maintain such close tolerance limits, manufacturing capabilities have to be enhanced, which certainly increases the manufacturing cost. The components manufactured have to undergo a closer scrutiny, which demands stringent inspection procedures and adequate instrumentation. This increases the cost of inspection. Hence, tolerance is a trade-off between the economical production and the accuracy required for proper functioning of the product. In fact, the tolerance limits specified for the components to be manufactured should be just sufficient to perform their intended functions.

3.3.3 Classification of Tolerance

Tolerance can be classified under the following categories:
1. Unilateral tolerance
2. Bilateral tolerance
3. Compound tolerance
4. Geometric tolerance

Unilateral Tolerance

When the tolerance distribution is only on one side of the basic size, it is known as unilateral tolerance. In other words, tolerance limits lie wholly on one side of the basic size, either above or below it. This is illustrated in Fig. 3.3(a). Unilateral tolerance is employed when precision fits are required during assembly. This type of tolerance is usually indicated when the mating parts are also machined by the same operator. In this system, the total tolerance as related to the basic size is in one direction only. Unilateral tolerance is employed in the drilling process wherein dimensions of the hole are most likely to deviate in one direction only, that is, the hole is always oversized rather than undersized. This system is preferred because the basic size is used for the GO limit gauge. This helps in standardization of the GO gauge, as holes and shafts of different grades will have the same lower and upper limits, respectively. Changes in the magnitude of the tolerance affect only the size of the other gauge dimension, the NOT GO gauge size.

Example $40 \, {}^{+0.02}_{+0.01}$, $40 \, {}^{+0.02}_{-0.00}$, $40 \, {}^{-0.01}_{-0.02}$, $40 \, {}^{+0.00}_{-0.02}$

Bilateral Tolerance

When the tolerance distribution lies on either side of the basic size, it is known as bilateral tolerance. In other words, the dimension of the part is allowed to vary on both sides of the basic

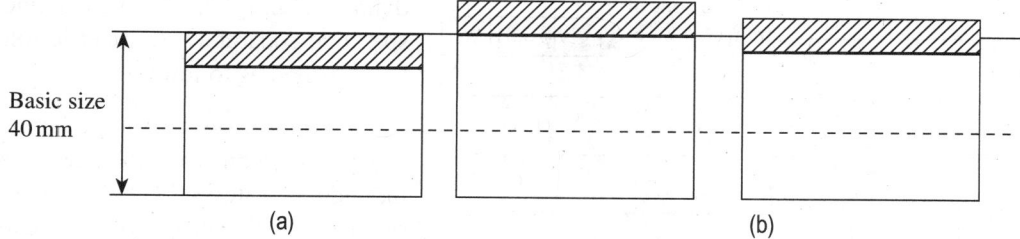

Fig. 3.3 Tolerances (a) Unilateral (b) Bilateral

size but may not be necessarily equally disposed about it. The operator can take full advantage of the limit system, especially in positioning a hole. This system is generally preferred in mass production where the machine is set for the basic size. This is depicted in Fig. 3.3(b). In case unilateral tolerance is specified in mass production, the basic size should be modified to suit bilateral tolerance.

Example
$$40 \pm 0.02, \ 40 \ ^{+0.02}_{-0.01}$$

Compound Tolerance

When tolerance is determined by established tolerances on more than one dimension, it is known as compound tolerance For example, tolerance for the dimension R is determined by the combined effects of tolerance on 40 mm dimension, on 60°, and on 20 mm dimension. The tolerance obtained for dimension R is known as compound tolerance (Fig. 3.4). In practice, compound tolerance should be avoided as far as possible.

Geometric Tolerance

Normally, tolerances are specified to indicate the actual size or dimension of a feature such as a hole or a shaft. In order to manufacture components more accurately or with minimum dimensional variations, the manufacturing facilities and the labour required become more cost intensive. Hence, it is essential for the manufacturer to have an in-depth knowledge of tolerances, to manufacture quality and reliable components economically. In fact, depending on the application of the end product, precision is engineered selectively. Therefore, apart from considering the actual size, other geometric dimensions such as roundness and straightness of a shaft have to be considered while manufacturing components. The tolerances specified should also encompass such variations. However, it is difficult to combine all errors of roundness, straightness, and diameter within a single tolerance on diameter. Geometric tolerance is defined as the total amount that the dimension of a manufactured part can vary. Geometric tolerance underlines the importance of the shape of a feature as against its size. Geometric dimensioning and tolerancing is a method of defining parts based on how they function, using standard symbols. This method is frequently used in industries. Depending on the functional requirements, tolerance on

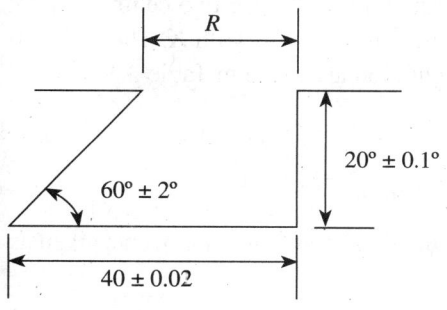

Fig. 3.4 Compound tolerance

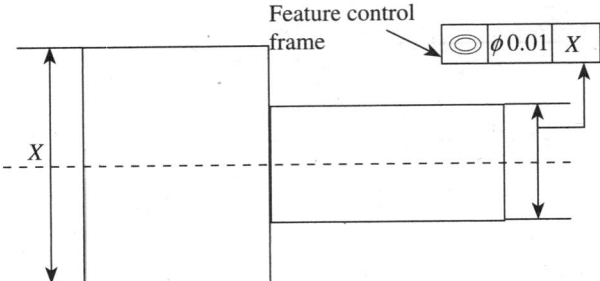

Fig. 3.5 Representation of geometric tolerance

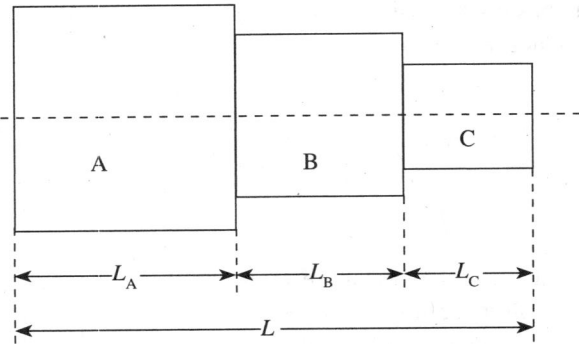

Fig. 3.6 Accumulation of tolerances

diameter, straightness, and roundness may be specified separately. Geometric tolerance can be classified as follows:

Form tolerances Form tolerances are a group of geometric tolerances applied to individual features. They limit the amount of error in the shape of a feature and are independent tolerances. Form tolerances as such do not require locating dimensions. These include straightness, circularity, flatness, and cylindricity.

Orientation tolerances Orientation tolerances are a type of geometric tolerances used to limit the direction or orientation of a feature in relation to other features. These are related tolerances. Perpendicularity, parallelism, and angularity fall into this category.

Positional tolerances Positional tolerances are a group of geometric tolerances that controls the extent of deviation of the location of a feature from its true position. This is a three-dimensional geometric tolerance comprising position, symmetry, and concentricity.

Geometric tolerances are used to indicate the relationship of one part of an object with another. Consider the example shown in Fig. 3.5. Both the smaller- and the larger-diameter cylinders need be concentric with each other. In order to obtain a proper fit between the two cylinders, both the centres have to be in line with each other. Further, perhaps both the cylinders are manufactured at different locations and need to be assembled on an interchangeable basis. It becomes imperative to indicate how much distance can be tolerated between the centres of these two cylinders. This information can be represented in the feature control frame that comprises three boxes. The first box on the left indicates the feature to be controlled, which is represented symbolically. In this example, it is concentricity. The box at the centre indicates the distance between the two cylinders that can be tolerated, that is, these two centres cannot be apart by more than 0.01 mm. The third box indicates that the datum is with X. The different types of geometric tolerances and their symbolic representation are given in Table 3.1.

Consider the example shown in Fig. 3.6.

Let $L_A = 30\,^{+0.02}_{-0.01}$ mm, $L_B = 20\,^{+0.02}_{-0.01}$ mm, and $L_C = 10\,^{+0.02}_{-0.01}$ mm.

The overall length of the assembly is the sum of the individual length of components given as

$$L = L_A + L_B + L_C$$
$$L = 30 + 20 + 10 = 60\,\text{mm}$$

Table 3.1 Symbolic representation of geometric tolerances

Type of geometric tolerance	Feature	Geometric characteristic	Definition	Symbol
Form tolerance	Independent	Straightness (two-dimensional)	Controls the extent of deviation of a feature from a straight line	
		Circularity (two-dimensional)	Exercises control on the extent of deviation of a feature from a perfect circle	
		Flatness (three-dimensional)	Controls the extent of deviation of a feature from a flat plane	
		Cylindricity (three-dimensional)	Controls the extent of deviation of a feature from a perfect cylinder	
	Related or single	Profile of a line (two-dimensional)	Controls the extent of deviation of an outline of a feature from the true profile	
		Profile of a surface (two-dimensional)	Exercises control on the extent of deviation of a surface from the true profile	
Orientation tolerance	Related	Perpendicularity (three-dimensional)	Exercises control on the extent of deviation of a surface, axis, or plane from a 90° angle	
		Parallelism (three-dimensional)	Controls the extent of deviation of a surface, axis, or plane from an orientation parallel to the specified datum	
		Angularity (three-dimensional)	Exercises controls on the deviation of a surface, axis, or plane from the angle described in the design specifications	
Position tolerance	Related	Position (three-dimensional)	Exercises control on the extent of deviation of location of a feature from its true position	
		Symmetry (three-dimensional)	Exercises control on extent of deviation of the median points between two features from a specified axis or centre plane	
		Concentricity (three-dimensional)	Exercises control on extent of deviation of the median points of multiple diameters from the specified datum axis	
Runout	Related	Circular runout (two-dimensional)	Exercises composite control on the form, orientation, and location of multiple cross-sections of a cylindrical part as it rotates	
		Total runout (three-dimensional)	Exercises simultaneous composite control on the form, orientation, and location of the entire length of a cylindrical part as it rotates	

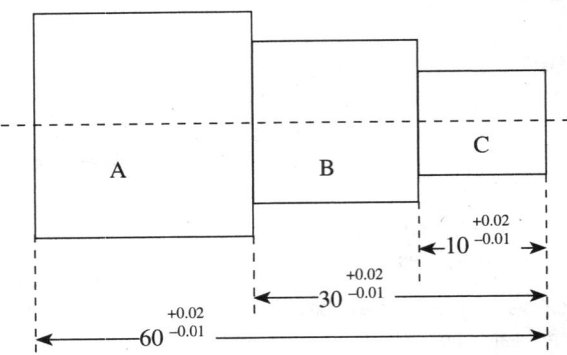

Fig. 3.7 Progressive dimensioning

Then, cumulative upper tolerance limit is 0.02 + 0.02 + 0.02 = 0.06 mm and cumulative lower limit = − 0.01 − 0.01 − 0.01 = −0.03 mm

Therefore, dimension of the assembled length will be = $60 \, ^{+0.06}_{-0.03}$ mm

It is essential to avoid or minimize the cumulative effect of tolerance build-up, as it leads to a high tolerance on overall length, which is undesirable. If progressive dimensioning from a common reference line or a baseline dimensioning is adopted, then tolerance accumulation effect can be minimized. This is clearly illustrated in Fig. 3.7.

3.4 MAXIMUM AND MINIMUM METAL CONDITIONS

Let us consider a shaft having a dimension of 40 ± 0.05 mm.

The maximum metal limit (MML) of the shaft will have a dimension of 40.05 mm because at this higher limit, the shaft will have the maximum possible amount of metal.

The shaft will have the least possible amount of metal at a lower limit of 39.95 mm, and this limit of the shaft is known as minimum or least metal limit (LML).

Similarly, consider a hole having a dimension of 45 ± 0.05 mm.

The hole will have a maximum possible amount of metal at a lower limit of 44.95 mm and the lower limit of the hole is designated as MML. For example, when a hole is drilled in a component, minimum amount of material is removed at the lower limit size of the hole. This lower limit of the hole is known as MML.

The higher limit of the hole will be the LML. At a high limit of 45.05 mm, the hole will have the least possible amount of metal. The maximum and minimum metal conditions are shown in Fig. 3.8.

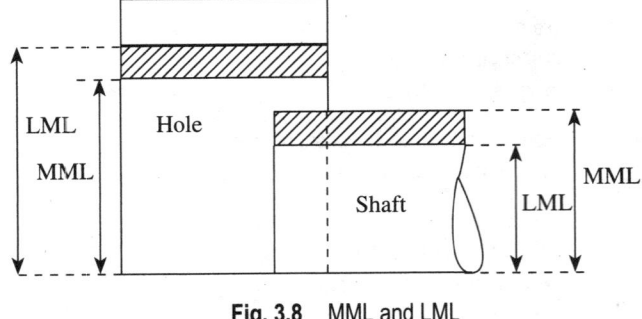

Fig. 3.8 MML and LML

3.5 FITS

Manufactured parts are required to mate with one another during assembly. The relationship between the two mating parts that are to be assembled, that is, the hole and the shaft, with respect to the difference in their dimensions before assembly is called a *fit*. An ideal fit is required for proper functioning of the mating parts. Three basic types of fits can be identified, depending on the actual limits of the hole or shaft:
1. Clearance fit

2. Interference fit

3. Transition fit

Clearance fit The largest permissible diameter of the shaft is smaller than the diameter of the smallest hole. This type of fit always provides clearance. Small clearances are provided for a precise fit that can easily be assembled without the assistance of tools. When relative motions are required, large clearances can be provided, for example, a shaft rotating in a bush. In case of clearance fit, the difference between the sizes is always positive. The clearance fit is described in Fig. 3.9.

Interference fit The minimum permissible diameter of the shaft exceeds the maximum allowable diameter of the hole. This type of fit always provides interference. Interference fit is a form of a tight fit. Tools are required for the precise assembly of two parts with an interference fit. When two mating parts are assembled with an interference fit, it will be an almost permanent assembly, that is, the parts will not come apart or move during use. To assemble the parts with interference, heating or cooling may be required. In an interference fit, the difference between the sizes is always negative. Interference fits are used when accurate location is of utmost importance and also where such location relative to another part is critical, for example, alignment of dowel pins. The interference fit is illustrated in Fig. 3.10.

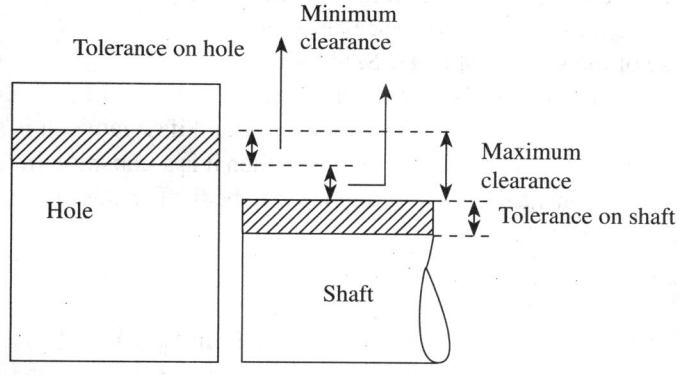

Fig. 3.9 Clearance fit

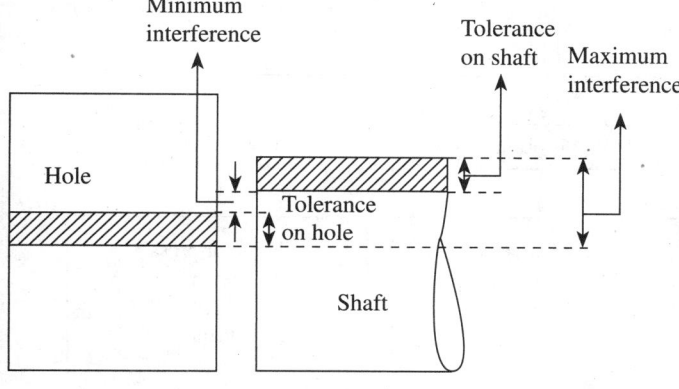

Fig. 3.10 Interference fit

Transition fit The diameter of the largest permissible hole is greater than the diameter of the smallest shaft and the diameter of the smallest hole is smaller than the diameter of the largest shaft. In other words, the combination of maximum diameter of the shaft and minimum diameter of the hole results in an interference fit, while that of minimum diameter of the shaft and maximum diameter of the hole yields a clearance fit. Since the tolerance zones overlap, this type of fit may sometimes provide clearance and sometimes interference, as depicted in Fig. 3.11. Precise assembly may be obtained with the assistance of tools, for example, dowel pins may be required in tooling to locate parts.

In a clearance fit, minimum clearance is the difference between minimum size of the hole, that is, low limit of the hole (LLH), and maximum size of the shaft, that is, high limit of the shaft (HLS), before assembly. In a transition or a

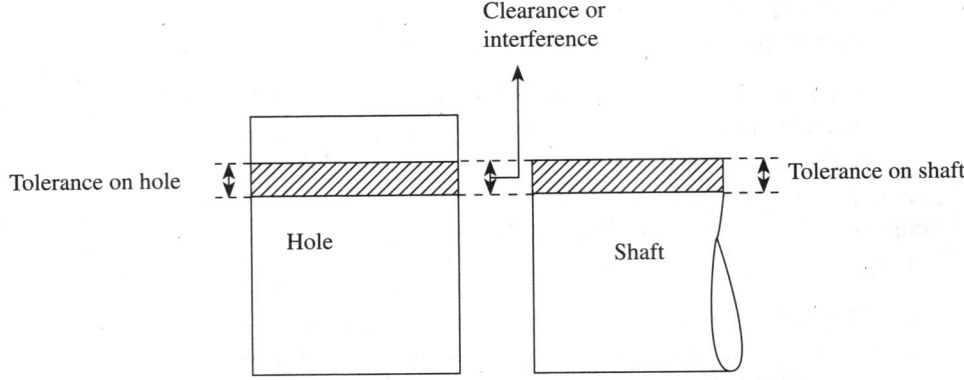

Fig. 3.11 Transition fit

clearance fit, maximum clearance is the arithmetical difference between the maximum size of the hole, that is, high limit of the hole (HLH), and the minimum size of the shaft, that is, low limit of the shaft (LLS), before assembly.

In an interference fit, minimum interference is the arithmetical difference between maximum size of the hole, that is, HLH, and minimum size of the shaft, that is, LLS, before assembly. In a transition or an interference fit, it is the arithmetical difference between minimum size of the hole, that is, LLH, and maximum size of the shaft, that is, HLS, before assembly.

Thus, in order to find out the type of fit, one needs to determine HLH − LLS and LLH − HLS. If both the differences are positive, the fit obtained is a clearance fit, and if negative, it is an interference fit. If one difference is positive and the other is negative, then it is a transition fit.

The three basic types of fits, clearance, transition, and interference, can be further classified, as shown in Fig. 3.12.

3.5.1 Allowance

An allowance is the intentional difference between the maximum material limits, that is, LLH and HLS (minimum clearance or maximum interference) of the two mating parts. It is the

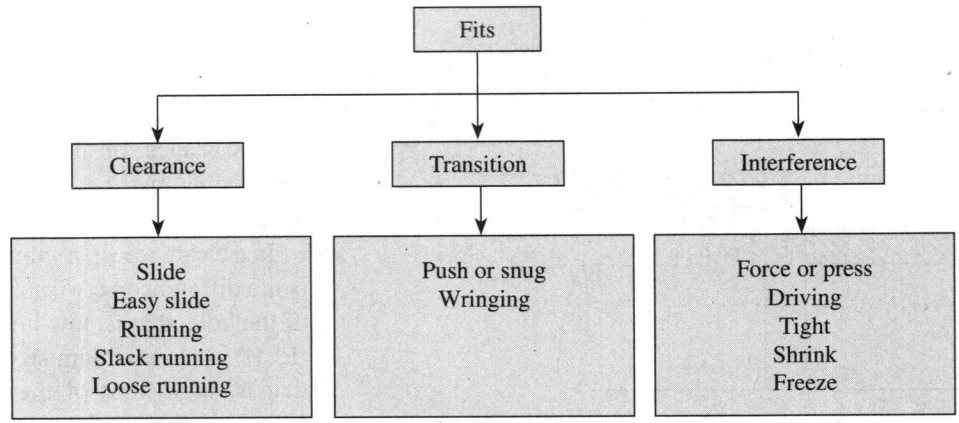

Fig. 3.12 Detailed classification of fits

Table 3.2 Examples of different types of fits

Description of fit	Class of fit	Application area
		Clearance fit
Slide	H7/h6	Sealing rings, bearing covers, movable gears in change gear trains, clutches, etc.
Easy slide	H7/g7	Lathe spindle, spigots, piston, and slide valves
Running	H8/f8	Lubricated bearings (with oil or grease), pumps and smaller motors, gear boxes, shaft pulleys, etc.
Slack running	H8/c11	Oil seals with metal housings, multi-spline shafts, etc.
Loose running	H8/d9	Loose pulleys, loose bearings with low revolution, etc.
		Interference fit
Force or press	H8/r6	Crankpins, car wheel axles, bearing bushes in castings, etc.
Driving	H7/s6	Plug or shaft slightly larger than the hole
Tight	H7/p6	Stepped pulleys on the drive shaft of a conveyor
Shrink	H7/u6, H8/u7	Bronze crowns on worm wheel hubs, couplings, gear wheels, and assembly of piston pin in IC engine piston
Freeze	H7/u6, H8/u7	Insertion of exhaust valve seat inserts in engine cylinder blocks and insertion of brass bushes in various assemblies
		Transition fit
Push or snug	H7/k6	Pulleys and inner ring of ball bearings on shafts
Wringing	H7/n6	Gears of machine tools

prescribed difference between the dimensions of the mating parts to obtain the desired type of fit. Allowance may be positive or negative. Positive allowance indicates a clearance fit, and an interference fit is indicated by a negative allowance.

Allowance = LLH − HLS

Table 3.2 gives examples of the classification of fits.

3.5.2 Hole Basis and Shaft Basis Systems

To obtain the desired class of fits, either the size of the hole or the size of the shaft must vary. Two types of systems are used to represent the three basic types of fits, namely clearance, interference, and transition fits. They are (a) hole basis system and (b) shaft basis system.

Although both systems are the same, hole basis system is generally preferred in view of the functional properties.

Hole Basis System

In this system, the size of the hole is kept constant and the shaft size is varied to give various types of fits. In a hole basis system, the fundamental deviation or lower deviation of the hole is zero, that is, the lower limit of the hole is the same as the basic size. The two limits of the shaft and the higher dimension of the hole are then varied to obtain the desired type of fit, as illustrated in Fig. 3.13.

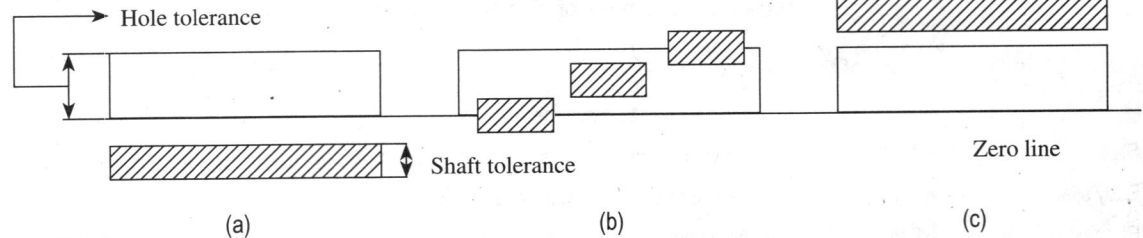

Fig. 3.13 Hole basis system (a) Clearance fit (b) Transition fit (c) Interference fit

This type of system is widely adopted in industries, as it is easier to manufacture shafts of varying sizes to the required tolerances. Standard size drills or reamers can be used to obtain a variety of fits by varying only the shaft limits, which leads to greater economy of production. The shaft can be accurately produced to the required size by standard manufacturing processes, and standard-size plug gauges are used to check hole sizes accurately and conveniently.

Shaft Basis System

The system in which the dimension of the shaft is kept constant and the hole size is varied to obtain various types of fits is referred to as shaft basis system. In this system, the fundamental deviation or the upper deviation of the shaft is zero, that is, the HLH equals the basic size. The desired class of fits is obtained by varying the lower limit of the shaft and both limits of the hole, as shown in Fig. 3.14.

This system is not preferred in industries, as it requires more number of standard-size tools such as reamers, broaches, and gauges, which increases manufacturing and inspection costs. It is normally preferred where the hole dimension is dependent on the shaft dimension and is used in situations where the standard shaft determines the dimensions of the mating parts such as couplings, bearings, collars, gears, and bushings.

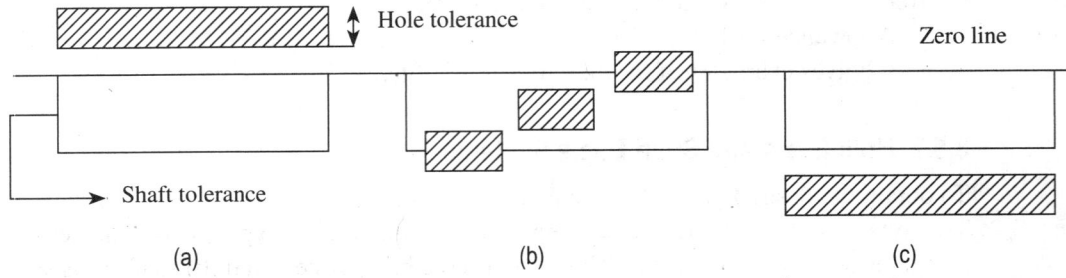

Fig. 3.14 Shaft basis system (a) Clearance fit (b) Transition fit (c) Interference fit

3.5.3 Numerical Examples

Example 3.1 In a limit system, the following limits are specified for a hole and shaft assembly:

Hole $= 30 \,^{+0.02}_{+0.00}$ mm and shaft $= 30 \,^{-0.02}_{-0.05}$ mm

Determine the (a) tolerance and (b) allowance.

Solution

(a) Determination of tolerance:

Tolerance on hole   = HLH – LLH

= 30.02 – 30.00 = 0.02 mm

Tolerance on shaft = HLS – LLS

= [(30 – 0.02) – (30 – 0.05)] = 0.03 mm

(b) Determination of allowance:

Allowance = Maximum metal condition of hole – Maximum metal condition of shaft

= LLH – HLS

= 30.02 – 29.98 = 0.04 mm

Example 3.2   The following limits are specified in a limit system, to give a clearance fit between a hole and a shaft:

Hole = $25\,^{+0.03}_{-0.00}$ mm and shaft = $25^{-0.006}_{-0.020}$ mm

Determine the following:

(a) Basic size

(b) Tolerances on shaft and hole

(c) Maximum and minimum clearances

Solution

(a) Basic size is the same for both shaft and hole.

(b) Determination of tolerance:

Tolerance on hole = HLH – LLH

= 25.03 – 25.00 = 0.03 mm

Tolerance on shaft = HLS – LLS

= [(25 – 0.006) – (25 – 0.020)] = 0.014 mm

Determination of clearances:

Maximum clearance = HLH – LLS

= 25.03 – 24.98 = 0.05 mm

Minimum clearance = LLH – HLS

= 25.00 – (25 – 0.006) = 0.06 mm

Example 3.3   Tolerances for a hole and shaft assembly having a nominal size of 50 mm are as follows:

Hole = $50\,^{+0.02}_{+0.00}$ mm and shaft = $50\,^{-0.05}_{-0.08}$ mm

Determine the following:

(a) Maximum and minimum clearances

(b) Tolerances on shaft and hole

(c) Allowance

(d) MML of hole and shaft

(e) Type of fit

Solution

(a) Determination of clearances:

Maximum clearance = HLH − LLS
$$= 50.02 - (50 - 0.08) = 0.10 \, mm$$

Minimum clearance = LLH − HLS
$$= 50.00 - (50 - 0.005) = 0.05 \, mm$$

(b) Determination of tolerance:

Tolerance on hole = HLH − LLH
$$= 50.02 - 50.00 = 0.02 \, mm$$

Tolerance on shaft = HLS − LLS
$$= [(50 - 0.05) - (50 - 0.08)] = 0.03 \, mm$$

(c) Determination of allowance:

Allowance = Maximum metal condition of hole − Maximum metal condition of shaft
$$= LLH - HLS$$
$$= 50.00 - (50 - 0.05) = 0.05 \, mm$$

(d) Determination of MMLs:

MML of hole = Lower limit of hole = 50.00 mm

MML of shaft = Higher limit of shaft = 50.00 − 0.05 = 49.05 mm

(e) Since both maximum and minimum clearances are positive, it can be conclude that the given pair has a clearance fit.

Example 3.4 A clearance fit has to be provided for a shaft and bearing assembly having a diameter of 40 mm. Tolerances on hole and shaft are 0.006 and 0.004 mm, respectively. The tolerances are disposed unilaterally. If an allowance of 0.002 mm is provided, find the limits of size for hole and shaft when (a) hole basis system and (b) shaft basis system are used.

Solution

(a) When hole basis system is used:

Hole size:

HLH = 40.006 mm

LLH = 40.000 mm

The allowance provided is +0.002 mm.

Therefore, HLS = LLH − Allowance
$$= 40.000 - 0.002 = 39.998 \, mm$$

LLS = HLS − Tolerance
$$= 39.998 - 0.004 = 39.994 \, mm$$

(b) When shaft basis system is used:

Shaft size:

HLS = 40.000 mm

LLS = 40.000 − 0.004 = 39.996 mm

The allowance provided is +0.002 mm.

Therefore, LLH = HLS + allowance
$$= 40.000 + 0.002 = 40.002 \, mm$$

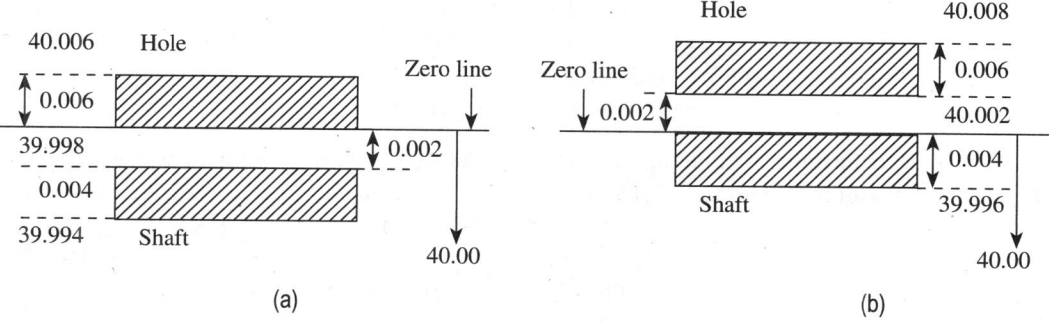

Fig. 3.15 Disposition of tolerances (a) Hole basis system (b) Shaft basis system

HLH = 40.002 + 0.006 = 40.008 mm

The disposition of tolerance for both hole basis and shaft basis systems are given in Fig. 3.15.

Example 3.5 For the following hole and shaft assembly, determine (a) hole and shaft tolerance and (b) type of fit.

Hole = $20 \, {}^{+0.025}_{+0.000}$ mm and shaft = $20 \, {}^{+0.080}_{+0.005}$ mm

Solution

(a) Determination of tolerance:

   Tolerance on hole = HLH − LLH

   = 20.025 − 20.00 = 0.025 mm

   Tolerance on shaft = HLS − LLS

   = 20.080 − 20.005 = 0.075 mm

(b) To determine the type of fit, calculate maximum and minimum clearances:

   Maximum clearance = HLH − LLS

   = 20.025 − 20.005 = 0.020 mm

(Clearance because the difference is positive)

Minimum clearance = LLH − HLS

   = 20.00 − 20.080 = −0.08 mm

(Interference because the difference is negative)

Since one difference is positive and the other negative, it can be concluded that the given hole and shaft pair has a transition fit.

Example 3.6 For the following hole and shaft assembly, determine (a) hole and shaft tolerance and (b) type of fit

Hole = $20 \, {}^{+0.05}_{+0.00}$ mm and shaft = $20 \, {}^{+0.08}_{+0.06}$ mm

Solution

(a) Determination of tolerance:

Tolerance on hole $= \text{HLH} - \text{LLH}$
$$= 20.05 - 20.00 = 0.05 \text{ mm}$$
Tolerance on shaft $= \text{HLS} - \text{LLS}$
$$= 20.08 - 20.06 = 0.02 \text{ mm}$$

(b) To determine the type of fit, calculate maximum and minimum clearances:
Maximum clearance $= \text{HLH} - \text{LLS}$
$$= 20.05 - 20.06 = -0.01 \text{ mm}$$
Minimum clearance $= \text{LLH} - \text{HLS}$
$$= 20.00 - 20.08 = -0.08 \text{ mm}$$

Since both differences are negative, it can be concluded that the given hole and shaft pair has an interference fit.

3.6 SYSTEM OF LIMITS AND FITS

The rapid growth of national and international trade necessitates the developments of formal systems of limits and fits, at the national and international levels. Economic success of most manufacturing industries critically depends on the conformity of the specifications of the products to international standards. The International Organization for Standardization (ISO) specifies the internationally accepted system of limits and fits. Indian standards are in accordance with the ISO.

The ISO system of limits and fits comprises 18 grades of fundamental tolerances to indicate the level of accuracy of the manufacture. These fundamental tolerances are designated by the letters IT followed by a number. The ISO system provides tolerance grades from IT01, IT0, and IT1 to IT16 to realize the required accuracy. The greater the number, the higher the tolerance limit. The choice of tolerance is guided by the functional requirements of the product and economy of manufacture. The degree of accuracy attained depends on the type and condition of the machine tool used. Table 3.3 gives the fundamental tolerance values required for various applications.

Tolerance values corresponding to grades IT5–IT16 are determined using the standard tolerance unit (i, in μm), which is a function of basic size.

Table 3.3 Tolerances grades for different applications

Fundamental tolerance	Applications
IT01–IT4	For production of gauges, plug gauges, and measuring instruments
IT5–IT7	For fits in precision engineering applications such as ball bearings, grinding, fine boring, high-quality turning, and broaching
IT8–IT11	For general engineering, namely turning, boring, milling, planning, rolling, extrusion, drilling, and precision tube drawing
IT12–IT14	For sheet metal working or press working
IT15–IT16	For processes such as casting, stamping, rubber moulding, general cutting work, and flame cutting

$$i = 0.453 \sqrt[3]{D} + 0.001D \text{ microns}$$

where D is the diameter of the part in mm. The linear factor $0.001D$ counteracts the effect of measuring inaccuracies that increase by increasing the measuring diameter. By using this formula, the value of tolerance unit 'i' is obtained for sizes up to 500 mm. D is the geometric mean of the lower and upper diameters of a particular diameter step within which the given or chosen diameter D lies and is calculated by using the following equation:

$$\sqrt{D_{max} \times D_{min}}$$

The various steps specified for the diameter steps are as follows:

1–3, 3–6, 6–10, 10–18, 18–30, 30–50, 50–80, 80–120, 120–180, 180–250, 250–315, 315–400, 400–500 500–630, 630–800, and 800–1000 mm.

Tolerances have a parabolic relationship with the size of the products. The tolerance within which a part can be manufactured also increases as the size increases. The standard tolerances corresponding to IT01, IT0, and IT1 are calculated using the following formulae:

> IT01: $0.3 + 0.008D$
>
> IT0: $0.5 + 0.012D$
>
> IT1: $0.8 + 0.020D$

The values of tolerance grades IT2–IT4, which are placed between the tolerance grades of IT1 and IT5, follow a geometric progression to allow for the expansion and deformation affecting both the gauges and the workpieces as dimensions increase. For the tolerance grades IT6–IT16, each grade increases by about 60% from the previous one, as indicated in Table 3.4.

Table 3.4 Standard tolerance units

Tolerance grade	IT6	IT7	IT8	IT9	IT10	IT11	IT12	IT13	IT14	IT15	IT16
Standard tolerance unit (i)	10	16	25	40	64	100	160	250	400	640	1000

The tolerance zone is governed by two limits: the size of the component and its position related to the basic size. The position of the tolerance zone, from the zero line (basic size), is determined by fundamental deviation. The ISO system defines 28 classes of basic deviations for holes and shafts, which are marked by capital letters A, B, C, ..., ZC (with the exception of I, L, O, Q, and W) and small letters a, b, c, ..., zc (with the exception of i, l, o, q, and w), respectively, as depicted in Fig. 3.16. Different combinations of fundamental deviations and fundamental tolerances are used to obtain various types of fits.

The values of these tolerance grades or fundamental deviations depend on the basic size of the assembly. The different values of standard tolerances and fundamental deviations can be obtained by referring to the design handbook. The choice of the tolerance grade is governed by the type of manufacturing process and the cost associated with it. From Fig. 3.16, a typical case can be observed in which the fundamental deviation for both hole H and shaft h having a unilateral tolerance of a specified IT grade is zero. The first eight designations from A (a) to H (h) for holes (shafts) are intended to be used in clearance fit, whereas the remaining designations, JS (js) to ZC (zc) for holes (shafts), are used in interference or transition fits. For JS, the two deviations are equal and given by $\pm IT/2$.

Consider the designation 40 H7/d9. In this example, the basic size of the hole and shaft is 40 mm. The nature of fit for the hole basis system is designated by H and the fundamental deviation of the hole is zero. The tolerance grade is indicated by IT7. The shaft has a d-type fit for which the fundamental deviation (upper deviation) has a negative value, that is, its dimension falls below the basic size having IT9 tolerance.

Depending on the application, numerous fits ranging from extreme clearance to extreme interference can be selected using a suitable combination of fundamental deviations and fundamental tolerances. From Fig. 3.16, it can be seen that the lower deviation for the holes 'A' to 'G' is above the zero line and that for 'K' to 'ZC' is below the zero line. In addition, it can be observed that for shafts 'a' to 'g', the upper deviation falls below the zero line, and for 'k' to 'zc' it is above the zero line.

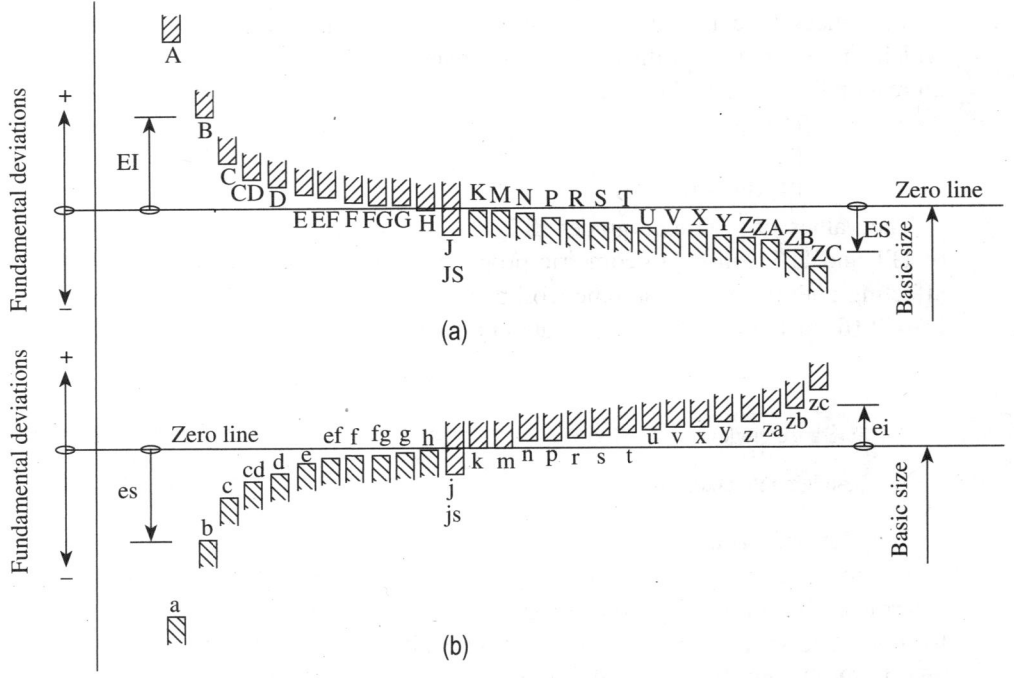

Fig. 3.16 Typical representation of different types of fundamental deviations
(a) Holes (internal features) (b) Shafts (external features)

It can be seen from Fig. 3.17 that 'EI' is above the zero line for holes 'A' to 'G', indicating positive fundamental deviation. In contrast, Fig. 3.18 shows that 'ei' is below the zero line for the shafts 'a' to 'g' and therefore the fundamental deviation is negative. In addition, from Figs 3.17 and 3.18, it can be observed that for holes 'K' to 'ZC', the fundamental deviation is negative ('EI' below the zero line), whereas for shafts 'k' to 'zc', it is positive ('ei' above the zero line).

It follows from Figs 3.17 and 3.18 that the values of 'ES' and 'EI' for the holes and 'es' and 'ei' for the shafts can be determined by adding and subtracting the fundamental tolerances, respectively. Magnitude and sign of fundamental deviations for the shafts, either upper deviation 'es' or lower deviation 'ei' for each symbol, can be determined

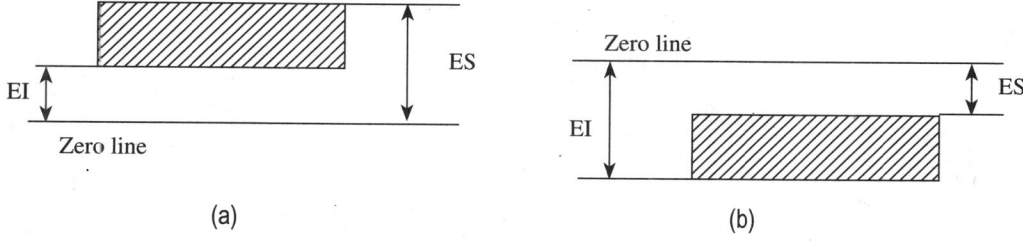

Fig. 3.17 Deviations for holes (a) Deviations for A to G (b) Deviations for K to Z_i

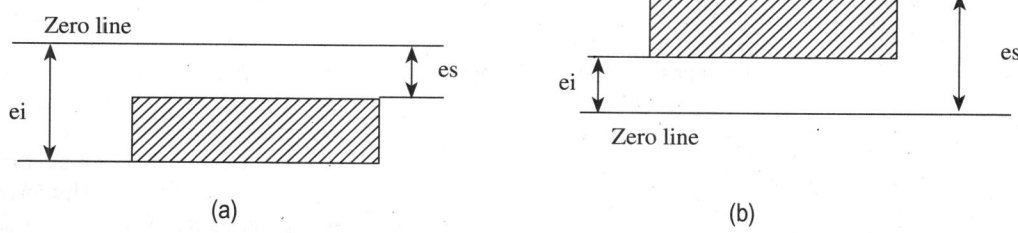

Fig. 3.18 Deviations for shafts (a) Deviations for a to g (b) Deviations for k to Z_i

from the empirical relationships listed in Tables 3.5 and 3.6, except for shafts j and js for which there is no fundamental deviation. The values given in Tables 3.5 and 3.6 are as per IS: 919. The

Table 3.5 Fundamental deviation formulae for shafts of sizes up to 500 mm

Upper deviation (es)		Lower deviation (ei)	
Shaft designation	In μm (for D in mm)	Shaft designation	In μm (for D in mm)
a	$= -(265 + 1.3D)$ for $D \leq 120$	j5–j8	No formula
		(For js two deviations are equal to IT ± 2)	
	$= -3.5D$ for $D > 120$	k4–k7	$= +0.6 \sqrt[3]{D}$
		For grades ≤k3 to ≥k8	$= 0$
b	$= -(140 + 0.85D)$ for $D \leq 160$	m	$= +(IT7 - IT6)$
	$= -1.8D$ for $D > 160$	n	$= +5D^{0.34}$
		p	$= + IT7 + 0$ to 5
c	$= -52D^{0.2}$ for $D + IT7 + 0$ to 40	r	$=$ Geometric mean of ei values for p and s
	$= -(95 + 0.8D)$ for $D > 40$		$= +IT8 + 1$ to 4 for $D \leq 50$
d	$= -16D^{0.44}$	s	$= +IT7 + 0.4D$ for $D > 50$
e	$= -11D^{0.41}$	t	$= +IT7 + 0.63D$
f	$= -5.5D^{0.41}$	u	$= +IT7 + D$
g	$= -2.5D^{0.34}$	v	$= +IT7 + 1.25D$

(Contd)

Table 3.5 (Contd)

		x	$= +IT7 + 1.6D$
		y	$= +IT7 + 2D$
		z	$= +IT7 + 2.5D$
h	$= 0$	za	$= +IT8 + 3.15D$
		zb	$= +IT9 + 4D$
		zc	$= +IT10 + 5D$

Table 3.6 Fundamental deviation formulae for holes of sizes up to 500 mm

For all deviations except the following:			General rule: Hole limits are identical with the shaft limits of the same symbol (letter and grade) but disposed on the other side of the zero line EI = upper deviation es of shaft of the same letter symbol but of opposite sign
For sizes above 3 mm	N	9 and coarser grades	ES = 0
	J, K, M, and N	Up to grade 8 inclusive	Special rule: ES = lower deviation ei of the shaft of the same letter symbol but one grade finer and of opposite sign, increased by the difference between the tolerances of the two grades in question
	P to ZC	Up to grade 7 inclusive	

other deviations of the holes and shafts can be determined using the following relationships that can be derived from Figs 3.17 and 3.18:

1. For holes A to G, EI is a positive fundamental deviation and EI = ES − IT.
2. For holes K to Zi, the fundamental deviation ES is negative and ES = EI − IT.
3. For shafts 'a' to 'g', the fundamental deviation es is negative and es = ei − IT.
4. For shafts 'k' to 'Zi', the fundamental deviation ei is positive, and ei = es − IT.

Production of large-sized components is associated with problems pertaining to manufacture and measurement, which do not exist in the case of smaller-sized components. As the size of the components to be manufactured increases, the difficulty in making accurate measurements also increases. Variations in temperature also affect the quality of measurement.

When the size of the components to be manufactured exceeds 500 mm, the tolerance grades IT01–IT5 are not provided, as they are considered to be too small. The fundamental tolerance unit in case of sizes exceeding 500 and up to 3150 mm is determined as follows:

$i = 0.004D + 2.1D$, where D is 0.001 mm.

The fundamental deviations for holes and shafts are given in Table 3.7, which are as per IS: 2101.

Table 3.7 Fundamental deviation for shafts and holes of sizes from above 500 to 3150 mm

Shafts			Holes			Formula for deviations in μm
Type	Fundamental deviation	Sign	Type	Fundamental deviation	Sign	(for D in mm)
d	es	−	D	EI	+	$16D^{0.44}$
e	es	−	E	EI	+	$11D^{0.41}$
f	es	−	F	EI	+	$5.5D^{0.41}$
g	es	−	G	EI	+	$2.5D^{0.34}$
h	es	No sign	H	EI	No sign	0
js	ei	−	JS	ES	+	$0.5IT\pi$
k	ei	−	K	ES	−	0
m	ei	+	M	ES	−	$0.024D + 12.6$
n	ei	+	N	ES	−	$0.04D + 21$
P	ei	+	P	ES	−	$0.072D + 37.8$
r	ei	+	R	ES	−	Geometric mean of the values for p and s or P and S
s	ei	+	S	ES	−	$IT7 + 0.4D$
t	ei	+	T	ES	−	$IT7 + 0.63D$
u	ei	+	U	ES	−	$IT7 + D$

3.6.1 General Terminology

The following are the commonly used terms in the system of limits and fits.

Basic size   This is the size in relation to which all limits of size are derived. Basic or nominal size is defined as the size based on which the dimensional deviations are given. This is, in general, the same for both components.

Limits of size   These are the maximum and minimum permissible sizes acceptable for a specific dimension. The operator is expected to manufacture the component within these limits. The maximum limit of size is the greater of the two limits of size, whereas the minimum limit of size is the smaller of the two.

Tolerance   This is the total permissible variation in the size of a dimension, that is, the difference between the maximum and minimum limits of size. It is always positive.

Allowance   It is the intentional difference between the LLH and HLS. An allowance may be either positive or negative.

Allowance = LLH − HLS

Grade   This is an indication of the tolerance magnitude; the lower the grade, the finer the tolerance.

Deviation   It is the algebraic difference between a size and its corresponding basic size. It may be positive, negative, or zero.

Upper deviation   It is the algebraic difference between the maximum limit of size and its corresponding basic size. This is designated as 'ES' for a hole and as 'es' for a shaft.

Lower deviation   It is the algebraic difference between the minimum limit of size and its corresponding basic size. This is designated as 'EI' for a hole and as 'ei' for a shaft.

Actual deviation   It is the algebraic difference between the actual size and its corresponding basic size.

Fundamental deviation   It is the *minimum* difference between the size of a component and its basic size. This is identical to the upper deviation for shafts and lower deviation for holes. It is the closest deviation to the basic size. The fundamental deviation for holes are designated by capital letters, that is, A, B, C, …, H, …, ZC, whereas those for shafts are designated by small letters, that is, a, b, c…, h…, zc. The relationship between fundamental, upper, and lower deviations is schematically represented in Fig. 3.19.

Zero line   This line is also known as the line of zero deviation. The convention is to draw the zero line horizontally with positive deviations represented above and negative deviations indicated below. The zero line represents the basic size in the graphical representation.

Shaft and hole   These terms are used to designate all the external and internal features of any shape and not necessarily cylindrical.

Fit   It is the relationship that exists between two mating parts, a hole and a shaft, with respect to their dimensional difference before assembly.

Maximum metal condition   This is the maximum limit of an external feature; for example, a shaft manufactured to its high limits will contain the maximum amount of metal. It is also the minimum limit of an internal feature; for example, a component that has a hole bored in it to its lower limit of size will have the minimum amount of metal removed and remain in its

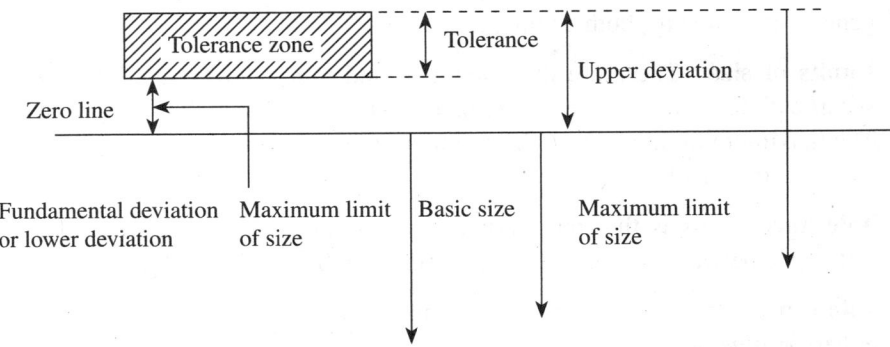

Fig. 3.19   Relationship between fundamental, upper, and lower deviations

maximum metal condition, (i.e., this condition corresponds to either the largest shaft or the smallest hole). This is also referred to as the GO limit.

Least metal condition   This is the minimum limit of an external feature; for example, a shaft will contain minimum amount of material, when manufactured to its low limits. It is also the maximum limit of an internal feature; for example, a component will have the maximum amount of metal removed when a hole is bored in it to its higher limit of size, this condition corresponds to either the smallest shaft or the largest hole. This is also referred to as the NO GO limit.

Tolerance zone   The tolerance that is bound by the two limits of size of the component is called the tolerance zone. It refers to the relationship of tolerance to basic size.

International tolerance grade (IT)   Tolerance grades are an indication of the degree of accuracy of the manufacture. Standard tolerance grades are designated by the letter IT followed by a number, for example, IT7. These are a set of tolerances that varies according to the basic size and provides a uniform level of accuracy within the grade.

Tolerance class   It is designated by the letter(s) representing the fundamental deviation followed by the number representing the standard tolerance grade. When the tolerance grade is associated with letter(s) representing a fundamental deviation to form a tolerance class, the letters IT are omitted and the class is represented as H8, f7, etc.

Tolerance symbols   These are used to specify the tolerance and fits for mating components. For example, in 40 H8f7, the number 40 indicates the basic size in millimetres; capital letter H indicates the fundamental deviation for the hole; and lower-case letter f indicates the shaft. The numbers following the letters indicate corresponding IT grades.

3.6.2 Limit Gauging

Eli Whitney, who is hailed as the father of the American system, won the first contract in 1798 for the production of muskets and developed the gauging system. Richard Roberts of the United Kingdom first used plug and collar gauges for dimensional control. In 1857, Joseph Whitworth demonstrated internal and external gauges for use with a shaft-based limit system.

As discussed in Section 3.1, in mass production, components are manufactured in accordance with the permissive tolerance limits, as suggested by the designer. Production of components within the permissive tolerance limits facilitates interchangeable manufacture. It is also essential to check whether the dimensions of the manufactured components are in accordance with the specifications or not. Therefore, it is required to control the dimensions of the components. Several methods are available to achieve the control on dimensions. Various precision measuring instruments can be used to measure the actual dimensions of the components, which can be compared with the standard specified dimensions to decide the acceptability of these components.

In mass production, where a large number of similar components are manufactured on an interchangeable basis, measuring the dimensions of each part will be a time-consuming and expensive exercise. In addition, the actual or absolute size of a component, provided that it is within the limits specified, is not of much importance because the permissible limits of variations in dimensions would have been specified by the designer in a logical manner, giving due consideration to the functional requirements. Therefore, in mass production, gauges can be used to check for the compliance of the limits of the part with the permissive tolerance limits, instead of measuring the actual dimensions. The term 'limit gauging' signifies the use of gauges for checking the limits of the components. Gauging plays an important role in the control of dimensions and interchangeable manufacture.

Limit gauges ensure that the components lie within the permissible limits, but they do not determine the actual size or dimensions. Gauges are scaleless inspection tools, which are used to check the conformance of the parts along with their forms and relative positions of the surfaces of the parts to the limits. The gauges required to check the dimensions of the components correspond to two sizes conforming to the maximum and minimum limits of the components. They are called GO gauges or NO GO or NOT GO gauges, which correspond, respectively, to the MML and LML of the component, as depicted in Figs 3.20 and 3.21. As discussed in section 3.4, MML is the lower limit of a hole and higher limit of the shaft and LML corresponds to the higher limit of a hole and lower limit of the shaft. The GO gauge manufactured to the maximum limit will assemble with the mating (opposed) part, whereas the NOT GO gauge corresponding to the low limit will not, hence the names GO and NOT GO gauges, respectively.

Practically, every gauge is a replica of the part that mates with the part for which the gauge has been designed. Consider an example of the manufacture of a cylinder that mates with a piston. The plug gauge, using which the cylinder bore is checked, is a copy of the opposed part (piston) as far as its form and size are concerned. When a gauge is designed as a replica

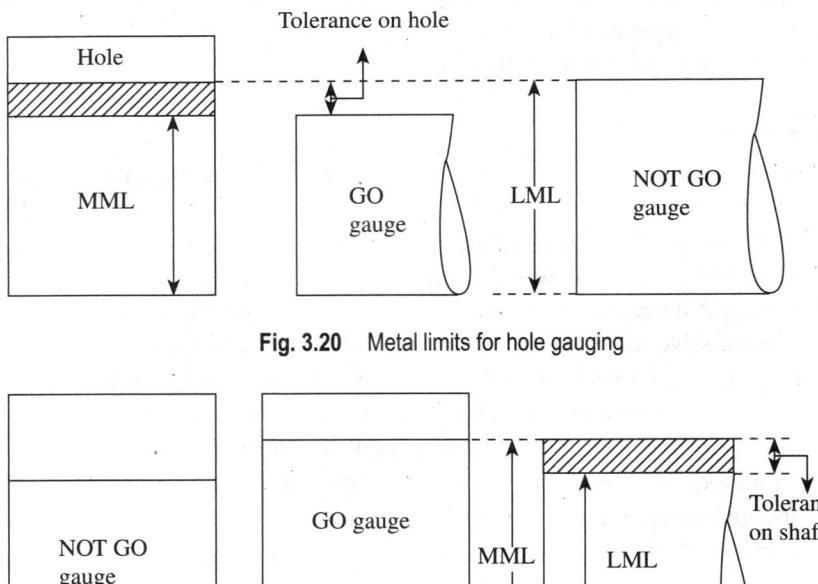

Fig. 3.20 Metal limits for hole gauging

Fig. 3.21 Metal limits for shaft gauging

of the mating (opposed) part so far as the dimension to be checked is concerned, it is known as 'standard gauge'. The main intention in the design of gauges is simplicity, which helps in making continuous and accurate measurements.

It is important to note that normally clearance fits are preferred for a majority of the assembly operations. Allowance or minimum clearance is determined by the algebraic difference of the MMLs of the mating parts. Therefore, for clearance fits, the MMLs of the mating parts become more critical than the LMLs. This assumes importance for the following two reasons:
1. MMLs are crucial for effective functioning of the parts.
2. If the MMLs slightly exceed the specified values then assembly itself becomes impossible.

As discussed earlier, for gauging the MMLs of the mating parts, GO gauges are used. Therefore, it becomes imperative that special attention needs to be given when GO gauges are designed for gauging these limits. Whenever the components are gauged for their MMLs, if the GO gauges fail to assemble during inspection, the components should not be accepted under any circumstances. The minimum limits in a clearance fit of a product are not so critical because even if they exceed the specified limits and the NOT GO gauge assembles, its acceptance may result in functional degradation and because of the reduced quality the useful life of the product may get affected. Hence, it becomes essential that more care is taken especially when GO gauges are used, when compared to NOT GO gauges during inspection.

3.6.3 Classification of Gauges

The detailed classification of the gauges is as follows:
1. Plain gauges
(a) According to their type:
   (i) Standard gauges
   (ii) Limit gauges

(b) According to their purpose:
   (i) Workshop
   (ii) Inspection
   (iii) Reference, or master, or control gauges

(c) According to the form of the tested surface:
   (i) Plug gauges for checking holes
   (ii) Snap and ring gauges for checking shafts

(d) According to their design:
   (i) Single- and double-limit gauges
   (ii) Single- and double-ended gauges
   (iii) Fixed and adjustable gauges

2. Adjustable-type gap gauges
3. Miscellaneous gauges
   (a) Combined-limit gauges
   (b) Taper gauges
   (c) Position gauges
   (d) Receiver gauges

(e) Contour gauges

(f) Profile gauges

3.6.4 Taylor's Principle

In 1905, William Taylor developed a concept relating to the gauging of components, which has been widely used since then. Since World War II, the term Taylor's principle has generally been applied to the principle of limit gauging and extensively used in the design of limit gauges. Prior to 1905, simple GO gauges were used. The components were carefully manufactured to fit the gauges. Since NOT GO gauges were not used, these components were without tolerance on their dimensions.

The theory proposed by Taylor, which is extensively used in the design of limit gauges, not only defines the function, but also defines the form of most limit gauges.

Taylor's principle states that the GO gauge is designed to check maximum metal conditions, that is, LLH and HLS. It should also simultaneously check as many related dimensions, such as roundness, size, and location, as possible.

The NOT GO gauge is designed to check minimum metal conditions, that is, HLH and LLS. It should check only one dimension at a time. Thus, a separate NOT GO gauge is required for each individual dimension.

During inspection, the GO side of the gauge should enter the hole or just pass over the shaft under the weight of the gauge without using undue force. The NOT GO side should not enter or pass.

The basic or nominal size of the GO side of the gauge conforms to the LLH or HLS, since it is designed to check maximum metal conditions. In contrast, the basic or nominal size of the NOT GO gauge corresponds to HLH or LLS, as it is designed to check minimum metal conditions.

It can be seen from Fig. 3.22 that the size of the GO plug gauge corresponds to the LLH and the NOT GO plug gauge to the HLH. Conversely, it can be observed from Fig. 3.23 that the GO snap gauge represents the HLS, whereas the NOT GO snap gauge represents the LLS.

It is pertinent to discuss here that since the GO plug is used to check more than one dimension of the hole simultaneously, the GO plug gauge must have a full circular section and must be of full length of the hole so that straightness of the hole can also be checked. During inspection, it can be ensured that if there is any lack of straightness or roundness of the hole a full entry of the GO plug gauge will not be allowed. Thus, it not only controls the diameter in any given cross-section but also ensures better bore alignment. However, it should be mentioned here that the

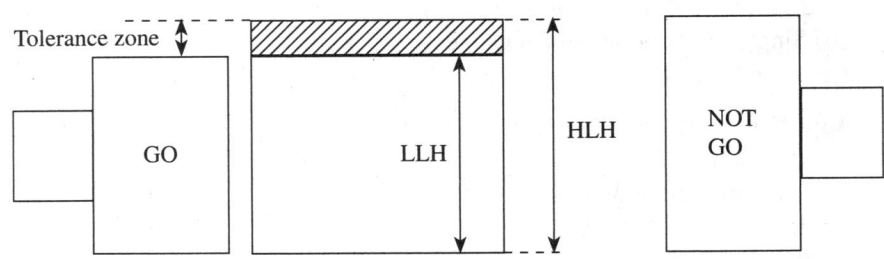

Fig. 3.22 GO and NOT GO limits of plug gauge

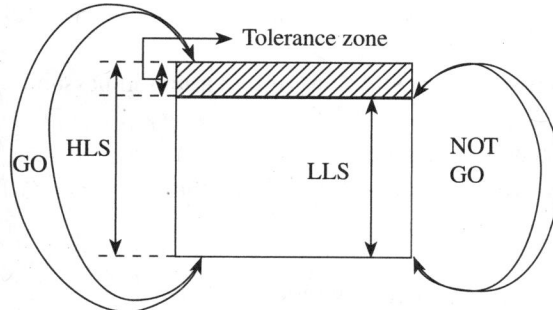

Fig. 3.23 GO and NOT GO limits of snap gauge

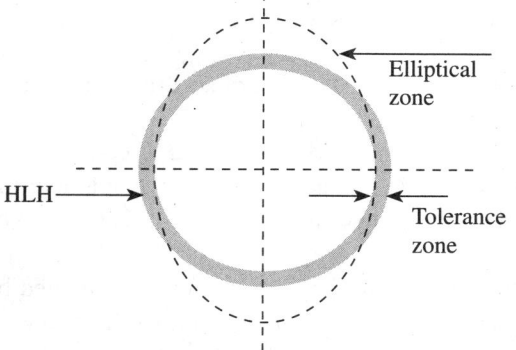

Fig. 3.24 Ovality in hole

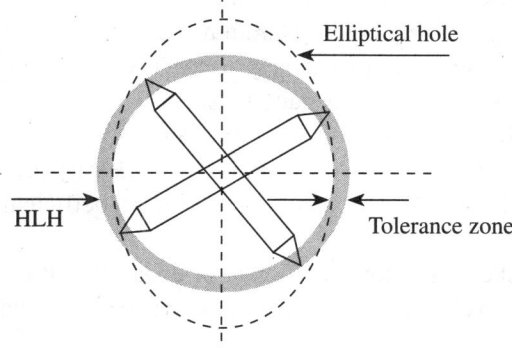

Fig. 3.25 Pin-type NOT GO to the check ovality

GO plug gauge cannot check the degree of ovality.

The short GO plug gauge, if used in inspection, will pass through all the curves and is hence not possible to identify defective parts. Therefore, in order to get good results, this condition has to be fulfilled during the inspection of the parts. The length of the plug should normally be more than 1.5 times the diameter of the hole to be checked. Compared to GO plug gauges, the NOT GO plug gauges are relatively shorter.

Let us consider the gauging of a cylindrical hole. A simple plug gauge is used to gauge this hole. During inspection, the GO gauge, which measures the minimum limit of the hole, enters but the NOT GO gauge, which corresponds to the maximum limit, does not. In this case, according to Taylor's theorem, the hole is considered to be within the specified limits and hence accepted. However, the shape of the hole has not been taken into account here. Most of the methods used in the manufacture of the holes are capable of producing truly circular holes. When these holes are gauged, they are accepted if they are within the specified tolerance zone. As long as there are no circularity errors, there is no problem. However, when the holes deviate from circularity, the problem starts. Consider an elliptical hole as shown in Fig. 3.24. It can be clearly observed from Fig. 3.24 that the NOT GO gauge does not enter because the minor axis of the ellipse is smaller than the HLH, which corresponds to the diameter of the NOT GO gauge. Therefore, even if the hole is slightly elliptical, the gauge does not take into account this variation of the hole shape and it still gets accepted as the GO gauge assembles and the NOT GO gauge does not.

One way of overcoming this problem is to employ the NOT GO, which is in the form of a pin. Any error in the circularity can easily be detected by placing the pin at different cross-sections of the hole, as depicted in Fig. 3.25. Hence, it can be said that Taylor's principle does not take care of the error of form, circularity, or straightness, and some modifications are needed to overcome these limitations.

3.6.5 Important Points for Gauge Design

The following points must be kept in mind while designing gauges:
1. The form of GO gauges should be a replica of the form of the opposed (mating) parts.

2. GO gauges enable several related dimensions to be checked simultaneously and hence are termed complex gauges.
3. During inspection, GO gauges must always be put into conditions of maximum impassability.
4. NOT GO gauges check a single element of feature at a time.
5. In inspection, NOT GO gauges must always be put into conditions of maximum passability.

3.6.6 Material for Gauges

The material used to manufacture gauges should satisfy the following requirements:
1. The material used in the manufacture of gauges should be hard and wear resistant for a prolonged life. This is one of the most important criteria that should be fulfilled.
2. It should be capable of maintaining dimensional stability and form.
3. It should be corrosion resistant.
4. It should be easily machinable, in order to obtain the required degree of accuracy and surface finish.
5. It should have a low coefficient of expansion, in order to avoid temperature effect.

High-carbon steel is the most suitable and inexpensive material used for manufacturing gauges. It can be heat treated suitably to provide stability and high hardness. It can easily be machined to a high degree of accuracy.

Mild steel gauges are the most suitable for larger gauges. They are suitably heat treated by carburizing to the required depth and then case hardened on their working surfaces to allow for final grinding and finishing. After hardening, internal stresses are relieved to improve stability.

Chromium-plated gauges are very popular and extensively used for gauging. Chromium plating makes the surface of the gauge very hard, and resistant to abrasion and corrosion. It is also very useful in reclaiming worn-out gauges. For gauging aluminium or other materials having an abrasive action, chromium-plated gauges are extensively used. The plug gauges employed for gauging have satellite ribs that are inserted in the gauging surface.

Glass gauges are not very popular although they have good wear and corrosion resistance properties. The problem with these gauges is that they either get damaged or are easily broken if dropped. They are not affected by changes in temperature and have very low coefficient of thermal expansion.

Although Invar, which contains 36% of nickel, has a low coefficient of expansion, it is not suitable over a long period. Elinvar has 42% of nickel, is more stable than Invar, and also has a low coefficient of expansion.

3.6.7 Gauge Tolerance (Gauge Maker's Tolerance)

We know that gauges have to be manufactured to their required dimensions corresponding to their maximum metal conditions. Gauges, like any other component, cannot be manufactured to their exact size or dimensions. In order to accommodate these dimensional variations, which arise due to the limitations of the manufacturing process, skill of the operator, etc., some tolerance must be allowed in the manufacture of gauges. Thus, the tolerance that is allowed in the manufacture of gauges is termed *gauge maker's tolerance* or simply *gauge tolerance*. Logically, gauge tolerance should be kept as minimum as possible; however, this increases the gauge manufacturing cost. There is no universally accepted policy for deciding the amount of tolerance to be provided on gauges. The normal practice is to take gauge tolerance as 10% of the work tolerance.

3.6.8 Wear Allowance

As discussed in Section 3.6.4, according to Taylor's principle, during inspection the NOT GO side should not enter or pass. The NOT GO gauge seldom engages fully with the work and therefore does not undergo any wear. Hence, there is no need to provide an allowance for wear in case of NOT GO gauges.

Taylor's principle also states that the GO side of the gauge should enter the hole or just pass over the shaft under the weight of the gauge without using undue force. During inspection, the measuring surfaces of the gauge constantly rub against the mating surfaces of the workpiece. Therefore, the GO gauges suffer wear on the measuring surfaces and thus lose their initial dimension. Hence, wear allowance is provided for GO gauges to extend their service life. As a consequence of this wear, the size of the GO plug gauge decreases while that of the ring or gap gauge increases. The wear allowance provided for the GO gauges are added in a direction opposite to wear. This allowance is added in for a plug gauge while subtracted for a ring or gap gauge. A wear allowance of 10% of gauge tolerance is widely accepted in industries. If the work tolerance of a component is less than 0.1 mm, no wear allowance on gauges is provided for that component, since a wear allowance of less than 0.001 mm will not have any practical effect on the gauges.

The allowance on new gauge is made by fixing the tolerance zone for the gauge from the MML of the work by an amount equal to the wear allowance. A new gauge is then manufactured within the limits specified by the tolerance zone for the gauge in this position. The gauge is then allowed to wear with use until its size coincides with the maximum material limit for the work.

3.6.9 Methods of Tolerance Specification on Gauges

By considering all the important points of gauge design, tolerances on gauges (snap and plug gauges) are specified by the following three methods.

First Method

This method of specifying tolerances on gauges was evolved during the developmental stages of gauges. In this system, the tolerance zones of workshop and inspection gauges are different and are made separately. Here the tolerances on the workshop gauge are set to fall inside the work tolerance, whereas the inspection gauge tolerances fall outside the work tolerance.

In workshop gauges, the tolerance on GO and NOT GO gauges is 10% of the work tolerance. For example, if the work tolerance is 10 units, then only 8 units will be left as the difference between maximum limit of GO gauge and minimum limit of NOT GO gauge, since these tolerances fall inside and are 1 unit each, as shown in Fig. 3.26. For inspection gauges, it can be seen that the tolerance on gauges is kept beyond 10% of work tolerance.

Disadvantages of workshop and inspection gauges:
1. Although some of the components are well within the work tolerance limits, they may be rejected under workshop gauges. These components again need to be checked by inspection gauges and may be accepted after that.

2. It may also happen that some of the components when tested by inspection gauges may get accepted.
3. As the tolerance zones are different, both workshop and inspection gauges have to be manufactured separately.

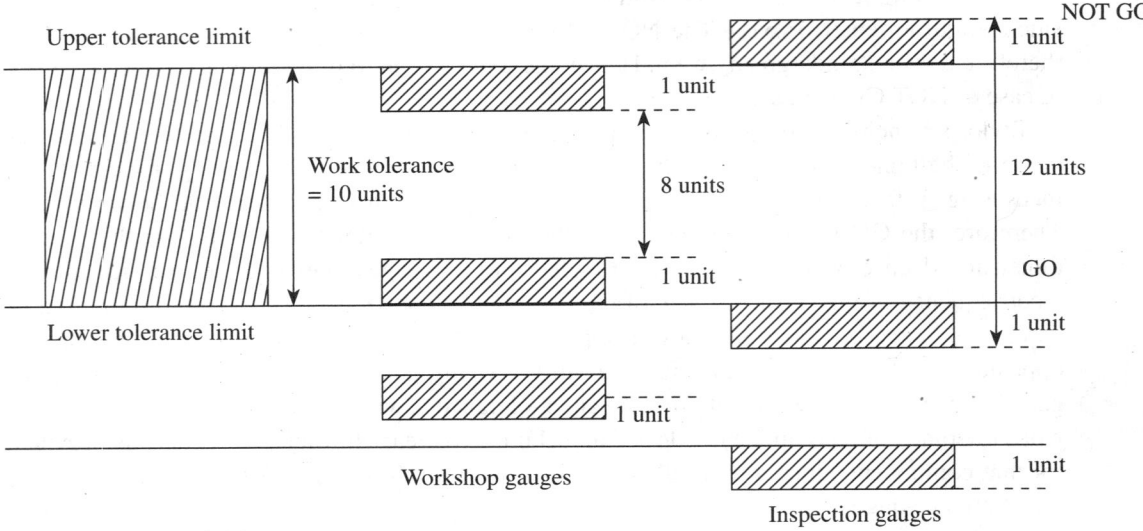

Fig. 3.26 Disposition of tolerances on workshop and inspection gauges

Revised System of Gauges

In this system, the tolerance zone on inspection gauges is reduced and the tolerance on workshop gauges is retained as before. The disadvantages of inspection gauges are reduced due to the reduction of tolerance on these gauges. It can be observed from Fig. 3.27 that instead of 120% of the tolerance provided in the first system for GO and NOT GO inspection gauges, in the revised system, 110% of the range of work tolerance is provided.

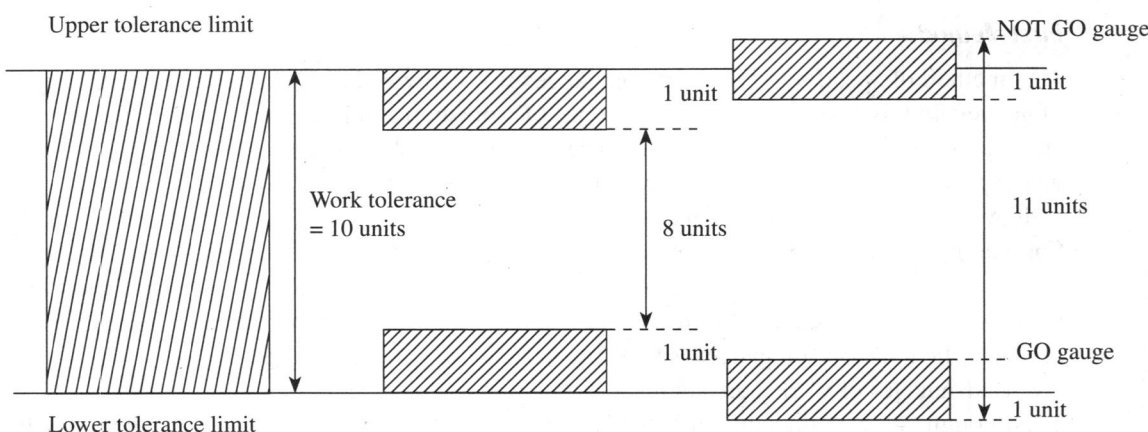

Fig. 3.27 Disposition of modified tolerances on workshop and inspection gauges

British System

The present British system is in accordance with Taylor's principle and is widely accepted in industries. This system works according to the following guidelines:

1. The specified tolerance should be wide and, at the same time, consistent with satisfactory functioning, economical production, and inspection.
2. During inspection, the work that lies outside the specified limits should not be accepted.

Thus, in this modern system, the same tolerance limits are specified on workshop and inspection gauges, and the same gauges can be used for both purposes.

It can be observed from Fig. 3.28 that the tolerance zone for the GO gauges is placed inside the work limits and that for the NOT GO gauges is outside the work limits.

A margin is provided in between the tolerance zone for the gauge and MML of the work to accommodate wear allowance for GO gauges. It is to be noted here that wear allowance should not be permitted beyond the MML of the work, especially when the limit is of critical importance. The magnitude of wear allowance is 10% of the gauge tolerance.

3.6.10 Numerical Examples

Example 3.7   Design the general type of GO and NOT GO gauges as per the present British system for a 40 mm shaft and hole pair designated as 40 H8/d9, given that

(a) $i = 0.453 \sqrt[3]{D} + 0.001D$
(b) 40 mm lies in the diameter range of 30–50 mm
(c) IT8 = 25i
(d) IT9 = 40i
(e) upper deviation of shaft = $-16D^{0.44}$
(f) wear allowance assumed to be 10% of gauge tolerance

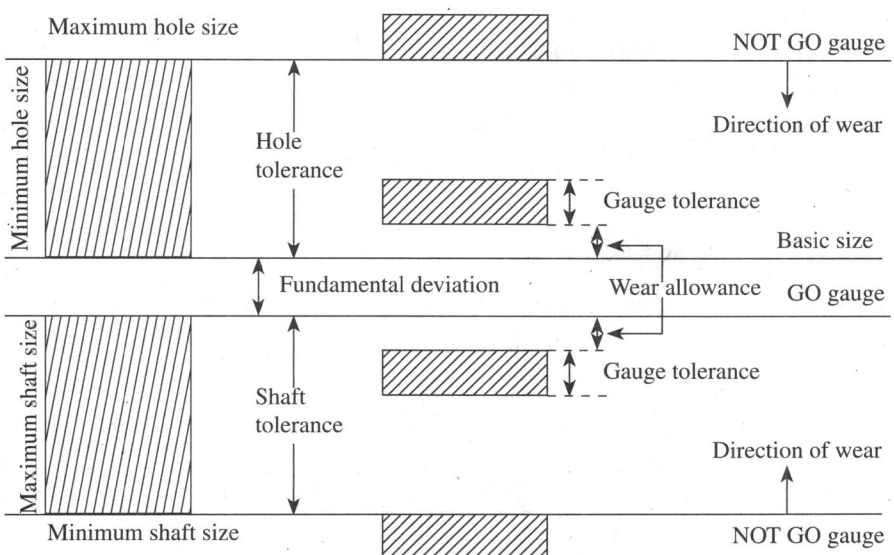

Fig. 3.28   Disposition of tolerances and allowances on gauges

Solution

The standard diameter steps for 40 mm shaft falls in the range of 30–50 mm.

D can be calculated using the equation $\sqrt{D_{max} \times D_{min}}$.
Therefore,

$$D = \sqrt{30 \times 50}$$
$$D = 38.7298 \text{ mm}$$

The value of fundamental tolerance unit is given by

$$i = 0.453 \sqrt[3]{D} + 0.001D$$
$$i = 0.453 \left(\sqrt[3]{38.7298} \right) + 0.001(38.7298) = 1.571 \,\mu\text{m}$$

For hole quality H8, the fundamental tolerance is $25i$.

$$25i = 25(1.571) = 39.275 \,\mu\text{m} = 0.039275 \text{ mm} \approx 0.039 \text{ mm}$$

For hole, the fundamental deviation is zero.
Hence, hole limits are as follows:

$$\text{LLH} = 40 \text{ mm}$$
$$\text{HLH} = 40.00 + 0.039 = 40.039 \text{ mm}$$
$$\text{Hole tolerance} = 40.039 - 40 = 0.039 \text{ mm}$$

For shaft quality $d9$, the fundamental tolerance is $40i$.

$$40i = 40(1.571) = 62.84 \,\mu\text{m} = 0.06284 \text{ mm} \approx 0.063 \text{ mm}$$

For d shaft, the fundamental deviation is given by $-16D^{0.44}$

Therefore, fundamental deviation $= -16(38.7298)^{0.44}$
$$= -79.9576 \,\mu\text{m} \approx -0.07996 \text{ mm} \approx -0.080 \text{ mm}$$

Hence, shaft limits are as follows:

$$\text{HLS} = 40 - 0.080 = 39.92 \text{ mm}$$
$$\text{LLS} = 40 - (0.080 + 0.063) = 39.857 \text{ mm}$$
Shaft tolerance $= 39.92 - 39.857 = 0.063 \text{ mm}$

Hence, the hole and shaft limits are as follows:

Hole $= 40 \,^{+0.039}_{+0.000}$ mm and shaft $= 40 \,^{-0.080}_{-0.143}$ mm

The disposition of tolerances is as shown in Fig. 3.29.

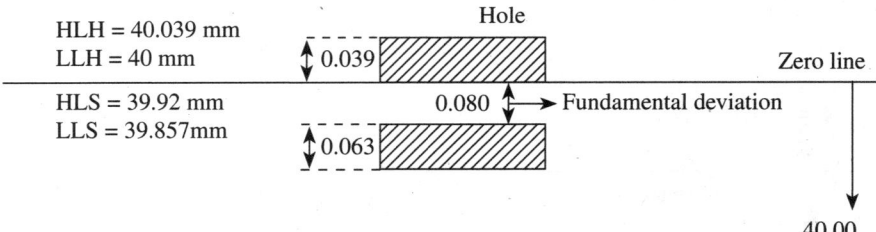

Fig. 3.29 Disposition of tolerances

Assuming gauge tolerance to be 10% of work tolerance,
gauge tolerance for hole = 10% of 0.039 = 0.0039 mm
Wear allowance for hole = 10% of Gauge tolerance; therefore,
Therefore, wear allowance for hole = 10% of 0.0039 = 0.00039 mm

Similarly,
Gauge tolerance for shaft = 10% of 0.063 = 0.0063 mm
Wear allowance for shaft = 10% of 0.0063 = 0.00063 mm

For hole:
The limits of GO Plug gauge are as follows:
Low limit = Basic size + Wear allowance
Low limit = 40.00 + 0.00039 = 40.00039 mm
High limit = Basic size + (Wear allowance + Gauge tolerance)
High limit = 40.00 + (0.00039 + 0.0039) mm
$\qquad$ = 40.00 + (0.00429) = 40.00429 mm

Limits of GO plug gauge = $40\ ^{+0.00429}_{+0.00039}$ mm
The limits of NOT GO Plug gauge are as follows:
Low limit = Basic size + Fundamental tolerance for hole
Low limit = 40.00 + 0.039= 40.039 mm
High limit = Basic size + (Fundamental tolerance for hole + Gauge tolerance)
High limit = 40.00 + (0.039 + 0.0039) mm
$\qquad$ = 40.00 + (0.0429) = 40.0429 mm

Limits of NOT GO plug gauge = $40\ ^{+0.0429}_{+0.0390}$ mm
For shaft:
The limits of GO snap gauge are as follows:
High limit = Basic size − (Fundamental deviation + Wear allowance)
High limit = 40.00 − (0.080 + 0.00063) mm
High limit = 40.00 − (0.08063) = 39.91937 mm
Low limit = Basic size − (Fundamental deviation + Wear allowance + Gauge tolerance)
Low limit = 40.00 − (0.080 + 0.00063 + 0.0063) mm
$\qquad$ = 40.00 − (0.08693) = 39.91307 mm

Limits of GO snap gauge = $40\ ^{-0.08063}_{-0.08693}$ mm
Limits of NOT GO snap gauge are as follows:
High limit = Basic size − (Fundamental deviation + Fundamental tolerance)
High limit = 40.00 − (0.080 + 0.063) mm
High limit = 40.00 − (0.143) = 39.857 mm
Low limit = Basic size − (Fundamental deviation + Fundamental tolerance + Gauge tolerance)
Low limit = 40.00 − (0.080 + 0.063 + 0.0063) mm
Low limit = 40.00 − (0.1493) = 39.8507 mm

Limits of NOT GO snap gauge = $40\ ^{-0.1430}_{-0.1493}$ mm

The disposition of gauge tolerances and wear allowance for the GO and NOT GO plug and snap gauge are schematically shown in Fig. 3.30.

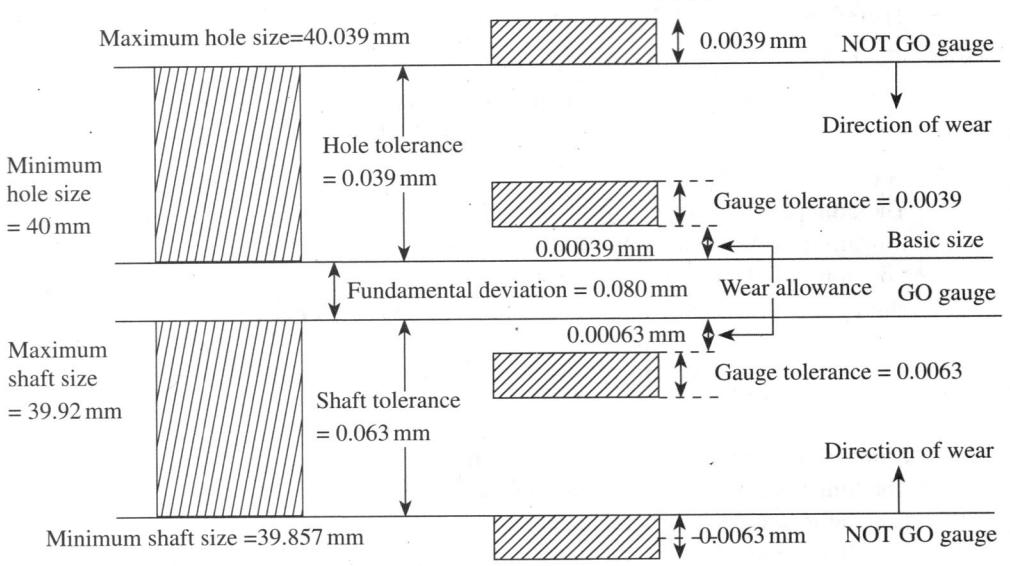

Fig. 3.30 Disposition of gauge tolerances and wear allowance

3.7 PLAIN PLUG GAUGES

Gauges that are used for gauging holes are known as plug gauges. Plain plug gauges are manufactured using hardened and stabilized wear-resistant steel. The gauging surface of the plug gauges are ground and lapped to make it smooth. The surface of these gauges is hardened to not less than 750 HV. Handles for these plug gauges may be made using any suitable steel. Handles made from light metal alloys can be used for heavy plain plug gauges. For smaller plain plug gauges, handles can be made using a suitable non-metallic material. Double-ended-type plug gauges for sizes up to 63 mm and single-ended-type gauges for sizes above 63 mm are recommended as standards. In order to protect the plug gauges against climatic conditions, they are normally coated with a suitable anticorrosive coating.

Plain plug gauges are designated by specifying

1. nominal size
2. 'GO' and 'NOT GO' on the GO and NOT GO sides, respectively
3. class of tolerance of the workpiece to be gauged
4. the manufacturer's name and trademark
5. marking the NOT GO side with a red colour band to distinguish it from the GO side, irrespective of whether the gauge is single or double ended

For example, if the plain plug is designated (according to Indian standard IS: 3484) as 'GO and NOT GO plain plug gauge 40 H7, IS: 3484', it means that it is a plain plug gauge for gauging a bore having a nominal size of 40 mm with a tolerance of H7 and is as per Indian standard.

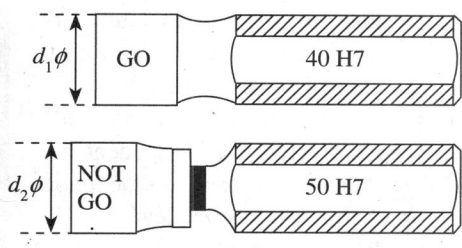

Fig. 3.31 Single-ended GO and NOT GO plug gauges

Single- and double-ended plug gauges are shown in Figs 3.31 and 3.32, respectively. The GO side is made to the lower limit of the hole represented by diameter d_1. The NOT GO side is made to a size equal to the upper limit of the hole represented by diameter d_2. A progressive type of plug gauge is shown in Fig. 3.33. In this type, both GO and NOT GO gauges are on the same side. This can be conveniently used to gauge smaller through holes. During gauging, the GO side of the gauge assembles to its full length. As the gauge is moved further, the NOT GO side obstructs the entry of the gauge into the hole.

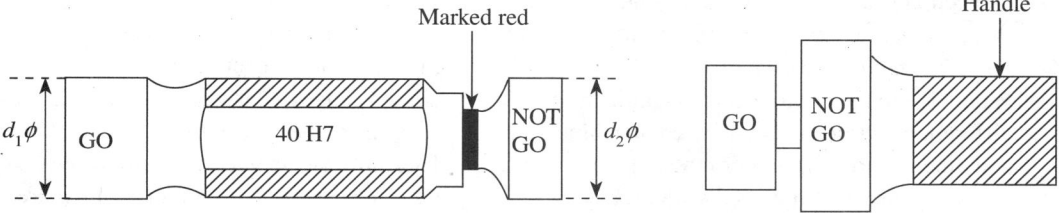

Fig. 3.32 Double-ended plug gauges

Fig. 3.33 Progressive plug gauges

3.8 SNAP GAUGES

Snap gauges are used to gauge the shaft to ascertain whether the dimensions of the shafts are well within the tolerance limits. These gauges like plug gauges are also manufactured using hardened and stabilized wear-resistant steel. The gauging surface is ground and lapped. Shafts of sizes ranging from 3 to 100 mm can conveniently be gauged using double-ended snap gauges, as shown in Fig. 3.34. For sizes over 100 and up to 250 mm, single-ended progressive-type snap gauges are used. A progressive snap gauge is schematically represented in Fig. 3.35.

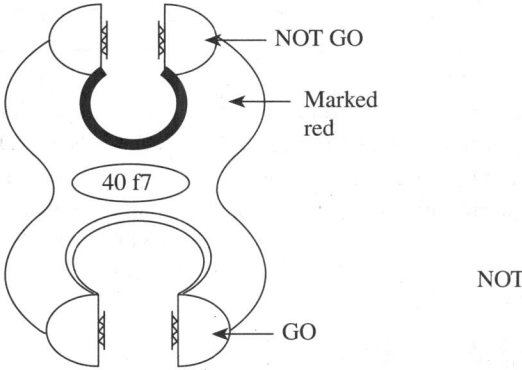

Fig. 3.34 Double-ended snap gauge

Fig. 3.35 Progressive snap gauge

A QUICK OVERVIEW

- No component can be manufactured precisely to a given dimension; it can only be made to lie between the upper and lower limits. The designer has to suggest these tolerance limits, which are

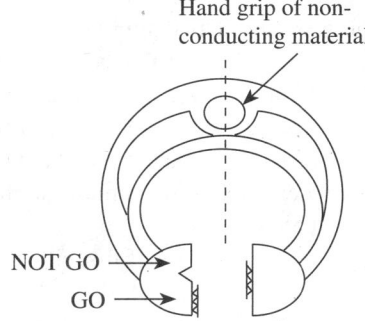

acceptable for each of the dimensions used to define shape and form, and ensure satisfactory operation in service. The difference between the upper and lower limits is termed permissive tolerance.

- The components of a product are manufactured in one or more batches by different persons on different machines and at different locations and are then assembled at one place. To achieve this, mass production becomes inevitable, but at the same time, adherence to the specified limits of accuracy is also important. Manufacture of components under such conditions is called *interchangeable manufacture*.
- When identical components, manufactured by different operators, using different machine tools and under different environmental conditions, are assembled and replaced without any further modification during the assembly stage, without affecting the functioning of the component, it is known as interchangeable assembly.
- When the tolerance distribution is only on one side of the basic size, it is known as unilateral tolerance, and when it lies on either side of the basic size, it is called bilateral tolerance.
- Fit is a relationship that results between the two mating parts to be assembled, that is, a hole and shaft, with respect to the difference in their dimensions before assembly.
- Depending on the actual limits of the hole or shaft, three basic types of fits can be identified,: clearance fit, interference fit, and transition fit.
- An allowance is the intentional difference between the MMLs, that is, LLH and HLS (minimum clearance or maximum interference) of the two mating parts.
- To obtain the desired class of fits, either the size of the hole or the size of the shaft must vary. Hole basis and shaft basis systems are used to represent the three basic types of fits.
- Fundamental deviation is the *minimum* difference between the size of a component and its basic size. This is identical to the upper deviation for shafts and lower deviation for holes. It is the closest deviation to the basic size. Fundamental tolerance is also called 'grade of tolerance'. In the Indian Standard system, there are 18 grades represented by number symbols for both hole and shaft. Fundamental deviation and fundamental tolerance indicate the type and quality of fit, respectively.
- Limit gauges ensure that the components lie within the permissible limits, but do not determine the actual size or dimensions. Gauges are scaleless inspection tools, which are used to check the conformance to the limits of the parts along with their forms and the relative positions of the surfaces of the parts.
- The gauges required to check the dimensions of the components correspond to two sizes conforming to the maximum and minimum limits of the components. They are called GO gauges or NO GO or NOT GO gauges, which correspond, respectively, to the MML and LML of the component.
- According to Taylor's principle, the GO gauge is designed to check maximum metal conditions. It should also simultaneously check as many related dimensions such as roundness, size, and location, as possible. The NOT GO gauge is designed to check minimum metal conditions. It should check only one dimension at a time. Thus, a separate NOT GO gauge is required for each individual dimension.

MULTIPLE-CHOICE QUESTIONS

1. Identify the correct feature control frame from the following:

   (a) ⊚ | $\phi 0.01$

   (b) ⊚ | $\phi 0.01$ | X

   (c) ⊚ | $\phi 0.01$ | X | XT

   (d) $\phi 0.01$ | ⊚

2. Which of the following does the symbol ⌒ indicate?
   (a) Profile of semicircle (c) Profile of line
   (b) Profile of surface (d) Profile of circle

3. Which of the following symbols indicates

circular runout?

(a)   (c) ◯

(b) ↗ (d) ↗↗

4. Which of the following represents the type of fit for a hole and shaft pair, given that

hole = $50^{+0.04}_{+0.00}$ mm and shaft = $50^{+0.060}_{+0.041}$ mm?

(a) Clearance fit (c) Loose fit
(b) Transition fit (d) Interference fit

5. How many grades of tolerances does the ISO system of limits and fits specify?

(a) 12 (b) 08 (c) 18 (d) 25

6. The most appropriate reason for specifying grades of tolerances in ISO system is

(a) to improve accuracy of manufacture
(b) for convenience
(c) for ease of manufacture of parts
(d) to help quality personnel accept more comp-onents

7. In a shaft basis system, the upper deviation of the size of shaft is

(a) 1 (c) not related to size
(b) less than 0 (d) 0

8. In the hole and shaft pair designation of 40 H7/ d9, the numbers 7 and 9 indicate

(a) nothing of importance
(b) tolerance grade
(c) accuracy of manufacture
(d) ease of assembly

9. NO GO gauges are designed

(a) for maximum passability
(b) for maximum impassability
(c) without any specified conditions
(d) without attaching any importance to them

10. An allowance is provided

(a) to help the operator
(b) to aid in production
(c) intentionally
(d) as permissive tolerance

11. MML corresponds to the

(a) higher limit of a hole and lower limit of the shaft
(b) lower limit of a hole and lower limit of the shaft
(c) higher limit of a hole and higher limit of the shaft
(d) lower limit of a hole and higher limit of the shaft

12. LML corresponds to the

(a) lower limit of a hole and higher limit of the shaft
(b) higher limit of a hole and lower limit of the shaft
(c) lower limit of a shaft and lower limit of the hole
(d) higher limit of a shaft and higher limit of the hole

13. Limit gauges are used to

(a) measure flatness of the component
(b) measure exact size of the component
(c) check if the component dimension lies within permissible limits
(d) measure surface roughness of the component

14. According to Taylor's principle, GO gauges are designed to check

(a) maximum metal condition
(b) minimum metal condition
(c) both of these
(d) none of these

15. The relationship that results between the two mating parts before assembly is called

(a) tolerance (c) limit
(b) allowance (d) fit

REVIEW QUESTIONS

1. Briefly explain the need to specify tolerance on components.
2. Define unilateral and bilateral tolerances. Give examples for each.
3. Explain why a unilateral tolerance system is generally preferred over bilateral system.
4. Explain the terms interchangeable manufacture and interchangeable assembly.
5. With an example, briefly explain the selective assembly approach.

6. Distinguish between tolerance and allowance.
7. Explain the following terms:
    (a) Limits
    (b) Fundamental deviation
    (c) Fundamental tolerance
8. Define fit and with the help of neat sketches, explain the different types of fits.
9. What are the essential conditions to obtain clearance and interference fits?
10. Differentiate between hole basis and shaft basis systems.
11. Explain the effect of work tolerance on manufacturing cost.
12. Explain the terms local interchangeability and universal interchangeability.
13. Explain why hole basis system is generally preferred.
14. With a neat sketch, discuss compound tolerance.
15. Define the following terms:
    (a) Basic size
    (b) Zero line
    (c) Tolerance zone
    (d) International tolerance grade
    (e) Tolerance class
    (f) Upper and lower deviations
16. What do you mean by accumulation of tolerances? Explain how it can be overcome.
17. Give a detailed classification of fits.
18. Discuss geometric tolerances.
19. Explain the different types of geometric tolerances and symbolically represent them.
20. Briefly explain the principle of limit gauging.
21. List the essential considerations made in the selection of materials for gauges.
22. Write a note on gauge materials.
23. Distinguish between a measuring instrument and a gauge.
24. State and explain Taylor's principle of gauge design.
25. Describe why a GO gauge should be of full form.
26. Classify gauges.
27. List the different points that have to be considered in the design of gauges.
28. List the guidelines followed in the British system.
29. Explain the different systems used to specify tolerances on gauges.
30. How are plain plug gauges designated?
31. Explain the term 'gauge maker's tolerance'.
32. Discuss why wear allowance should be provided to gauges.
33. With a neat sketch, explain double-ended plug gauge.
34. Give a sketch of and explain a snap gauge.
35. Describe a progressive plug gauge with a neat diagram.
36. With the help of neat sketches, discuss metal limits for hole and shaft gauging.
37. Explain why special attention should be given to GO gauges compared to NOT GO gauges during the design of gauges.

PROBLEMS

1. The tolerances for a hole and shaft assembly having a nominal size of 50 mm are as follows:

    Hole = $40 \, ^{+0.021}_{+0.000}$ mm and shaft = $40 \, ^{-0.040}_{-0.075}$ mm

    Determine
    (a) maximum and minimum clearances
    (b) tolerances on shaft and hole
    (c) allowance
    (d) MML of hole and shaft
    (e) type of fit
2. A shaft is manufactured within the specified limits of 30.02 and 29.98 mm. Find the high and low limits of the bush to give a maximum clearance of 0.10 mm and minimum clearance of 0.02 mm.
3. Calculate the limits, tolerances, and allowances on a 25 mm shaft and hole pair designated H7/g6 to get a precision fit. The fundamental tolerance is calculated by the following equation:

    $$i = 0.453 \sqrt[3]{D} + 0.001D$$

    The following data is given:
    (a) Upper deviation of shaft = $-2.5D^{0.34}$
    (b) 25 mm falls in the diameter step of 18–30 mm
    (c) IT7 = 16i

(d) IT6 = 10i

(e) Wear allowance = 10% of gauge tolerance
In addition, determine the maximum and minimum clearance.

4. Design the general type of GO and NO GO gauge for components having 30 H7/f8 fit. Given that
   (a) $i = 0.453 \sqrt[3]{D} + 0.001D$
   (b) upper deviation of 'f' shaft = $-5.5D^{0.41}$
   (c) 30 mm falls in the diameter step of 18–30 mm
   (d) IT7 = 16i
   (e) IT8 = 25i

(f) wear allowance = 10% of gauge tolerance

5. Design a general type of GO and NO GO gauge for components having 50 H7/d9 fit. The fundamental tolerance is calculated by the following equation:

$$i = 0.453 \sqrt[3]{D} + 0.001D$$

The following data is given:
   (a) Upper deviation of shaft = $-16D^{0.44}$
   (b) 50 mm falls in the diameter step of 30–50 mm
   (c) IT7 = 16i
   (d) IT9 = 40i
   (e) Wear allowance = 10% of gauge tolerance

ANSWERS

Multiple-choice Questions

1. (b)	2. (c)	3. (b)	4. (d)	5. (c)	6. (a)	7. (d)	8. (b)
9. (a)	10. (c)	11. (d)	12. (b)	13. (c)	14. (a)	15. (d)	

Problems

1. (a) Maximum clearance = 0.096 mm,
   Minimum clearance = 0.040 mm
   (b) Tolerance on hole = 0.021 mm,
   Tolerance on shaft = 0.035 mm
   (c) Allowance = 0.040 mm
   (d) Maximum metal limit of hole = 40.00 mm,
   Maximum metal limit of shaft = 39.96 mm
   (e) Clearance fit

2. Hole = $30^{+0.08}_{-0.04}$ mm

3. Hole tolerance = 0.021 mm,
   Shaft tolerance = 0.0132 mm,
   Maximum clearance = 0.0412 mm,
   Minimum clearance = 0.007 mm,
   Allowance = 0.007 mm,
   Hole = $25^{-0.021}_{-0.000}$ mm, Shaft = $25^{-0.0070}_{-0.0202}$ mm

4. $i = 1.316$ microns, Hole tolerance = 0.021 mm,
   Fundamental deviation = -0.0200 mm,
   Shaft tolerance = 0.033 mm,
   Hole = $30^{+0.021}_{+0.000}$ mm and Shaft = $30^{-0.020}_{-0.053}$ mm,
   Limits of GO Plug gauge = $30^{+0.00231}_{+0.00021}$ mm,
   Limits of NOT GO Plug gauge = $30^{+0.0231}_{+0.0210}$ mm,
   Limits of GO Snap gauge = $30^{-0.02033}_{-0.02363}$ mm,
   Limits of NOT GO Snap gauge = $30^{-0.0530}_{-0.0563}$ mm

5. $i = 1.571$ microns, Hole tolerance = 0.025 mm,
   Fundamental deviation = -0.080 mm,
   Shaft tolerance = 0.063 mm,
   Hole = $50^{+0.025}_{+0.000}$ mm, Shaft = $50^{-0.080}_{-0.143}$ mm,
   Limits of GO Plug gauge = $40^{+0.00275}_{+0.00025}$ mm,
   Limits of NOT GO Plug gauge = $50^{+0.0275}_{+0.0250}$ mm,
   Limits of GO Snap gauge = $50^{-0.08063}_{-0.08693}$ mm,
   Limits of NOT GO Snap gauge = $50^{-0.1430}_{-0.1493}$ mm

CHAPTER 4

Linear Measurement

After studying this chapter, the reader will be able to

- understand the basic principles of design of linear measuring instruments
- explain the use of 'datum planes' in dimensional measurement
- appreciate the advantages offered by scaled instruments in contrast to a simple steel rule, and discuss the various applications and limitations of their variants
- elucidate the vernier family of instruments for linear measurement and read vernier instruments
- describe how Abbe's law applies to micrometer measurement
- explain how the micrometer principle has developed into a family of diverse instruments
- utilize the digital electronic instruments in linear measurement
- explain the use of slip gauges, and their manufacture and calibration

4.1 INTRODUCTION

In Chapter 2, the methods by which engineering standards of length are established were discussed. Both direct and indirect linear measuring instruments conform to these established standards of length and provide convenient means for making accurate and precise linear measurements. Vernier calliper and vernier micrometer are the most widely used linear measuring instruments in machine shops and tool rooms. Measuring instruments are designed either for line measurements (e.g., steel rule or vernier calliper) or for end measurements in order to measure the distance between two surfaces using an instrument (e.g., screw gauge). Callipers and dividers, which are also linear measurement devices, are basically *dimension transfer instruments*. They will not directly provide the measurement of length on a scale. Quality of measurement not only depends on the accuracy of these instruments, but also calls for application of certain simple principles to be followed during measurements. Illustrations are given throughout this chapter, especially on the latter issue, to highlight that care should be exercised for the proper use of linear measuring instruments.

Most people's first contact with linear measurement is with a steel rule or a tape measure. However, today's engineer has a choice of a wide range of instruments—from purely mechanically operated instruments to digital electronics instruments. One has to consider only the nature of application and cost of measurement to decide which instrument is the best for an application. This chapter covers a broad range of linear measurement instruments, from a simple steel rule to digital callipers and micrometers. However, many of these instruments, such as depth gauge and height gauge, need to be used with a datum to ensure accuracy of measurements. The foundation for all dimensional measurements is the 'datum plane', the most important ones being the surface plate and the V-block. Constructions of the surface plate and V-block are also explained with illustrations.

4.2 DESIGN OF LINEAR MEASUREMENT INSTRUMENTS

The modern industry demands manufacture of components and products to a high degree of dimensional accuracy and surface quality. Linear measurement instruments have to be designed to meet stringent demands of accuracy and precision. At the same time, the instruments should be simple to operate and low priced to make economic sense for the user. Proper attachments need to be provided to make the instrument versatile to capture dimensions from a wide range of components, irrespective of the variations in cross-sections and shapes. The following points highlight important considerations that have to be addressed in the design of linear measurement instruments:

1. The measuring accuracy of line-graduated instruments depends on the original accuracy of the line graduations. Excessive thickness or poor definition of graduated lines affects the accuracy of readings captured from the instrument.
2. Any instrument incorporating a scale is a suspect unless it provides compensation against wear.
3. Attachments can enhance the versatility of instruments. However, every attachment used along with an instrument, unless properly deployed, may contribute to accumulated error. Wear and tear of attachments can also contribute to errors. Use attachments when their presence improves reliability more than their added chance for errors decreasing it.
4. Instruments such as callipers depend on the feel of the user for their precision. Good quality of the instrument promotes reliability, but it is ultimately the skill of the user that ensures accuracy. Therefore, it is needless to say that proper training should be imparted to the user to ensure accurate measurements.
5. The principle of alignment states that the line of measurement and the line of dimension being measured should be coincident. This principle is fundamental to good design and ensures accuracy and reliability of measurements.
6. Dial versions of instruments add convenience to reading. Electronic versions provide digital readouts that are even easier to read. However, neither of these guarantees accuracy and reliability of measurements unless basic principles are adhered to.
7. One important element of reliability of an instrument is its *readability*. For instance, the smallest division on a micrometer is several times larger than that on a steel rule of say 0.1 mm resolution, which is difficult to read. However, the micrometer provides better least count, say up to 0.01 mm, compared to the same steel rule. Therefore, all other things being equal, a micrometer is more reliable than even a vernier scale. However, micrometers have a lesser range than verniers.

8. If cost is not an issue, digital instruments may be preferred. The chief advantage of the electronic method is the ease of signal processing. Readings may be directly expressed in the required form without additional arithmetic. For example, they may be expressed in either metric or British units, and can also be stored on a memory device for further use and analysis.

9. Whenever a contact between the instrument and the surface of the job being measured is inevitable, the contact force should be optimum to avoid distortion. The designer cannot leave the fate of the instrument on the skill of the user alone. A proper device like a *ratchet stop* can limit the contact force applied on the job during measurements, thereby avoiding stress on the instrument as well as distortion of the job.

4.3 SURFACE PLATE

In Section 4.2, we understood that every linear measurement starts at a reference point and ends at a measured point. This is true when our basic interest is in measuring a single dimension, length in this case. However, the foundation for all dimensional measurements is the 'datum plane', the most important one being the surface plate. A surface plate is a hard, solid, and horizontal flat plate, which is used as the reference plane for precision inspection, marking out, and precision tooling set-up (Fig. 4.1). Since a surface plate is used as the datum for all measurements on a job, it should be finished to a high degree of accuracy. It should also be robust to withstand repeated contacts with metallic workpieces and not be vulnerable to wear and tear.

The history of surface plates can be traced to the early 19th century when Richard Robert invented the planer in 1817, which was presumably the first machine tool that used a flat surface. He showed a way of duplicating flat surfaces with a high degree of accuracy, and the world of sliding motions and flat surfaces was born. However, the surface plates used by Roberts were of quite low accuracy compared to today's standards. One should credit the contribution of Sir Joseph Whitworth, a leading name in metrology, who recognized the lack of understanding of the concept of flatness at that time and devised a methodology in the year 1840 for generating a flat surface by the 'three-plate method'. This method is being used even today to manufacture surface plates, although better and modern methods of fabricating surface plates are becoming increasingly popular. In this method, three cast iron plates with ribbed construction (for rigidity) are rough machined along their edges and top surfaces. The plates are kept in the open for normalizing for about a year. Natural changes in temperature relieve the internal stresses. The plates are then finish-machined to a high degree of accuracy and are marked #1, #2, and #3, and applied with a coating of Prussian blue. In a six-step process, the surfaces of two of the plates are placed in contact in a particular order and the blued portions are scraped. The pairing of the plates is varied in a pre-planned sequence, which ensures that all three surfaces match to a high degree, thereby ensuring accurate flat surfaces.

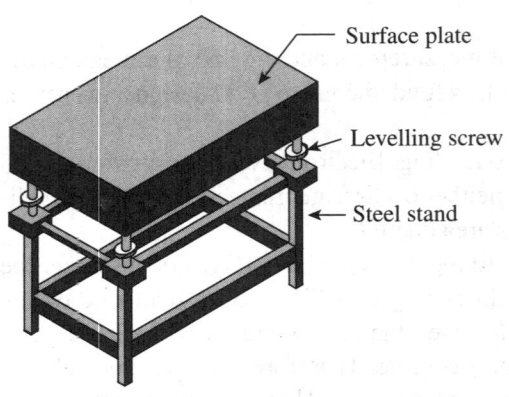

Fig. 4.1 Surface plate

Surface plate

Levelling screw

Steel stand

The surface plates are made either from cast iron or from granite. Even though granite surface plates are perceived to be superior, cast iron surface plates are still in wide use. In fact, a cast iron surface plates is used as a tool for lapping granite surface plates to the required degree of accuracy. Cast iron allows itself to be impregnated with the lapping media over a large flat surface. In the following paragraphs, we will look into the construction and use of cast iron and granite surface plates in greater detail.

Cast Iron Surface Plates

Despite a drop in their usage, cast iron surface plates still retain popularity as surface masters. They are made of either plain or alloyed close-grained cast iron, reinforced with ribs to provide strength against bending or buckling. IS2285-1991 specifies the composition, size, and cross-sectional details of ribs and thickness of plates. The plates are manufactured in three grades, namely grade 0, grade I, and grade II. While grade 0 and grade I plates are hand scraped to achieve the required degree of flatness, grade II plates are precision machined to the required degree of accuracy. Table 4.1 illustrates some of the standard sizes of surface plates as per IS 2285-1991.

Table 4.1 Cast iron surface plate specifications as per IS 2285-1991

Size (mm)	Maximum deviation from flatness in microns			Approximate weight (kg)
	Grade 0	Grade I	Grade II	
300 × 300	4	7	15	21
400 × 400	4.5	9	17	50
450 × 300	4	8	16	39
450 × 450	4.5	9	18	62
600 × 450	5	10	20	79
630 × 400	5	10	20	96
600 × 600	5	10	20	128
630 × 630	5	10	21	156
900 × 600	6	12	23	204
1500 × 1200	8	16	33	986

Smaller-size plates are usually provided with a handle. All surface plates need to be provided with protective covers when not in use. Fabricated heavy angle iron stands with levelling screws provide convenient working height for surface plates. As per the proven practice devised by Sir Whitworth, surface plates are fabricated in sets of three. Cast iron is dimensionally more stable over time compared to granite plates. Unlike granite, it also has uniform optical properties with very small light penetration depth, which makes it a favourable material for certain optical applications. One significant drawback of cast iron is its higher coefficient of thermal expansion, which makes it unsuitable for applications involving large variations in temperature.

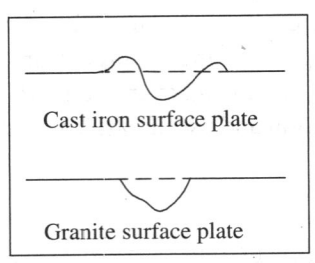
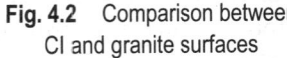

Fig. 4.2 Comparison between CI and granite surfaces

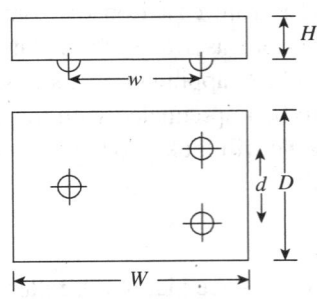

Fig. 4.3 Dimension parameters used in Table 4.2

Granite Surface Plates

In recent times, granite has replaced cast iron as the preferred material for surface plates. Most surface plates are made of black granite, while pink granite is the next preferred choice. Granite has many advantages over cast iron. Natural granite that is seasoned in the open for thousands of years is free from warp age or deterioration. It is twice as hard as cast iron and not affected by temperature changes. It is not vulnerable to rusting and is non-magnetic. As shown in Fig. 4.2, it is free from burrs or protrusions because of its very fine grain structure.

Table 4.2 lists some of the standard sizes of surface plates as per IS 2285-1991. Figure 4.3 illustrates the meaning of dimension parameters used in Table 4.2. The plates are manufactured in three grades, namely grade 00, grade 0, and grade 1. Flatness for grade 00 surface plates range from 0.002 to 0.007 mm, whereas that for grade 0 plates varies from 0.003 to 0.014 mm. Grade 1 plate is the coarsest, with flatness ranging from 0.005 to 0.027 mm.

Table 4.2 Sizes of granite surface plates as per IS 2285-1991

Grade	S. no.	Dimensions $W \times D \times H$ (mm)	Flatness (mm)	w (mm)	d (mm)	Mass (kg)
00	1	300 × 300 × 100	0.002	240	240	27
	2	450 × 300 × 100	0.002	390	240	40
	3	750 × 500 × 130	0.003	630	420	146
	4	1500 × 1000 × 200	0.004	1100	700	900
	5	3000 × 2000 × 500	0.007	2000	1500	9000
0	1	300 × 300 × 100	0.003	240	240	27
	2	450 × 300 × 100	0.003	390	240	40
	3	750 × 500 × 130	0.005	630	420	146
	4	1500 × 1000 × 200	0.008	1100	700	900
	5	3000 × 2000 × 500	0.014	2000	1500	9000
1	1	300 × 300 × 100	0.005	240	240	27
	2	450 × 300 × 100	0.006	390	240	40
	3	750 × 500 × 130	0.009	630	420	146
	4	1500 × 1000 × 200	0.016	1100	700	900
	5	3000 × 2000 × 500	0.027	2000	1500	9000

Glass Surface Plates

Glass is an alternative material for surface plates. It was used during World War II when material and manufacturing capacity were in short supply. Glass can be ground suitably and has the benefit that it chips rather than raising a burr, which is a problem in cast iron surface plates.

4.4 V-BLOCKS

V-blocks are extensively used for inspection of jobs with a circular cross section. The major purpose of a V-block is to hold cylindrical workpieces to enable measurement. The cylindrical surface rests firmly on the sides of the 'V', and the axis of the job will be parallel to both the base and the sides of the V-block. Generally, the angle of the V is 90°, though an angle of 120° is preferred in some cases. It is made of high-grade steel, hardened above 60 Rc, and ground to a high degree of precision. V-blocks are manufactured in various sizes ranging from 50 to 200 mm. The accuracy of flatness, squareness, and parallelism is within 0.005 mm for V-blocks of up to 150 mm length, and 0.01 mm for those of length between 150 and 200 mm (Fig. 4.4).

V-blocks are classified into two grades, grade A and grade B, according to IS: 2949-1964, based on accuracy. Grade A V-blocks have minimum departure from flatness (up to 5 μm for 150 mm length) compared to grade B V-blocks.

There are many variants of V-blocks, such as V-blocks with clamp, magnetic V-block, and cast iron V-block. Figure 4.5 illustrates a V-block with a stirrup clamp. It is convenient for clamping the job onto the V-block, so that measurements can be made accurately. Another popular type of V-block is the magnetic V-block, shown in Fig. 4.6. The magnetic base sits on a flat surface, preferably on a surface plate. The base and two sides are energized for gripping onto a flat surface and a 'vee'slot enables the device to grip the job firmly with a circular cross section. A push-button control turns the permanent magnetic field on and off, thereby enabling the attachment or detachment of the V-block to a flat surface. All three magnetic surfaces are carefully ground and, when switched on, all three magnetic surfaces are activated simultaneously. Magnetic V-blocks are used in tool rooms for drilling and grinding round jobs.

4.5 GRADUATED SCALES

We often use the words 'rule' and 'scale' to mean the simple devices that we have been using since primary-school geometry class. However, there is a clear difference in the actual meaning of these two familiar words. A scale is graduated in *proportion* to a unit of length. For example,

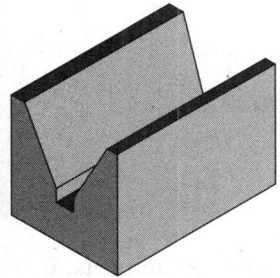

Fig. 4.4 V-block

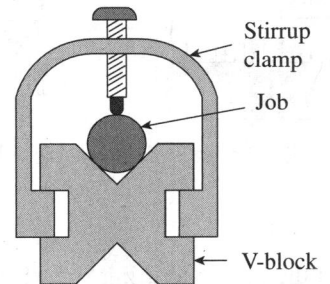

Fig. 4.5 V-block with a stirrup clamp

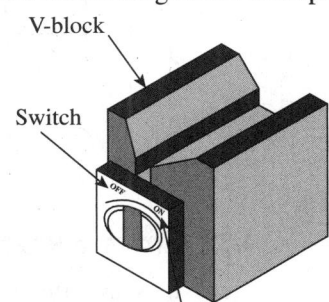

Fig. 4.6 Magnetic V-block

the divisions in an architect's scale, illustrated in Fig. 4.7, represent feet and inches, while the plumber's scale would have divisions in terms of 1/32th or 1/64th of an inch. The divisions of a rule, on the other hand, are the unit of length, its divisions, and its multiples. Typically, the rules with which we are familiar have graduations (in centimetres, millimetres, or inches) and their decimal divisions throughout the length.

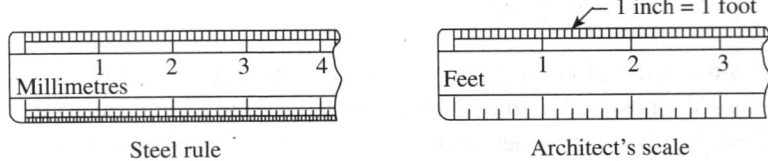

Fig. 4.7 Illustration of the difference between a rule and a scale

Steel rules are most popular in metrology applications since they are more accurate and durable compared to rules made from other materials such as wood or plastic. While rigid rules can be used for laying out lines on a job, flexible steel rules can also be used to measure surfaces with circular profiles. Steel rules are either stamped or cut from a roll of spring steel. The graduations are photo-engraved and tempered with satin chrome finish for good readability. The ruler can be 150, 300, 500, or 1000 mm long; 19 or 25 mm wide; and 1. mm thick. The finer sub-divisions may be marked either throughout the length of the scale or in only a part of its length.

The use of steel rule requires consideration of the relationship between the reference point and the measured point. Figure 4.8 illustrates the preferred way of choosing the reference point for making a measurement. A graduated line on the rule, rather than an edge of the

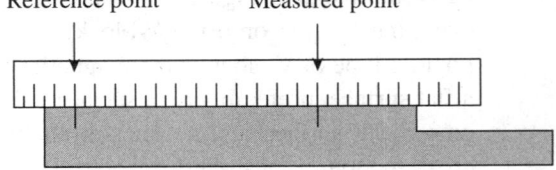

Fig. 4.8 Illustration of reference and measured points

rule, is selected as the reference point. This method improves the accuracy of measurement considerably, even though a little effort is required to align carefully, the reference and measured points. It is recommended not to use the edge of the rule as the reference point, as the edge is subjected to wear and tear and worn-out corners may contribute to error in measurements. Sometimes an attachment such as a hook or a knee is used to facilitate measurement, as shown in Fig. 4.9.

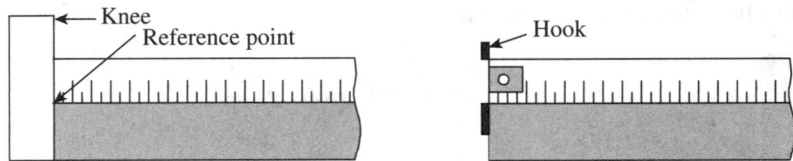

Fig 4.9 Correct ways of using a rule with attachments

4.5.1 Errors in Measurements

Steel rules are often used on the manufacturing shop floors for making measurements. Even though the least count of rules is limited to 1 mm or at the most to 0.5 mm, the user should never be lax in the usage of any measuring device, however simple or trivial it may be. Therefore, one should be aware of the two common errors that can creep up in measurements involving rules. The first one is an inherent error because of poor quality of the rule. This can be attributed to either poor quality of the material of the rule or poor workmanship in the manufacture of the rule. One can avoid this by purchasing a good-quality rule from a standard vendor.

The second error may creep up because of wrong usage of a rule and the observational error. The rule should be properly aligned with the job being measured, ensuring that the reference and measured points are set accurately. The rule should not be twisted or bent in the process of measurement. The major observational error occurs because of parallax error. This is illustrated in Fig. 4.10. Parallax is the apparent shift in the position of an object caused by the change of position of the observer. If an observer views the scale along the direction B or C, the line of sight would be such that there is an apparent shift in the recorded reading by a division or two, as apparent from Fig. 4.10. The more the shift of the eye, from a vertical position right above the measured point, the more pronounced the error. It is needless to say that parallax error can be avoided if the observer recognizes this typical error and takes care to align his/her eyesight in the direction A, shown in Fig. 4.10.

Steel rules come in various sizes and shapes, depending upon the requirements of the component being measured. Accordingly, there are narrow rules, flexible fillet rules, short rules with holders, angle rules, measuring tapes, pocket tapes, and so on (Fig. 4.11). Narrow rules, fillet rules, and angle rules are used for measuring the inside of small holes, narrow slots, and grooves. Short rules with holders are convenient for measuring within recesses. Short rule is an extension of steel rule and obeys the same principle For very precise measurements, temperature expansion and contraction must be considered.

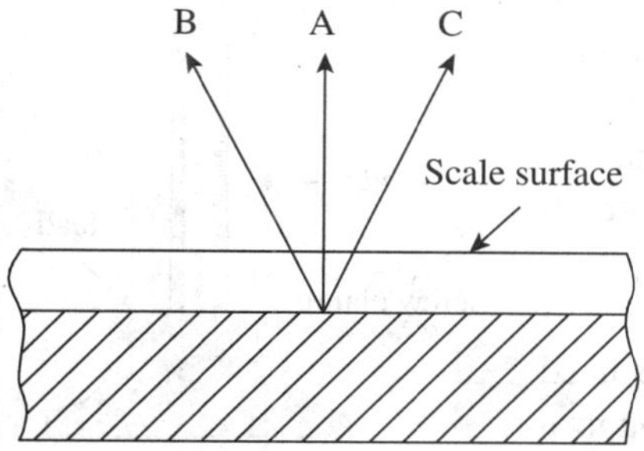

Fig. 4.10 Parallax error that can be minimized by direct eye measurements

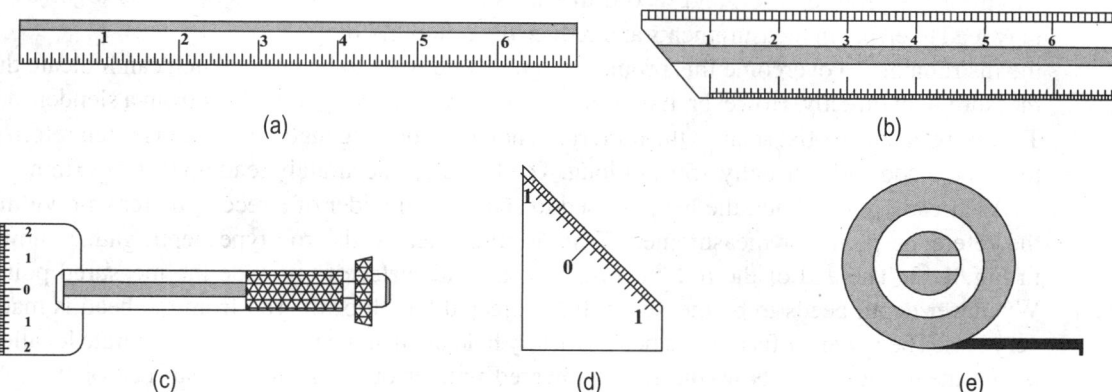

(a) (b) (c) (d) (e)

Fig. 4.11 Types of steel rules (a) Narrow tempered steel rule (b) Flexible fillet rule (c) Short rule with holder (d) Angle rule (e) Steel measuring tape

4.6 SCALED INSTRUMENTS

Rules are useful for many shop floor measurements. However, measurements of certain components require some mechanical means to either hold the measuring device steadily against the component being measured or capture the reading, which can be read at leisure. Another important advantage of a scaled instrument is that the least count of measurement can be improved greatly compared to an ordinary steel rule. Most of the modern scaled instruments provide digital display, which comes with a high degree of magnification. Measurements can be made up to micron accuracy. This section presents three scaled instruments, namely depth gauge, combination set, and callipers, which are necessary accessories in a modern metrology laboratory.

4.6.1 Depth Gauge

Depth gauge is the preferred instrument for measuring holes, grooves, and recesses. It basically consists of a graduated rod or rule, which can slide in a T-head (simply called the head) or stock. The rod or rule can be locked into position by operating a screw clamp, which facilitates accurate reading of the scale. Figure 4.12 illustrates a depth gauge, which has a graduated rule to read the measurement directly. The head is used to span the shoulder of a recess, thereby providing the reference point for measurement. The rod or rule is pushed into the recess until it bottoms. The screw clamp helps in locking the rod or rule in the head. The depth gauge is then withdrawn, and reading is recorded in a more convenient position. Thus, depth gauge is useful for measuring inaccessible points in a simple and convenient manner.

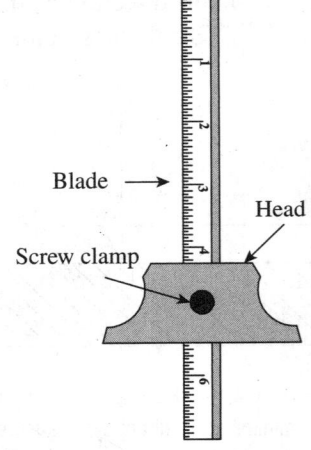

Fig. 4.12   Depth gauge

As already pointed out, either rods or rules can be used in depth gauges for the purpose of measurement. Although a slender rod can easily transfer measurements from narrow and inaccessible holes and recesses, the instrument cannot directly display the reading. One has to use another rule to measure the length of the protruded rod and record the measurement. This may lead to errors in measurements and reduce the reliability of the instrument. To overcome this problem, a graduated rod can be used, which can indicate the measurement directly. However, it is somewhat difficult to read graduations from a slender rod. Therefore, a narrow flat scale is the preferred choice for depth gauges. The rule is often referred to as the blade and is usually 150 mm long. The blade can accurately read up to 1 or ½ mm.

As already pointed out, the head is used to span the shoulder of a recess, thereby providing the reference point for measurement. This is illustrated in the rod-type depth gauge shown in Fig. 4.13. The end of the rod butts against the end surface to provide the measured point. Whenever depth needs to be measured, the projected length of the rod from the head is made very less. The lower surface of the head is firmly held against the job to ensure accurate location of the measured point. Now the rod is lowered until it butts against the surface of the job, thereby marking the measured point. The screw clamp is tightened, the instrument is slowly taken out, and the depth of the hole is read in a convenient position. This method is preferred

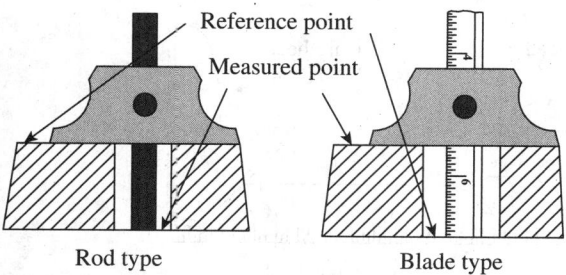

Fig. 4.13 Measured and reference points in depth gauges

for narrow recesses and holes. To summarize, the depth gauge is first positioned against the reference point, followed by the capture of the measured point in order to complete the measurement process.

Sometimes, it becomes necessary to alter the reference and measured points to suit the requirement, as illustrated by the blade-type depth gauge in Fig. 4.13. If the hole is large enough for visually positioning the blade of the depth gauge, the preferred method is to first locate the end of the blade against the lower surface of the hole. The blade is extended from the head, the instrument is brought close to the job, and the end of the blade is butted against the lower surface of the hole. This establishes the reference point for measurement. Now, the head is lowered and the lower surface of the head is made to butt against the top of the job, as shown in Fig. 4.10. The surface of the head provides the measured point. The screw clamp is now tightened and the measurement recorded.

Although depth gauge provides an easy and convenient method for measuring depths of holes and recesses, it has the following limitations:

1. The job size is limited by the width of the head of the depth gauge. Usually, the maximum width of the hole that can be spanned is about 50 mm.
2. The base of the head should be perpendicular to the line of measurement. Otherwise, the line of measurement will be skewed, resulting in erroneous readings.
3. The end of the blade must butt against the desired reference. This will be rather difficult to achieve, especially in blind holes.
4. The end of the blade and the lower surface of the head are always in contact with the job being measured. Therefore, these surfaces will undergo wear and tear. The instrument should be periodically checked for accuracy and replaced if the wear amounts to one graduation line of the instrument.

4.6.2 Combination Set

A combination set has three devices built into it: a combination square comprising a square head and a steel rule, a protractor head, and a centre head. While the combination square can be used as a depth or height gauge, the protractor head can measure the angles of jobs. The centre head comes in handy for measuring diameters of jobs having a circular cross section. The combination set is a useful extension of steel rule. This non-precision instrument is rarely used in any kind of production inspection. However, it is frequently used in tool rooms for tool and die making, pattern making, and fabrication of prototypes. It is a versatile and interesting instrument that has evolved from a try-square, which is used for checking squareness between two surfaces.

The graduated steel rule is grooved all along its length. The groove enables the square head to be moved along the length of the rule and fixed at a position by tightening the clamp screw provided on the square head. The square head along with the rule can be used for measuring heights and depths, as well as inside and outside squaring operations. The blade of the graduated

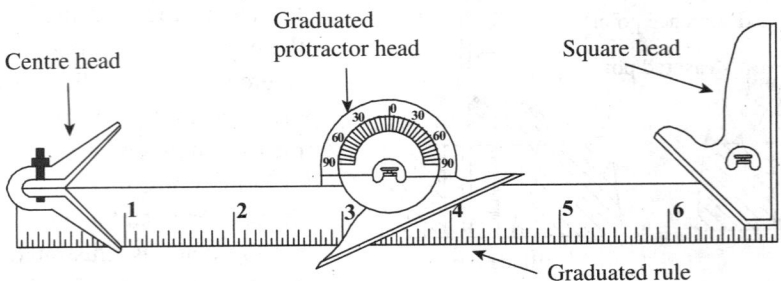

Fig. 4.14 Combination set

protractor head can be swivelled to any angle, which enables the measurement of angles on jobs. The protractor can also be moved along the scale and fixed at a convenient point. Protractors of some combination sets are provided with a spirit level for the purpose of levelling a surface. The centre head attachment is used with the rule to locate the centre of bar stocks. The illustration in Fig. 4.14 shows how each of these attachments are integrated in the combination set.

Square Head

The square head along with the graduated rule on the combination set provides an easy way of measuring heights and depths. While the square head provides a right angle reference, the rule provides a means for directly taking the readings. However, a primary requirement is that the square head can be used only against a flat reference surface. Figure 4.15 illustrates a typical method of measuring height using the combination set. The square head is firmly held against a flat surface of the job, and the rule is lowered until it touches the reference point at the bottom of the job, as shown in the figure. The rule can be locked in this position, and the reading noted down in a convenient position. Attachments are available to mark the measured point with reference to the end of the steel rule. The range of measurement can also be extended by using attachments. In some instruments, the square head is provided with a spirit level, which can be used to test the surfaces for parallelism. A scribing point is provided at the rear of the base in some instruments for scribing purposes.

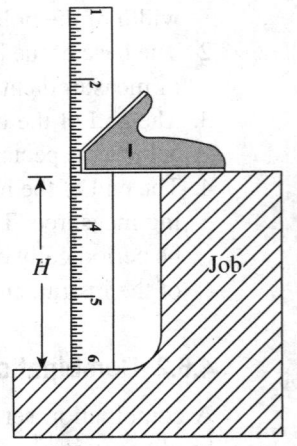

Fig. 4.15 Combination square as a height gauge

Protractor Head

Figure 4.16 illustrates the use of the protractor head on the combination set. This head comprises a rotatable turret within a stock. The turret has an angular scale graduated in degrees. Similar to the square head, the protractor head can also slide along the rule. The blade of the protractor is held firmly against the job and the angle can be directly read from the scale. A spirit level provided on the protractor head can be used for the purpose of levelling a surface. The protractor can also be used to determine the deviation of angle on the job from the desired one. The protractor is first set to the correct angle and locked in position. Now it is held against

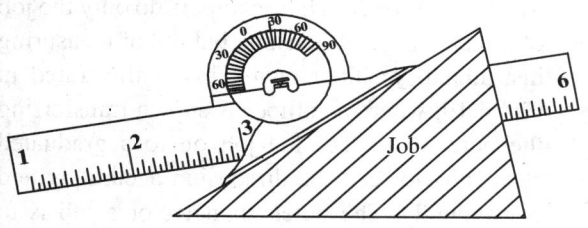

Fig. 4.16 Use of a protractor head for angle measurement

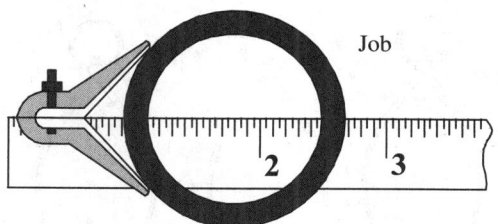

Fig. 4.17 Use of a centre head for the measurement of diameter

the surface of the job for which angle is to be measured. Any deviation from the desired angle can be checked by inserting angle gauges (feeler gauges) in the gap between the blade of the protractor and the job.

Centre Head

The centre head attachment is used along with the steel rule to locate the centre of a bar stock or a circular job. It can be seen from Fig. 4.17 that one edge of the steel rule bisects the V-angle of the centre head. Therefore, it lies on the centre line of any circular job held against the centre head. The diameter of the job can be directly read on the graduated scale, which is useful for marking the centre of the job by using a scriber. The V between the two blades of the centre head facilitates accurate positioning of circular jobs, which greatly improves measurement accuracy when compared to taking readings directly using a rule held against the job. The latter method is completely manual and depends on the skill of the person taking the reading, and therefore is highly prone to error.

Combination sets are available in various sizes. The length of the steel rule ranges from 150 to 600 mm. The scales are graduated in mm and 0.5 mm, and the graduations are available on both sides of the scale. Considering two sets of graduations on one side of the rule, it is possible to have four scales with different graduation schemes and least count. This enhances the versatility of the instrument considerably.

4.6.3 Callipers

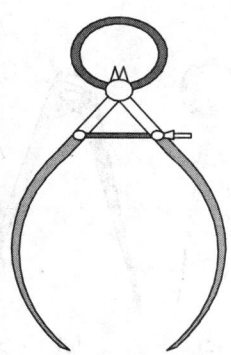

Fig. 4.18 Callipers, the original transfer instruments

There are many jobs whose dimensions cannot be accurately measured with a steel rule alone. A typical case in point is a job with a circular cross section. An attempt to take measurement using a steel rule alone will lead to error, since the steel rule cannot be positioned diametrically across the job with the required degree of accuracy. One option is to use the combination set. However, callipers are the original transfer instruments to transfer such measurements on to a rule (Fig. 4.18). They can easily capture the diameter of a job, which can be manually identified as the maximum distance between the legs of the calliper that can just slide over the diameter of the job. Even though callipers are rarely used in production inspection, they are widely used in tool room and related work.

Busch defines *callipers as instruments that physically duplicate the separation between the reference point and measured point of any dimension*

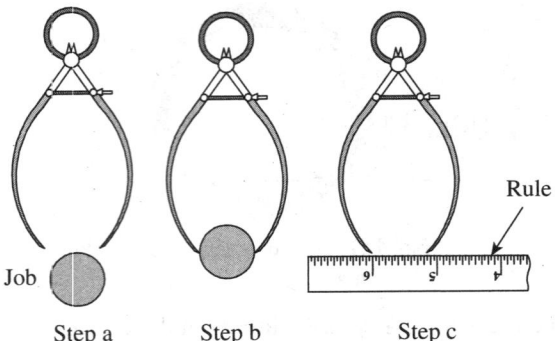

Fig. 4.19 Calliper being used to transfer a dimension from a job to a rule

within their range. Thus, callipers do only the job of transferring a dimension, but not of measuring instruments on their own. This is illustrated in Fig. 4.19, where a calliper is shown transferring the outer diameter of a job on to a graduated steel rule, to read the dimension accurately and conveniently. The outer diameter of a job is to be measured (Step a). Aligning the ends of the two legs of the calliper to a feature of the part being measured, like the one shown in Fig. 4.19, is accomplished quite easily (Step b) because the calliper provides for easy flexing of the two legs and a means of locking them into position whenever required. Now, simply laying the ends of the calliper on a steel rule facilitates easy measurement of the dimension in question (Step c). Thus, as the definition stated earlier mentions, physical duplication of the separation of reference and measured points is accomplished with a high degree of accuracy.

Callipers are available in various types and sizes. The two major types are the *firm joint calliper* and the *spring calliper*. A firm joint calliper, as the name itself suggests, can hold the position of two legs opened out to a particular degree unless moved by a certain force. This is possible because of higher friction developed at the joint between the two legs of the calliper. They are adjusted closely for size by gentle tapping of a leg. A locknut is needed to lock the calliper in a particular position. On the other hand, a spring calliper can hold a particular position due to the spring pressure acting against an adjusting nut. This permits a very careful control, and no lock is needed. Figure 4.20 illustrates the classification of callipers. Callipers are manufactured in a large number of sizes. They are designated not by their measurement ranges, but by the length of their legs, which range from 50 to 500 mm.

Figure 4.21 illustrates the different types of callipers. These are all simple callipers, whose ends are adjustable to transfer a measurement from the job to a steel rule. Although a member of the calliper family, a divider classified under callipers is simply referred to as a divider.

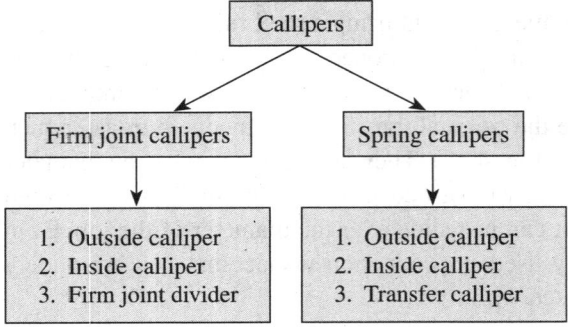

Fig. 4.20 Classification of callipers

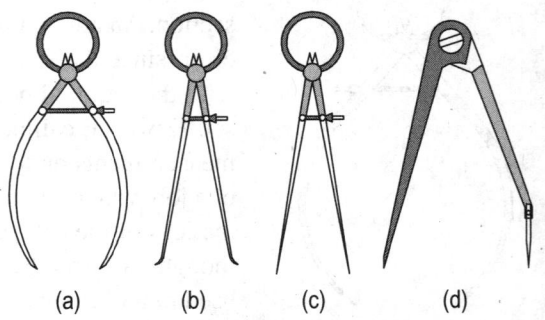

Fig. 4.21 Types of callipers (a) Outside calliper (b) Inside calliper (c) Divider (d) Hermaphrodite calliper

Outside calliper   As the name suggests, an outside calliper is used to measure the outside or external dimensions. While taking measurements, the legs of the calliper should be set gently against the job either by a small physical force in case of a firm joint calliper or by an adjustment of the nut in case of a spring calliper. The user should ensure that a firm contact is established between the calliper and the job before transferring the measurement to a rule. One leg of the calliper is firmly held against a graduation on the steel rule to establish the reference point. The other leg is now gently transferred on to the rule to capture the measured point. The legs of callipers are made of alloy steel, with the measuring points being heat treated to withstand wear and tear. The legs are of a rectangular cross section, and should be free from cracks and any other type of flaw for longevity. The springs in spring callipers are made of spring steel, which is hardened and tempered. The spring force is adjusted by a knurled nut operating on a precision machined screw.

Inside calliper   As illustrated in Fig. 4.21, the measuring ends of the legs of an inside calliper are shaped in the form of a hook. While taking measurements, the calliper legs are initially in a folded condition, which are inserted into the component. Now, the legs are stretched out to butt against the surface of the job. The calliper is carefully rocked over the centre. The feel provided as they pass the centre is the limit of their sensitivity.

Divider   As with callipers, the primary use of dividers is to transfer measurements. However, they are used with line references rather than with surface references. Dividers are used for the following tasks: (a) transferring a dimension of a job to a rule for measurement, (b) transferring a dimension from one part to another part, and (c) transferring a dimension from a rule to a job for layout of the part. Scribing arcs during layout work is another chief use of dividers.

Hermaphrodite calliper   It is essentially a scribing tool comprising one divider leg and one calliper leg. The scriber can be mounted by means of a locknut, as shown in Fig. 4.21. The chief advantage of a hermaphrodite calliper is that a scriber of any required shape and size can be fitted to it and used.

The proper use of the inside and outside callipers depends to a large extent on the skill of the person taking measurements. Measuring with a calliper consists of adjusting the opening so that its reference points duplicate the features of the job being measured. In other words, there is no other provision in a calliper that helps in its alignment than the reference points. As illustrated in Fig. 4.22, the greatest accuracy is achieved in case of callipers when the line of measurement coincides with a plane perpendicular to the job. The divider provides the best accuracy when the measurements are taken from well-marked lines, as shown in Fig. 4.22. Many a time measurements need to be taken between edges, in which case care must be exercised in ascertaining the proper way of taking measurements.

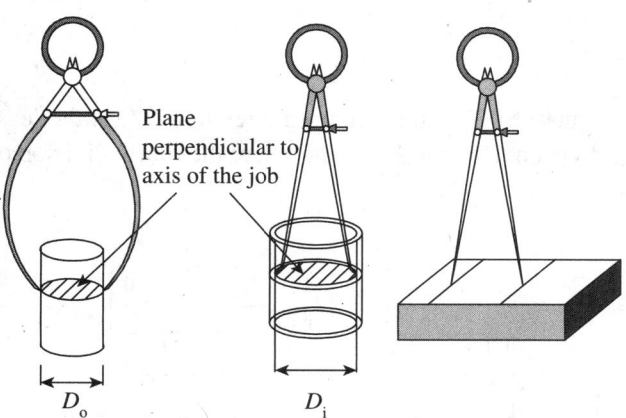

Fig. 4.22   Measurement using callipers (a) Outside calliper (b) Inside calliper (c) Divider

4.7 VERNIER INSTRUMENTS

The instruments discussed in this chapter until now can be branded 'non-precision' instruments, not for their lack of precision but for their lack of amplification. A steel rule can measure accurately up to 1 mm or at best up to 0.5 mm. It is not sensitive to variations in dimensions at much finer levels because of the inherent limitation in its design. On the other hand, vernier instruments based on the vernier scale principle can measure up to a much finer degree of accuracy. In other words, they can amplify finer variations in dimensions and can be branded as 'precision' instruments.

The vernier scale was invented in its modern form in 1631 by the French mathematician Pierre Vernier (1580–1637). Vernier instruments are being used for more than two centuries. The American, Joseph Brown, is credited with the invention of the vernier calliper. As is perhaps known to a student, a vernier scale provides a least count of up to 0.01 mm or less, which remarkably improves the measurement accuracy of an instrument. It has become quite common in the modern industry to specify dimensional accuracy up to 1 μm or less. It is the responsibility of an engineer to design and develop measuring instruments that can accurately measure up to such levels.

It will not be out of place here to briefly brush up our memory of the basic principles of a vernier scale. A vernier scale comprises two scales: the main scale and the vernier scale. Consider the scale shown in Fig. 4.23. Let us say that the main scale has graduations in millimetres up to a minimum division of 1 mm. The vernier scale also has graduations, having 10 equal divisions. In this example, 10 vernier scale divisions (VSDs) equal nine main scale divisions (MSDs). Obviously, the value of one VSD is less than one MSD. Such a vernier scale is called a *forward vernier*. On the other hand, suppose 10 VSDs equal 11 MSDs, the value of one VSD is more than that of one MSD. Such a vernier scale is called the *backward vernier*.

Calculation of least count   The minimum length or thickness that can be measured with a vernier scale is called the *least count*. For a forward vernier shown in Fig. 4.23,

N VSD = $(N-1)$ MSD

1 VSD = $(N-1)/N$ MSD

Least count = 1 MSD – 1 VSD

Therefore, Least count = 1 MSD – $(N-1)/N$ MSD

Least count = $[1- (N-1)/N]$ MSD

Least count = 1 MSD/N

Total reading = MSR + (VC × LC), where MSR is the main scale reading, LC is the least count, and VC is the vernier coinciding division. Refer to Fig. 4.24 where the fourth division of the vernier coincides with a division on the main scale.

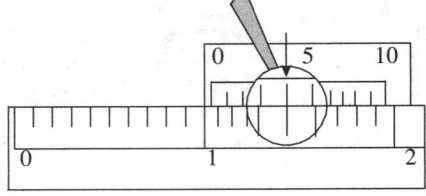

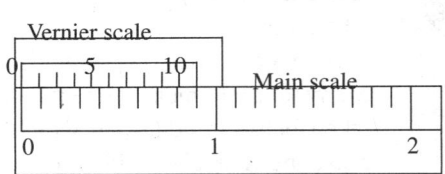

Fig. 4.23   Principle of a vernier scale

Fig. 4.24   Fourth division of vernier coinciding with a division on the main scale

Least count = 1 MSD/N = 1 mm/10 = 0.1 mm

Therefore, total reading = 1 + (4 × 0.1) = 1.4 mm

Digital read-out instruments and dial callipers are rapidly replacing vernier instruments. However, the principle of vernier measurement is basic to metrology, and the use of vernier instruments comprising vernier calliper, vernier depth gauge, vernier height gauge, vernier micrometers, etc., is still widespread in the industry. One can argue that anyone who can measure reliably with the vernier instruments can use the digital versions with equal reliability without any additional training. Even though the use of vernier instruments is not frequent for production inspection, they continue to play an important role in tool room and laboratory work. Production inspectors prefer limit gauges and comparators, which can speed up the inspection process considerably.

4.7.1 Vernier Calliper

A vernier calliper consists of two main parts: the main scale engraved on a solid L-shaped frame and the vernier scale that can slide along the main scale. The sliding nature of the vernier has given it another name—*sliding calliper*. The main scale is graduated in millimetres, up to a least count of 1 mm. The vernier also has engraved graduations, which is either a forward vernier or a backward vernier. The vernier calliper is made of either stainless steel or tool steel, depending on the nature and severity of application.

Figure 4.25 illustrates the main parts of a vernier calliper. The L-shaped main frame also serves as the fixed jaw at its end. The movable jaw, which also has a vernier scale plate, can slide over the entire length of the main scale, which is engraved on the main frame or the beam. A clamping screw enables clamping of the movable jaw in a particular position after the jaws have been set accurately over the job being measured. This arrests further motion of the movable jaw, so that the operator can note down the reading in a convenient position. In order to capture a dimension, the operator has to open out the two jaws, hold the instrument over the job, and slide the movable jaw inwards, until the two jaws are in firm contact with the job. A fine adjustment screw enables the operator to accurately enclose the portion of the job where measurement is required by applying optimum clamping pressure. In the absence of the fine adjustment screw, the operator has to rely on his careful judgement to apply the minimum force that is required to close the two jaws firmly over the job. This is easier said than done, since any excessive application of pressure increases wear and tear of the instrument and may also cause damage to delicate or fragile jobs. The two jaws are shaped in such a manner that

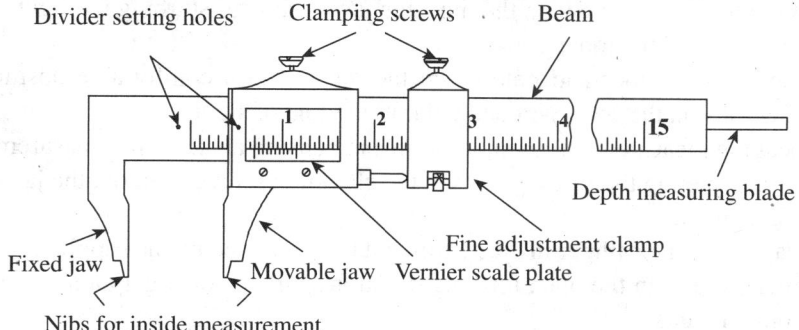

Fig. 4.25 Main parts of a vernier calliper

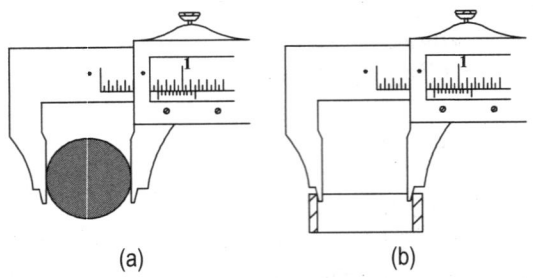

Fig. 4.26 Measurement of dimensions (a) Outside dimension (b) Inside dimension

they can be used to measure both inside and outside dimensions. Notice the nibs in Fig. 4.25, which can be used to measure inside dimension. Figure 4.26 illustrates the method of measuring inside and outside dimensions using a vernier calliper. Whenever the vernier slides over the main frame, a depth-measuring blade also slides in and out of the beam of the calliper. This is a useful attachment for measuring depths to a high degree of accuracy. Divider setting holes are provided, which enable the use of a divider to aid the measurement process.

Measuring a diameter is easier than measuring between flat surfaces, because the diameter is the greatest distance separating the reference and the measured points. Compared to the measurement between flat surfaces, the area of contact between the calliper and the job is much lesser in diameter measurement. Therefore, the resultant force acting either on the job or on the jaws of the calliper is lesser, with the result that there is no deformation or buckling of the jaws. This not only improves the accuracy of measurement, but also reduces the wear and tear of the instrument. Whether the measurement is done for the inside diameter or outside diameter, the operator has to rely on his/her feel to judge if proper contact is made between the measured surfaces and also that excessive force is not exerted on the instrument or the job. Continued closing of the calliper will increase the springing. High gauging pressure causes rapid wear of the jaws, burnishes the part (localized hardening of metal), and may cause damage to the calliper. The following guidelines are useful for the proper use of a vernier calliper:

1. Clean the vernier calliper and the job being measured thoroughly. Ensure that there are no burrs attached to the job, which could have resulted from a previous machining operation.
2. When a calliper's jaws are fully closed, it should indicate zero. If it does not, it must be recalibrated or repaired.
3. Loosen the clamping screw and slide the movable jaw until the opening between the jaws is slightly more than the feature to be measured.
4. Place the fixed jaw in contact with the reference point of the feature being measured and align the beam of the calliper approximately with the line of measurement.
5. Slide the movable jaw closer to the feature and operate the fine adjustment screw to establish a light contact between the jaws and the job.
6. Tighten the clamp screw on the movable jaw without disturbing the light contact between the calliper and the job.
7. Remove the calliper and note down the reading in a comfortable position, holding the graduations on the scale perpendicular to the line of sight.
8. Repeat the measurement a couple of times to ensure an accurate measurement.
9. After completing the reading, loosen the clamping screw, open out the jaws, and clean and lubricate them.
10. Always store the calliper in the instrument box provided by the supplier. Avoid keeping the vernier calliper in the open for long durations, since it may get damaged by other objects or contaminants.
11. Strictly adhere to the schedule of periodic calibration of the vernier calliper.

According to IS: 3651-1974,vernier callipers are of three types: type A, type B, and type C. While all the three types have the scale on the front of the beam, type A vernier scale has jaws on both sides for external and internal measurements, along with a blade for depth measurement. Type B, shown in Fig. 4.25, is provided with jaws on one side only for both external and internal measurements. Type C has jaws on both sides for making the measurements. However, the jaws have knife edge faces for marking purpose. The recommended measuring ranges for vernier callipers are 0–125, 0–200, 0–250, 0–300, 0–500, 0–750, 0–1000, 750–1500, and 750–2000 mm.

Dial Calliper

A vernier calliper is useful for accurate linear measurements. However, it demands basic mathematical skill on the part of the user. One should be able to do simple calculations involving MSD, vernier coinciding division, and least count, in order to compute the measured value of a dimension. In addition, considerable care should be exercised in identifying the coinciding vernier division. These problems can be offset by using a dial calliper (Fig. 4.27).

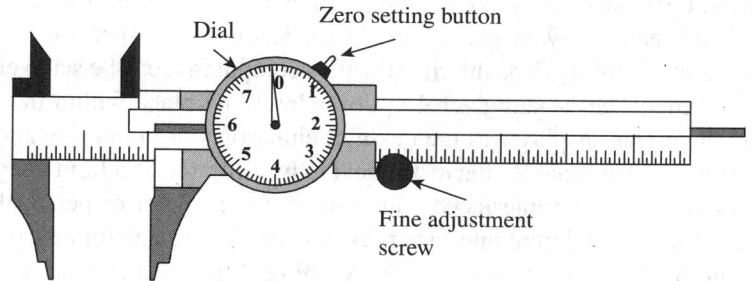

Fig. 4.27 Dial calliper

In a dial calliper, the reading can be directly taken from a dial gauge that is attached to the calliper. The dial gauge has its own least count, which is clearly indicated on the face of the dial. By multiplying the value of the reading indicated by the least count, one can calculate the measured value easily. A small but precise pair of rack and pinion drives a pointer on a circular scale. This facilitates direct reading without the need to read a vernier scale. Typically, the pointer undergoes one complete rotation per centimetre or per millimetre of linear measurement. This measurement should be added to the main scale reading to get the actual reading. A dial calliper also eliminates parallax error, which is associated with a conventional vernier calliper.

A dial calliper is more expensive than the vernier calliper. In addition, the accuracy of the reading mechanism of the dial calliper is a function of length of travel, unlike the vernier calliper that has the same accuracy throughout its length. A dial calliper is also subject to malfunctioning because of the delicate nature of the dial mechanism.

Electronic Digital Calliper

An electronic digital calliper is a battery-operated instrument that displays the reading on a liquid crystal display (LCD) screen. The digital display eliminates the need for calculations and provides an easier way of taking readings. Figure 4.28 illustrates the main parts of an electronic digital calliper.

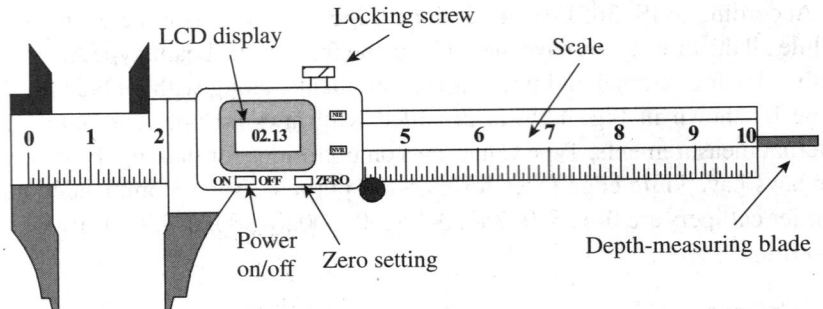

Fig. 4.28 Electronic digital calliper

The LCD display is turned on or off with a button. In order to initialize the instrument, the external jaws are brought together until they touch and the 'zero button' is pressed to set the reading to zero. The digital calliper can now be used to measure a linear dimension. Some digital callipers can be switched between centimetres or millimetres, and inches. Digital callipers are made of stainless steel and are generally available in three sizes: 150, 200, and 300 mm.

The two greatest advantages of an electronic digital calliper are its electronic calculator functions and capability to be interfaced with a computer. It can be set to either metric or British system of units. The 'floating zero' option allows any place within the scale range to be set to zero. The digital display will then exhibit either plus or minus deviations of the jaw from a reference value. This enables the instrument to be also used as a limit gauge. More importantly, a digital calliper can be interfaced with a dedicated recorder or personal computer through a serial data cable. The digital interface provides secured storage for a series of readings, thereby improving the reliability of the records. It can be connected to a printer to provide a printed record or can be directly interfaced with a computer of a statistical control system.

4.7.2 Vernier Depth Gauge

Section 4.7.1 already highlighted the construction and working principle of a depth gauge. However, even though it is simple to operate, it cannot take measurements finer than 1 mm accuracy. A vernier depth gauge is a more versatile instrument, which can measure up to 0.01 mm or even finer accuracy. Figure 4.29 illustrates the constructional features of a vernier depth gauge. The lower surface of the base has to butt firmly against the upper surface of the hole or recess whose depth is to be measured. The vernier scale is stationary and screwed onto the slide, whereas the main scale can slide up and down. The nut on the slide has to be loosened to move the main scale. The main scale is lowered into the hole or recess, which is being measured. One should avoid exerting force while pushing the scale against the surface of the job being measured, because this will not only result in the deformation of the scale resulting in erroneous measurements, but also accelerate the wear and tear of the instrument. This problem is eliminated thanks to the fine adjustment clamp provided with the instrument. A fine adjustment wheel will rotate the fine adjustment screw, which in turn will cause finer movement of the slide. This ensures firm but delicate contact with the surface of the job.

Vernier depth gauges can have an accuracy of up to 0.01 mm. Periodic cleaning and lubrication are mandatory, as the main scale and fine adjustment mechanism are always in motion in the process of taking measurements.

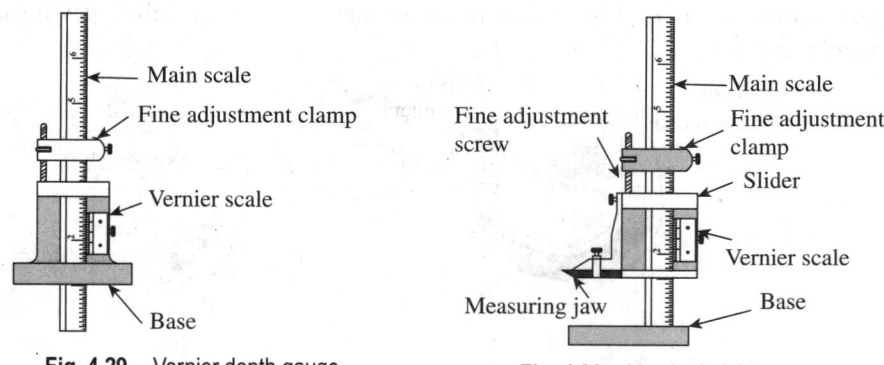

Fig. 4.29 Vernier depth gauge

Fig. 4.30 Vernier height gauge

4.7.3 Vernier Height Gauge

In a vernier height gauge, as illustrated in Fig. 4.30, the graduated scale or bar is held in a vertical position by a finely ground and lapped base. A precision ground surface plate is mandatory while using a height gauge. The feature of the job to be measured is held between the base and the measuring jaw. The measuring jaw is mounted on a slider that moves up and down, but can be held in place by tightening of a nut. A fine adjustment clamp is provided to ensure very fine movement of the slide in order to make a delicate contact with the job. Unlike in depth gauge, the main scale in a height gauge is stationary while the slider moves up and down. The vernier scale mounted on the slider gives readings up to an accuracy of 0.01 mm.

Vernier height gauges are available in sizes ranging from 150 to 500 mm for precision tool room applications. Some models have quick adjustment screw release on the movable jaw, making it possible to directly move to any point within the approximate range, which can then be properly set using the fine adjustment mechanism. Vernier height gauges find applications in tool rooms and inspection departments. Modern variants of height gauges such as optical and electronic height gauges are also becoming increasingly popular.

4.8 MICROMETER INSTRUMENTS

The word 'micrometer' is known by two different meanings. The first is as a unit of measure, being one thousandth of a millimetre. The second meaning is a hand-held measuring instrument using a screw-based mechanism. The word *micrometer* is believed to have originated in Greece, the Greek meaning for this word being *small*. The first ever micrometer screw was invented by William Gascoigne of Yorkshire, England, in the 17th century and was used in telescopes to measure angular distances between stars. The commercial version of the micrometer was released by the Browne & Sharpe Company in the year 1867. Obviously, micrometer as an instrument has a long and cherished history in metrological applications. There have been many variants of the instrument, and modern industry makes use of highly sophisticated micrometers, such as digital micrometers and laser scan micrometers. A micrometer can provide better least counts and accuracy than a vernier calliper. Better accuracy results because of the fact that the line of measurement is in line with the axis of the instrument, unlike the vernier calliper that does not conform to this condition. This fact is best explained by *Abbe's principle*, which states that '*maximum accuracy may be obtained only when the standard is in line with the axis of the part being measured'*. Figure 4.31 illustrates the relevance of Abbe's law for micrometers and vernier callipers.

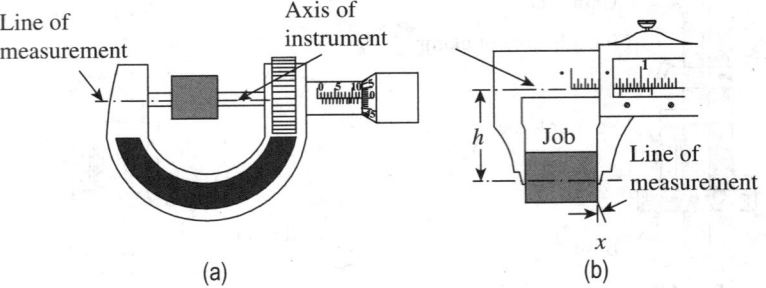

Fig. 4.31 Conformity to Abbe's law (a) Micrometer (b) Vernier calliper

In case of a micrometer, the axis of the job being measured is in line with the line of measurement of the instrument, as illustrated in Fig. 4.31(a). In case of a vernier calliper, for the reading to be accurate, the beam would have to be perfectly straight and the two jaws perfectly at 90° to it. However, this is rarely the case. There is always some lack of straightness of the beam, and the jaws may not be perfectly square with the beam. With continuous usage and wear and tear, the jaws will develop more and more play (Play refers to uncontrolled movements due to slip of one part over the other.) because of repeated sliding movements. Therefore, a certain amount of angular error, marked as x in Fig. 4.31(b), will always be present. This angular error also depends on how far the line of measurement is from the axis of the instrument. The higher the value of this separation h, the greater will be the angular error. We can therefore conclude that *the degree to which an instrument conforms to Abbe's law determines its inherent accuracy.*

4.8.1 Outside Micrometer

Figure 4.32 illustrates the details of an outside micrometer. It consists of a C-shaped frame with a stationary anvil and a movable spindle. The spindle movement is controlled by a precision ground screw. The spindle moves as it is rotated in a stationary spindle nut. A graduated scale is engraved on the stationary sleeve and the rotating thimble. The zeroth mark on the thimble will coincide with the zeroth division on the sleeve when the anvil and spindle faces are brought together. The movement of the screw conforms to the sets

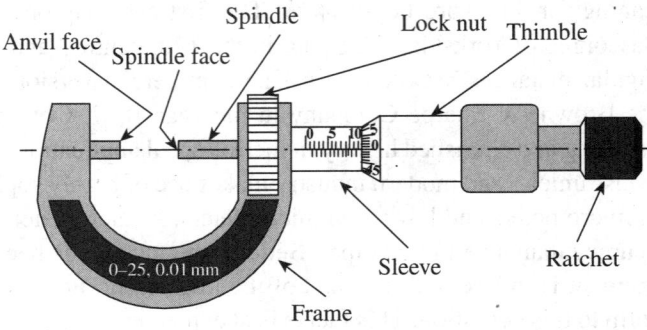

Fig. 4.32 Outside micrometer

of graduations. The locknut enables the locking of the spindle while taking a reading. The ratchet ensures a 'feel' while taking a reading and prevents application of excessive force on the job. The ranges of micrometers are normally 0–25, 25–50, or 0–50 mm. The maximum range of micrometers is limited to 500 mm.

A micrometer is made of steel or cast steel. The measuring faces are hardened to about 60–65 HRC since they are in constant touch with metallic jobs being measured. If warranted, the faces are also tipped with tungsten carbide or a similar material to prevent rapid wear. The anvil is ground and lapped to a high degree of accuracy. The material used for thimble and ratchet should be wear-resistant steel.

Micrometers with metric scales are prevalent in India. The graduations on the sleeve are in millimetres and can be referred to as the main scale. If the smallest division on this scale reads 0.5 mm, each revolution of the thimble advances the spindle face by 0.5 mm. The thimble, in turn, will have a number of divisions. Suppose the number of divisions on the thimble is 50, then the least count of the micrometer is 0.5/50, that is, 0.01 mm. Figure 4.33 illustrates how the micrometer scale is read when a job is held between the anvil face and the spindle face.

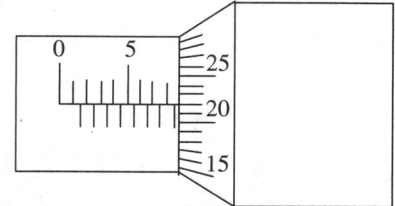

Fig. 4.33 Reading an outside micrometer

In this example, the main scale reading is 8.5 mm, which is the division immediately preceding the position of the thimble on the main scale. As already pointed out, let us assume the least count of the instrument to be 0.01 mm. The 22nd division on the thimble is coinciding with the reference line of the main scale. Therefore, the reading is as follows:

$$8.5 + 22\,(0.01)\,mm = 8.72\,mm$$

Thus, a micrometer is a simple instrument to use. However, there are two precautions to be observed while reading a micrometer. The thimble must be read in the correct direction. The other precaution concerns the zero position on the thimble. When passing the index line on the main scale, there is a chance to read an extra 0.5 mm. This is caused by the fact that the next main scale graduation has begun to show but has not yet fully appeared. This is avoided by being careful to read only full divisions on the barrel. Assuming that these simple precautions are adhered to, a micrometer has many advantages over other linear measurement instruments. It has better readability than a vernier scale and there is no parallax error. It is small, lightweight, and portable. It retains accuracy over a longer period than a vernier calliper and is less expensive. On the flip side, it has a shorter measuring range and can only be used for end measurement.

Types of Micrometers

A micrometer is a versatile measuring instrument and can be used for various applications by simply changing the anvil and the spindle face. For example, the anvil may be shaped in the form of a V-block or a large disk. Figure 4.34 shows a few variants, namely the disk micrometer, screw thread micrometer, dial micrometer, and blade micrometer. The following paragraphs briefly highlight the use of each type of micrometer in metrology applications:

Disk micrometer   It is used for measuring the distance between two features with curvature. A tooth span micrometer is one such device that is used for measuring the span between the two teeth of a gear. Although it provides a convenient means for linear measurement, it is prone to error in measurement when the curvature of the feature does not closely match the curvature of the disk.

Screw thread micrometer   It measures pitch diameters directly. The anvil has an internal 'vee', which fits over the thread. Since the anvil is free to rotate, it can accommodate any rake range of thread. However, interchangeable anvils need to be used to cover a wide range of thread pitches. The spindle has a conical shape and is ground to a precise dimension.

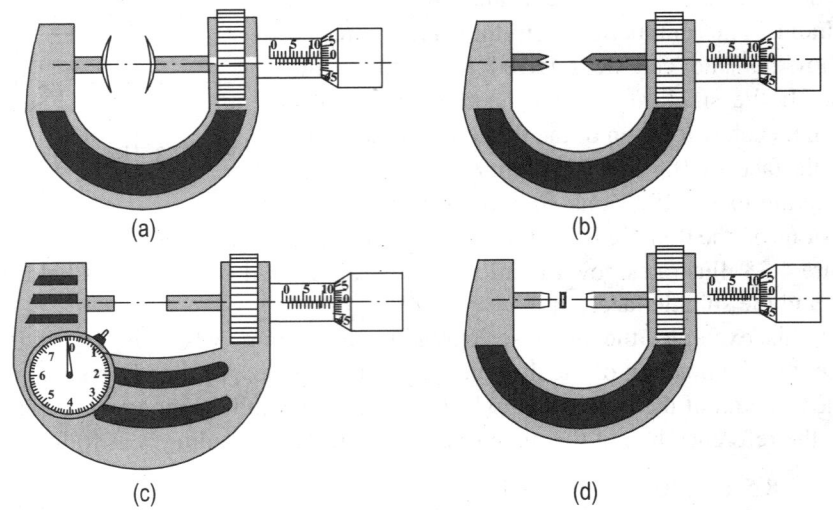

Fig. 4.34   Types of micrometers (a) Disk type (b) Screw thread type (c) Dial type (d) Blade type

Dial micrometer   The dial indicator fixed to the frame indicates the linear displacement of a movable anvil with a high degree of precision. It is especially useful as a comparator for GO/NO-GO judgement in mass production. The dial micrometer normally has an accuracy of 1 µm and repeatability of 0.5 µm. Instruments are available up to 50 mm measuring distance, with a maximum measuring force of 10 N. The dial tip is provided with a carbide face for a longer life.

Blade micrometer   The anvil and spindle faces are in the form of narrow blades and useful for measuring narrow grooves, slots, keyways, and recesses. The blade thickness is around 0.75–1 mm. The spindle does not rotate when the movable blade is moving along the measuring axis. Due to the slender nature of the instrument and non-turning spindle working against a rotating screw, it is vulnerable to rapid wear and tear and needs careful use and maintenance.

Universal micrometer   It has interchangeable anvils such as flat, spherical, spline, disk, or knife edge. It is called universal because of its modular design. The micrometer fitted with the required accessories can function as an outside micrometer, a depth micrometer, a step micrometer, etc.

4.8.2 Vernier Micrometer

A micrometer that we considered hitherto can provide an accuracy of at best 0.01 mm or 10 μm. Placing a vernier scale on the micrometer permits us to take readings up to the next decimal place. In other words, one can accurately measure up to 1 μm or 0.001 mm, which is an excellent proposition for any precision workmanship. As illustrated in Fig. 4.35, in addition to the barrel and thimble scales, a vernier scale is provided next to the barrel scale. Divisions on this vernier scale have to be read in conjunction with the barrel scale to provide the next level of discrimination in readings. The vernier scale consists of a number of equally spaced lines, which are numbered from 0 to 5 or 10, depending on the scale.

The principle of measurement of a vernier micrometer is very similar to that of a vernier calliper. If a division on the thimble is exactly coinciding with the reference line (line marked 0 in the vernier scale in Fig. 4.35) of the vernier scale, the reading is taken in a way similar to an ordinary micrometer explained earlier. However, if none of the divisions on the thimble coincide with the reference line, we need to examine which division on the thimble coincides with one of the divisions on the vernier scale. Hence, an additional step is involved in the calculation since the vernier reading should be taken into account.

Refer to Fig. 4.36, which shows a sample reading. In this case, the thimble has crossed the 12.5 mm mark on the barrel scale. None of the divisions on the thimble coincides with the zeroth line on the vernier scale, that is, the reference line on the barrel. However, the reference line is between the 24th and 25th divisions on the thimble. Suppose the thimble has 50 divisions, and five divisions on the vernier scale correspond to six divisions on the thimble, we can calculate the least count of the instrument as follows.

If one complete rotation of the thimble moves it by 0.5 mm on the barrel scale, the least count of the micrometer scale is $0.5/50 = 0.01$ mm.

Since five divisions on the vernier scale correspond to six divisions on the thimble, the least count of the vernier scale is equal to $0.01/5 = 0.002$ mm. In Fig. 4.36, the fourth division on the vernier scale is coinciding with a division on the thimble.

Therefore, the reading is $12.5 + 24 (0.01) + 4 (0.002) = 12.748$ mm.

Guidelines for Use of Micrometers

1. Before placing the micrometer on the job being measured, bring it near the desired opening. Do this by rolling the thimble along the hand but not by twirling. Hold the micrometer

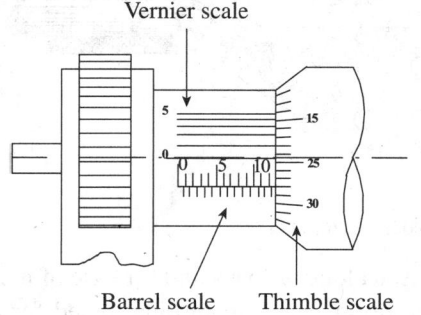

Fig. 4.35 Vernier micrometer

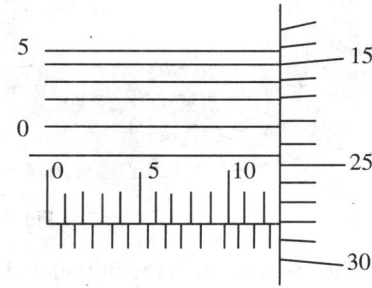

Fig. 4.36 Reading a vernier scale

firmly with one hand, and use the feel of the hand to ensure that the axis of the micrometer is perpendicular to the reference plane of the job. Close the micrometer using the ratchet stop until it disengages with a click.

2. Even though a micrometer can be locked in position by tightening the clamp ring (locknut) and used as a snap gauge for inspection purposes, it is not basically designed for this role. Locking the spindle movement and forcing the measuring faces over the job result in sliding friction, which accelerates wear on the contact surfaces as well as on the micrometer screw.

3. The locknut is a memory device. It retains the reading so that it can be read in a convenient position. However, avoid tightening the locknut when the spindle is withdrawn. Doing so will injure the clamping mechanism.

4. It is not wise to buy a micrometer that does not have a controlled force feature. Excessive force while closing the measuring faces over the job will result in rapid wear and tear of the instrument. A ratchet stop acts as an overriding clutch that holds the gauging force at the same amount for each measurement regardless of the differences in manual application of force.

5. While measuring the diameter of a cylindrical part, rock the cylinder to find the maximum opening that provides the desired feel.

6. Do not expect the micrometer to guarantee reliable measurement if it is (a) dirty; (b) poorly lubricated; (c) poorly adjusted; or (d) closed too rapidly.

7. At the end of each day, the micrometer should be wiped clean, visually inspected, oiled, and replaced in its case to await the next use.

4.8.3 Digital Micrometer

The 'multifunction' digital micrometer is becoming very popular in recent times. The readings may be processed with ease. The push of a button can convert a reading from decimal to inch and vice versa. Any position of the spindle can be set to zero and the instrument can be used to inspect a job within a specified tolerance. The instrument can be connected to a computer or a printer. Most instruments can record a series of data and calculate statistical information such as mean, standard deviation, and range (Fig. 4.37).

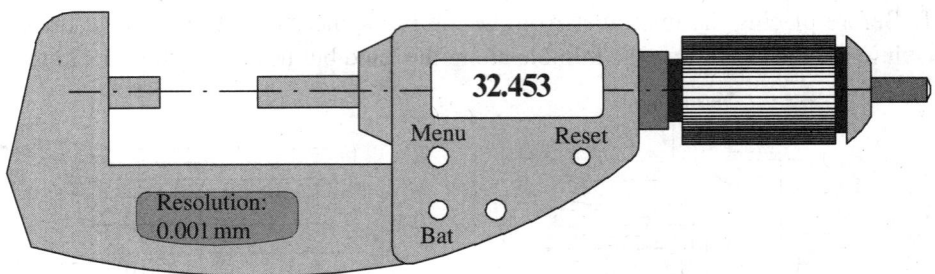

Fig. 4.37   Digital micrometer

The instrument is recommended to be used along with a stand for ease of measurement. The spindle is made of stainless steel and measuring faces are carbide tipped for a longer life. A locking clamp ensures locking of spindle at any desired setting. A constant and low measuring force is ensured by the thimble mechanism. Most of the instruments have a least count of

0.001 mm. An LCD screen displays the reading with absolute linear scale with simple digimatic data collection for personal computer (SPC) data output. An easy push button control is provided to choose the various functions of the instrument. The push buttons controlling the various functions are as follows:

1. ON/OFF: To power the instrument on or off
2. IN/MM: To select either inch or metric system of measurement
3. ZERO: To set the screen display to zero at any desired position
4. HOLD: To hold the measurement taken until the push button is operated again
5. ORIGIN: To set the minimum value for the micrometer depending upon its size; micrometer count commences from this value
6. Alarm indicator: To indicate low voltage and counting value composition error

4.8.4 Inside Micrometer Calliper

The inside micrometer calliper is useful for making small measurements from 5 to 25 mm. In this instrument, unlike a regular micrometer, the axis of the instrument does not coincide with the line of measurement. In addition, unlike the outside micrometer where there is a surface contact between the job and the instrument, the contact between the job and the instrument is line contact. The *nibs*, as the contacts are called, are ground to a small radius. As a necessity, this radius has to be smaller than the smallest radius the instrument can measure. Therefore, all measurements are made with line contacts.

As illustrated in Fig. 4.38, the movable jaw can be moved in and out by the rotation of the thimble. One complete rotation of the thimble moves it by one division on the barrel scale. A locknut can be operated to hold the position of the movable jaw for ease of noting down a reading. While taking measurements, it needs to be rocked and centralized to assure that the axis of the instrument is parallel to the line of measurement. This makes the instrument prone to rapid wear. It is therefore needless to say that the instrument needs to be checked and calibrated regularly.

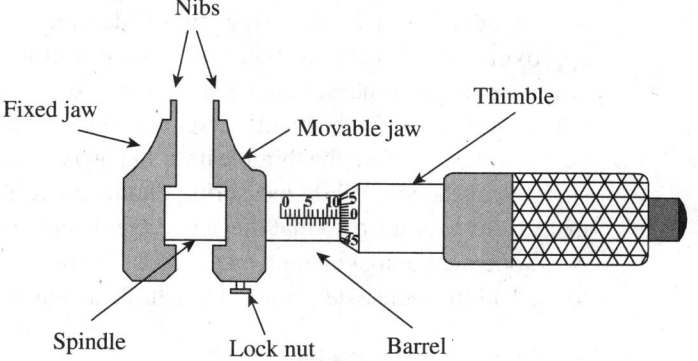

Fig. 4.38 Inside micrometer calliper

4.8.5 Inside Micrometer

This instrument perfectly complies with Abbe's law. The axis of an inside micrometer is also its line of measurement. It is useful for measuring the inside diameter of cylinders, rings, and other machine parts. The inside micrometer set has several accessories, which have to be assembled together for taking the readings. The main unit is the measuring head, which has a thimble that moves over a barrel, same as in the case of an outside micrometer. Graduated scales are provided on the barrel and thimble, which give readings up to an accuracy of 0.01 mm, but with a limited range. The rear end of the measuring head has a contact surface, whereas extension rods of various lengths can be fitted to the front end of the measuring head. A set of extension rods are provided with the instrument to cover a wide range of measurements. The rod ends are spherical and present nearly point contact to the job being measured. A chuck attached to the spindle facilitates the attachment of extension rods. Using a particular extension rod, the distance between contact surfaces can be varied by rotating the thimble up to the range of the micrometer screw. Higher diameters and distances can be measured using longer extension rods. Figure 4.39 illustrates the construction details of an inside micrometer.

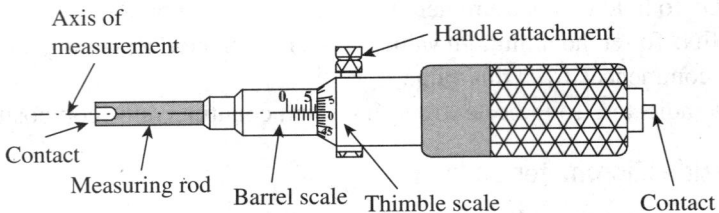

Fig. 4.39 Inside micrometer

The inside micrometer is more common than the inside micrometer calliper because of the flexibility it affords to measure a large range of dimensions. A range of 25 mm above the length of extension rods is commonly used in shops. A standard set will have five extension rods of ranges 50–75, 75–100, 100–125, 125–150, 150–175, and 175–200 mm. In addition to extension rods, a spacing collar (usually 12.5 mm in length) is provided for smaller adjustments in the range of measurements. The micrometer is also provided with a detachable handle for easier handling of the instrument.

The best practice for using an inside micrometer is to first measure the dimension approximately by using a steel rule. Then select a suitable extension rod that can adequately cover the range of measurements. Insert the extension rod into the chuck and set the instrument to read zero. Now fix the handle and lower the instrument into the gap where the dimension is to be measured. Operate the thimble until the two contact surfaces establish a firm contact with the surfaces of the job. While measuring diameters, it is always recommended to lightly move the instrument to and fro so that the actual diameter is sensed by the person taking measurements. Now, operate the locknut and take out the instrument. The micrometer reading has to be added to the length of the extension rod to get the actual reading.

4.8.6 Depth Micrometer

An alternative to vernier depth gauge is the depth micrometer. In fact, most shop floor engineers vouch for its superiority over vernier depth gauges because of its greater measuring range, better reliability, and easier usability. One peculiarity of this instrument is that it reads in reverse from other micrometers. Looking from the ratchet side, a clockwise rotation moves the spindle downwards, that is, into the depth of the job being measured. Therefore, the entire barrel scale is visible when the tip of the measuring rod is in line with the bottom surface of the base. As the measuring rod advances into the depths, the thimble will move over the barrel scale. Reliable measurements of up to 0.01 mm are possible with this instrument. Figure 4.40 illustrates the parts of a depth micrometer. The bottom flat surface of the base butts over the reference plane on the job, and the micrometer scale directly gives the depth of the measuring rod tip from the reference plane.

The head movement of the depth micrometer is usually 25 mm. Inter-changeable measuring rods, similar to an inside micrometer discussed in the previous section, provide the required measuring range for the instrument. Measuring rods of up to 250 mm length are used in a standard set.

4.8.7 Floating Carriage Micrometer

A floating carriage micrometer, sometimes referred to as an effective diameter-measuring micrometer, is an instrument that is used for accurate measurement of 'thread plug gauges'. Gauge dimensions such as outside diameter, pitch diameter, and root diameter are measured with the help of this instrument. All these dimensions have a vital role in thread plug gauges, since the accuracy and interchangeability of the component depend on the gauges used. To reduce the effect of slight errors in the micrometer screws and measuring faces, this micrometer is basically used as a comparator. Figure 4.41 illustrates a floating carriage micrometer.

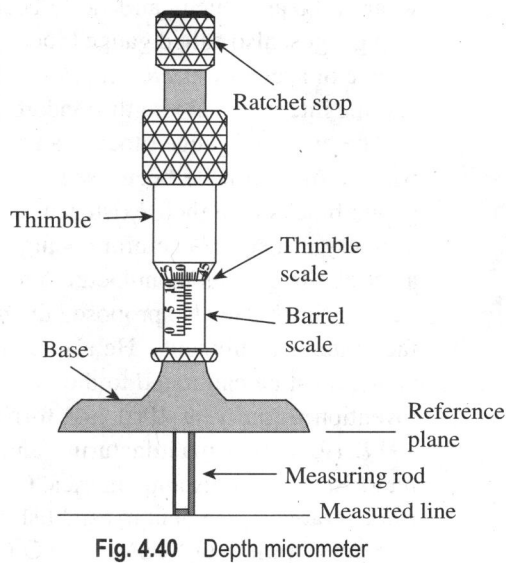

Fig. 4.40 Depth micrometer

The carriage has a micrometer with a fixed spindle on one side and a movable spindle with the micrometer on the other side. The carriage moves on a finely ground 'V' guide way or an antifriction guide way to facilitate movement in a direction parallel to the axis of the plug gauge mounted between the centres. The micrometer has a non-rotary spindle with a least count of up to 0.001 or 0.002 mm. The instrument is very useful for thread plug gauge manufacturers, in gauge calibration laboratories (established under NABL accreditation), and in standard rooms where in-house gauge calibration is carried out.

Fig. 4.41 Floating carriage micrometer

4.9 SLIP GAUGES

Hitherto we have seen instruments such as vernier calliper, depth gauge, and micrometer, which can facilitate measurement to a fairly high degree of accuracy and precision. All these measurements involve line standards. The accuracy of these instruments depends on the accuracy of the workmanship involved in their manufacture. Any minor misalignment or error in a screw can lead to errors in measurement. Repetitive use of a screw or joint results in rapid wear and tear, which can lead to accumulation of errors in measurement within a short time. Slip gauges, also called gauge blocks, can counter some of these limitations and provide a high degree of accuracy as *end standards*. In fact, slip gauges are a direct link between the measurer and the international length standards.

The origin of gauge blocks can be traced to the 18th century Sweden, where 'gauge sticks' were known to have been used in machine shops. However, the modern-day slip gauges or gauge blocks owe their existence to the pioneering work done by C.E. Johansson, a Swedish armoury inspector. Therefore, gauge blocks are also known as *Johansson gauges*. He devised a set of slip gauges manufactured to specific heights with a very high degree of accuracy and surface finish. He also proposed the method of 'wringing' slip gauges to the required height to facilitate measurements. He also emphasized that the resulting slip gauges, to be of universal value, must be calibrated to the international standard. Johansson was granted a patent for his invention in the year 1901 and formed the Swedish company CE Johansson AB in the year 1917. He started manufacturing and marketing his gauge blocks to the industry, and found major success in distant America. One of his customers was Henry Ford with whom he signed a cooperative agreement to establish a gauge making shop at his Cadillac automobile company. The development of 'GO' and 'NO-GO' gauges also took place during this time.

Figure 4.42 illustrates the functional features of a slip gauge. It is made of hardened alloy steel having a 30 mm × 10 mm cross section. Steel is the preferred material since it is economical and has the same coefficient of thermal expansion as a majority of steel components used in production. Hardening is required to make the slip gauge resistant to wear. Hardening is followed by stabilizing at a sub-zero temperature to relieve stresses developed during heat treatment. This is followed by finishing the measuring faces to a high degree of accuracy, flatness, and surface finish. The height of a slip gauge is engraved on one of the rectangular faces, which also features a symbol to indicate the two measured planes. The length between the measuring surfaces, flatness, and surface conditions of measuring faces are the most important requirements of slip gauges.

Carbide gauge blocks are used for their superior wear resistance and longer life. They also have low coefficient of thermal expansion. However, they are quite expensive and used when rapid wear of gauges is to be avoided.

Several slip gauges are combined together temporarily to provide the end standard of a specific length. A set of slip gauges should enable the user to stack them together to provide an accuracy of up to one-thousandth of a millimetre or better. In order to achieve this, individual gauges must be available in dimensions needed to achieve any combination within the available number of gauges. The surfaces of neighbouring slip gauges should stick so close together that there should not be any scope for even a layer of air to be trapped between them, which can add error to the final reading. For this to happen, there should be absolute control over the form, flatness, parallelism, surface finish, dimensional stability of material, and homogeneity of gauging surfaces. While building slip gauges to the required height, the surfaces of slip gauges are pressed

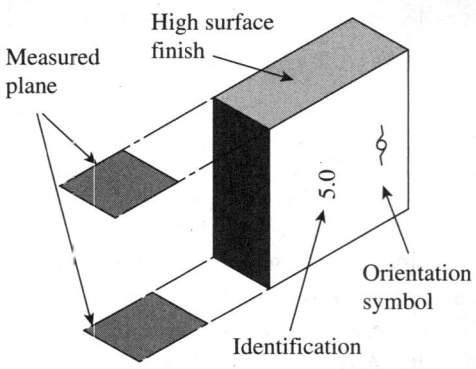

Fig. 4.42 Functional features of a slip gauge

into contact by imparting a small twisting motion while maintaining the contact pressure. The slip gauges are held together due to molecular adhesion between a liquid film and the mating surfaces. This phenomenon is known as 'wringing'.

4.9.1 Gauge Block Shapes, Grades, and Sizes

Slip gauges are available in three basic shapes: rectangular, square with a central hole, and square without a central hole. Rectangular blocks are the most commonly used since they can be used conveniently where space is restricted and excess weight is to be avoided. Square slip gauges have larger surface area and lesser wear rate because of uniform distribution of stresses during measurements. They also adhere better to each other when wrung together. Square gauge blocks with central holes permit the use of tie rods, which ensure that the built-up slip gauges do not fall apart.

Slip gauges are classified into grades depending on their *guaranteed* accuracy. The grade defines the type of application for which a slip gauge is suited, such as inspection, reference, or calibration. Accordingly, slip gauges are designated into five grades, namely grade 2, grade 1, grade 0, grade 00, and inspection grade.

Grade 2 This is the workshop-grade slip gauge. Typical uses include setting up machine tools, milling cutters, etc., on the shop floor.

Grade 1 This grade is used for tool room applications for setting up sine bars, dial indicators, calibration of vernier, micrometer instruments, and so on.

Grade 0 This is an inspection-grade slip gauge. Limited people will have access to this slip gauge and extreme care is taken to guard it against rough usage.

Grade 00 This set is kept in the standards room and is used for inspection/calibration of high precision only. It is also used to check the accuracy of the workshop and grade 1 slip gauges.

Calibration grade This is a special grade, with the actual sizes of slip gauges stated on a special chart supplied with the set of slip gauges. This chart gives the exact dimension of the slip gauge, unlike the previous grades, which are presumed to have been manufactured to a set tolerance. They are the best-grade slip gauges because even though slip gauges are manufactured using precision manufacturing methods, it is difficult to achieve 100% dimensional accuracy. Calibration-grade slip gauges are not necessarily available in a set of preferred sizes, but their sizes are explicitly specified up to the third or fourth decimal place of a millimetre.

Many other grading standards are followed for slip gauges, such as JIS B 7506-1997 (Japan), DIN 861-1980 (Germany), ASME (USA), and BS 4311:Part 1:1993 (UK). Most of these standards assign grades such as A, AA, AAA, and B. While a grade B may conform to the workshop-grade slip gauge, grades AA and AAA are calibration and reference grades, respectively.

Slip gauges are available in standard sets in both metric and inch units. In metric units, sets of 31, 48, 56, and 103 pieces are available. For instance, the set of 103 pieces consists of the following:
1. One piece of 1.005 mm
2. 49 pieces ranging from 1.01 to 1.49 mm in steps of 0.01 mm

3. 49 pieces ranging from 0.5 to 24.5 mm in steps of 0.5 mm
4. Four pieces ranging from 25 to 100 mm in steps of 25 mm

A set of 56 slip gauges consists of the following:
1. One piece of 1.0005 mm
2. Nine pieces ranging from 1.001 to 1.009 mm in steps of 0.001 mm
3. Nine pieces ranging from 1.01 to 1.09 mm in steps of 0.01 mm
4. Nine pieces ranging from 1.0 to 1.9 mm in steps of 0.1 mm
5. 25 pieces ranging from 1 to 25 mm in steps of 1.0 mm
6. Three pieces ranging from 25 to 75 mm in steps of 25 mm

Generally, the set of slip gauges will also include a pair of tungsten carbide protection gauges. These are marked with letter 'P', are 1 or 1.5 mm thick, and are wrung to the end of the slip gauge combination. They are used whenever slip gauges are used along with instruments like sine bars, which are made of metallic surfaces that may accelerate the wear of regular slip gauges. Wear blocks are also recommended when gauge block holders are used to hold a set of wrung gauges together. The purpose of using a pair of wear blocks, one at the top and the other at the bottom of the stack, is to ensure that major wear is concentrated over the two wear gauges, which can be economically replaced when worn out. This will extend the useful life of the set of slip gauges.

4.9.2 Wringing of Slip Gauges

Wringing is the phenomenon of adhesion of two flat and smooth surfaces when they are brought into close contact with each other. The force of adhesion is such that the stack of a set of blocks will almost serve as a single block, and can be handled and moved around without disturbing the position of individual blocks. More importantly, if the surfaces are clean and flat, the thin layer of film separating the blocks will also have negligible thickness. This means that stacking of multiple blocks of known dimensions will give the overall dimension with minimum error.

Wringing Phenomenon

When two surfaces are brought into contact, some amount of space exists between them. This is because of surface irregularities and presence of dirt, oil, grease, or air pockets. Let us assume that the two surfaces are perfectly flat with highly finished surfaces, free from dirt and oil, and firmly pressed together. Now the air gap becomes so small that it acts in the same way as a liquid film. The thickness of this film can be as low as 0.00001 mm. Now a question arises as to why the blocks stick together so firmly that even a high magnitude of force acting perpendicular to their surfaces will not be able to separate them. A combination of two factors appears to ensure this high adhesion force. First, as shown in Fig. 4.43, an atmospheric force of 1 bar is acting in the direction shown by the two arrows. This is contributing to the adhesion of the surfaces of the two slip gauges.

Secondly, the surfaces are in such close proximity that there is molecular adhesion of high magnitude that creates a high adhesion force. Since the slip gauge surfaces undergo lapping as a super finishing operation, material removal takes place at the molecular level. Since some molecules are lost during the lapping operation, the material is receptive to molecules of the mating surface, which creates high molecular adhesion.

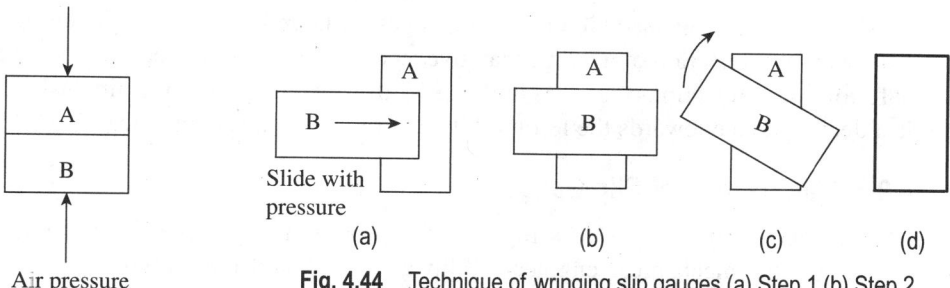

Fig. 4.43 Wringing phenomenon

Fig. 4.44 Technique of wringing slip gauges (a) Step 1 (b) Step 2 (c) Step 3 (d) Step 4

These two factors collectively ensure adhesion of slip gauges with minimum air gap between them. Therefore, a stack of slip gauges will have a length equal to the sum of the individual heights.

Technique of Wringing Slip Gauges

The ability of a given gauge block to wring is called *wringability*; it is defined as 'the ability of two surfaces to adhere tightly to each other in the absence of external means'. The minimum conditions for wringability are a surface finish of 0.025 µm or better, and a flatness of at least 0.13 µm. Wringing of slip gauges should be done carefully and methodically because a film of dust, moisture, or oil trapped between gauges will reduce the accuracy of measurement. The first step is to clean the slip gauges immediately before wringing because any gap in time will allow dust and moisture to settle on the gauges. A very fine hairbrush can be used to clean them. Some people are under the false notion that a thin film of oil should always be applied to the gauge surfaces before wringing. Most often, the application of oil itself may introduce unwanted dust and oil in between the gauges. The need for additional oil film is felt for worn out gauges where there is reduced metal-to-metal contact resulting in poor molecular adhesion. The following are the preferred steps in the wringing of slip gauges:

1. Clean slip gauge surfaces with a fine hairbrush (camel hairbrushes are often recommended) and a dry pad.
2. Overlap gauging surfaces by about one-fourth of their length, as shown in Fig. 4.44(a).
3. Slide one block perpendicularly across the other by applying moderate pressure. The two blocks should now form the shape as shown in Fig. 4.44(b).
4. Now, gently rotate one of the blocks until it is in line with the other block, as in Fig. 4.44(c) and (d).

Combining Slip Gauges

As pointed out earlier, gauge blocks are available in standard sets of 31, 48, 56, and 103 pieces. While figuring out the slip gauges that are required to make up a given dimension, a procedure must be followed to save time and, more importantly, to ensure that a minimum number of gauges are used. Please remember that more the number of gauges used, more is the separation of gauges by a thin film, which can cumulatively contribute to substantial error. In addition, the accuracy up to which a dimension can be built depends on the gauge that can give accuracy up to the last possible decimal place. For instance, while the 103-piece set can give an accuracy of up to 0.005 mm, the 56-piece set can give up to 0.005 mm.

Thus, whenever we need to build slip gauges to the required height/dimension, the primary concern is the selection of a gauge that gives the dimension to the required decimal place. This is followed by selection of gauges in the order in which they meet the dimension from the next last decimal place towards the left until the entire selection is complete.

4.9.3 Manufacture of Slip Gauges

Manufacturing slip gauges to a high degree of accuracy and precision is one of the major challenges for mechanical engineers. Slip gauges should meet high standards of accuracy, flatness, parallelism, and surface quality. There are several recommended methods of manufacturing slip gauges. Internationally acclaimed methods are the ones recommended by the United States bureau of Standards and the German (Zeiss) method. In India, slip gauges are manufactured as per the guidelines recommended by the National Physical Laboratory (NPL). This method is discussed in the following paragraphs.

Steel blanks containing 1% carbon, 1.8% chromium, and 0.4% manganese are sawn from steel bars such that the blanks are oversized by about 0.5 mm on all sides. They are then hardened and stabilized by an artificial seasoning process. Now, the blanks are loaded on the magnetic chuck of a precision grinding machine, and all the faces are ground in the first setting. In the second setting, the gauges are reversed and ground to within a 0.05 mm size. Grinding operation is followed by lapping operation using a lapping chuck. NPL recommends that eight gauges be lapped at a time by an arrangement shown in Fig. 4.45.

Preliminary lapping operation ensures that all the eight blanks become parallel to about 0.0002 mm and sizes are reduced to an accuracy of 0.002 mm. Final lapping is carried out as per the layout shown in Fig. 4.45(a). The blanks are then reversed and opposite faces lapped again until they lie on one true plane. Now, the positions of blanks are interchanged as per the second arrangement shown in Fig. 4.45(b), and another round of lapping is carried out until all faces lie on one true plane. At this stage, the blanks would have achieved a high degree of dimensional accuracy and parallelism.

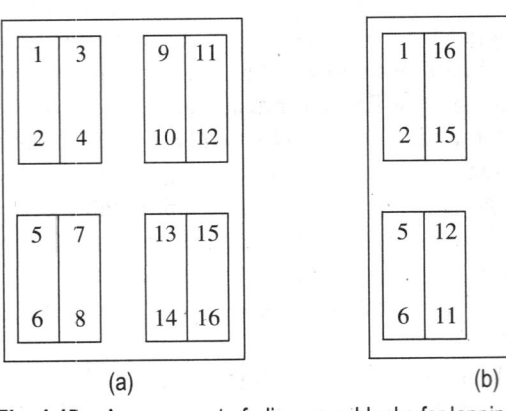

(a)   (b)

Fig. 4.45 Arrangement of slip gauge blanks for lapping operation
(a) First arrangement (b) Second arrangement

4.9.4 Calibration of Slip Gauges

Slip gauges are calibrated by direct comparison with calibration grade gauges using a comparator. Slip gauges need to be calibrated at regular intervals, since even a slightly worn out slip gauge can create havoc at the quality control stage, resulting in increased scrap losses. NPL has recommended schedules for the calibration of different grades of slip gauges. Notwithstanding regular calibration schedules, a slip gauge is a candidate for recalibration under the following conditions:

1. Visual inspection shows wear beyond permissible level.

2. Wringing becomes difficult.

3. An unusual increase in rejections occurs during quality inspection.

Working slip gauge blocks are calibrated using master sets. The master gauges, in turn, are calibrated by grand masters, which are maintained by the National Bureau of Standards. In addition, usually all manufacturers of gauge blocks provide calibration services. In most of the advanced countries, there are independent metrology laboratories that mainly deal with providing calibration services. Such a service is conspicuous by its absence in India.

It is of academic interest to know the different types of comparators used for calibrating slip gauges. The popular ones are the Brook-level comparator, Eden-Rolt millionth comparator, and the NPL-Hilger interferometer. The working principle of the Brook-level comparator is explained to give flair of the significance of instrumentation for calibration of slip gauges.

Brook-level Comparator

The Brook-level comparator works in a way similar to a spirit level, but is more sophisticated and highly sensitive. It compares differences in the levels of two objects to a sub-micron-level accuracy. The comparator comprises a sensitive level mounted in an aluminium casing. The underside of the casing has two steel balls, each of which rests on top of the two objects whose heights are to be compared. Whenever there is a variation in height, the bubble moves over a graduated scale, which has a resolution of $1\,\mu m$ or $0.1\,\mu m$. The slip gauge to be calibrated (test gauge) and the reference calibration gauge (standard) are kept on top of a surface plate. It is needless to say that the surfaces of the two slip gauges and the surface plate should be thoroughly cleaned and positioned so that there is no additional layer of dust, oil, etc., which will reduce the accuracy. The Brook-level comparator is kept on top of this pair of slip gauges such that one steel ball rest on the test gauge and the other on the standard, as shown in Fig. 4.46. The reading on the graduated scale on the top face of the comparator is noted. Let us say this reading is 12.2.

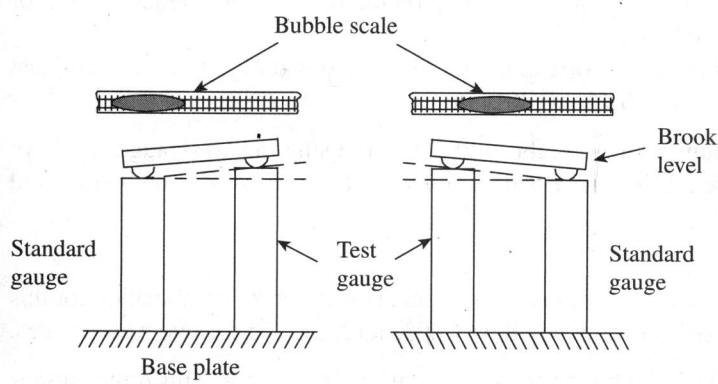

Fig. 4.46   Brook-level comparator

Now, the comparator is lifted up and the surface plate is given a 180° rotation. The Brook-level comparator is again lowered onto the two slip gauges, such that the two contact points are reversed. The new reading is noted down. Let us say the second reading is 12.0. If the resolution of the comparator is 0.00001 mm, then the difference in length is calculated as follows:

$$\text{Difference in length} = (0.0001) \times (12.2 - 12.0)/2 = 0.00001\,\text{mm}$$

4.10 NUMERICAL EXAMPLES

Example 4.1   In a vernier calliper, the main scale reads in millimetres with a least count of 0.1 mm. Ten divisions on the vernier correspond to nine divisions of the main scale. Determine the least count of the calliper.

Solution

1 MSD = 0.1 mm

10 VSD = 9 MSD

Therefore, least count = 1 MSD − 1 VSD = 1 MSD − 9/10 MSD = 1/10 (0.1) mm = 0.01 mm

Example 4.2 The main scale in a vernier instrument is graduated in millimetres, with the smallest division being 1 mm. Ten divisions on the vernier scale correspond to nine divisions on the main scale. Answer the following questions:

(a) Is the vernier scale a forward vernier or a backward vernier?

(b) What is the least count of the instrument?

(c) If the main scale reads 13 mm and the fifth division on the vernier scale coincides with a division on the main scale, what is the value of the dimension being measured?

Solution

(a) Since the length of one MSD is higher than the length of one VSD, the scale is a forward vernier scale.

(b) Least count = 1 MSD − 1 VSD = (1 − 9/10) MSD = 0.1 mm

(c) The dimension is = 13 + 5 ´ 0.1 = 13.5 mm

Example 4.3 The barrel scale of a vernier micrometer has graduations in millimetres, with each division measuring 1 mm. The thimble scale has 50 equal divisions, and one complete rotation of the thimble moves it by one division over the barrel scale. If five divisions on the vernier scale attached to the micrometer correspond to six divisions on the thimble, calculate the least count of the instrument.

Solution

If one complete rotation of the thimble moves it by 1 mm on the barrel scale, the least count of the micrometer scale is 1/50 = 0.02 mm.

Since five divisions on the vernier scale correspond to six divisions on the thimble, the least count of the vernier scale is equal to 0.02/5 = 0.001 mm.

Example 4.4 Slip gauges have to be built up to a height of 26.125 mm using the 103-gauge set. Give the selection of slip gauges if wear blocks of 1.5 mm thickness are to be used at the bottom and top of the stack.

Solution

We need to deduct 3 mm from 26.125 mm, that is, 23.125 mm, and select the combination for this height. The selection of gauges in this example will be 1.005, 1.12, 21, and the two wear gauges.

Example 4.5 It is required to build a 0.95 mm height for some calibration purpose. This dimension is less than the least gauge in the 103-gauge set, that is, the 1.005 mm gauge. How can you meet the requirement?

Solution

In such cases, it is necessary to use a combination that is slightly greater than 0.95 mm and subtract a second combination to get the required result.

Accordingly, we can first choose the 1.005 mm gauge, which is in excess of 0.95 mm by 0.055 mm. Addition of 0.055 to 1.005 mm gives a value of 1.06 mm. Now,

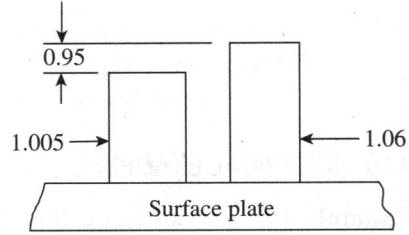

Fig. 4.47 Indirect combination of slip gauges

for the second combination, select the 1.06 mm gauge. If the two gauges are kept on a surface plate, their difference will give the required value of 0.95 mm, as shown in Fig. 4.47. This can be measured by using a suitable comparator of high accuracy.

A QUICK OVERVIEW

- The foundation for all dimensional measurements is the 'datum plane', the most important one being the surface plate. A surface plate is a hard, solid, and horizontal flat plate, which is used as the reference plane for precision inspection, marking out, and precision tooling set-up. V-blocks are extensively used for inspection of jobs with circular cross sections.
- Most of the basic linear measurements are derived from the simple steel rule. However, scaled instruments provide the most convenient means for linear measurement in contrast to steel rules, which are limited by accuracy, resolution, and mechanism to ensure ease of measurement.
- Callipers are the original transfer instrument to transfer such measurements onto a rule. They ensure physical duplication of the separation of reference point and measured point with a high degree of accuracy.
- The principle of vernier measurement is basic to metrology and the use of vernier instruments comprising vernier calliper, vernier depth gauge, vernier height gauge, vernier micrometers, etc., is still widespread in the industry. However, digital read-out instruments and dial callipers are rapidly replacing vernier instruments. The advantages of vernier instruments are their long measuring range and convenience. Their chief disadvantages are reliance on the 'feel' of the user and susceptibility to misalignment.
- A micrometer can provide better least counts and accuracy than a vernier calliper. Better accuracy results from the fact that the line of measurement is in line with the axis of the instrument, unlike in a vernier calliper, which does not conform to this condition. This fact is best explained by *Abbe's principle*, which states that *maximum accuracy may be obtained only when the standard is in line with the axis of the part being measured*.
- A floating carriage micrometer, sometimes referred to as an effective diameter-measuring micrometer, is an instrument that is used for accurate measurement of 'thread plug gauges'. Gauge dimensions such as outside diameter, pitch diameter, and root diameter are measured with the help of this instrument.
- Slip gauges provide the most practical standard for linear measurement. They are readily available and a small number of them, in a standard set, can be used to build thousands of dimensions. The errors involved with them are known and therefore predictable.

MULTIPLE-CHOICE QUESTIONS

1. The foundation for all dimensional measurements is the
   (a) datum plane   (c) datum point
   (b) datum line   (d) geometry of work piece
2. As per the practice devised by Sir Whitworth, the surface plates are fabricated in sets of
   (a) two   (c) four
   (b) three   (d) ten
3. Granite surface plates are free from burrs or protrusions because of
   (a) lower coefficient of expansion
   (b) light weight
   (c) very fine grain structure
   (d) none of these
4. The preferred instrument for measuring holes, grooves, and recesses is
   (a) plain scale   (c) slip gauge
   (b) vernier calliper (d) depth gauge

5. The centre head attachment on a combination set is used to
   (a) measure angles
   (b) measure height and depth
   (c) measure distance between centres
   (d) locate the centre of a bar stock or a circular job

6. A scribing tool comprising one divider leg and one calliper leg is known by the name
   (a) hermaphrodite calliper
   (b) hermasonic calliper
   (c) transfer calliper
   (d) firm joint calliper

7. _____ vernier calliper has jaws on both sides for making measurements and the jaws have knife edge faces for marking purpose.
   (a) Type A   (c) Type C
   (b) Type B   (d) Type D

8. The degree to which an instrument conforms to the _____ law determines its inherent accuracy.
   (a) Moore's law   (c) Johnson's law
   (b) Abbe's law   (d) Mikelson's law

9. Micrometer measuring faces are tipped with _____ to prevent rapid wear.
   (a) aluminium   (c) molybdenum oxide
   (b) chromium   (d) tungsten carbide

10. Gauge blocks owe their existence to the pioneering work done by
    (a) Johansson   (c) Taylor
    (b) Abbe   (d) None of these

11. Which of the following best describes wringing of slip gauges?
    (a) Squeezing the oil out from between two gauges
    (b) Causing blocks to adhere by molecular attraction
    (c) Effect of atmospheric pressure
    (d) Process of removing minute burrs

12. Which of the following is preferred for selecting a combination of slip gauges?
    (a) Selective halving
    (b) Writing out all combinations and selecting the closest one
    (c) Averaging
    (d) Determining the fewest number of blocks

13. Which of the following statements with reference to a micrometer is false?
    (a) It is not as precise as a vernier calliper.
    (b) It can be used for end measurement only.
    (c) It has no parallax error.
    (d) It has a shorter measuring range compared to a vernier calliper.

14. Which of the following part features is easiest to measure with a vernier calliper?
    (a) Large distances between outside planes
    (b) Heights from a surface plate
    (c) Cylindrical features
    (d) Concave features

15. Slip gauges marked with the letter 'P' are
    (a) wear gauges
    (b) preferred gauges
    (c) discarded gauges
    (d) inspection grade gauges

REVIEW QUESTIONS

1. Why are callipers and dividers called dimension transfer instruments?
2. List the important considerations for the design of linear measurement instruments.
3. When do you prefer cast iron surface plates over granite surface plates and vice versa?
4. What is the main purpose of a 'V-block'? What is the basis for their classification?
5. How is a scale different from a rule?
6. What are the common errors associated with measurements using a steel rule?
7. How do you measure the depth of a hole or recess using a depth gauge? What are the limitations of this instrument?
8. How do you employ a combination set to measure the following?
   (a) Height
   (b) Angle of a surface
   (c) Centre of a bar stock
9. Give a classification of callipers. Illustrate with sketches the different types of callipers.
10. Differentiate using sketches the measurements of inside and outside dimensions using a vernier calliper.

11. Discuss the guidelines to be followed for the proper use of a vernier calliper.
12. Explain the working principle of a dial calliper.
13. What are the major benefits of an electronic digital calliper?
14. Explain how a micrometer conforms to Abbe's law.
15. Write a note on the types of micrometers.
16. Differentiate between an inside micrometer and an inside micrometer calliper.
17. What is the application of a floating carriage micrometer?
18. Why are slip gauges called 'Johansson gauges'?
19. Explain the phenomenon involved in 'wringing' of slip gauges.
20. What is the significance of calibration of slip gauges?

PROBLEMS

1. In a vernier calliper, the main scale reads in millimetres with a least count of 0.5 mm. Twenty divisions on the vernier correspond to 19 divisions on the main scale. Determine the least count of the calliper.

2. The main scale in a vernier instrument is graduated in millimetres, with the smallest division being 1 mm. Ten divisions on the vernier scale correspond to 11 divisions on the main scale. Answer the following questions:
   (a) Is the vernier scale a forward vernier or backward vernier?
   (b) What is the least count of the instrument?
   (c) If the main scale reads 12 mm and the 10th division on the vernier scale coincides with a division on the main scale, what is the value of the dimension being measured?

3. The barrel scale of a vernier micrometer has graduations in millimetres, with each division measuring 1 mm. The thimble scale has 100 equal divisions, and one complete rotation of the thimble moves it by one division over the barrel scale. If 10 divisions on the vernier scale attached to the micrometer corresponds to 11 divisions on the thimble, calculate the least count of the instrument.

4. A selection of slip gauges is required to build a height of 48.155 mm. Propose the best combination of gauges using the 103-gauge set.

5. It is required to set a height difference of 0.45 mm to set the sine bar to a known angle. How will you select the combination of slip gauges to set the sine bar using a set of 56 slip gauges? (Hint: Since the required value is less than the dimension of the minimum available gauge, it is required to select combinations on either sides of the sine bar, so that the difference in height should give the required value.)

6. For the case described in problem 4, what will be the combination of slip gauges if two protection gauges of 1.5 mm thickness need to be used?

ANSWERS

Multiple-choice Questions

1. (a)   2. (b)   3. (c)   4. (d)   5. (d)   6. (a)   7. (c)   8. (b)
9. (d)   10. (a)   11. (b)   12. (d)   13. (a)   14. (c)   15. (a)

Problems

1. 0.025 mm   2. (a) Backward vernier scale   (b) 0.1 mm   (c) 13 mm
3. 0.01 mm   4. 1.005 + 1.15 + 21 + 25
5. First combination: 1 + 1.1   Second combination: 1.05 + 1.5
6. First combination: 1.5 + 1 + 1.1 + 1.5   Second combination: 1.5 + 1.05 + 1.5 + 1.5

Angular Measurement

After studying this chapter, the reader will be able to

• understand the basic requirements of angular measurement in the industry and the variety of instruments available at our disposal
• elucidate the basic principle of a protractor and its extension as the universal bevel protractor, which is an indispensable part of a metrology laboratory
• measure angles using the sine principle and explain the use of sine bar, sine block, sine plate, and sine centre
• use angle gauges to set them accurately to the required angle
• appreciate the importance of 'bubble instruments' such as the conventional spirit level and clinometers in angular measurement
• explain the principles of optical measurement instruments, the most popular ones being the autocollimator and the angle dekkor

5.1 INTRODUCTION

Length standards such as foot and metre are arbitrary inventions of man. This has necessitated the use of wavelength of light as a reference standard of length because of the difficulty in accurately replicating the earlier standards. On the other hand, the standard for angle, which is derived with relation to a circle, is not man-made but exists in nature. One may call it degree or radian, but the fact remains that it has a direct relation to a circle, which is an envelope of a line moving about one of its ends. Whether one defines a circle as the circumference of a planet or path of an electron around the nucleus of an atom, its parts always bear a unique relationship.

The precise measurement of angles is an important requirement in workshops and tool rooms. We need to measure angles of interchangeable parts, gears, jigs, fixtures, etc. Some of the typical measurements are tapers of bores, flank angle and included angle of a gear, angle made by a seating surface of a jig with respect to a reference surface, and taper angle of a jib. Sometimes, the primary objective of angle measurement is not to measure angles. This may sound rather strange, but this is the case in the assessment of alignment of machine parts. Measurement of straightness, parallelism, and flatness of machine parts requires highly sensitive instruments like autocollimators. The angle reading from such an instrument is a measure of the error of alignment.

There are a wide range of instruments, starting from simple scaled instruments to sophisticated types that use laser interferometry techniques. The basic types are simple improvisations of a protractor, but with better discrimination (least count), for example, a vernier protractor. These instruments are provided with a mechanical support or a simple mechanism to position them accurately against the given workpiece and lock the reading. A spirit level has universal applications, not only in mechanical engineering but also in civil engineering construction for aligning structural members such as beams and columns. Instruments employing the basic principle of a spirit level but with higher resolution, such as conventional or electronic clinometers, are popular in metrology applications. By far, the most precise instruments are collimators and angle dekkors, which belong to the family of instruments referred to as *optical tooling*. This chapter deals with some of the popular angle measurement devices that are widely used in the industry.

5.2 PROTRACTOR

A simple protractor is a basic device used for measuring angles. At best, it can provide a least count of 1° for smaller protractors and ½° for large ones. However, simple though it may be, the user should follow the basic principles of its usage to measure angles accurately. For instance, the surface of the instrument should be parallel to the surface of the object, and the reference line of the protractor should coincide perfectly with the reference line of the angle being measured. Positioning of the protractor and observation of readings should be performed with care to avoid parallax error.

Similar to a steel rule, a simple protractor has limited usage in engineering metrology. However, a few additions and a simple mechanism, which can hold a main scale, a vernier scale, and a rotatable blade, can make it very versatile. A universal bevel protractor is one such instrument that has a mechanism that enables easy measurement and retention of a reading. A vernier scale improves the least count substantially. Additional attachments enable the measurement of acute and obtuse angles with ease and thereby justify its name as the universal bevel protractor. It can measure the angle enclosed by bevelled surfaces with ease and hence the name.

In fact, if one traces the history of development of angle-measuring devices, the bevel protractor preceded the universal bevel protractor. The earliest bevel protractor had a simple mechanism that facilitated rotation of measuring blades and locked them in place. It had a scale graduated in degrees on which the measurements could be directly read. However, these instruments have largely been replaced by universal bevel protractors and the older types are not being used in metrology applications now. Therefore, we shall directly go to a discussion on the universal bevel protractor.

5.2.1 Universal Bevel Protractor

The universal bevel protractor with a 5' accuracy is commonly found in all tool rooms and metrology laboratories. Figure 5.1 illustrates the construction of a universal bevel protractor. It has a base plate or stock whose surface has a high degree of flatness and surface finish. The stock is placed on the workpiece whose angle is to be measured. An adjustable blade attached

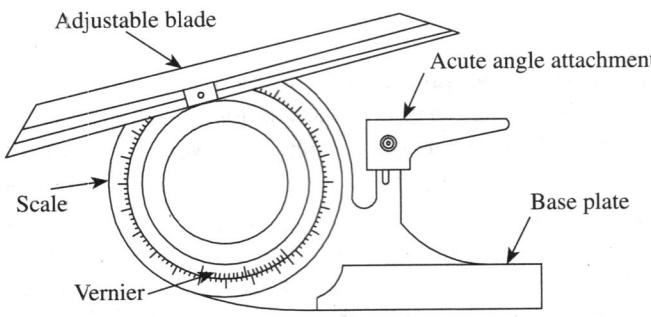

Fig. 5.1 Universal bevel protractor

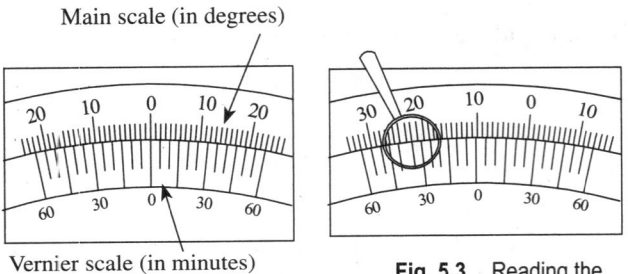

Main scale (in degrees)

Vernier scale (in minutes)

Fig. 5.2 Divisions on the vernier scale

Fig. 5.3 Reading the vernier scale

to a circular dial is made to coincide with the angular surface. It can be swivelled to the required angle and locked into position to facilitate accurate reading of the circular scale that is mounted on the dial. The main scale on the dial is graduated in degrees and rotates with the rotation of the adjustable blade. A stationary vernier scale mounted close to the dial, as shown in Fig. 5.1, enables measurements to a least count of 5' or less. An acute angle attachment is provided for the measurement of acute angles.

The main scale on the dial is divided into four quadrants, each measuring 90°. Each division on this scale reads 1°. The degrees are numbered from 0 to 90 on either side of the zeroth division. The vernier scale has 24 divisions, which correspond to 46 divisions on the main scale. However, the divisions on the vernier scale are numbered from 0 to 60 on either side of the zeroth division, as shown in Fig. 5.2.

Calculation of Least Count

Value of one main scale division = 1°

24 vernier divisions correspond to 46 main scale divisions. From Fig. 5.2, it is clear that one vernier division equals 1/12th of 23°. Let us assume that the zeroth division on both the main and the vernier scales are lined up to coincide with each other. Now, as the dial rotates, a vernier division, starting from the fifth minute up to the 60th minute, progressively coincides with a main scale division until the zeroth division on the vernier scale moves over the main scale by 2°.

Therefore, the least count is the difference between one vernier division and two main scale divisions, which is 1/12° or 5'.

Reading Vernier Scales

Consider the situation shown in Fig. 5.3. The zeroth division of the vernier scale is just past the 10° division on the main scale. The seventh division, marked as the 35' division, on the left-hand side of the vernier scale coincides with a division on the main scale. Therefore, the reading in this case is 10°35'.

Sometimes, confusion arises regarding the direction in which the vernier has to be read. This confusion may crop up for the aforementioned example also. It is possible that a division on the right side of zero on the vernier scale may be coinciding with a division on the main scale (dial scale). In order to eliminate this confusion, we follow a simple rule. *Always read the vernier from zero in the same direction that you read the dial scale.* In the given example, the 10th division on the dial, which is close to the zeroth division on the vernier, is to the left of the zeroth division on the dial scale. In other words, the dial scale is being read in

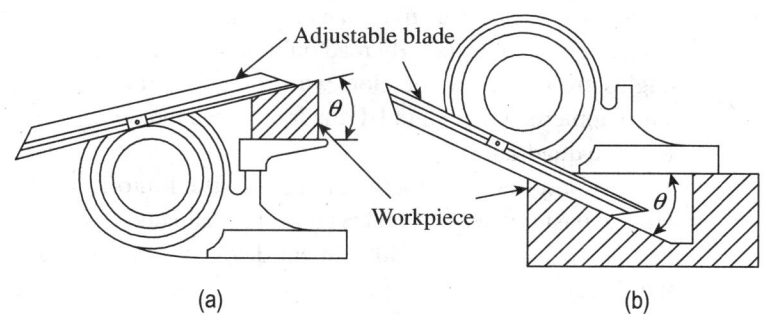

(a) (b)

Fig. 5.4 Measurement of angles using bevel protractor (a) Acute angle attachment (b) Inside bevelled face angle measurement

the leftward or anticlockwise direction. Therefore, the vernier should also be read towards the left of the vernier zero division. Figure 5.4 illustrates the use of a bevel protractor for measurement of angles. While Fig. 5.4(a) illustrates the use of acute angle attachment, Fig. 5.4(b) shows how the angle of an inside bevelled face can be measured.

Angles and their Supplements

Since a universal bevel protractor can measure both acute and obtuse angles, care should be exercised to clearly differentiate between the angle being indicated on the scale and its supplement. The dial gauge is graduated from 0° to 90° in four quadrants, as shown in Fig. 5.5. Figure 5.5(a) illustrates the orientation of the blade with respect to the base when the protractor is set to 90°. The zeroth division on the vernier coincides with the 90° division on the dial scale.

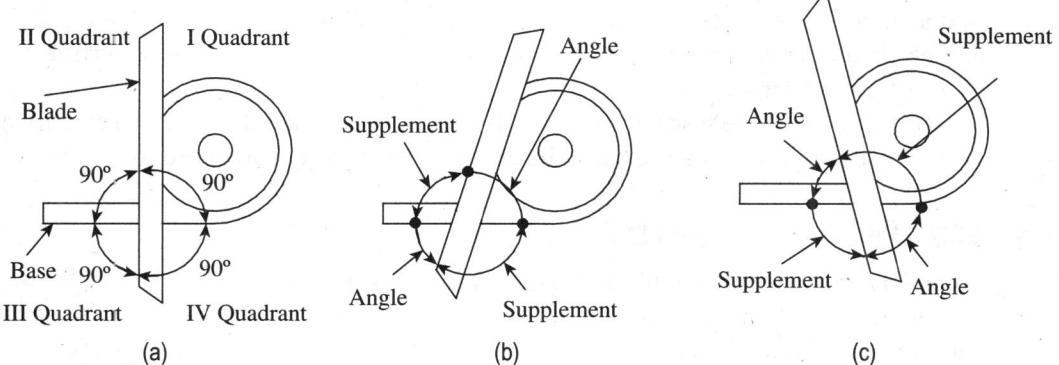

(a) (b) (c)

Fig. 5.5 Angles and their supplements (a) Blade oriented with base (b) Blade turned clockwise (c) Blade turned counterclockwise

Suppose the blade is turned clockwise, as in Fig. 5.5(b), the angles read directly are those that are formed from the blade to the base counterclockwise. Thus, if the angles of a work part are being measured in quadrant I or III, the angles can be read directly from the scale. On the other hand, if the angles of a work part are being measured in quadrant II or IV, the actual angles are given by their supplements. In other words, the value of the angle is obtained by subtracting the angle indicated by the scale from 180°. Both these angles, obviously, are obtuse angles.

Further, Fig. 5.5(c) illustrates a situation when the blade is turned counterclockwise. Here also, angles can be directly read only in two quadrants, namely the second and the fourth. These angles are formed in the clockwise direction from the blade to the base, and are acute angles. The supplements in I and III quadrants give the obtuse angles of the work parts, which are held in these quadrants.

Bevel protractors in general are classified into four types: A, B, C, and D. Types C and D are the basic types, with the dial scale graduated in degrees. They are neither provided with a vernier scale nor a fine adjustment device. Types A and B are provided with a vernier scale, which can accurately read up to 5'. While type A has both an acute angle attachment and a fine adjustment device, type B does not have either of them.

A bevel protractor is a precision angle-measuring instrument. To ensure an accurate measurement, one should follow these guidelines:

1. The instrument should be thoroughly cleaned before use. It is not recommended to use compressed air for cleaning, as it can drive particles into the instrument.
2. It is important to understand that the universal bevel protractor does not essentially measure the angle on the work part. It measures the angle between its own parts, that is, the angle between the base plate and the adjustable blade. Therefore, one should ensure proper and intimate contact between the protractor and the features of the part.
3. An easy method to determine if the blade is in contact with the work part is to place a light behind it and adjust the blade so that no light leaks between the two.
4. It should always be ensured that the instrument is in a plane parallel to the plane of the angle. In the absence of this condition, the angle measured will be erroneous.
5. The accuracy of measurement also depends on the surface quality of the work part. Burrs and excessive surface roughness interfere with the intimate contact between the bevel protractor and the work part, leading to erroneous measurements.
6. One should be careful to not slide the instrument over hard or abrasive surfaces, and not over-tighten clamps.
7. Before replacing the instrument in its case, it has to be wiped with a clean and dry cloth, a thin rust-preventing coating has to be applied, and moving parts need to be lubricated.

5.2.2 Optical Bevel Protractor

An optical protractor is a simple extension of the universal bevel protractor. A lens in the form of an eyepiece is provided to facilitate easy reading of the protractor scale. Figure 5.6 illustrates the construction details of an optical bevel protractor. The blade is clamped to the dial by means of a blade clamp. This enables fitting of blades of different lengths, depending on the work part being measured. In a protractor without a vernier, the dial scale reading can be

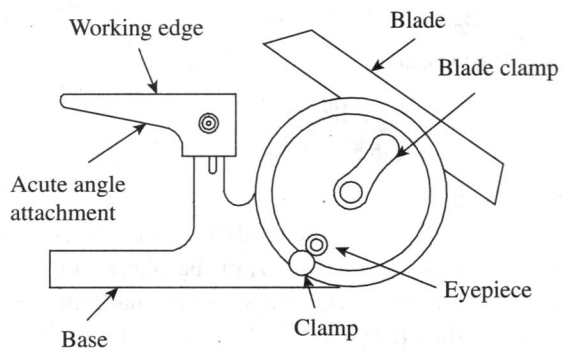

Fig. 5.6 Optical bevel protractor

directly read through the eyepiece. In vernier protractors, the eyepiece is attached on top of the vernier scale itself, which together move as a single unit over the stationary dial scale. The eyepiece provides a magnified view of the reading for the convenience of the user.

Most of the universal protractors in use are of this type. An acute angle attachment is provided to facilitate measurement of acute angles on work parts. A clamp is provided to lock the reading, so that it can be read and recorded at a convenient position by the user.

5.3 SINE BAR

A sine bar is used to measure angles based on the sine principle. Its upper surface forms the hypotenuse of a triangle formed by a steel bar terminating in a cylinder near each end. When one of the cylinders, called a roller, is resting on a flat surface, the bar can be set at any desired angle by simply raising the second cylinder. The required angle is obtained when the difference in height between the two rollers is equal to the sine of the angle multiplied by the distance between the centres of the rollers. Figure 5.7 illustrates the construction details of a sine bar.

Sine bars are made of corrosion-resistant steel, and are hardened, ground, and stabilized. The size is specified by the distance between the centres of the cylinders, which is 100, 200, or 300 mm. The upper surface has a high degree of flatness of up to 0.001 mm for a 100 mm length and is perfectly parallel to the axis joining the centres of the two cylinders. The parallelism of upper surface with the datum line is of the order of 0.001 mm for a 100 mm length. Relief holes are sometimes provided to reduce the weight of the sine bar. This by itself is not a complete measuring instrument. Accessories such as a surface plate and slip gauges are needed to perform the measurement process. Figure 5.8 illustrates the application of a sine rule for angle measurement.

The sine of angle θ formed between the upper surface of a sine bar and the surface plate (datum) is given by

$$\text{Sin}\,(\theta) = h/L$$

where h is the height difference between the two rollers and L is the distance between the centres of the rollers.

Therefore, $h = L\,\text{Sin}\,(\theta)$

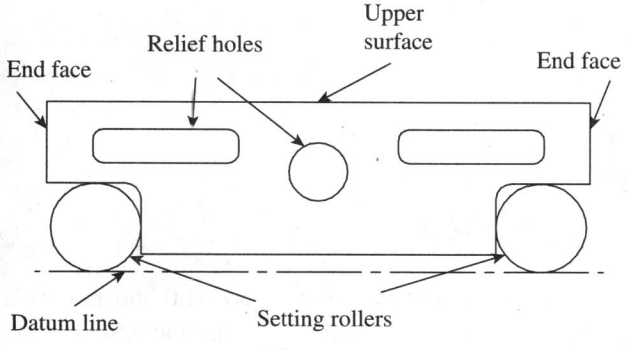

Fig. 5.7 Sine bar

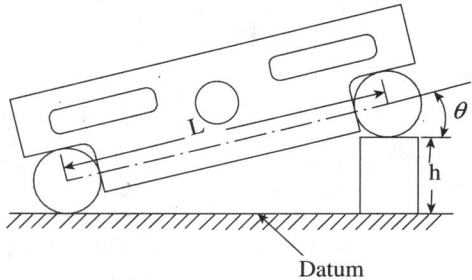

Fig. 5.8 Sine rule

5.3.1 Setting Sine Bars to Desired Angles

By building slip gauges to height h and placing the sine bar on a surface plate with one roller on top of the slip gauges, the upper surface can be set to a desired angle with respect to the surface plate. The set-up is easier because the cylinders are integral parts of the sine bar and no separate clamping is required. No measurement is required between the cylinders since this is a known length. It is preferable to use the sine bar on a grade A surface plate. In addition, it is desirable to support both rollers on gauge blocks so that the minute irregularities of the surface plate may be eliminated.

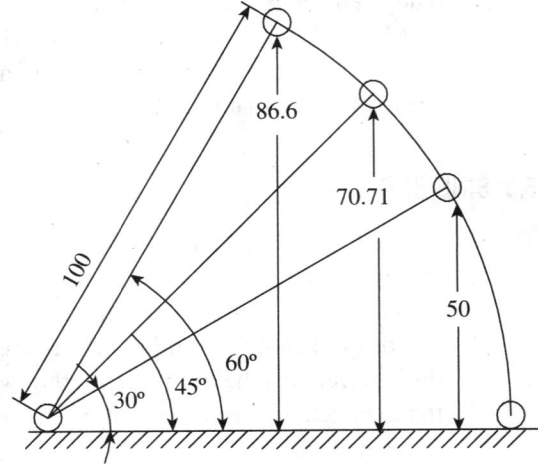

Fig. 5.9 Change in sine value with change in angle

The question often asked is about the maximum angle that can be set using a sine bar. The standard response is 45°. At higher angles, errors due to the distance between the centres of the rollers and gauge blocks get magnified. This is explained in the following example.

Let us assume a 100 mm-long sine bar and calculate the heights to set it at 30°, 45°, and 60°, as shown in Fig. 5.9. The heights are respectively 53, 73.71, and 89.6 mm for angles 30°, 60°, and 90°.

Assume there is an error of +0.1 mm in the height, h. Table 5.1 illustrates the error in the measured angles. The actual angle in 'degrees' is given by $\sin^{-1}(h/L)$.

Table 5.1 Relationship between error and angle being set

Angle to be set (degrees)	Length of sine bar (mm)	Height of slip gauges (mm)	Actual angle (degrees)	Error in measurement (degrees)
30	100.0	50	30	0.066
		50.1	30.066	
45	100.0	70.71	45	0.080
		70.81	45.080	
60	100.0	86.6	60	0.112
		86.7	60.112	

Accounting for an error of 0.1 mm in height h, the angular error for 30° is 0.066 mm. This error increases to 0.08 mm for 45° and jumps to 0.112 mm for 60°. A similar increase in error is observed if the distance between the centres of the rollers of the sine bar has an error. This is the primary reason why metrologists avoid using sine bars for angles greater than 45°.

5.3.2 Measuring Unknown Angles with Sine Bar

A sine bar can also be used to measure unknown angles with a high degree of precision. The angle of the work part is first measured using an instrument like a bevel protractor. Then, the work part is clamped to the sine bar and set on top of a surface plate to that angle using slip gauges, as shown in Fig. 5.10 (clamping details are not shown in the figure).

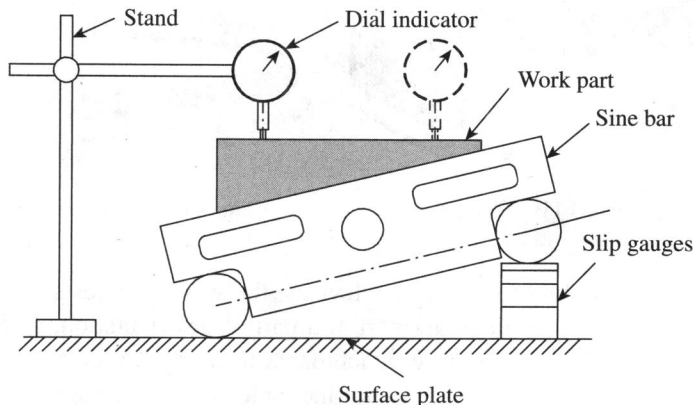

Fig. 5.10 Measurement of unknown angles

A dial gauge fixed to a stand is brought in contact with the top surface of the work part at one end and set to zero. Now, the dial indicator is moved to the other end of the work part in a straight line. A zero reading on the dial indicator indicates that the work part surface is perfectly horizontal and the set angle is the right one. On the other hand, if the dial indicator shows any deviations, adjustments in the height of slip gauges is necessary to ensure that the work part surface is horizontal. The difference in height corresponding to the dial gauge reading is incorporated in the slip gauges, and the procedure is repeated until the dial indicators show zero deviation. The actual angle is calculated using the total height of the slip gauges.

Instead of a dial gauge, a high-amplification comparator can be used for better accuracy. Whether setting a sine bar to a known angle or for measuring unknown angles, a few guidelines should be followed to ensure proper usage of the instrument:

1. It is not recommended to use sine bars for angles greater than 45° because any error in the sine bar or height of slip gauges gets accentuated.
2. Sine bars provide the most reliable measurements for angles less than 15°.
3. The longer the sine bar, the better the measurement accuracy.
4. It is preferable to use the sine bar at a temperature recommended by the supplier. The accuracy of measurement is influenced by the ambient temperature.
5. It is recommended to clamp the sine bar and the work part against an angle plate. This prevents misalignment of the workpiece with the sine bar while making measurements.
6. One should always keep in mind that the sine principle can be put to use provided the sine bar is used along with a high-quality surface plate and set of slip gauges.

5.3.3 Sine Blocks, Sine Plates, and Sine Tables

A *sine block* is a sine bar that is wide enough to stand unsupported (Fig. 5.11). If it rests on an integral base, it becomes a *sine plate* (Fig. 5.12). A sine plate is wider than a sine block. A heavy-duty sine plate is rugged enough to hold work parts for machining or inspection of angles. If a sine plate is an integral part of another device, for example, a machine tool, it is called a *sine table*. However, there is no exact dividing line between the three.

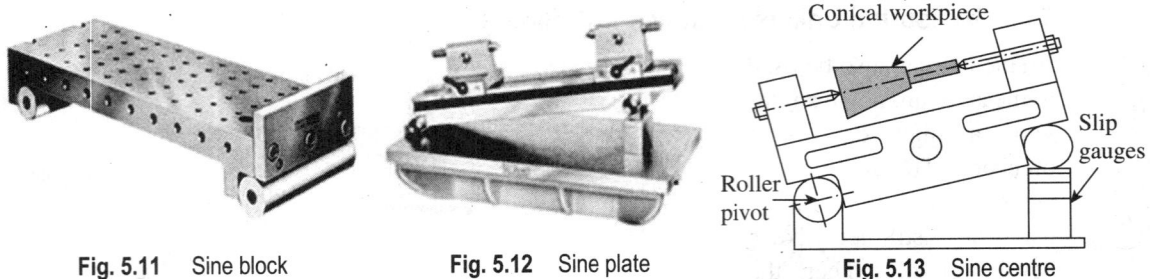

Fig. 5.11 Sine block   **Fig. 5.12** Sine plate   **Fig. 5.13** Sine centre

In all these three devices, the work part rests on them. They are often used like a fixture to keep the work part at a particular orientation, so that the required angle is machined. The instruments have attachments to raise and lock the block to the required angle, and to also fasten work parts. The sine table is the most rugged device, which may be swung to any angle from 0° to 90° by pivoting about the hinged end.

There are many instances where compound angles need to be machined or inspected. While simple angles lie on one plane, compound angles of a surface lie on more than one plane. In a surface formed by the intersections of planes, the angles on the surface planes are called *face angles*. A compound sine plate can conveniently measure or set itself to this face angle. In a typical compound sine plate, there are two sine plates: a base plate creates one plane, while the top plate creates the second plane. Compound sine plates are usually used for finishing operations, for example, a finish grinding operation.

5.3.4 Sine Centre

A sine centre provides a convenient means of measuring angles of conical workpieces that are held between centres, as shown in Fig. 5.13. One of the rollers is pivoted about its axis, thereby allowing the sine bar to be set to an angle by lifting the other roller. The base of the sine centre has a high degree of flatness, and slip gauges are wrung and placed on it to set the sine bar at the required angle.

Conical workpieces that need to be inspected are placed between the centres. The sine centre is used for measuring angles up to 60°. The procedure for measuring angles is very similar to the one described in Section 5.3.2. A dial gauge clamped to a stand is set against the conical workpiece. The sine bar is set to an angle such that the dial gauge registers no deviation when moved from one end of the workpiece to the other. The angle is determined by applying the sine rule.

5.4 ANGLE GAUGES

Angle gauges, which are made of high-grade wear-resistant steel, work on a principle similar to slip gauges. While slip gauges can be built to give linear dimensions, angle gauges can be built to give the required angle. The gauges come in a standard set of angle blocks that can be wrung together in a suitable combination to build an angle. C.E. Johansson, who developed slip gauges, is also credited with the invention of angle gauge blocks. However, the first set of a combination of angle gauges was devised by Dr G.A. Tomlinson of the National Physical

Laboratory, UK, in the year 1939, which provided the highest number of angle combinations. His set of 10 blocks can be used to set any angle between 0° and 180° in increments of 5'.

At the outset, it seems improbable that a set of 10 gauges is sufficient to build so many angles. However, angle blocks have a special feature that is impossible in slip gauges—the former can be subtracted as well as added. This fact is illustrated in Fig. 5.14.

This illustration shows the way in which two gauge blocks can be used in combination to generate two different angles. If a 5° angle block is used along with a 30° angle block, as shown in Fig. 5.14(a), the resulting angle is 35°. If the 5° angle block is reversed and combined with the 30° angle block, as shown in Fig. 5.14(b), the resulting angle is 25°. Reversal of an angle block subtracts itself from the total angle generated by combining other angle blocks. This provides the scope for creating various combinations of angle gauges in order to generate angles that are spread over a wide range by using a minimum number of gauges.

Angle gauges are made of hardened steel, which is lapped and polished to a high degree of accuracy and flatness. The gauges are about 75 mm long and 15 mm wide, and the two surfaces that generate the angles are accurate up to ±2". The gauges are available in sets of 6, 11, or 16. Table 5.2 provides the details of individual blocks in these sets.

Most angles can be combined in several ways. However, in order to minimize error, which gets compounded if the number of gauges used is increased, it is preferable to use the least number of angle gauge blocks. The set of 16 gauges forms all the angles between 0° and 99° in 1" steps—a total of 3,56,400 combinations! The laboratory master-grade set has an accuracy of one-fourth of a second. While the inspection-grade set has an accuracy of ½", the tool room-grade set has an accuracy of 1".

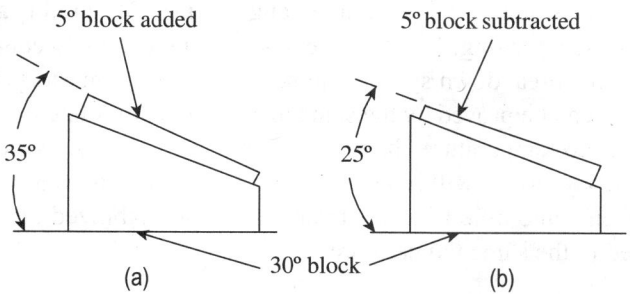

Fig. 5.14   Angle gauge block (a) Addition (b) Subtraction

Table 5.2   Angle gauge block sets

Smallest increment by which any angle can be produced	Number of individual blocks contained in the set	Detailed listing of the blocks composing the set
1°	6	Six blocks of 1°, 3°, 5°, 15°, 30°, and 45°
1'	11	Six blocks of 1°, 3°, 5°, 15°, 30°, and 45° Five blocks of 1', 3', 5', 20', and 30'
1"	16	Six blocks of 1°, 3°, 5°, 15°, 30°, and 45° Five blocks of 1', 3', 5', 20', and 30' Five blocks of 1", 3", 5", 20", and 30"

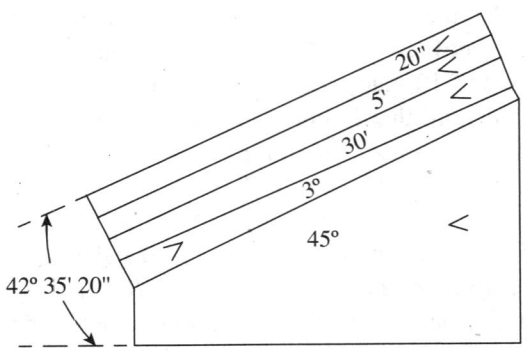

Fig. 5.15 Combination of angle gauges for 42°35'20"

The illustrations in Fig. 5.15 show how angle gauges can be combined to provide the required angles. It may be noted that each angle gauge is engraved with the symbol '<', which indicates the direction of the included angle. Obviously, when the angles of the gauges need to be added up, the symbol < of all gauges should be in line. On the other hand, whenever an angle gauge is required to be subtracted from the combination, the gauge should be wrung such that the symbol < is in the other direction.

Let us consider an angle 42°35'20", which is to be built using the 16-gauge set. Starting from degrees, the angle of 42° can be built by subtracting a 3° block from a 45° block. The angle of 35' can be obtained by combining a 30' gauge with a 5' gauge. A 20" gauge is readily available. The resulting combination is shown in Fig. 5.15.

It can be noticed in this combination that except for the 3° angle gauge, all other gauges are added. Accordingly, the 3° gauge is reversed and wrung with the other gauges, as shown in the figure. The 'wringing' method is the same as that of slip gauges explained in Chapter 4. After wringing, the entire combination is placed on a surface plate and the edges are properly aligned to facilitate measurement.

From the calibration point of view, it is much easier to calibrate angle gauge blocks compared to slip gauges. This is due to the fact that an angle being measured is a portion of a full circle and is, therefore, self-proving. For instance, each of three exactly equal portions of 90° must equal 30°. Thus, the breakdown system can be used to create masters of angle measuring, and each combination can be proved by the same method. In addition, the accuracy of angle gauges is not as sensitive to temperature changes as that of slip gauges. For example, a gauge block manufactured at, say, 30 °C will retain the same angle when used at 40 °C, assuming that the readings are taken some time after the temperature has stabilized and the whole body of the gauge is exposed to the same temperature.

5.4.1 Uses

Angle gauges are used for measurement and calibration purposes in tool rooms. It can be used for measuring the angle of a die insert or for inspecting compound angles of tools and dies. They are also used in machine shops for either setting up a machine (e.g., the revolving centre of a magnetic chuck) or for grinding notches on a cylindrical grinding machine. The illustration in Fig. 5.16 highlights the inspection of a compound angle by using angle gauge blocks. In this case, a surface is inclined in two planes at an angle of 90° to each other. Angle gauges offer a simple means for inspecting such a compound angle by using a dial gauge mounted on a stand.

Figure 5.16 shows a workpiece with a compound angle. Let us assume that the back angle (α) is 15°20' and the side angle (β) is 5°. In order to set the workpiece to 15°20', two angle gauge blocks of 15° and 20' are selected and wrung together. This combination is placed on a surface plate, and the workpiece is positioned on top of the angle gauges. Now, the dial indicator

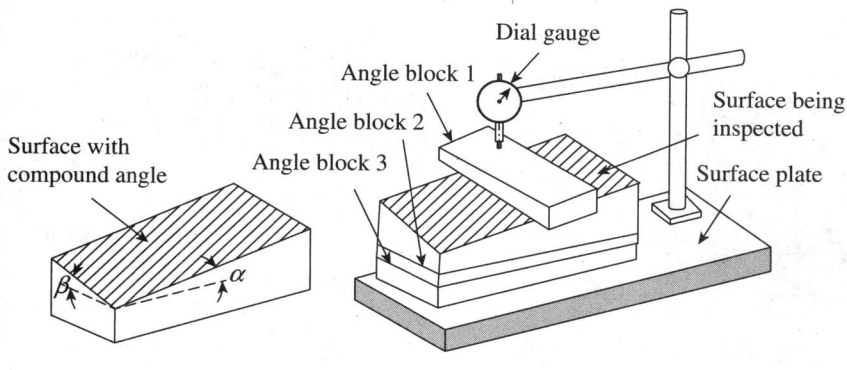

Fig. 5.16 Inspection of a compound angle using angle gauges

reading is taken along a longitudinal direction. If the reading remains zero, it indicates a conformance of the angle α; in this case, the back angle is 15°20'. Then, a 5° angle block is selected and is positioned across the workpiece in the transverse direction, as shown in the figure.

Correctness of the compound angle is then easily determined by running the dial indictor across the top surface of the 5° gauge in the transverse direction. A zero reading on the dial gauge indicates the conformance of angle β, that is, 5°. Thus, in a single setting, the compound angles can be inspected using simple devices.

5.4.2 Manufacture and Calibration

Angle gauges are quite often used as the standard for calibration purpose, and therefore, should be manufactured with a high degree of accuracy and precision. Steel blanks are cut to the required shape and machined to the nearest size and shape. Before subjecting them to finish machining, the blanks are heat treated to impart the required hardness. The heat treatment process involves quenching and tempering, followed by a stabilizing process. Now, the gauges are ground and lapped using a sine table. The advantage of using a sine table is that it can ensure that a precise angle is generated without needing to use a custom-made jig or fixture. The non-gauging sides of an angle block are ground and lapped to uniform thickness. This is followed by lapping of the gauging faces. It is important to ensure that the gauging faces are perfectly square to the non-gauging sides. The gauges are then inspected using an autocollimator or interferometry technique to ensure that they meet the accuracy requirements.

The newly manufactured or in-use angle gauges are subjected to calibration to ensure accuracy. One of the popular calibration methods for angle gauges is the interferometry method, which is quite a simple and accurate way of calibrating angle gauges.

The angle gauge, which needs to be calibrated, is carefully placed on a steel platen. The platen is nothing but a surface plate with a degree of flatness. An optical flat is positioned above the angle gauge at some angle, as shown in Fig. 5.17. A monochromatic light source is provided for the optical flat, so that fringe patterns are seen on both the platen and the angle gauge. Assuming that the angle gauge and the platen surfaces are inclined in one plane only, the fringes are straight, parallel, and equally spaced. However, the pitch of the two sets of fringes is different; those on the angle gauge have a lesser pitch. For a distance 'l' measured across the fringes, if 'p' is the number of fringes on the platen and 'q' the number of fringes on the angle gauge, then,

$$\theta = \theta_1 + \theta_2 = \lambda\,(p - q)/2l$$

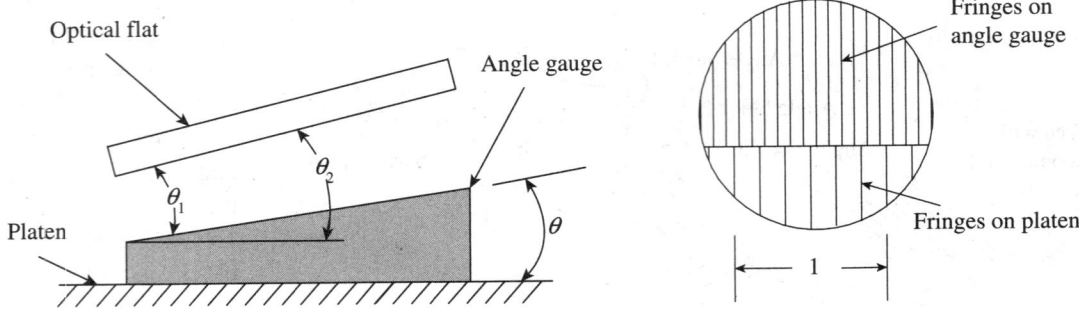

Fig. 5.17 Calibration of angle gauges by interferometry

where λ is the wavelength of light. In this method, it is possible to calibrate angle gauges up to 0.1" of an arc.

5.4.3 True Square

A true square is used as a companion tool along with an angle gauge block set. It is available for both tool room and laboratory master sets. As the name itself suggests, it is a square piece made of hardened and wear-resistant steel. All faces of the true square are precisely at a 90° angle to adjacent gauging surfaces. It has a high degree of optical flatness and parallelism to be used with autocollimators. The main advantage of a true square is that it extends the range of the angle block set to 360°, be it in degree, minute, or second steps. Figure 5.18 illustrates the shape of a true square.

5.5 SPIRIT LEVEL

A spirit level is a basic 'bubble instrument', which is widely used in engineering metrology. It is derived from the practice in cold western countries. To combat freezing, the tubes were filled with 'spirits of wine', hence the general term spirit level. Spirit level, as you are aware, is an angular measuring device in which the bubble always moves to the highest point of a glass vial. The details of a typical spirit level are shown in Fig. 5.19. The base, called the reference plane, is seated on the machine part for which straightness or flatness is to be determined. When the base is horizontal, the bubble rests at the centre of the graduated scale, which is engraved on the glass. When the base of the spirit level moves out of the horizontal, the bubble shifts to the highest point of the tube. The position of the bubble with reference to the scale is a measure of the angularity of the machine part. This scale is calibrated to directly provide the reading in minutes or seconds. A cross test level provided at a right angle to the main bubble scale indicates the inclination in the other plane. A screw adjustment is provided to set the bubble to zero by referencing with a surface plate.

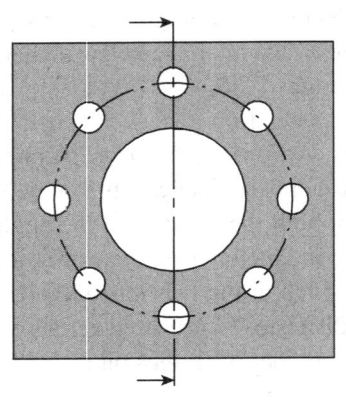

Fig. 5.18 True square

The performance of the spirit level is governed by the

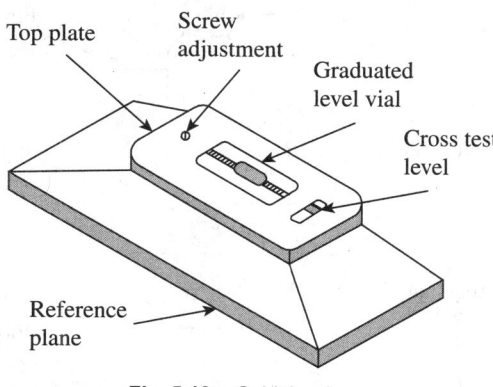

Fig. 5.19 Spirit level

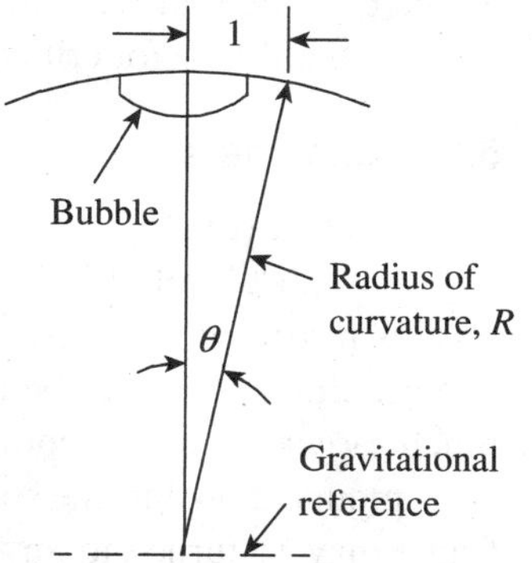

Fig. 5.20 Relationship between radius of curvature and bubble scale reading

geometrical relationship between the bubble and the two references. The first reference is the effect of gravity acting at the centre of the bubble. The second is the scale against which the bubble position is read. The sensitivity of the spirit level depends on the radius of curvature of the bubble, which is formed against the inside surface of the glass vial, and the base length of its mount.

Assuming that a level has graduations on the vial with a least count of 1 mm and a radius of curvature, R, as shown in Fig. 5.20, the angle $\theta = l/R$ (since θ is very small).

If the graduations are at a 2 mm interval and represent a tilt of 10", then the following can be concluded:

$$\theta^c = 10 \times \pi/(180 \times 3600)$$

Therefore, $R = 41{,}273.89$ mm or 41.274 m approximately.

If the base length is 250 mm, the height h, to which one end must be raised for a 2 mm bubble movement, is given by the following relation:

$$\theta^c = h/250$$

Therefore, $h = 0.012$ mm.

It is obvious from these computations that sensitivity of the instrument depends on the radius of curvature of the bubble tube and the base length. Sensitivity can be increased by either increasing the radius of curvature or reducing the base length. The most useful sensitivity for precision measurement is 10" per division.

The main use of a spirit level is not for measuring angles, but for measuring alignment of machine parts and determining flatness and straightness. Typically, the level is stepped along the surface in intervals of its own base length, the first position being taken as the datum. The heights of all other points are determined relative to this datum. One should always keep in mind that the accuracy of a given spirit level depends on the setting of the vial relative to the base. There is bound to be a certain amount of error in the setting of the vial. In order to minimize this error, the following procedure is recommended while using a spirit level for precision measurement:

1. Take readings from both ends of the vial.
2. Reverse the base of the spirit level.
3. Repeat readings from both ends.

4. Average the four readings.

5. Repeat all steps for critical cases.

5.5.1 Clinometer

A clinometer (Fig. 5.21) is a special case of a spirit level. While the spirit level is restricted to relatively small angles, clinometers can be used for much larger angles. It comprises a level mounted on a frame so that the frame may be turned to any desired angle with respect to a horizontal reference. Clinometers are used to determine straightness and flatness of surfaces. They are also used for setting inclinable tables on jig boring machines and angular jobs on surface grinding machines. They provide superior accuracy compared to ordinary spirit levels.

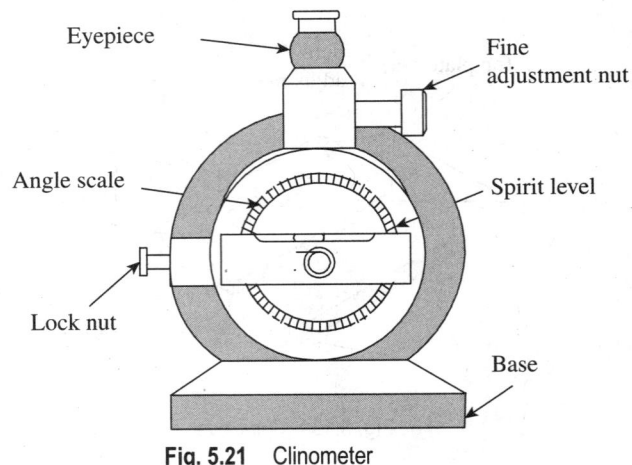

Fig. 5.21 Clinometer

To measure with clinometers, the base is kept on the surface of the workpiece. The lock nut is loosened, and the dial comprising the circular scale is gently rotated until the bubble in the spirit level is approximately at the centre. Now, the lock nut is tightened and the fine adjustment nut is operated until the bubble is exactly at the centre of the vial scale. The reading is then viewed through the eyepiece. Most clinometers in a metrology laboratory provide readings up to an accuracy of 1'. Precision clinometers can be used if the accuracy requirement is up to 1".

A recent advancement in clinometers is the electronic clinometer. It consists of a pendulum whose displacement is converted into electrical signals by a linear voltage differential transformer. This provides the advantage of electronic amplification. It is powered by an electronic chip that has a provision for recording and data analysis. Electronic clinometers have a sensitivity of 1". A major advantage of these clinometers is that the readings settle down within 1 second in contrast to the mechanical type, which requires a couple of seconds for the reading to settle down.

5.6 OPTICAL INSTRUMENTS FOR ANGULAR MEASUREMENT

Chapter 7 is devoted to the discussion of optical measurements. Nevertheless, it would be appropriate to highlight the four principles that govern the application of optics in metrology here. The most vital one is *magnification*, which provides visual enlargement of the object. Magnification enables easy and accurate measurement of the attributes of an object. The second one is *accuracy*. A monochromatic light source provides the absolute standard of length, and therefore, ensures a high degree of accuracy. The third principle is one of *alignment*. It uses light rays to establish references such as lines and planes. The fourth, and a significant one, is the principle of *interferometry*, which is a unique phenomenon associated with light. These principles have driven the development of a large number of measuring instruments and comparators. This section is devoted to two such instruments, which are most popular in angular measurement, namely the autocollimator and the angle dekkor.

An autocollimator is an optical instrument that is used to measure small angles with very high sensitivity. It has a wide variety of applications, including precision alignment, detection of angular movement, verification of angle standards, etc. The user has a wide choice of instruments from the conventional visual collimator to the digital and laser autocollimators. An angle dekkor is a particular type of autocollimator that is essentially used as a comparator. Although it is not as precise an instrument as an autocollimator, it has a wide field of applications for general angular measurements, as angular variations are directly read without the operation of a micrometer.

5.6.1 Autocollimator

It is a special form of telescope that is used to measure small angles with a high degree of resolution. It is used for various applications such as precision alignment, verification of angle standards, and detection of angular movement, among others. It projects a beam of collimated light onto a reflector, which is deflected by a small angle about the vertical plane. The light reflected is magnified and focused on to an eyepiece or a photo detector. The deflection between the beam and the reflected beam is a measure of the angular tilt of the reflector. Figure 5.22 illustrates the working principle of an autocollimator.

The reticle is an illuminated target with a cross-hair pattern, which is positioned in the focal plane of an objective lens. A plane mirror perpendicular to the optical axis serves the purpose of reflecting an image of the pattern back on to the observation point. A viewing system is required to observe the relative position of the image of the cross-wires. This is done in most of the autocollimators by means of a simple eyepiece. If rotation of the plane reflector by an angle θ results in the displacement of the image by an amount d, then, $d = 2f\theta$, where f is the focal length of the objective lens.

It is clear from this relationship that the sensitivity of an autocollimator depends on the focal length of the objective lens. The longer the focal length, the larger the linear displacement for a given tilt of the plane reflector. However, the maximum reflector tilt that can be accommodated is consequently reduced. Therefore, there is a trade-off between sensitivity and measuring range. The instrument is so sensitive that air currents between the optical path and the target mirror can cause fluctuations in the readings obtained. This effect is more severe when the distance between the two increases. Therefore, an autocollimator is housed inside a sheet-metal or a PVC plastic casing to ensure that air currents do not hamper measurement accuracy.

Autocollimators may be classified into three types:
1. Visual or conventional autocollimator
2. Digital autocollimator
3. Laser autocollimator

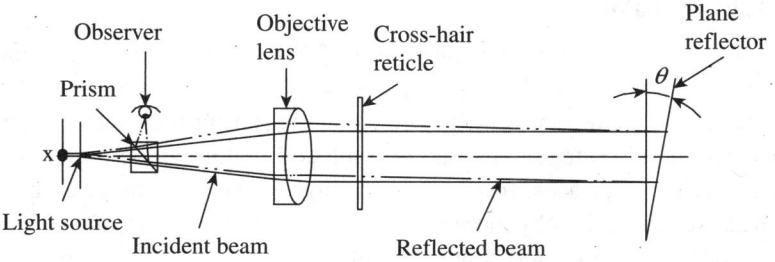

Fig. 5.22 Principle of autocollimation

Visual Autocollimator

In this type of autocollimator, the displacement of the reflected image is determined visually. A pinhole light source is used, whose reflected image is observed by the operator through an eyepiece. Visual collimators are typically focused at infinity, making them useful for both short and long-distance measurements. The plane reflector is one of the vital parts of an autocollimator, because a mirror that is not flat will defocus the return image, resulting in poor definition of the image. High-quality mirrors with a flatness tolerance of 1 μm per 100 mm are used. Most visual collimators have a resolution of 3–5" over a distance of 1.5 m.

The following are some of the typical applications of visual autocollimators:
1. Determination of angular measurements up to 3"
2. Determination of straightness of machine guideways
3. Determination of parallelism of slide movements with respect to guideways
4. Flatness estimation of machine tables, surface plates, etc.
5. Verification of right angle prisms for angular errors
6. Angle comparisons of reflecting surfaces

Digital Autocollimator

A digital autocollimator uses an electronic photo detector to detect the reflected light beam. A major advantage of this type of collimator is that it uses digital signal processing technology to detect and process the reflected beam. This enables the filtering out of stray scattered light, which sharpens the quality of the image. The illuminated target reticle slit is imaged back in its own plane through the objective lens and reflecting mirror. It is then re-imaged onto a vibrating slit by means of a relay lens. A photocell positioned behind the vibrating slit generates an output, which captures both the magnitude and the direction of rotation of the mirror from a central null position. These instruments have a resolution of up to 0.01 arc-second and a linearity of 0.1%. Since the output is digital in nature, it can be transferred to a data acquisition system, thereby facilitating storage and further processing of data. Another major advantage is that it can also measure angles of dynamic systems to a high degree of resolution, thanks to high sampling rates of digital electronic systems.

The following are some of the applications of a digital autocollimator:
1. Angular measurement of static as well as dynamic systems
2. Alignment and monitoring of robotic axes
3. Verification of angular errors of rotary tables, indexing heads, and platforms of machine parts
4. Remote monitoring of alignment of large mechanical systems

Laser Autocollimator

Laser autocollimators represent the future of precision angle measurement in the industry. Superior intensity of the laser beam makes it ideal for the measurement of angles of very small objects (1 mm in diameter) as well as for long measuring ranges that extend to 15 m or more. Another marked advantage is that a laser autocollimator can be used for the measurement of non-mirror-quality surfaces. In addition, the high intensity of the laser beam creates ultra-low noise measurements, thereby increasing the accuracy of measurement.

TL40 and TL160 lasers are popular in autocollimators. Table 5.3 gives a comparative picture of the capabilities of these two lasers.

Table 5.3 Comparison of TL40 and TL160 laser collimators

Specification	TL40 laser	TL160 laser
Maximum measuring range (arc-seconds in each axis)	±3600	±600
Maximum resolution (arc-second)	0.1	0.1
Maximum working distance (m)	1	15
Measurement linearity (%)	0.5	0.5
Maximum data bandwidth (Hz)	1000	1000
Dicde laser light source	670 nm, class II	670 nm, class II

You are perhaps aware of the power of laser and impracticability of direct viewing of laser beam unlike the conventional source of light. Therefore, an alternative method has to be developed to capture the readings of the instrument. Figure 5.23 illustrates the construction details of a laser collimator.

All the components of a laser collimator are housed in a precisely machined barrel. The line of sight, in this case, laser beam, is precisely cantered, thanks to the holding and supporting fixture. A laser beam is produced by a continuous wave plasma tube. It is coherent and has a diameter of approximately 8–10 mm. The optical axis is exactly in line with the mechanical axis of the barrel. The rear lens along with the spatial filter ensures a sharp beam. The laser beam is aimed at the target. The instrument barrel, in addition to the laser emitter, contains a beam splitter and an array of photoelectric sensors. These are arranged in such a manner that the sensors do not receive laser emissions but receive only the return beam from the mirror.

Unlike in the visual autocollimator, in a laser collimator, the internal target is a bi-cell array of sensors. Two sensors are provided to measure displacement in each axis. The sensor output is converted to angular displacements by a logic circuit. Most laser collimators have a resolution of ±3 arc-seconds.

5.6.2 Autocollimator Applications

Autocollimators are used for the measurement of straightness and flatness of machine parts and accessories such as guideways, machine tables, surface plates as well as for the assessment of

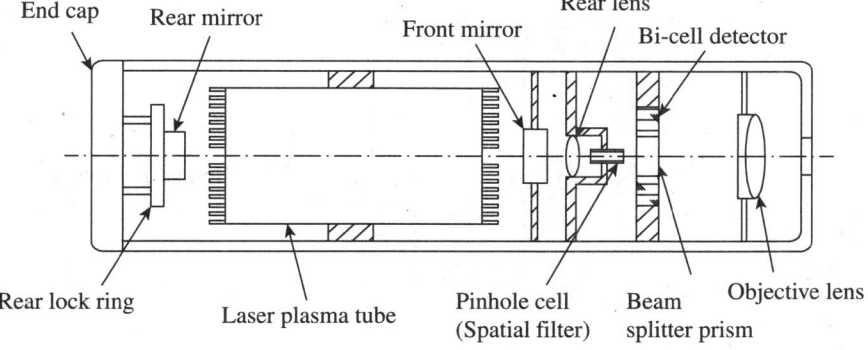

Fig. 5.23 Laser collimator

parallelism of machine slide movement with respect to guideways. Angle gauges can also be calibrated using an autocollimator. Most of these measurement techniques are discussed later in Chapter 10. However, determination of straightness using an autocollimator is discussed in this section to highlight its use in metrology.

Checking the straightness of machine guideways is one of the most frequent uses of an autocollimator. The measurement set-up is shown in Fig. 5.24. The autocollimator is kept aligned with the guideway for which straightness is to be determined. It is mounted on a levelling base. The levelling base facilitates alignment of the optical axis in line with the surface being measured. It incorporates spring-loaded clamps and a circular bubble level to help in perfect alignment.

A mirror carriage is another important accessory for an auto-collimator. It has a polished and clear re-flecting surface, which is perfectly square with the base. The base of the standard length sits on the machine guideway or a similar

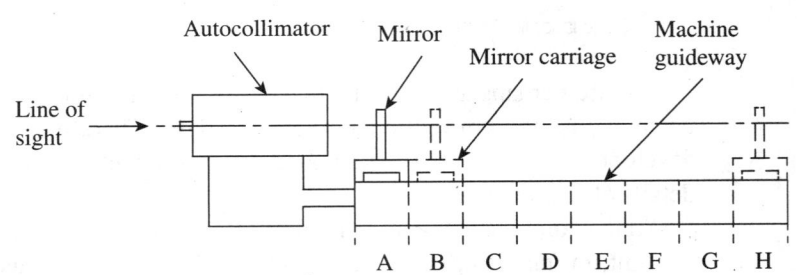

Fig. 5.24 Straightness measurement using an autocollimator

surface for which straightness is being ascertained. Markings can be made on the machine sur-face to step off equal lengths by shifting the mirror carriage successively; whenever the base of the carriage is not straight with respect to the axis of the autocollimator, the mirror will have a small tilt with respect to the optical axis of the autocollimator. This results in the measurement of the tilt angle by the autocollimator.

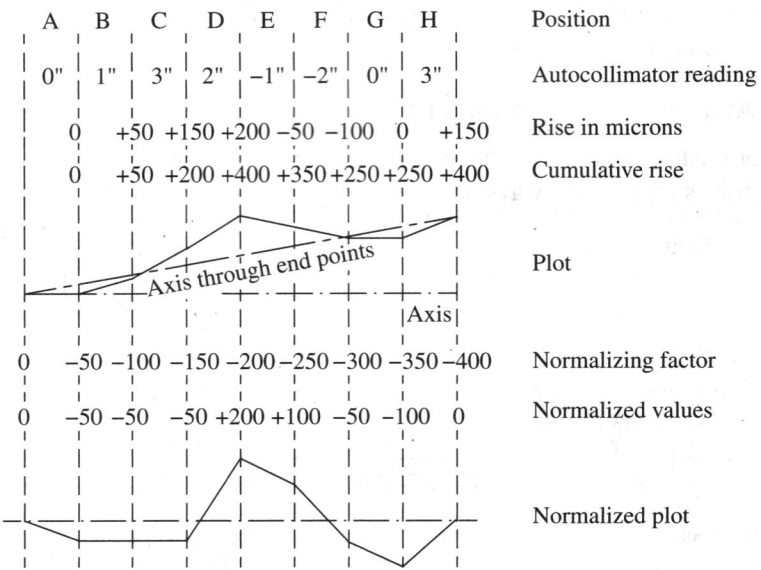

Fig. 5.25 Straightness plot

The first step is to convert deviation from straightness of guideway into successive tilts of the mirror carriage. At each position of the mirror carriage, readings are taken from the autocollimator. These readings are converted into a profile plot, as shown in Fig. 5.25. This plot clearly illustrates the deviation of the guideway from straightness, but is somewhat difficult to interpret. A plot that is created with respect to the two end points is more useful for interpretation of the deviation. Accordingly, a normalizing factor, which is nothing but the ratio of cumulative rise to a number of positions, is added to each reading. A new plot with normalized values is drawn, wherein the profile plot is realigned with respect to the two end points. This plot illustrates, in graphical form, the deviation of the guideway from straightness. In fact, the maximum deviation of the plot from the axis is a measure of straightness. Thus, an autocollimator provides a quick and accurate means for ascertaining straightness of a guideway.

5.6.3 Angle Dekkor

An angle dekkor is a small variation of the autocollimator. This instrument is essentially used as a comparator and measures the change in angular position of the reflector in two planes. It has an illuminated scale, which receives light directed through a prism. The light beam carrying the image of the illuminated scale passes through the collimating lens, as shown in Fig. 5.26, and falls onto the reflecting surface of the workpiece. After getting reflected from the workpiece, it is refocused by the lens in field view of the eyepiece. While doing so, the image of the illuminated scale would have undergone a rotation of 90° with respect to the optical axis. Now, the light beam will pass through the datum scale fixed across the path of the light beam, as shown in Fig. 5.26. When viewed through the eyepiece, the reading on the illuminated scale measures angular deviations from one axis at 90° to the optical axis, and the reading on the fixed datum scale measures the deviation about an axis mutually perpendicular to this.

The view through the eyepiece, which gives the point of intersection of the two scales, is shown in Fig. 5.27. The scales usually measure up to an accuracy of 1'. This reading actually indicates changes in angular position of the reflector in two planes. In other words, the initial

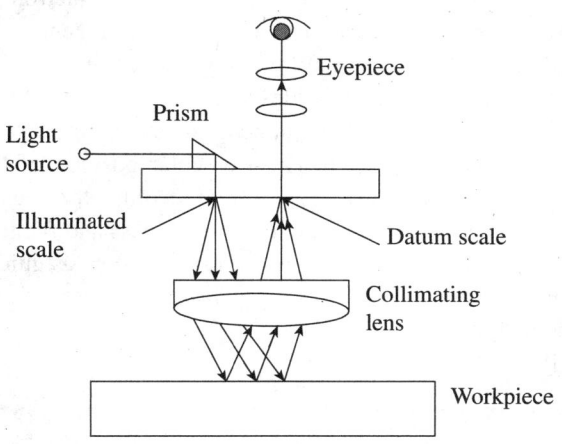

Fig. 5.26   Angle dekkor

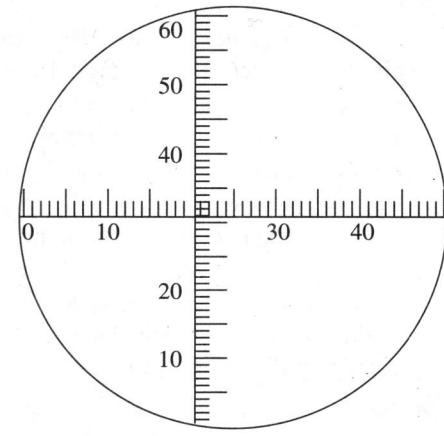

Fig. 5.27   Intersection of two scales

reading of the angle dekkor corresponds to the reading on the two scales before shifting the position of the reflector. After the reflector undergoes an angular tilt, the second reading is noted down by recording the point of intersection on both scales. The difference in readings on the two scales indicates the tilt of the reflector in two planes at 90° to each other.

The optical system in an angle dekkor is enclosed in a tube, which is mounted on an adjustable bracket. It has a wide range of applications, as angular variations can be directly read through the eyepiece of the instrument. Some of the typical applications are as follows:

1. Measurement of sloping angle of V-blocks
2. Calibration of taper gauges
3. Measurement of angles of conical parts
4. Measurement of angles of work part surfaces, which are simultaneously inclined in two planes
5. Determination of a precise angular setting for machining operations, for example, milling a slot at some precise angle to a previously machined datum surface.

A QUICK OVERVIEW

- The precise measurement of angles is an important requirement in workshops and tool rooms. We need to measure angles of interchangeable parts, gears, jigs, fixtures, etc. There are a wide range of instruments, starting from simple scaled instruments to sophisticated types, which use laser interferometry techniques that can be used for angle measurement.

- The universal bevel protractor with a 5' accuracy is commonly found in all tool rooms and metrology laboratories. Since the universal bevel protractor can measure both acute and obtuse angles, care should be exercised to clearly differentiate between the angle being indicated on the scale and its supplement.

- The sine bar is used to measure angles based on the sine principle. Simple variants of the sine bar such as sine block, sine plate, sine table, and sine centre have many applications in metrology.

- Angle gauges, which are made of high-grade wear-resistant steel, work in a manner similar to slip gauges. While slip gauges can be built to give linear dimensions, angle gauges can be built

to give the required angle. The gauges come in a standard set of angle blocks that can be wrung together in a suitable combination to build an angle.

- The main use of spirit level-based instruments like clinometers is not for measuring angles per se. They are used for measuring alignment of machine parts and determination of flatness and straightness.

- Optical instruments enable very precise measurement of angles. The four principles that govern the application of optics in metrology are magnification, accuracy, alignment, and interferometry. While the autocollimator can measure small angles with very high sensitivity, an angle dekkor can be used as a comparator to measure the change in angular position of the object in two planes. An autocollimator is used for the measurement of straightness and flatness of machine parts and accessories such as guideways, machine tables, and surface plates, among others.

MULTIPLE-CHOICE QUESTIONS

1. While measuring the angle of a workpiece using a universal bevel protractor in clockwise direction, in which quadrants can the angle be read directly from the scale?
   (a) Quadrants I and II
   (b) Quadrants I and III
   (c) Quadrant I only
   (d) All the quadrants

2. Which of the following statements is false with respect to a bevel protractor?
   (a) If the angle of a work part is being measured in II quadrant, the actual angle is given by the supplement.
   (b) If the angle of a work part is being measured in IV quadrant, the actual angle is given by the supplement.
   (c) The angle measured in II quadrant is always an obtuse angle.
   (d) The angle measured in IV quadrant is always an acute angle.

3. Which type of bevel protractor has a vernier scale as well as an acute angle attachment?
   (a) Type A   (c) Type C
   (b) Type B   (d) Type D

4. The purpose of providing relief holes in sine bars is to
   (a) improve accuracy   (c) reduce weight
   (b) improve precision   (d) reduce wear

5. The maximum angle that can be set using a sine bar is limited to
   (a) 15°   (c) 45°
   (b) 30°   (d) 60°

6. Which of the following statements is true?
   (a) The longer the sine bar, the better the accuracy.
   (b) The shorter the sine bar, the better the accuracy.
   (c) Accuracy of a sine bar does not depend on an ambient temperature.
   (d) A sine bar cannot measure unknown angles.

7. What is the minimum number of angle gauges required to set any angle between 0° and 180° in increments of 5'?
   (a) 8   (c) 12

   (b) 10   (d) 15

8. Which of the following statements is true?
   (a) An angle dekkor is a small variation of a clinometer.
   (b) An angle dekkor is a small variation of an autocollimator.
   (c) An angle dekkor is a small variation of a sine bar.
   (d) None of these

9. Which of the following instruments is capable of measuring compound angles?
   (a) Sine centre
   (b) Compound sine plate
   (c) Compound surface plate
   (d) All of these

10. Which of the following is correct if the 5° angle block is reversed and combined with the 30° angle block?
    (a) The resulting angle becomes 25°.
    (b) The resulting angle becomes 35°.
    (c) The resulting angle remains 30°.
    (d) Such a combination is not possible.

11. It is much easier to calibrate angle gauge blocks compared to slip gauges, because
    (a) angle gauges are made of superior material
    (b) there are better instruments to calibrate angles rather than linear dimensions
    (c) the angle being measured is always a portion of a triangle comprising 180°
    (d) the angle being measured is a portion of a full circle and is, therefore, self-proving

12. Which of the following can extend the range of the angle block set to 360°?
    (a) True square   (c) Combination set
    (b) Try square   (d) Sine plate

13. The performance of the spirit level is governed by the geometrical relationship between the bubble and
    (a) a single datum
    (b) its top plate
    (c) two references
    (d) three references along mutually per-pendicular directions

14. The sensitivity of a spirit level depends on

(a) the width of the bubble
(b) accuracy of base plate
(c) both (a) and (b)
(d) the radius of curvature of the bubble tube
15. Precision measurement of angles of non-mirror-

quality surfaces is possible only with
(a) a laser autocollimator
(b) a visual autocollimator
(c) an angle dekkor
(d) any of these

REVIEW QUESTIONS

1. What is the chief difference between length standard and angle standard?
2. What is the working principle of a universal bevel protractor? What are the precautions to be taken while using it?
3. The vernier scale in a bevel protractor is read in the same direction as the dial. Why?
4. How does an optical bevel protractor differ from a mechanical type?
5. What is the basic difference between sine bars, sine plates, and sine tables?
6. In a sine bar, when should the set-up be made for the complement of an angle?
7. Discuss the essential requirements for maintaining accuracy in the construction of a sine bar.
8. How is a sine bar specified?
9. What is the maximum recommended angle to which a sine bar can be set? Discuss the relationship between the angle being set and the error of measurement for a sine bar.
10. Explain how conical workpieces are inspected on a sine centre.
11. Why are angle gauges also referred to as

'Tomlinson gauges'?
12. Using the set of 16 angle gauges, how can we set the angle 20°40'10"?
13. The compound angle of a forging die needs to be inspected. What is the procedure of inspection using angle gauges? Discuss with a simple sketch.
14. The main aim of clinometers is not to measure angles, but to measure alignment, straightness, and parallelism. Clarify.
15. Describe with a sketch the principle behind the working of an autocollimator.
16. Discuss any two important uses of an auto-collimator in the industry.
17. Give the differences among the following: visual collimator, digital collimator, and laser collimator.
18. Explain with the help of a plot how straightness of a machine guideway is assessed using an autocollimator.
19. How does an angle dekkor differ from an autocollimator?
20. Discuss the applications of an angle dekkor in metrology.

Answers to Multiple-choice Questions

1. (b)   2. (d)   3. (a)   4. (c)   5. (c)   6. (a)   7. (b)   8. (b)
9. (b)   10. (a)   11. (d)   12. (a)   13. (c)   14. (d)   15. (a)

Comparators

After studying this chapter, the reader will be able to

- appreciate the difference between direct measurement and comparison measurement
- understand the functional requirements of comparators
- explain the basic principles of construction and operations of various types of comparators such as mechanical, optical, electrical, and electronic comparators
- describe the functions of various attachments that can be used with different comparators to enhance their functional aspects
- elucidate the basic measurement principles of comparators
- discuss the applications and limitations of various comparators

6.1 INTRODUCTION

All measurements require the unknown quantity to be compared with a known quantity, called the *standard*. A measurement is generally made with respect to time, mass, and length. In each of these cases, three elements are involved: the unknown, the standard, and a system for comparing them. In Chapter 4 we came across linear measurement instruments, such as verniers and micrometers, in which standards are in-built and calibrated. Hence, these instruments enable us to directly measure a linear dimension up to the given degree of accuracy.

On the other hand, in certain devices the standards are separated from the instrument. It compares the unknown length with the standard. Such measurement is known as *comparison measurement* and the instrument, which provides such a comparison, is called a *comparator*. In other words, a comparator works on relative measurement. It gives only dimensional differences in relation to a basic dimension or master setting. Comparators are generally used for linear measurements, and the various comparators currently available basically differ in their methods of amplifying and recording the variations measured.

Figure 6.1 illustrates the difference between direct and comparison measurements. As can be seen in the figure, a calibrated standard directly gives the measured value in case of direct measurement. On the other hand, a comparator has to be set to a reference value (usually zero setting) by employing a standard. Once it is set to this reference value, all subsequent readings indicate the deviation from the standard. The deviation can be read or recorded by means of a display or recording unit, respectively. Accuracy of direct measurement depends on four factors: accuracy of the standard, accuracy of scale, least count of the scale, and accuracy of reading the scale. The last factor is the human element, which depends on the efficiency with which the scales are read and the accurate interpretation of the readings.

Accuracy of comparison measurement primarily depends on four factors: accuracy of the standard used for setting the comparator, least count of the standard, sensitivity of the comparator, and accuracy of reading the scale. In contrast to direct measurement, the role of the sensing element is significant in a comparator. The sensitivity of the comparator to sense even a minute variation in the measured value is equally important. The variation in the measured value may be in terms of change in displacement, pressure, fluid flow, temperature, and so on.

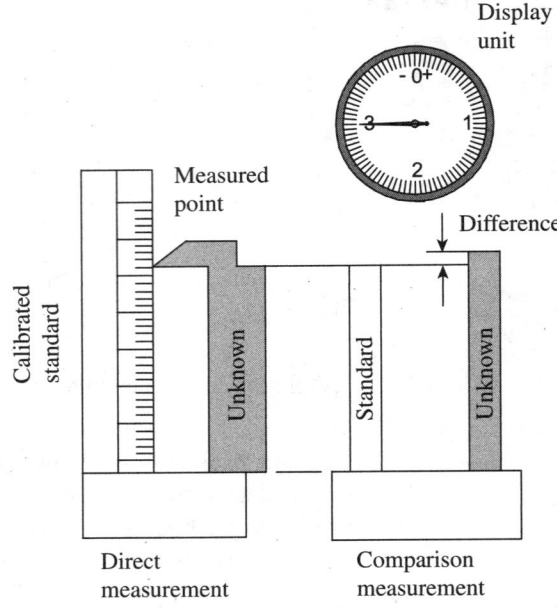

Fig. 6.1 Direct measurement versus comparison measurement

6.2 FUNCTIONAL REQUIREMENTS

A comparator has to fulfil many functional requirements in order to be effective in the industry. It should not only provide a high degree of accuracy and precision but also be convenient for use. It should withstand the rough and tough operating environment on the shop floor and also have good sensitivity to detect minute changes in the parameter being measured. We can summarize the major requirements of a comparator as follows:

1. A comparator should have a high degree of accuracy and precision. We can safely say that in general, comparison measurement provides better accuracy and precision than direct measurement. In *direct measurement*, precision is dependent on the least count of the scale and the means for reading it. In *comparison measurement*, it is dependent on the least count of the standard and the means for comparing. Accuracy, in contrast, is dependent on other factors, geometrical considerations being the most important of them. Direct measurement instruments such as vernier calliper and micrometer have the standard built into it, with the result that measurement is done by the *displacement method*. It is the relationship between the distance displaced and a standard that constitutes the measurement. On the other hand, comparison measurement uses the *interchange method* for measurement. In this method, both ends of the unknown feature are compared with both ends of the standard at the same time. This enables comparators to have a more favourable geometry, which gives scope for better accuracy.

2. The scale should be linear and have a wide range. Since a comparator, be it mechanical, pneumatic, or electrical, has a means of amplification of signals, linearity of the scale within the measuring range should be assured.

3. A comparator is required to have high amplification. It should be able to amplify changes in the input value, so that readings can be taken and recorded accurately and with ease. Amplification demands use of more number of linkages in a mechanical system and a more elaborate circuit in an electrical system. This puts load on the system, resulting in the system being unable to sense small changes in the input signal. Therefore, one has to strike a compromise between the two. Alternately, the designer can be biased in favour of one at the cost of the other, depending on the major objective of measurement.

4. A comparator should have good resolution, which is the least possible unit of measurement that can be read on the display device of the comparator. Resolution should not be confused with readability, the former being one among many factors that influence the latter. Other factors include size of graduations, dial contrast, and parallax.

5. There should be a provision incorporated to compensate for temperature effects.

6. Finally, the comparator should be versatile. It should have provisions to select different ranges, attachments, and other flexible means, so that it can be put to various uses.

6.3 CLASSIFICATION OF COMPARATORS

We can classify comparators into mechanical device and electrical device on the basis of the means used for comparison. In recent times, engineers prefer to classify comparators as low- and high-amplification comparators, which also reflect the sophistication of the technology that is behind these devices. Accordingly, we can draw the following classification.

With respect to the principle used for amplifying and recording measurements, comparators are classified as follows:
1. Mechanical comparators
2. Mechanical–optical comparators
3. Electrical and electronic comparators
4. Pneumatic comparators
5. Other types such as projection comparators and multi-check comparators

Each of these types of comparators has many variants, which provide flexibility to the user to make an appropriate and economical selection for a particular metrological application.

6.4 MECHANICAL COMPARATORS

Mechanical comparators have a long history and have been used for many centuries. They provide simple and cost-effective solutions. The skills for fabricating and using them can be learnt relatively easily compared to other types of comparators. The following are some of the important comparators in metrology.

6.4.1 Dial Indicator

The dial indicator or the dial gauge is one of the simplest and the most widely used comparator.

It is primarily used to compare workpieces against a master. The basic features of a dial gauge consist of a body with a circular graduated dial, a contact point connected to a gear train, and an indicating hand that directly indicates the linear displacement of the contact point. The contact point is first set against the master, and the dial scale is set to zero by rotating the bezel. Now, the master is removed and the workpiece is set below the contact point; the difference in dimensions between the master and the workpiece can be directly read on the dial scale. Dial gauges are used along with V-blocks in a metrology laboratory to check the roundness of components. A dial gauge is also part of standard measuring devices such as bore gauges, depth gauges, and vibrometers. Figure 6.2 illustrates the functional parts of a dial indicator.

The contact point in a dial indicator is of an interchangeable type and provides versatility to the instrument. It is available as a mounting and in a variety of hard, wear-resistant materials. Heat-treated steel, boron carbide, sapphire, and diamond are some of the preferred materials. Although flat and round contact points are commonly used, tapered and button-type contact points are also used in some applications. The stem holds the contact point and provides the required length and rigidity for ease of measurement. The bezel clamp enables locking of the dial after setting the scale to zero. The scale of the dial indicator, usually referred to as dial, provides the required least count for measurement, which normally varies from 0.01 to 0.05 mm. The scale has a limited range of linear measurements, varying from 5 to 25 mm. In order to meet close least count, the dial has to be large enough to improve readability.

The dials are of two types: *continuous* and *balanced*. A continuous dial has graduations starting from zero and extends to the end of the recommended range. It can be either clockwise or anti-clockwise. The dial corresponds to the unilateral tolerance of dimensions. On the other hand, a balanced dial has graduations marked both ways of zero. This dial corresponds to the use of bilateral tolerance. Figure 6.3 illustrates the difference between the two types of dials.

Metrological features of a dial indicator differ entirely from measuring instruments such as slide callipers or micrometers. It measures neither the actual dimension nor does it have a reference point. It measures the amount of deviation with respect to a standard. In other words, we measure not length, but *change in length*. In a way, this comparison measurement is *dynamic*, unlike direct measurement, which is *static*. Obviously, the ability to detect and measure the change is the *sensitivity* of the instrument.

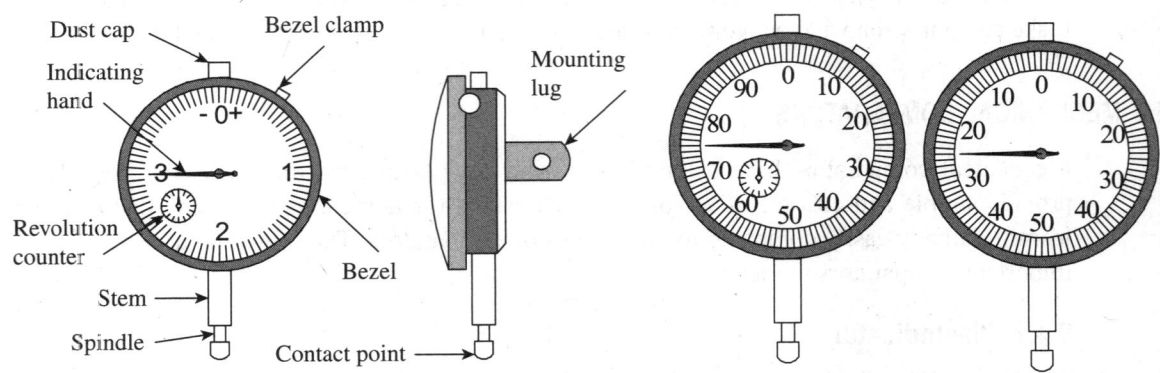

Fig. 6.2 Functional parts of a dial indicator

Fig. 6.3 Method for designating numbers

Working Mechanism of Dial Indicators

Figure 6.4 illustrates the mechanism used in a dial indicator in order to achieve high magnification using a set of gears and pinions. The plunger and spindle are usually one piece. The spindle attached to the bottom of the rack is the basic sensing element. A coil spring resists the measurement movement and thereby applies the necessary gauging pressure. Thus, the application of gauging pressure is built into the mechanism rather than leaving it to the technician. It also returns the mechanism to the 'at-rest' position after each measurement.

The plunger carries a rack, which meshes with a gear (marked gear A in the figure). A rack guide prevents the rotation of the plunger about its own axis. A small movement of the plunger causes the rack to turn gear A. A larger gear, B, mounted on the same spindle as gear A, rotates by the same amount and transfers motion to gear C. Attached to gear C is another gear, D, which meshes with gear E. Gear F is mounted on the same spindle as the indicator pointer. Thus, the overall magnification obtained in the gear train A–B–C–D–E is given by $T_D/T_E \times T_B/T_C$, where T_D, T_E, T_B, and T_C are the number of teeth on the gears D, E, B, and C, respectively. The magnification is further enhanced at the tip of the pointer, depending on the length of the pointer. A hair spring loads all the gears in the train against the direction of gauging movement. This eliminates backlash that would be caused by gear wear. The gears are precision cut and usually mounted on jewelled bearings.

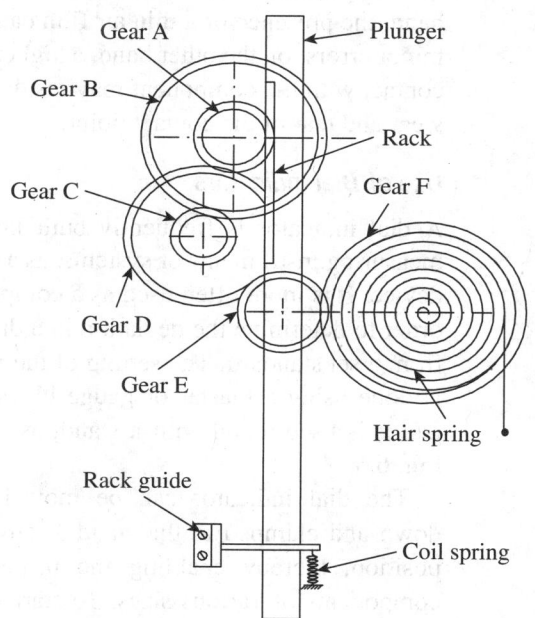

Fig. 6.4 Working mechanism of a dial indicator

Contact Points

Dial indicators are versatile instruments because their mountings adapt them to many methods of support. Interchangeable contact points adapt them to varied measurement situations. Contact points are available in various hard and wear-resisting materials such as boron carbide, sapphire, and diamond. Contact points made of hardened steel are also often used. Figure 6.5 illustrates some of the standard contact points.

The standard or spherical contact point is the most preferred one because it presents point contact to the mating surface irrespective of whether it is flat or cylindrical. However, care must be taken to pass them through the centre line of the spindle. The highest reading will be the diameter. It becomes less reliable when gauging spherical components because sphere-to-sphere contact makes the highest point of contact difficult to find. Another limitation is that it can take only limited gauging pressure, as high gauging pressure will leave an indent on the workpiece. A button-type contact point can be used if light contact pressure on smaller components is required.

A tapered point is convenient for component surfaces that cannot be accessed by either standard or flat contact points. The use of contact points on spherical surfaces presents some problems. Only a flat point is suitable in such cases. It gives reliable readings for cylindrical surfaces too. Paradoxically, flat contact points are not preferred for flat surfaces. On the one hand, the presence of a thin air film can lead to minor errors; on the other hand, a higher area of contact with the component may result in rapid wear and tear of the contact point.

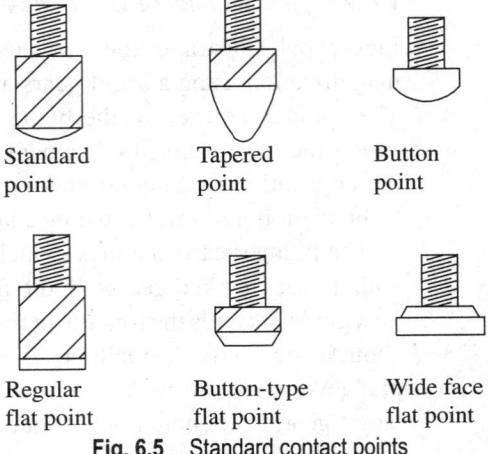

Fig. 6.5 Standard contact points

Use of Dial Indicators

A dial indicator is frequently built into other measuring instruments or systems, as a read-out device. It is more often used as a comparator in order to determine the deviation in a dimension from a set standard. The setting of the indicator is done using a master or gauge block. A dial gauge is used along with a stand, as shown in Fig. 6.6.

The dial indicator can be moved up and down and clamped to the stand at any desired position, thereby enabling the inspection of components of various sizes. To start with, the indicator is moved up and the standard is placed on the reference surface, while ensuring that the spindle of the indicator does not make contact with the standard. Next, the stand clamp is loosened and the spindle of the indicator is gently lowered onto the surface of the standard such that the spindle is under the required gauge pressure. Now, the indicator is held in position by tightening the stand clamp. The bezel clamp is loosened, the bezel is rotated, and the reading is set to zero. The dial indicator should be set to a dimension that is approximately in the centre of the spread over which the actual object size is expected to vary.

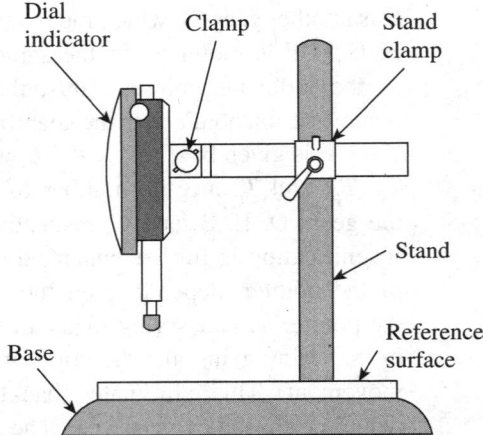

Fig. 6.6 Dial indicator mounted on a stand

Once the zero setting is done, the standard is gently taken out by hand and the workpieces are gently inserted below the spindle, one by one. Most of the dial indicators are provided with a plunger lifting lever, which provides a slight upward motion of the spindle and enables inserting and withdrawing of workpieces, without causing damage to the indicator mechanism. Now, the difference in height between the standard and the workpiece is read from the dial gauge scale. The following guidelines are recommended for the proper use of dial indicators:

1. A dial indicator is a delicate instrument as the slender spindle can be damaged easily. The user should avoid sudden contact with the workpiece surface, over-tightening of contact points, and side pressure.

2. Any sharp fall or blow can damage the contact points or upset the alignment of bearings, and hence should be avoided.

3. Standard reference surfaces should be used. It is not recommended to use non-standard attachments or accessories for reference surfaces.

4. The dial indicator should be cleaned thoroughly before and after use. This is very important because unwanted dust, oil, and cutting fluid may seep inside the instrument and cause havoc to the maze of moving parts.

5. Periodic calibration of the dial gauge is a must.

6.4.2 Johansson Mikrokator

The basic element in this type of comparator is a light pointer made of glass fixed to a thin twisted metal strip. Most of us, during childhood, would be familiar with a simple toy having a button spinning on a loop of string. Whenever the loop is pulled outwards, the string unwinds, thereby spinning the button at high speed. This type of comparator, which was developed by the Johansson Ltd Company of USA, uses this principle in an ingenious manner to obtain high mechanical magnification. The basic principle is also referred to as the 'Abramson movement' after H. Abramson who developed the comparator.

The two halves of the thin metal strip, which carries the light pointer, are twisted in opposite directions. Therefore, any pull on the strip will cause the pointer to rotate. While one end of the strip is fixed to an adjustable cantilever link, the other end is anchored to a bell crank lever, as shown in Fig. 6.7. The other end of the bell crank lever is fixed to a plunger. Any linear motion of the plunger will result in a movement of the bell crank lever, which exerts either a push or a pull force on the metal strip. Accordingly, the glass pointer will rotate either clockwise or anticlockwise, depending on the direction of plunger movement. The comparator is designed in such a fashion that even a minute movement of the plunger will cause a perceptible rotation of the glass pointer. A calibrated scale is employed with the pointer so that any axial movement of the plunger can be recorded conveniently. We can easily see the relationship of the length and width of the strip with the degree of amplification.

Thus, $d\theta/dl \propto l/nw^2$, where $d\theta/dl$ is the amplification of the mikrokator, l is the length of the metal strip measured along the neutral axis, n is the number of turns on the metal strip, and w is the width of the metal strip.

It is clear from the preceding equation that magnification varies inversely with the number of turns and width of the metal strip. The lesser the number of turns and thinner the strip, the higher is the magnification. On the other hand, magnification varies directly with the length of the metal strip. These three parameters are varied optimally to get a compact but robust instrument. A

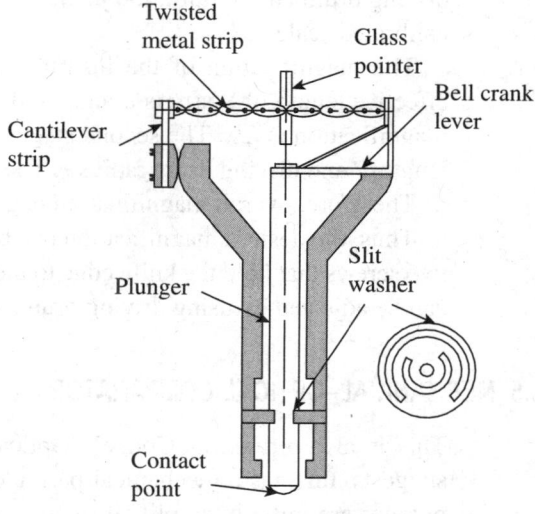

Fig. 6.7 Johansson mikrokator

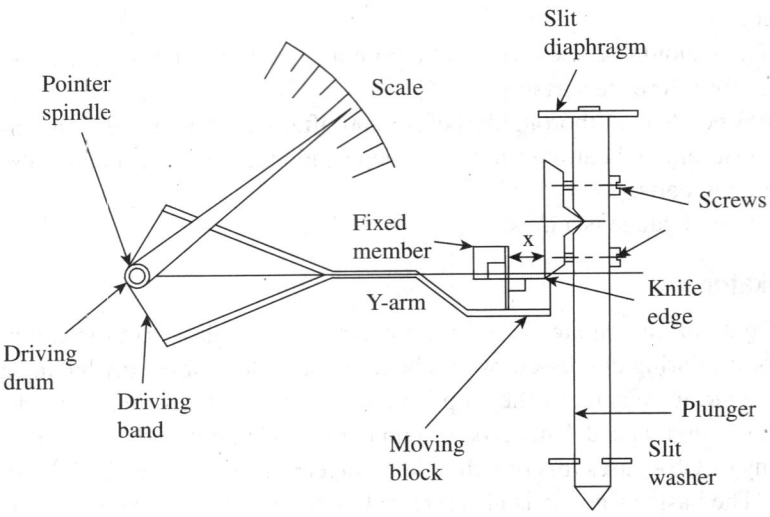

Fig. 6.8 Sigma mechanical comparator

pull on the metal strip subjects it to tensile force. In order to prevent excessive stress on the central portion of the metal strip, perforations are made in the strip, which can be noticed in Fig. 6.7. A slit washer is provided to arrest the rotation of the plunger along its axis.

6.4.3 Sigma Comparator

It is a simple but ingenious mechanical comparator developed by the Sigma Instrument Company, USA. A linear displacement of a plunger is translated into the movement of a pointer over a calibrated scale. Figure 6.8 illustrates the working parts of a Sigma mechanical comparator.

The plunger is the sensing element that is in contact with the work part. It moves on a slit washer, which provides frictionless linear movement and also arrests rotation of the plunger about its axis. A knife edge is screwed onto the plunger, which bears upon the face of the moving member of a cross-strip hinge. This unit comprises a fixed member and a moving block, connected by thin flexible strips at right angles to each other. Whenever the plunger moves up or down, the knife edge drives the moving member of the cross-strip hinge assembly. This deflects an arm, which divides into a 'Y' form. The extreme ends of this Y-arm are connected to a driving drum by means of phosphor-bronze strips. The movement of the Y-arm rotates the driving drum and, in turn, the pointer spindle. This causes the movement of the pointer over a calibrated scale.

The magnification of the instrument is obtained in two stages. In the first stage, if the effective length of Y-arm is L and the distance from the hinge pivot to the knife edge is x, then magnification is L/x. The second stage of magnification is obtained with respect to the pointer length R and driving drum radius r. The magnification is given by R/r.

Therefore, overall magnification is given by $(L/x) \times (R/r)$.

Thus, the desired magnification can be obtained by adjusting the distance x by operating the two screws that hold the knife edge to the plunger. In addition, the second level of magnification can be adjusted by using driving drums of different radii (r).

6.5 MECHANICAL–OPTICAL COMPARATOR

This is also termed as Cooke's Optical Comparator. As the name of the comparator itself suggests, this has a mechanical part and an optical part. Small displacements of a measuring plunger are initially amplified by a lever mechanism pivoted about a point, as shown in Fig. 6.9. The mechanical system causes a plane reflector to tilt about its axis. This is followed by

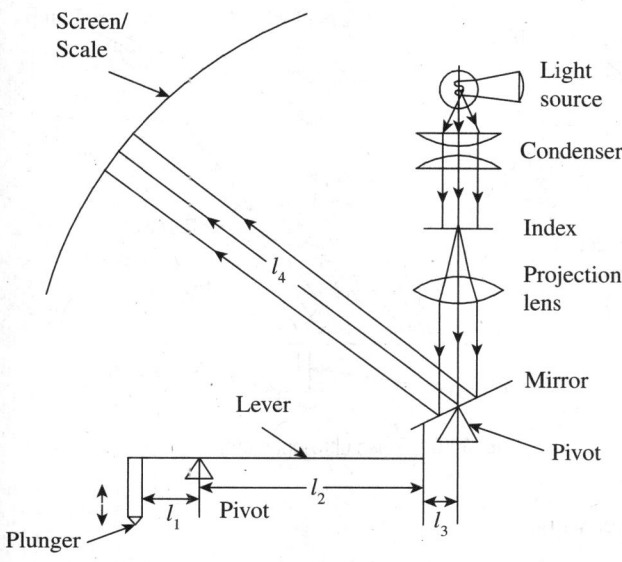

Fig. 6.9 Principle of a mechanical optical comparator

a simple optical system wherein a pointed image is projected onto a screen to facilitate direct reading on a scale.

The plunger is spring loaded such that it is biased to exert a downward force on the work part. This bias also enables both positive and negative readings, depending on whether the plunger is moving up or down. The scale is set to zero by inserting a reference gauge below the plunger. Now, the reference gauge is taken out and the work part is introduced below the plunger. This causes a small displacement of the plunger, which is amplified by the mechanical levers. The amplified mechanical movement is further amplified by the optical system due to the tilting of the plane reflector. A condensed beam of light passes through an index, which normally comprises a set of cross-wires. This image is projected by another lens onto the plane mirror. The mirror, in turn, reflects this image onto the inner surface of a ground glass screen, which has a scale. The difference in reading can be directly read on this calibrated screen, which provides the linear difference in millimetres or fractions of a millimetre. Optical magnifications provide a high degree of precision in measurements due to the reduction of moving members and better wear-resistance qualities.

With reference to Fig. 6.9, mechanical amplification = l_2/l_1 and optical amplification = 2 (l_4/l_3).

The multiplication factor 2 figures in the optical amplification because if the mirror is tilted by $\theta°$, then the image is tilted by $2\theta°$ over the scale. Thus, the overall magnification of the system is given by $2 \times (l_4/l_3) \times (l_2/l_1)$.

6.5.1 Zeiss Ultra-optimeter

The Zeiss ultra-optimeter is another mechanical optical comparator that can provide higher magnification than the simple mechanical optical comparators explained in Section 6.4. This magnification is made possible by the use of two mirrors, which create double reflection of light. Figure 6.10 illustrates the working principle of the Zeiss ultra-optimeter.

It is preferable to have a monochromatic light source passing through a condenser lens followed by an index that carries the image of two cross-wires onto a tilting mirror (marked mirror 1 in the figure). Mirror 1 reflects the image onto mirror 2 (kept parallel to it), which is again reflected to mirror 1. After the reflection from three surfaces in succession, the light rays pass through an objective lens. The magnified image is formed at the eyepiece after passing through a transparent graticule. The graticule has a scale that enables the reading of linear displacement of the plunger.

The working of a comparator is quite similar to the one explained in Section 6.5. A movement of the plunger corresponds to the change in linear dimension of a work part with respect to a standard. The plunger movement tilts mirror 1, which moves the image of cross-wires over the scale. The scale thus directly provides the linear deviation and thereby provides a convenient means for inspection of work parts. The entire set-up is enclosed in a PVC enclosure and tubings. In order to set the instrument to zero, a screw is provided to move the projected image of the graticule over the scale. Subsequent readings are either plus or minus values, depending on whether a dimension is larger or smaller than the set value, respectively.

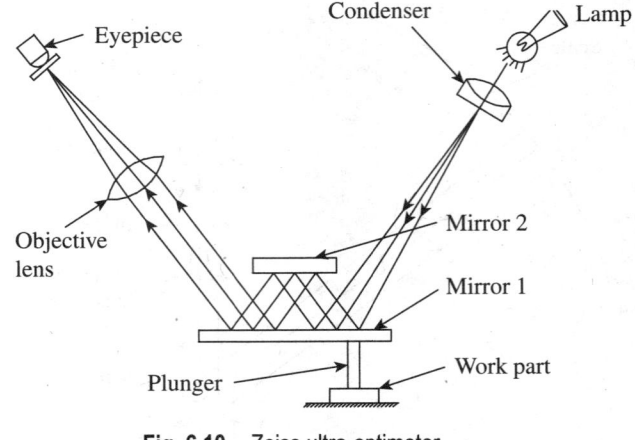

Fig. 6.10   Zeiss ultra-optimeter

6.5.2 Optical Projector

An optical projector is a versatile comparator, which is widely used for inspection purpose. It is especially used in tool room applications. It projects a two-dimensional magnified image of the workpiece onto a viewing screen to facilitate measurement. It comprises three main elements: the projector itself comprising a light source and a set of lens housed inside the enclosure, a work table to hold the workpiece in place, and a transparent screen with or without a chart gauge for comparison or measurement of parts.

Figure 6.11 illustrates the various parts of an optical projector. The workpiece to be inspected is mounted on a table such that it is in line with the light beam coming from the light source. The table may be either stationary or movable. In most projectors, the table can be moved in two mutually perpendicular directions in the horizontal plane. The movement is effected by operating a knob attached with a double vernier micrometer, which can provide a positional accuracy of up to 5 µm or better. The light beam originating from the lamp is condensed by means of a condenser and falls on the workpiece. The image of the workpiece is carried by the light beam, which passes through a projection lens. The projection lens magnifies the image, which falls on a highly polished mirror kept at an angle. The reflected light beam carrying the image of the workpiece now falls on a transparent screen. Selecting high-quality optical elements and a lamp, and mounting them at the right location will ensure a clear and sharp image, which, in turn, will ensure accuracy in measurement.

The most preferred light source is the tungsten filament lamp, although mercury or xenon lamps are also used sometimes. An achromatic collimator lens is placed in the path of a light beam coming from the lamp. The collimator lens will reorient the light rays into a parallel beam large enough in diameter to provide coverage of the workpiece. Mounting and adjustment of the lamp are critical to assure proper positioning of the filament with respect to the optical axis.

The collimated beam of light passes across the area the workpiece is positioned on the work table. Care should be taken to ensure that the contour of the workpiece that is of interest is directly in line with the light beam. The distance of the table from the projection lens should be such that it matches with the focal length of the lens, in order to ensure a sharp image. The table can be either stationary or movable. The movable tables are designed to generally travel in two mutually perpendicular directions in the horizontal plane. The table moves on anti-friction guide-ways and is controlled by the knob of a double vernier micrometer. This micrometer provides an accurate means of measuring the dimensions of the workpiece.

The light beam, after passing through the projection lens, is directed by a mirror onto the viewing screen. Screens are made of glass, with the surface facing the operator, ground to a very fine grain

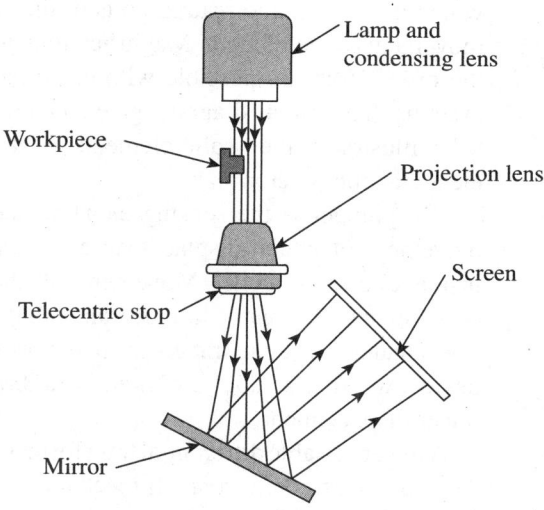

Fig. 6.11 Optical projector

size. The location of the screen should be such that it provides an accurate magnification and perfectly conforms to the measurement indicated by the micrometer. A reticle attached to the end of the projection lens provides images of two mutually perpendicular cross-wires, which can be used for the purpose of measurement. Many projector screens can also be rotated about the centre, thereby enabling measurement of angular surfaces also.

The following are the typical applications of profile projectors:
1. Inspection of elements of gears and screws
2. Measurement of pitch circle diameters of holes located on components
3. Measurement of unusual profiles on components such as involute and cycloidal, which are difficult to measure by other means
4. Measurement of tool wear (Drawing of a tool to scale is made on a tracing sheet. The tracing sheet is clamped on to the screen. Now, the used tool is fixed on the table and the image is projected to the required magnification. Using a pencil, one can easily trace the actual profile of the tool on to the tracing sheet. This image superimposed on the actual drawing is useful for measuring the tool wear.)

6.6 Electrical Comparators

Electrical and electronic comparators are in widespread use because of their instantaneous response and convenience in amplifying the input. An electronic comparator, in particular, can achieve an exceptionally high magnification of the order of 10^5:1 quite easily. Electrical and electronic comparators mainly differ with respect to magnification and type of output. However, both rely on mechanical contact with the work to be measured.

Electrical comparators generally depend on a Wheatstone bridge circuit for measurement. A direct current (DC) circuit comprising four resistors, two on each arm, is balanced when the ratios of the resistances in the two arms are equal. Displacement of the sensing element, a plunger, results in an armature connected to one of the arms of the bridge circuit to cause an imbalance in the circuit. This imbalance is registered as an output by a galvanometer, which is calibrated to read in units of linear movement of the plunger. Magnifications of the order $10^4:1$ are possible with electrical systems. The block diagram given in Fig. 6.12 illustrates the main elements of an electrical comparator.

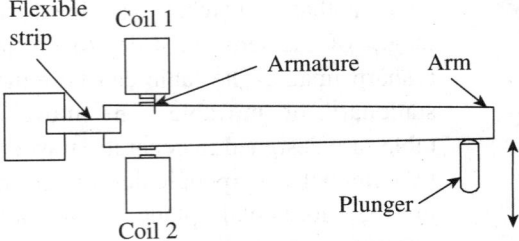

Fig. 6.12 Elements of an electrical comparator

The plunger is the sensing element, the movement of which displaces an armature inside a pair of coils. Movement of the armature causes change in inductance in the two coils, resulting in a net change in inductance. This change causes an imbalance in the bridge circuit, resulting in an output. The output display device, whether analog or digital, is calibrated to show the readings in units of length, that is, linear displacement.

A linear variable differential transformer (LVDT) is one of the most popular electromechanical devices used to convert small mechanical displacements (of the order of a few millimetres or fractions of a millimetre) into amplified electrical signals.

6.6.1 Linear Variable Differential Transformer

An LVDT provides an alternating current (AC) voltage output proportional to the relative displacement of a transformer core with respect to a pair of electrical windings. It provides a high degree of amplification and is very popular because of its ease of use. Moreover, it is a non-contact-type device, where there is no physical contact between the plunger and the sensing element. As a consequence, friction is avoided, resulting in better accuracy and long life for the comparator. It can be conveniently packaged in a small cartridge. Figure 6.13 illustrates the construction of an LVDT.

An LVDT produces an output proportional to the displacement of a movable core within the field of several coils. As the core moves from its 'null' position, the voltage induced by the coils change, producing an output representing the difference in induced voltage. It works on the mutual inductance principle. A primary coil and two secondary coils, identical to each other, are wound on an insulating form, as shown in Fig. 6.13. An external AC power source is applied to the primary coil and the two secondary coils are connected together in phase opposition. In order to protect the device from humidity, dust, and magnetic influences, a shield of ferromagnetic material is spun over the metallic end washers. The magnetic core is made of an alloy of nickel and iron.

The motion of the core varies the mutual inductance of secondary coils. This change in inductance determines the electrical voltage induced from the primary coil to the secondary coil. Since the secondary coils are in series, a net differential output results for any given position of the core. Figure 6.14 illustrates the characteristic curve of an LVDT. This curve shows the relationship between the differential output voltage and the position of the core with respect to the coils. It can be seen from this graph that if the core is centred in the middle of the two secondary windings, then the voltage induced in both the secondary coils will be equal in magnitude but opposite in phase, and the net output will be zero.

An output voltage is generated when the core moves on either side of the null position. Theoretically, output voltage magnitudes are the same for equal core displacements on either

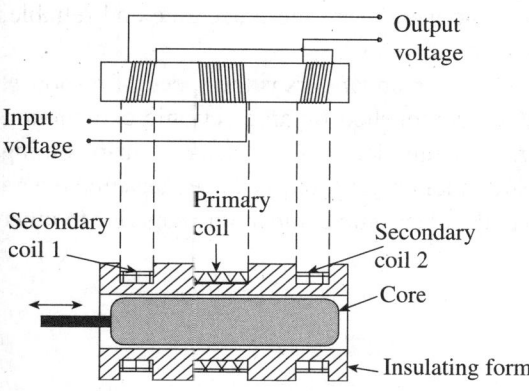

Fig. 6.13 Construction details of an LVDT

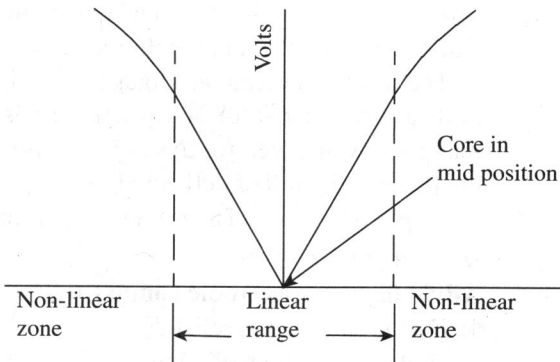

Fig. 6.14 Characteristic curve of an LVDT

side of the null balance. However, the phase relation existing between power source and output changes 180° through the null. Therefore, it is easy, through phase determination, to distinguish between outputs resulting from displacements on either side of the null. For such displacements, which are within the linear range of the instrument, output voltage is a linear function of core displacement. However, as Fig. 6.14 indicates, the linear range of the instrument is limited. Care should be taken to ensure that the actual measurement ranges are limited to the linear range of the LVDT.

Sensitivity of an LVDT is stated in terms of millivolts output per volt input per 1 mm core displacement. The per-volt input voltage refers to the exciting voltage that is applied to the circuit. Sensitivity varies from 0.1 to 1.5 mV for a range varying from 0.01 to 10 mm of core displacement. Sensitivity is directly proportional to excitation voltage, frequency of input power, and number of turns on the coils. An LVDT enjoys several distinct advantages compared to other comparators.

Advantages of LVDTs

1. It directly converts mechanical displacement into a proportional electrical voltage. This is unlike an electrical strain gauge, which requires the assistance of some form of elastic member.
2. It cannot be overloaded mechanically. This is because the core is completely separated from the remainder of the device.
3. It is highly sensitive and provides good magnification.
4. It is relatively insensitive to temperature changes.
5. It is reusable and economical to use.

The only disadvantage of an LVDT is that it is not suited for dynamic measurement. Its core has appreciable mass compared, for example, to strain gauges. The resulting inertial effects may lead to wrong measurements.

6.6.2 Electronic Comparator

Generally, electrical and electronic comparators differ with respect to magnification and type of output. However, both rely on the mechanical contact with the work to be measured. While the electronic comparator is more complex, advances in integrated circuits have reduced the size and power consumption of the equipment. Electronic gauges are more accurate and reliable, which has made them the preferred choice in many applications.

The most significant advantage offered by electronic comparators is the speed of response. A measurement rate of 500 per minute is easily accomplished by an electronic comparator, making it well suited for *dynamic measurement*. For example, the thickness of a strip coming out of a rolling mill or deflection of a machine part under varying loads can be measured over a given period of time. The following advantages make electronic comparators superior to other types of comparators.

Advantages of electronic comparators
1. High accuracy and reliability
2. High sensitivity in all ranges
3. High speed of response
4. Easy provision for multiple amplification ranges
5. Versatility (a large number of measurement situations can be handled with standard accessories)
6. Easy integration into an automated system

Continuous output of an electronic comparator can be achieved in numerous ways. However, the basic principle involved is that the AC voltage applied to the gauge head is altered by deflection of a mechanical spindle. The spindle deflection is in response to the size of the workpiece being measured. The following are some of the means employed for electronic gauging.

Methods of electronic gauging
1. One of the earliest electronic comparators employed a movable, spindle-actuated armature whose change in position caused an imbalance of a bridge circuit. This imbalance resulted in the movement of a meter hand.
2. A second system uses a movable coil (primary). As this coil is linearly displaced with respect to two stationary secondary coils, there is a change in output voltage. This change in voltage, in turn, displaces a meter hand accordingly. Magnification of up to 100,000× can be achieved in this way.
3. Many electronic gauges use the LVDT principle to achieve high magnification.
4. Another electronic system employs a capacitive-type gauge head. In this type of comparator, electrical energy is applied to two metal plates, one of them movable with the spindle. As the air gap between the plates changes, the change in capacitance results in an imbalance of the bridge circuit. This changes the output voltage, which is further amplified electronically and displayed in either analog or digital form.
5. Movement at the probe tip actuates the inductance transducer, which is supplied with an AC from an oscillator. Movement of the probe changes oscillator frequency, which is demodulated and fed to an electronic comparator circuit. The change in frequency with respect to the preset frequency is a measure of linear displacement.

While the foregoing methods are employed in various electronic gauges with different levels of sophistication, the 'Sigma electronic comparator' is quite popular in metrology laboratories. It uses the principle explained in method 5.

Sigma Electronic Comparator

Figure 6.15 illustrates the components of a Sigma electronic comparator.

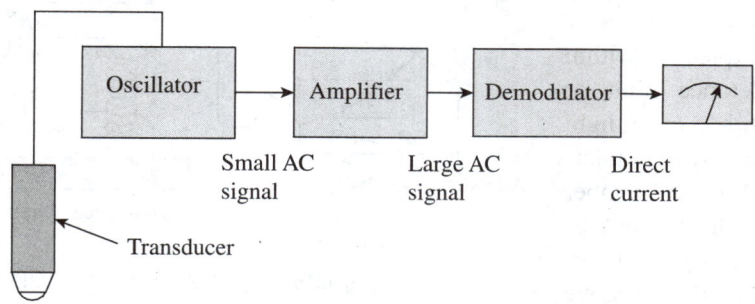

Fig. 6.15   Components of an electronic comparator

The movement at the probe tip actuates the inductance transducer, which is supplied with an AC source from the oscillator. The transducer converts this movement into an electrical signal, which is then amplified and fed via an oscillator to the demodulator. The current, in DC form, then passes to the meter and the probe tip movement is displayed as a linear measurement over a circular scale. Various measuring and control units can be incorporated, which provide for a wide range of single or multiple measurements to be made simultaneously. Using various adaptors to suit the work, the comparator can be put to many applications such as external and internal gauging, flatness testing, thickness gauging, and tube wall thickness.

Figure 6.16 illustrates the various parts of a Sigma comparator, which is commercially available. The set-up consists of a transducer stand and a display unit. The transducer stand consists of a mounting arrangement for the plunger, which moves inside a ball bushing free of friction. The plunger housing is fixed to a horizontal platform, which can be moved up or down, thanks to a nut-and-screw arrangement. The platform can be raised to the required height by loosening the nut and clamped in position by tightening the nut. Once the main nut is tightened, there may be a small shift in the position of the plunger, which can be made up by operating the fine adjustment knob. The plunger is held against a light spring load to ensure that it makes a firm contact with the workpiece while the reading is being taken.

The display unit comprises all the electronics. It consists of a needle moving over a circular scale, several knobs for range selection, zero setting and other adjustments, and light indicators to display the inspection results. To start with, the standard, which may be a master component or a slip gauge, is loaded below the plunger and a light contact is made. The appropriate range is selected. The range may vary from micron to millimetre levels. The user has to select the range depending on the level of tolerance required. Now, the zero setting knob is operated to set the scale to read zero.

In most of the instruments, there is a provision to set either the unilateral or the bilateral tolerance in the display unit. A couple of knobs are provided, which can be operated to set the tolerance. The procedure advocated by the manufacturer has to be diligently followed to set these values accurately. Now the standard is taken out and the workpieces are inserted below the plunger one after the other, in order to carry out inspection. The deviation from the set value can be directly read on the scale. More importantly, the light indication on the display unit conveys whether the workpiece is within the tolerance limit or not. While a green light indicates 'within toler-ance', red and amber lights indicate 'out of tolerance'. Out of toler-ance on the positive side is indicated by the amber light, which means that the workpiece has excess material and can be further processed and brought to within tolerance limits. Out of tolerance on the negative side is indicated by the red light, which means that the workpiece has less material than even the minimum material condition and has to be scrapped.

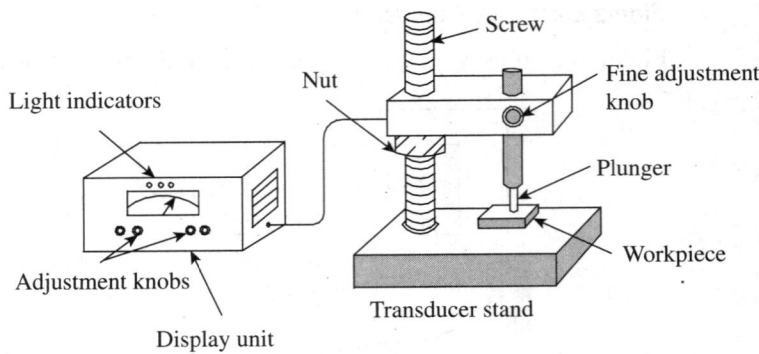

Fig. 6.16 Sigma electronic comparator

The Sigma electronic comparator (Fig. 6.16) is extremely popular in inspection processes because of the following reasons:
1. It is easy to use and provides a convenient means of measurement.
2. It has a high degree of accuracy and repeatability.
3. It has a provision to set several ranges of tolerances very easily.
4. Light indications on its display unit enable fast inspection, since the inspector of components does not have to refer to the scale every time.
5. It can be easily integrated with a computer or micro-controller. Therefore, inspection data can be recorded for further analysis.

6.7 PNEUMATIC COMPARATORS

Pneumatic comparators use air as a means of measurement. The basic principle involved is that changes in a calibrated flow respond to changes in the part feature. This is achieved using several methods and is referred to as pneumatic gauging, air gauging, or pneumatic metrology. Since a pneumatic gauge lends itself to the gauging of several features at once, it has become an indispensable part of production inspection in the industry. It is possible to gauge length, diameter, squareness, parallelism, taper, concentricity, etc., using a simple set-up. For instance, if one is inspecting the bore of an engine cylinder, it is also possible to assess its size, taper, camber, and straightness in the same setting.

Pneumatic metrology is quite popular because of several advantages: absence of metal-to-metal contact, higher amplification, and low cost. Absence of metal-to-metal contact between the gauge and the component being inspected greatly increases the accuracy of measurement. The gauge also has greater longevity because of a total absence of wearable parts. Amplification may be increased without much reduction in range, unlike mechanical or electronic instruments. However, similar to electronic comparators, amplification is achieved by application of power from an external source. Hence, a pneumatic comparator does not depend on the energy imparted to the pick-up element by contact with the component being inspected. Table 6.1 presents functional and metrological features of pneumatic comparators.

Table 6.1 Functional and metrological features of pneumatic comparators

Functional features	Metrological features
1. No wearing of parts	1. Non-contact inspection of work parts
2. Rapid response	2. Minimum gauging force
3. Remote positioning of gauge heads	3. Both variable inspection (variability in size) and attribute inspection (GO and NO GO type) possible
4. Self-cleansing of heads and parts	
5. No hysteresis	4. High range of amplification
6. Scope for inspecting diverse part features	5. Suited for varied inspections such as length, position, and surface topography
7. Compact size of gauge head	

Pneumatic comparators are best suited for inspecting multiple dimensions of a part in a single setting ranging from 0.5 to 1000 mm. It is also amenable for on-line inspection of parts on board a machine tool or equipment. Based on the type of air gauge circuit, pneumatic gauges can be classified as *free flow gauges* and *back pressure gauges*. The back pressure gauge was developed first, but the free flow gauge is in greater use.

6.7.1 Free Flow Air Gauge

This uses a simple pneumatic circuit. Compressed air with a pressure in the range 1.5–2 bar is passed through a tapered glass column that contains a small metal float. The air then passes through a rubber or plastic hose and exits to the atmosphere through the orifice in the gauging head. Since the gauging head is inserted inside the work part that is being inspected, there is a small clearance between the gauging head and the work part. This restricts the flow of air, thereby changing the position of the float inside the tapered glass column. The set-up is illustrated in Fig. 6.17. Compressed air from the factory line is filtered and reduced to the required pressure. A shut-off valve is provided to ensure shut-off of air supply when not in use. Air bleed and zero adjustment screws are provided to facilitate calibration of the gauge. The gauge head is mounted onto a handle, which provides a convenient way of handling the gauge head during inspection.

As mentioned, the amount of clearance between the gauge head and the work part determines the rate of air flow in the glass column, which in turn regulates the position of the float. Figure 6.18 illustrates the relationship between the clearance and the flow rate. It is clear from the graph that the flow rate increases with the increase in the clearance. The curve has a linear portion, which is used for the purpose of measurement. This linearity in the gauging range permits dimensional variation to be accurately measured up to 1 µm. A calibrated scale enables the reading to be directly read from the scale. Amplification of up to 100,000:1 has been built into these gauges. High amplification and long range permit easier and accurate readings.

Another significant advantage of this type of pneumatic gauge is that it is relatively free from operator error. Gauge readings will be uniform from operator to operator because they do

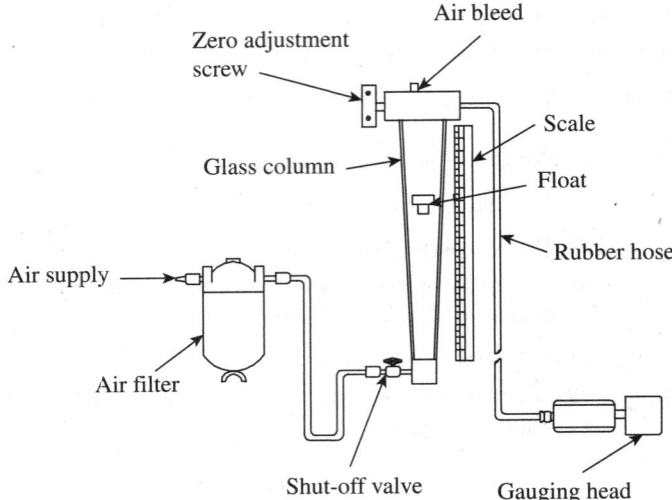

Fig. 6.17 Free flow air gauge

not depend on a high degree of operator skill or sense of feel during gauging. A typical gauge head has two orifices diametrically opposite each other, as shown in Fig. 6.19. If the spindle of the gauge head is moved to one side, the air flow is decreased; however, the air flow through the diametrically opposite orifice increases by an equal amount. The air gaps in position 1 are a on the left side and b on the right side, with b being more pronounced than a. In position 2, while the gauging head has slightly shifted towards the right, the air gap c on the left side is more pronounced than d, the air gap on the right side. While a lesser air gap puts a restriction on air flow through one orifice, the corresponding increase in the air gap for the other orifice results in a balancing act, which ensures that the net air flow rate remains constant. Thus, regardless of small variation in positioning the gauge head inside the work part, the same reading is ensured on the scale.

It will be interesting to the reader to probe the actual mechanics involved in measurement, which is based on the position of the float in the tapered glass column. The float takes up a position in the tapered tube such that the air velocity through the 'annulus' created by the float and the tube is constant. The air then escapes through the gauging orifice. In order to use the gauge as a comparator, the user uses a master gauge of known dimension and geometric form, and sets the float to a reference value by adjusting the air flow rate. In other words, the air gauge is set to a *datum* rate of air flow through the system. Now, the master gauge is taken out and the gauge head is inserted into the work part being inspected. Any variation in the dimension of the work part will produce a variation in the rate of flow through the system. This is reflected in the change in height of the float in the glass column, and the difference in dimension can be directly read on the graduated scale.

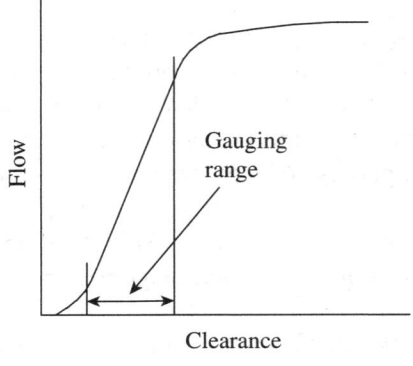

Fig. 6.18 Flow–Clearance curve

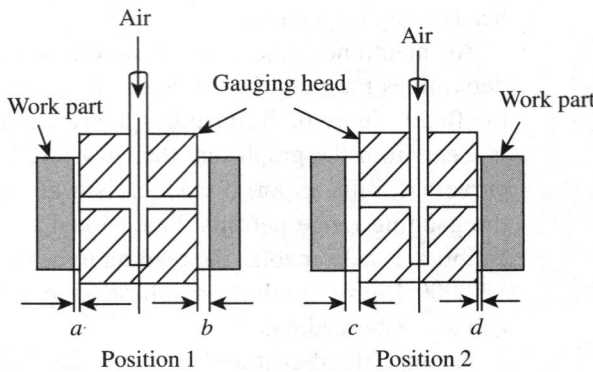

Fig. 6.19 Air gaps in air gauging

6.7.2 Back Pressure Gauge

This system uses a two-orifice arrangement, as shown in Fig. 6.20. While the orifice O_1 is called the *control orifice*, the orifice O_2 is referred to as the *measuring orifice*. The measuring head gets compressed air supply at a constant pressure P, which is called the *source pressure*. It passes through the control orifice into an intermediate chamber. Air exits the measuring head through the measuring orifice. While the size of the control orifice remains constant, the effective size of the measuring orifice varies because of the gap d between the measuring orifice and the work surface. Depending on the gap d, the *back pressure* P_b changes, thereby providing a means for measuring dimension d.

The indicating device is essentially a pressure gauge or a manometer, which can be calibrated to read in terms of the linear deviation. By suitably selecting the values of O_1, O_2, and P, the pressure P_b may be made to vary linearly for any change in gap d. Figure 6.21 shows the characteristic curve of a back pressure gauge. Assuming that the areas of control orifice and measuring orifice are A_1 and A_2 respectively, the relationship between the ratio of back pressure to source pressure and the ratio of the areas of control orifice to measuring orifice is almost linear for P_b/P values from 0.5 to 0.8. This range is selected for the design of the back pressure gauge.

Figure 6.22 illustrates the construction details of a back pressure gauge. Compressed air is filtered and passed through a pressure regulator. The regulator reduces the pressure to about 2 bar. The air at this reduced pressure passes through the control orifice and escapes to the atmosphere through the orifice of the measuring head.

Alternatively, the air pressure can be reduced and maintained at a constant value by passing it through a dip tube into a water chamber, the pressure setting being done by the head of water displaced. The excess air is let off to the atmosphere.

Depending on the clearance between the measuring head and the work part surface, back pressure is created in the circuit, which, as already pointed out, has a direct relationship with the effective area of the measuring orifice. Various transducers are available to display the linear gap between the measuring head and the work part. In the set-up shown in Fig. 6.22, the back pressure is let into a bourdon tube, which undergoes deflection depending on the magnitude of air pressure. This deflection of the bourdon tube is amplified by a lever and gear arrangement, and indicated on a dial. The reading can be directly taken from a calibrated scale, as shown

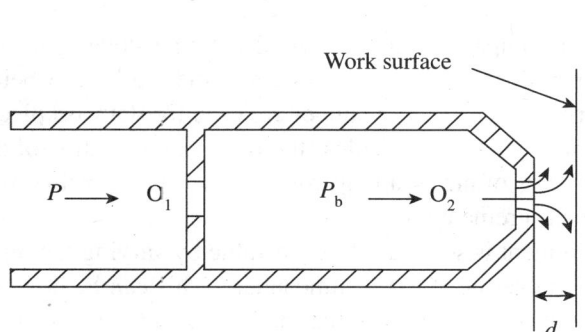

Fig. 6.20 Working principle of a back pressure gauge

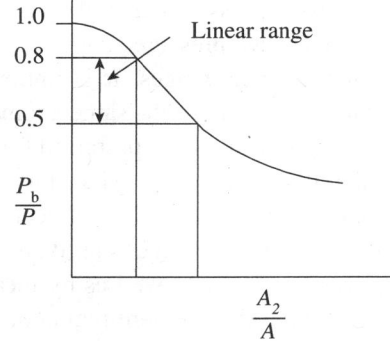

Fig. 6.21 Characteristic curve of a back pressure gauge

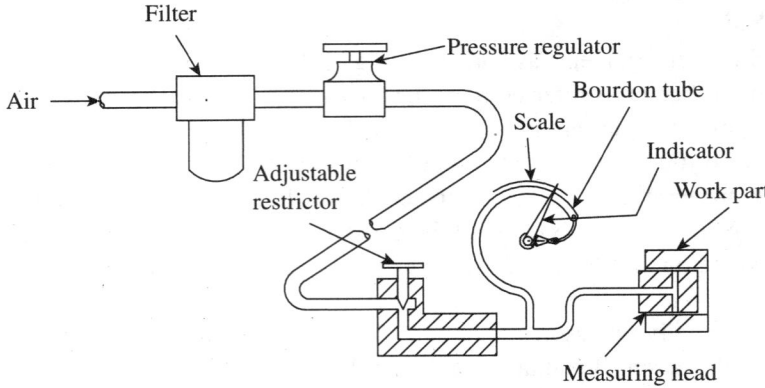

Fig. 6.22 Back pressure gauge

in the figure. Magnification of up to 7500:1 can be achieved by employing the bourdon tube principle. Readings up to 0.01 mm is common in most of the gauges.

There are several other methods of transducing the change in back pressure into the displacement of a dial gauge pointer over a scale. Some of the commonly used methods are discussed here:

Water column back pressure gauge   In this gauge, the back pressure is indicated by the head of water displaced in the manometer tube. The tube is graduated linearly to show the changes in pressure resulting from the changes in the gap d. Amplification of up to 50,000 can be obtained in this system. In some gauges, the back pressure actuates a relay whose output is read by a well-type water manometer or a dial indicator.

Differential back pressure gauge   In this system, regulated air passes through two channels—one going to the bellows cavity and gauge head, and the other partially exhausting to the atmosphere for zero setting and terminating in the bellows. Restricting air flow at the measuring head causes a pressure differential that is registered by a dial indicator.

Venturi back pressure gauge   A venturi tube is placed between the air supply and the measuring head. The restriction of air flow at the measuring head causes a difference in the two sections of the venturi tube. This results in a higher rate of flow in the narrower channel. The pressure difference actuates bellows, which is amplified by mechanical means resulting in the movement of the dial indicator over a calibrated scale.

Most of these gauges are also equipped with electrical contacts and relays to activate signal lights when preset dimensional limits are reached. This can speed up the inspection process since the operator need not observe the readings displayed on a scale. The response speed of a back pressure systems is less than that of a free flow system because there is a time lag before the back pressure can build up.

The back pressure gauge is essentially a comparator, and the initial setting is done by means of reference gauges. It is important for both the reference gauge and the workpiece being inspected to have the same geometric form. Therefore, slip gauges are used for flat workpieces and ring gauges are preferred for cylindrical workpieces. Unless there is a conformation of the geometric form, the expansion characteristics of air escaping from the measuring orifice will change, resulting in loss of accuracy in measurements.

The master gauge is used and the instrument is set to a reference value by varying the input pressure of air as well as by means of variable bleed to the atmosphere. This can be done by operating the pressure regulator. Air pressure is adjusted so that the instrument is set to some datum value on the scale. Now, the reference gauge is taken out and the workpiece is introduced

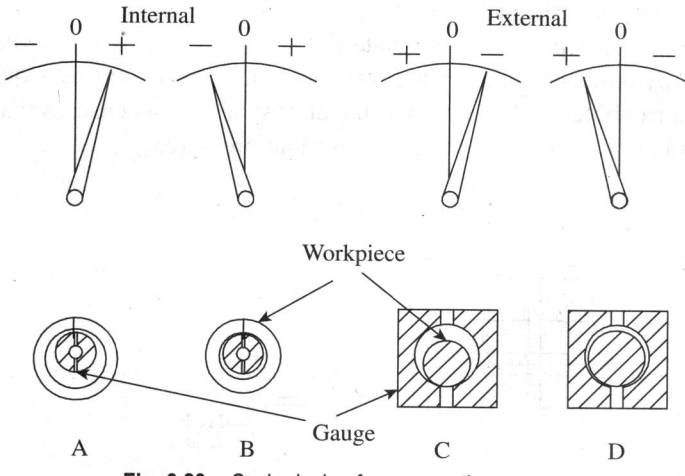

Fig. 6.23 Scale design for pneumatic gauges

with the measuring gauge. The deviation in dimension can be directly read on the scale.

Scale Selection

The general rule for selection of a scale for pneumatic comparators is to select a scale that will contain the tolerance of the dimension to be measured entirely. This is a tricky thing to balance, because closer tolerances, on the one hand, and readability, on the other hand, may result in the need to have quite long scales, which may become unwieldy to use. Whenever space is limited, scale divisions must be crowded together, affecting readability. The designer has to strike a compromise between the two. In any case, separate scales are needed for internal and external measurements because the relationship between flow rate and plus–minus sign is reversed. Figure 6.23 illustrates the design of scales for a bourdon-tube-type back pressure gauge.

In case A, the gauging head is in a large hole. There is lesser restriction on the air escaping through the measuring orifice. It races through the system, resulting in a greater increase in the radius of curvature of the bourdon tube and carries the needle to the plus side of the scale. In case B, the same element is closely confined by a smaller hole. The flow is restricted, resulting in lesser flow rate of air. Therefore, the radius of curvature of the bourdon tube reduces, resulting in a negative value being registered on the scale. Thus, the gauge registers a plus reading when the hole is large and a minus reading when the hole is small.

In cases C and D, the matter is reversed; the part feature is a diameter and the jets are in a ring around it. The flow is rapid when the feature is small, a reverse of the previous case. In case C, the smaller part results in a larger air flow. On the other hand, a larger part shown in case D restricts flow. Thus, for external measurement, minus represents a large flow and plus represents a small flow. This difference between inside and outside measurements exists for all measurements. Only pneumatic comparators permit quick scale changes to facilitate measurement. The actual reading of the scales is quite simple. They are all marked with amplification, least count, and range. In order to select the proper scale, the user should decide on the sensitivity and magnification required for a particular inspection.

6.7.3 Solex Pneumatic Gauge

This air gauge has been developed and marketed by Solex Air Gauges Ltd, USA, and is one of the most popular pneumatic comparators in the industry. The Solex pneumatic gauge is generally used for the inspection of internal dimensions, although it is also used for external measurements with suitable attachments. Figure 6.24 illustrates the construction details of this comparator. Compressed air is drawn from the factory air supply line, filtered, and regulated to a pressure of about 2 bar. Air will now pass through a dip tube immersed in a glass water tank. The position of the dip tube in terms of depth H will regulate the effective air pressure in the system at the input side. Extra air, by virtue of a slightly higher supply air pressure, will leak out of the water tank in the form of air bubbles and escape into the atmosphere. This ensures that the air moving towards the control orifice will be at a desired constant pressure.

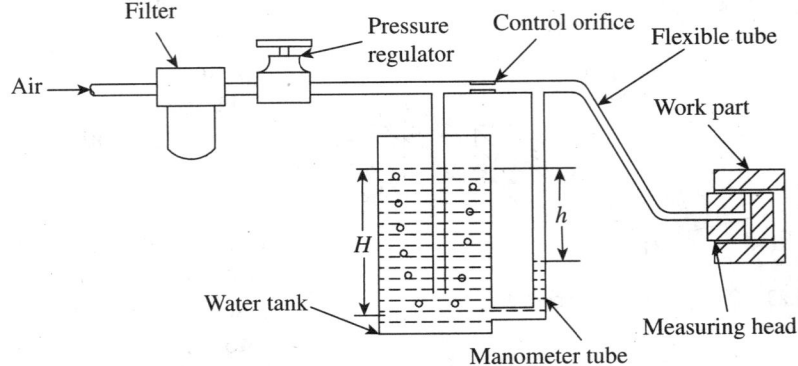

Fig. 6.24 Solex pneumatic gauge

The air at a reduced pressure then passes through the control orifice and escapes from the measuring orifice in the measuring head. Based on the clearance between the work part and the measuring orifice, a back pressure is created, which results in the head of water being displaced in the manometer tube. As we have already seen, within a limited measuring range, change in pressure varies linearly with change in internal dimension of the work part. Therefore, the change in linear dimension can be directly read from a linearly calibrated scale. The Solex comparator has a high degree of resolution, and variation in dimension up to a micrometre can be determined easily. Amplification of up to 50,000 is obtainable in this gauge.

6.7.4 Applications of Pneumatic Comparators

Pneumatic gauging is one of the widely used methods for inspection of holes. While it comprises relatively simple elements such as air filters, glass columns, manometer tubes, and bourdon tubes, the inspection can be carried out with an accuracy up to 1 μm. The gauging elements can be adapted to measure nearly any feature of the hole, including diameter, roundness, squareness, and straightness. Figure 6.25 illustrates the use of a single-jet nozzle, which can be used to carry out a variety of inspections.

The gauging element in pneumatic metrology can be classified into three types: type 1, type 2, and type 3. In type 1, the hole being measured is the exit nozzle of the gauging element. This (as illustrated in Fig. 6.26a) is only suitable for inside measurement and is used when the cross-sectional area is to be controlled rather than the shape. Typical applications include inspection of automobile cylinder bores, nozzle of carburettor, etc.

The gauging element in type 2 is illustrated in Fig. 6.26(b). In this case, an air jet not in contact with the part is the gauging element. The rate of flow of air depends on the cross-sectional area of the nozzle and the clearance between the nozzle and the part features. In other words, it is basically an air jet placed close to the part.

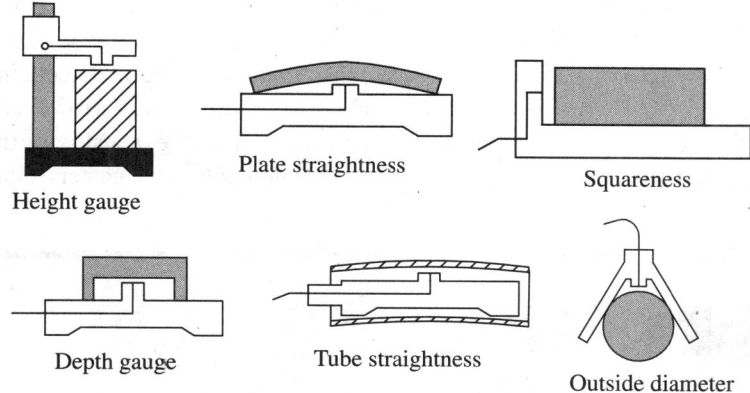

Height gauge

Plate straightness

Squareness

Depth gauge

Tube straightness

Outside diameter

Fig. 6.25 Use of a single-jet nozzle for inspection

In type 3, the air jet is mechanically actuated by contact with the part. This is more suited for attribute inspection (GO and NO GO type). It is compact and can replace an LVDT. It incorporates an air valve that changes the air flow in proportion to the linear change. This is often used interchangeably with an electronic gauge head.

The pneumatic gauging head may have one or more measuring orifices. Accordingly, a gauging head with a single orifice results in the indicator needle moving to either the positive or the negative side, depending on the variation in gap between the orifice and the work part. However, two opposing orifices in the measuring head can provide differential measurement. The clearance with respect to both the orifices will get added up, resulting in an equivalent gap. By rotating the measuring head, characteristics, for example, out-of-roundness can be reliably measured. Figure 6.27 illustrates four types of gauging heads with one, two, three, and four measuring orifices. Table 6.2 lists the typical applications of each.

Table 6.2 Applications of multiple-orifice gauging heads

Type	No. of orifices	Applications
A	1	Checking concentricity, location, squareness, flatness, straightness, length, and depth
B	2	Checking inside diameter, out-of-roundness, bell-mouth, and taper
C	3	Checking triangular out-of-roundness (lobbing)
D	4	Furnishing average diameter readings in a single setting

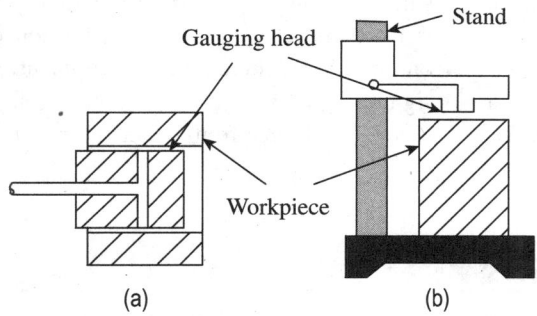

Gauging head

Stand

Workpiece

(a)

(b)

Fig. 6.26 Types of pneumatic gauging elements
(a) Type 1 (b) Type 2

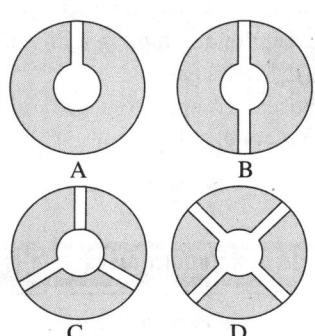

A

B

C

D

Fig. 6.27 Types of gauging heads

Pneumatic comparators are always preferred for the inspection of small holes (<15 mm). Precision up to 10 μm is obtained easily. Pneumatic comparators are preferred for larger holes as well, because they provide a number of desirable features such as high amplification, excellent readability, non-contact operation, and absence of mechanical parts, among others. However, pneumatic comparators suffer from three limitations: short measurement range, sensitivity to surface finish of work parts, and the need for expensive gauging elements and masters that offset the low instrument cost.

A QUICK OVERVIEW

- A comparator works on the principle of relative measurement. It gives only dimensional differences in relation to a basic dimension or master setting.
- In *direct measurement*, precision is dependent on the least count of the scale and the means for reading it. In *comparison measurement*, it is dependent on the least count of the standard and the means for comparing.
- Comparators are classified as mechanical comparator, mechanical–optical comparator, electrical and electronic comparator, pneumatic comparator, and other types such as projection comparators and multi-check comparators.
- Optical comparators provide a high degree of precision in measurements due to reduction of moving members and better wear-resistance qualities.
- Electronic comparators are in widespread use because they give responses instantaneously and it is convenient to amplify the input. An electronic comparator in particular can achieve an exceptionally high magnification of the order of $10^5:1$ quite easily.
- Electrical comparators generally depend on a Wheatstone bridge circuit for measurement.

Displacement of the sensing element, a plunger, results in an armature connected to one of the arms of the bridge circuit to cause imbalance in the circuit. This imbalance is registered as an output by a galvanometer, which is calibrated to read in units of linear movement of the plunger.
- An LVDT directly converts mechanical displacement into a proportional electrical voltage. This is unlike an electrical strain gauge, which requires the assistance of some form of an elastic member.
- The Sigma electronic comparator is extremely popular in inspection processes because of advantages such as ease of usage, high degree of accuracy and repeatability, provision for setting tolerances during inspection, and ease of integration into a computer-integrated manufacturing system.
- A pneumatic comparator is best suited for inspecting multiple dimensions of a part in a single setting ranging from 0.5 to 1000 mm. It is also amenable to on-line inspection of parts on board a machine tool or equipment. Another significant advantage of a pneumatic gauge is that it is relatively free from operator error.

MULTIPLE-CHOICE QUESTIONS

1. In comparison measurement, precision depends on _____.
   (a) least count of the comparator
   (b) least count of the standard
   (c) least count of the scale of the instrument
   (d) all of these

2. In case of a comparator, measurement is done by
   (a) displacement method
   (b) interchange method
   (c) direct method
   (d) Parkinson method

3. A balanced dial in a dial gauge has
   (a) graduations in both metric and British systems
   (b) graduations starting from zero and extending to the end of the recommended range
   (c) graduations marked both ways of zero
   (d) graduations in logarithmic scale

4. How is the precision of a dial comparator determined?
   (a) By comparison of the readings with slip gauges
   (b) By dispersal of a series of readings
   (c) From the manufacturer's specifications
   (d) From the distance between dial graduations

5. _____ contact point is most preferred in dial gauges, since it presents point contact to the mating surface irrespective of whether it is flat or cylindrical.
   (a) Spherical
   (b) Flat
   (c) Tapered
   (d) Button

6. The basic principle of the Johansson mikrokator is based on
   (a) Johansson movement
   (b) Abbey movement
   (c) Abraham movement
   (d) Abramson movement

7. With respect to the Johansson mikrokator, which of the following statements is true?
   (a) Magnification varies inversely with the number of turns and width of the metal strip.
   (b) Magnification varies directly with the number of turns and width of the metal strip.
   (c) The more the number of turns of the strip, the higher the magnification.
   (d) The thicker the strip, the higher the magnification.

8. In a Sigma mechanical comparator, magnification is obtained
   (a) in a single stage
   (b) in two stages
   (c) in three stages
   (d) depending on the manufacturer's instruction

9. Double reflection of light using a pair of mirrors is a unique feature of
   (a) an autocollimator
   (b) a clinometer
   (c) a Zeiss ultra-optimeter
   (d) an optical projector

10. Which of the following comparators is best suited for inspection of small gears and screws?
    (a) Autocollimator
    (b) Profile projector
    (c) Johansson mikrokator
    (d) Zeiss ultra-optimeter

11. Which of the following comparators is non-contact type?
    (a) Johansson mikrokator
    (b) Mechanical optical comparator
    (c) Sigma comparator
    (d) LVDT

12. An LVDT works on the principle of
    (a) mutual inductance
    (b) mutual capacitance
    (c) mutual resistance
    (d) magnetic induction

13. Which of the following gauges is relatively free from operator's error?
    (a) Sigma mechanical comparator
    (b) Zeiss ultra-optimeter
    (c) Pneumatic gauge
    (d) All of these

14. Which of the following comparators can give amplification of up to 50,000?
    (a) LVDT
    (b) Solex pneumatic gauge
    (c) Dial gauge
    (d) Sigma electronic comparator

15. How many orifices will you recommend for pneumatic gauging of triangular out-of-roundness (lobbing)?
    (a) One
    (b) Two
    (c) Three
    (d) Four

REVIEW QUESTIONS

1. Compare comparison measurement with direct measurement.
2. Define a comparator. Discuss the functional requirements of a comparator.
3. Give the classification of comparators.
4. With the help of a neat sketch, explain the functional parts of a dial indicator.
5. Explain the working mechanism of a dial indicator.
6. Write a note on the various contact points used in a dial indicator.
7. What are the guidelines for the proper use of dial indicators?
8. Explain the factors that influence amplification in a Johansson mikrokator.
9. With the help of a neat sketch, explain the working principle of a Sigma mechanical comparator.
10. Explain the optical system in a mechanical optical comparator. What are its advantages when compared to a mechanical comparator?
11. Explain the function of a Zeiss ultra-optimeter.
12. Discuss the various metrological applications of an optical projector.
13. What is an LVDT? Explain its working principle. Discuss the characteristic curve of an LVDT with a sketch.
14. What are the major advantages of electronic comparators that have made them the first choice in inspection metrology?
15. Explain the working principle of a Sigma electronic comparator.
16. Discuss the functional and metrological features of pneumatic comparators.
17. Explain how the relationship between the clearance (between gauge head and work part) and air flow rate enables measurement in a free-flow-type pneumatic gauge.
18. A pneumatic gauge is relatively free from operator error. How do you justify this statement?
19. Explain the working principle of a pneumatic back pressure gauge. Discuss the relevance of the characteristic curve in measurement.
20. With the help of a figure, explain the working principle of a Solex pneumatic gauge.
21. Discuss the major applications of pneumatic gauges.
22. The operation of instruments requires energy. What is the source of energy for dial indicators, pneumatic comparators, and electrical comparators?
23. What is the most important way in which an electronic comparator differs from a mechanical comparator?
24. What is a balanced bridge? What is its significance in electrical comparators?
25. What are the limitations of pneumatic metrology?

Answers to Multiple-choice Questions

1. (b)	2. (b)	3. (c)	4. (b)	5. (a)	6. (d)	7. (a)	8. (b)
9. (c)	10. (b)	11. (d)	12. (a)	13. (c)	14. (b)	15. (c)	

Optical Measurement and Interferometry

After studying this chapter, the reader will be able to

- understand the basic principles of optical measurement
- explain the construction, measurement, and applications of optical instruments such as tool maker's microscope, optical projector, and optical square
- describe the phenomenon of interference and the formation of fringe bands
- elucidate how fringe bands are manipulated for linear measurement
- discuss the instrumentation available for making measurements using interference technique

7.1 INTRODUCTION

Today, it is an accepted fact that light waves provide the best standard for length. The significance of light waves as the length standard was first explored by Albert A. Michelson and W.L. Worley, although indirectly. They were using an interferometer to measure the path difference of light that passed through a tremendous distance in space. In their experiment, they measured the wavelength of light in terms of metre, the known standard then. They soon realized that the reverse was more meaningful—it made more sense to define a metre in terms of wavelengths of light. This aspect was soon recognized, as scientists began to understand that the wavelength of light was stable beyond any material that had hitherto been used for the standard. Moreover, they realized that light was relatively easy to reproduce anywhere.

Optical measurement provides a simple, easy, accurate, and reliable means of carrying out inspection and measurements in the industry. This chapter provides insights into some of the important instruments and techniques that are widely used. Although an autocollimator is an important optical instrument that is used for measuring small angles, it is not discussed here, as it has already been explained in Chapter 5.

Since optical instruments are used for precision measurement, the projected image should be clear, sharp, and dimensionally accurate. The design of mechanical elements and electronic controls should be compatible with the main optical system. In general, an optical instrument should have the following essential features:

1. A light source

2. A condensing or collimating lens system to direct light past the work part and into the optical system
3. A suitable stage or table to position the work part, the table preferably having provisions for movement in two directions and possibly rotation about a vertical axis
4. The projection optics comprising lenses and mirrors
5. A viewing screen or eyepiece to receive the projected image
6. Measuring and recording devices wherever required

When two light waves interact with each other, the wave effect leads to a phenomenon called *interference* of light. Instruments designed to measure interference are known as *interferometers*. Application of interference is of utmost interest in metrology. Interference makes it possible to accurately compare surface geometry with a master, as in the case of optical flats. Microscopic magnification enables micron-level resolution for carrying out inspection or calibration of masters and gauges. Lasers are also increasingly being used in interferometers for precision measurement. The first part of this chapter deals with a few prominent optical instruments such as the tool maker's microscope and optical projector. The latter part deals with the principle of interferometry and related instrumentation in detail.

7.2 OPTICAL MEASUREMENT TECHNIQUES

We are quite familiar with the most common application of optics, namely *microscope*. Biologists, chemists, and engineers use various types of microscopes, wherein the primary requirement is visual magnification of small objects to a high degree with an additional provision for taking measurements. Optical magnification is one of the most widely used techniques in metrology. However, optics has three other principal applications: in *alignment*, in *interferometry*, and as an absolute *standard* of length. An optical measurement technique to check alignment employs light rays to establish references such as lines and planes. Interferometry uses a phenomenon of light to facilitate measurements at the micrometre level. Another significant application of optics is the use of light as the absolute standard of length, which was discussed in Chapter 2.

7.2.1 Tool Maker's Microscope

We associate microscopes with science and medicine. It is also a metrological tool of the most fundamental importance and greatest integrity. In addition to providing a high degree of magnification, a microscope also provides a simple and convenient means for taking readings. This enables both absolute and comparative measurements. Let us first understand the basic principle of microscopy, which is illustrated in Fig. 7.1.

A microscope couples two stages of magnification. The *objective lens* forms an image of the workpiece at I_1 at the *stop*. The stop frames the image so that it can be enlarged by the *eyepiece*. Viewed through the eyepiece, an enlarged virtual image I_2 is obtained. Magnification at each stage multiplies. Thus, a highly effective magnification can be achieved with only moderate magnification at each stage.

Among the microscopes used in metrology, we are most familiar with the tool maker's microscope. It is a multifunctional device that is primarily used for measurement on factory shop floors. Designed with the measurement of workpiece contours and inspection of surface

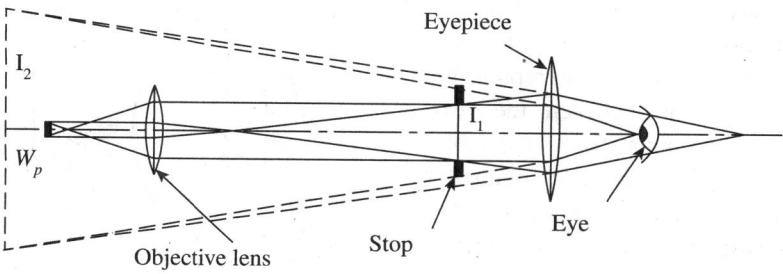

Fig. 7.1 Principle of microscopy

features in mind, a tool maker's microscope supports a wide range of applications from shop floor inspection, and measurement of tools and machined parts to precision measurement of test tools in a measuring room. The main use of a tool maker's microscope is to measure the shape, size, angle, and position of small components that fall under the microscope's measuring range. Figure 7.2 illustrates the features of a typical tool maker's microscope.

It features a vertical supporting column, which is robust and carries the weight of all other parts of the microscope. It provides a long vertical working distance. The workpiece is loaded on an *XY* stage, which has a provision for translatory motion in two principal directions in the horizontal plane. Micrometers are provided for both *X* and *Y* axes to facilitate linear measurement to a high degree of accuracy. The entire optical system is housed in the measuring head. The measuring head can be moved up and down along the supporting column and the image can be focused using the focusing knob. The measuring head can be locked into position by operating the clamping screw. An angle dial

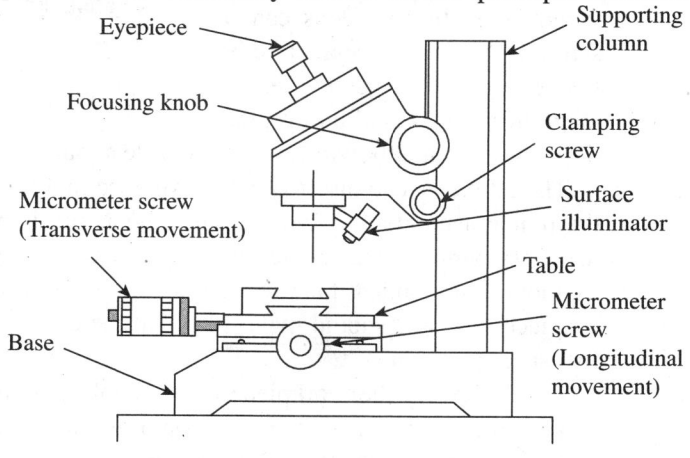

Fig. 7.2 Tool maker's microscope

built into the eyepiece portion of the optical tube allows easy angle measurement. A surface illuminator provides the required illumination of the object, so that a sharp and clear image can be obtained.

The element that makes a microscope a measuring instrument is the *reticle*. When the image is viewed through the eyepiece, the reticle provides a reference or datum to facilitate measurement. Specialized reticles have been developed for precise setting. A typical reticle has two 'cross-wires', which can be aligned with a reference line on the image of the workpiece. In fact, the term 'cross-wire' is a misnomer, because modern microscopes have cross-wires etched on glass. Figure 7.3 illustrates the procedure for linear measurement. A measuring point on the workpiece is aligned with one of the cross-wires and the reading R_1 on the microscope is noted down. Now, the *XY* table is moved by turning the micrometer head, and another measuring point is aligned with the same cross-wire. The reading, R_2 is noted down. The difference between the two readings represents the dimension between the two measuring points. Since the table can be moved in two mutually perpendicular directions (both in the longitudinal as well as transverse directions) using the micrometers, a precise measurement can be obtained. In some tool maker's microscopes, instead of a micrometer head, vernier scales are provided for taking readings.

Table 7.1 gives the details of lenses available in a 'Mitutoyo' tool maker's microscope. While the eyepiece is inserted in an eyepiece mount, the objective lens can be screwed into the optical tube. For example, an objective lens of magnification 2× and an eyepiece

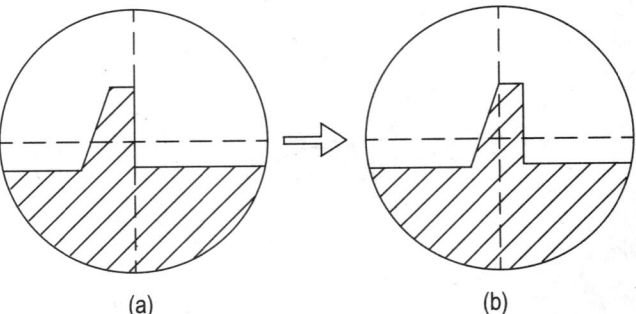

Fig. 7.3 Alignment of cross-wires with the measuring point (a) Reading R_1 (b) Reading R_2

Table 7.1 Lenses used in the Mitutoyo tool maker's microscope

Lens	Magnification	Working distance (mm)
Eyepiece	10×	–
	20×	–
Objective lens	2×	65
	5×	33
	10×	14

of magnification 20× will together provide a magnification of 40×.

The reticle is also inserted in the eyepiece mount. A positioning pin is provided to position the reticle accurately. A dioptre adjustment ring is provided in the eyepiece mount to bring the cross-wires of the reticle into sharp focus. The measuring surface is brought into focus by moving the optical tube up and down, with the aid of a focusing knob. Looking into the eyepiece, the user should make sure that the cross-wires are kept in ocular focus during the focusing operation.

Positioning of the workpiece on the table is extremely important to ensure accuracy in measurement. The measuring direction of the workpiece should be aligned with the traversing direction of the table. While looking into the eyepiece, the position of the eyepiece mount should be adjusted so that the horizontal cross-wire is oriented to coincide with the direction of the table movement. Now, the eyepiece mount is firmly secured by tightening the fixing screws. The workpiece is placed/clamped on the table and the micrometer head turned to align an edge of the workpiece with the centre of the cross-wires. Then, the micrometer is operated and the moving image is observed to verify whether the workpiece pavement is parallel to the measuring direction. By trial and error, the user should ensure that the two match perfectly.

Most tool maker's microscopes are provided with a surface illuminator. This enables the creation of a clear and sharp image. Out of the following three types of illumination modes that are available, an appropriate mode can be selected based on the application:

Contour illumination   This type of illumination generates the contour image of a workpiece, and is suited for measurement and inspection of workpiece contours. The illuminator is equipped with a green filter.

Surface illumination   This type of illumination shows the surface of a workpiece, and is used in the observation and inspection of workpiece surfaces. The angle and orientation of the illuminator should be adjusted so that the workpiece surface can be observed under optimum conditions.

Simultaneous contour and surface illuminations   Both contour and surface of a workpiece can be observed simultaneously.

Some of the latest microscopes are also provided with angle dials to enable angle measurements. Measurement is done by aligning the same cross-wire with two edges of the workpiece, one after the other. An angular vernier scale, generally with a least count of 6^1, is used to take the readings.

Applications of Tool Maker's Microscope

1. It is used in shop floor inspection of screw threads, gears, and other small machine parts.
2. Its application includes precision measurement of test tools in tool rooms.
3. It helps determine the dimensions of small holes, which cannot be measured with micrometers and callipers.
4. It facilitates template matching inspection. Small screw threads and involute gear teeth can be inspected using the optional template reticles.
5. It enables inspection of tapers on small components up to an accuracy of 6^1.

7.2.2 Profile Projector

The profile projector, also called the optical projector, is a versatile comparator, which is widely used for the purpose of inspection. It is especially used in tool room applications. It projects a two-dimensional magnified image of the workpiece onto a viewing screen to facilitate measurement. A profile projector is made up of three main elements: the projector comprising a light source and a set of lens housed inside an enclosure, a work table to hold the workpiece in place, and a transparent screen with or without a chart gauge for comparison or measurement of parts. A detailed explanation of a profile projector is given in Section 6.5.2 of Chapter 6, and the reader is advised to refer to it.

7.2.3 Optical Squares

An optical square is useful in turning the line of sight by 90° from its original path. Many optical instruments, especially microscopes, have this requirement. An optical square is essentially a pentagonal prism (pentaprism). Regardless of the angle at which the incident beam strikes the face of the prism, it is turned through 90° by internal reflection. Unlike a flat mirror, the accuracy of a pentaprism is not affected by the errors present in the mounting arrangement. This aspect is illustrated in Figs 7.4 and 7.5. It can be seen from Fig. 7.4 that

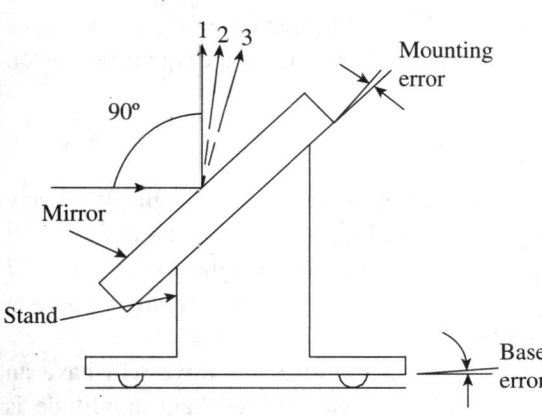

1 — Reflected ray without errors
2 — Reflected ray due to mounting error
3 — Reflected ray due to base error

Fig. 7.4   Mirror reflecting light by 90°

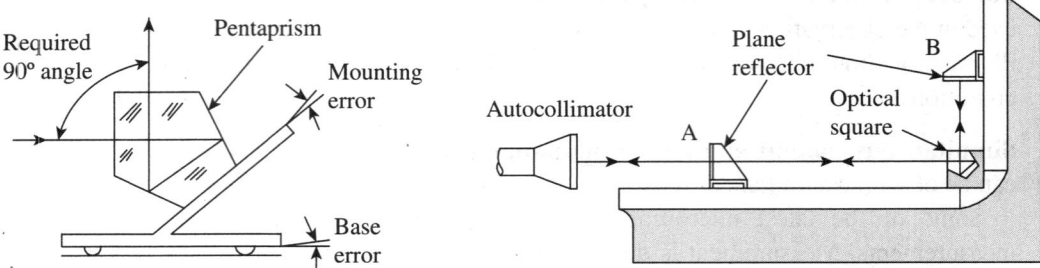

Fig. 7.5   Optical square   **Fig. 7.6**   Use of an optical square to test squareness

a mirror is kept at an angle of 45° with respect to the incident ray of light, so that the reflected ray will be at an angle of 90° with respect to the incident ray. It is observed that any error in the mounting of the mirror or in maintaining its base parallel, in a fixed reference, to the beam is greatly magnified by the optical lever effect. These two errors in combination may even be greater than the workpiece squareness error.

This problem may be overcome by using an optical square. Figure 7.5 illustrates the optical path through an optical square. The incident ray is reflected internally from two faces and emerges from the square at exactly 90° to the incident light. This is a remarkable property. Any slight deviation or misalignment of the prism does not affect the right angle movement of the light ray.

Optical squares are of two types. One type is fitted into instruments like telescopes, wherein an optical square is factory-fitted to ensure that the line of sight is perpendicular to the vertex. The second type comes with the necessary attachments for making adjustments to the line of sight. This flexibility allows optical squares to be used in a number of applications in metrology. Figure 7.6 illustrates the use of an optical square to test the squareness of machine slideways.

Squareness of the vertical slideway with respect to a horizontal slideway or bed is of utmost importance in machine tools. The test set-up requires an autocollimator, plane reflectors, and an optical square. It is necessary to take only two readings, one with the reflector at position A and a second at position B, the optical square being set down at the intersection of the two surfaces when the reading at B is taken. The difference between the two readings is the squareness error.

7.3 OPTICAL INTERFERENCE

A ray of light is composed of an infinite number of waves of equal wavelength. We know that the value of the wavelength determines the colour of light. For the sake of simplicity, let us consider two waves, having sinusoidal property, from two different light rays. Figure 7.7 illustrates the combined effect of the two waves of light. The two rays, A and B, are in phase at the origin O, and will remain so as the rays propagate through a large distance.

Suppose the two rays have amplitudes y_A and y_B, then the resultant wave will have an amplitude $y_R = y_A + y_B$. Thus, when the two rays are in phase, the resultant amplitude is maximum and the intensity of light is also maximum. However, if the two rays are out of phase, say by an amount δ, then the resultant wave will have an amplitude $y_R = (y_A + y_B) \cos \delta/2$. It is clear that the combination of the two waves no longer produces maximum illumination.

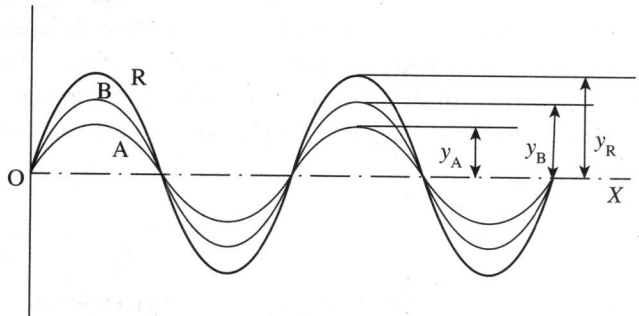

Fig. 7.7 Two waves of different amplitudes that are in phase

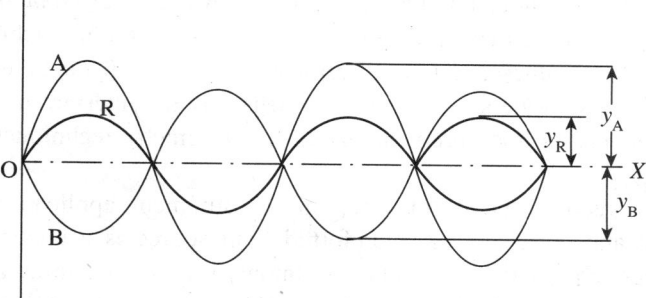

Fig. 7.8 Two waves of different amplitudes, out of phase by 180°

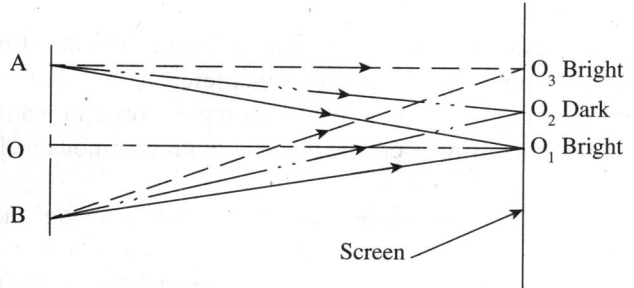

Fig. 7.9 Formation of fringes

Consider the case where the phase difference between the two waves is 180°. The amplitude of the resulting wave, which is shown in Fig. 7.8, is the algebraic sum of y_A and y_B. The corollary is that if y_A and y_B are equal, then y_R will be zero since cos(180/2) is zero. This means that complete *interference* between two waves having the same wavelength and amplitude produces darkness.

One of the properties of light is that light from a single source can be split into two component rays. Observing the way in which these components recombine shows us that the wave length of light can be used for linear measurement. The linear displacement δ between the wavelengths of the two light rays results in maximum interference when $\delta = \lambda/2$, where λ is the wavelength of light.

Now in what way is this property going to help us in taking linear measurements? Figure 7.9 illustrates how the property of interference of light can be used for linear measurement. Let us consider two monochromatic light rays from two point sources, A and B, which have the same origin. The light rays are made to fall on a flat screen that is placed perpendicular to the axis OO₁. The axis OO₁ is in turn perpendicular to the line joining the two point sources, A and B. Since both rays originate from the same light source, they are of the same wavelength. Let us also assume that the distances OA and OB are equal.

Now, consider convergence of two rays at point O₁ on the screen. Since the distances AO₁ and BO₁ are equal, the two rays are in phase, resulting in maximum illumination at point O₁. On the other hand, at point O₂, the distance BO₂ is longer than the distance AO₂. Therefore, by the time the two rays arrive at point O₂, they are out of phase. Assuming that the phase difference $\delta = \lambda/2$, where λ is the wavelength of light, complete interference occurs, forming a dark spot.

At point O₃ on the screen, the distance BO₃ is longer than AO₃. If the difference between the two distances, that is, BO₃ − AO₃, is equal to an even number of half wavelengths, the two light rays arriving at O_3 will be in phase, leading to the formation of a bright spot. This process repeats on either side of O_1 on the screen, resulting in the formation of alternate dark and bright areas. This pattern of alternate bright and dark areas is popularly known as fringes. The dark areas will occur whenever the path difference of A and B amounts to an odd number of half wavelengths, and the bright areas will occur when the path difference amounts to an even number of half wavelengths.

7.4 INTERFEROMETRY

It is now quite obvious to the reader that the number of fringes that appear in a given length on the screen is a measure of the distance between the two point light sources and forms the basis for linear measurement. This phenomenon is applied for carrying out precise measurements of very small linear dimensions, and the measurement technique is popularly known as *interferometry*. This technique is used in a variety of metrological applications such as inspection of machine parts for straightness, parallelism, and flatness, and measurement of very small diameters, among others. Calibration and reference grade slip gauges are verified by the interferometry technique. The instrument used for making measurements using interferometry technique is called an *interferometer*.

A variety of light sources are recommended for different measurement applications, depending on convenience of use and cost. The most preferred light source is a tungsten lamp with a filter that transmits monochromatic light. Other commonly used light sources are mercury, mercury 198, cadmium, krypton 86, thallium, sodium, helium, neon, and gas lasers. Among all the isotopes of mercury, mercury 198 is one of the best light sources, producing rays of sharply defined wavelength. In fact, the wavelength of mercury 198 is the international secondary standard of length.

Krypton-86 light is the basis for the new basic international standard of length. The metre is defined as being exactly 1,650,763.73 wavelengths of this light source, measured in vacuum. Gas lasers comprising a mixture of neon and helium produce light that is far more monochromatic than all the aforementioned sources. Interference fringes can be obtained with enormous path differences, up to 100 million wavelengths.

While optical flats continue to be the popular choice for measurement using the interferometry technique, a host of other instruments, popularly known as interferometers, are also available. An interferometer, in other words, is the extension of the optical flat method. While interferometers have long been the mainstay of dimensional measurement in physical sciences, they are also becoming quite popular in metrology applications. While they work according to the basic principle of an optical flat, which is explained in the Section 7.4.1 they provide additional conveniences to the user. The mechanical design minimizes time-consuming manipulation. The instrument can be fitted with additional optical devices for magnification, stability, and high resolution. In recent times, the use of lasers has greatly extended the potential range and resolution of interferometers.

7.4.1 Optical Flats

The most common interference effects are associated with thin transparent films or wedges bounded on at least one side by a transparent surface. Soap bubbles, oil films on water, and optical flats fall in this category. The phenomenon by which interference takes place is readily described in terms of an optical flat, as shown in Fig. 7.10.

An optical flat is a disk of high-quality glass or quartz. The surface of the disk is ground and lapped to a high degree of flatness. Sizes of optical flats vary from 25 to 300 mm in diameter, with a thickness ranging from 25 to 50 mm. When an optical flat is laid over a flat reflecting surface, it orients at a small angle θ, due to the presence of an air cushion between the two surfaces. This is illustrated in Fig. 7.10. Consider a ray of light from a monochromatic light source falling on the upper surface of the optical flat at an angle. This light ray is partially reflected at point 'a'. The remaining part of the light ray passes through the transparent glass material across the air gap and is reflected at point 'b' on the flat work surface. The two reflected components of the light ray are collected and recombined by the eye, having travelled two different paths whose length differs by an amount 'abc'.

If 'abc' = λ/2, where λ is the wavelength of the monochromatic light source, then the condition for complete interference has been satisfied. The difference in path length is one-half the wavelength, a perfect condition for total interference, as explained in Section 7.3. The eye is now able to see a distinct patch of darkness termed a fringe. Next, consider another light ray from the same source falling on the optical flat at a small distance from the first one. This ray gets reflected at points 'd' and 'e'. If the length 'def' equals 3λ/2, then total interference occurs again and a similar fringe is seen by the observer. However, at an intermediate point between the two fringes, the path difference between two reflected portions of the light ray will be an even number of half wavelengths. Thus, the two components of light will be in phase, and a light band will be seen at this point.

To summarize, when light from a monochromatic light source is made to fall on an optical flat, which is oriented at a very small angle with respect to a flat reflecting surface, a band of alternate light and dark patches is seen by the eye. Figure 7.11 illustrates the typical fringe pattern seen on a flat surface viewed under an optical flat. In case of a perfectly flat surface, the fringe pattern is regular, parallel, and uniformly spaced. Any deviation from this pattern is a measure of error in the flatness of the surface being measured.

Fringe patterns provide interesting insights into the surface being inspected. They reveal surface conditions like contour lines on a map. Figure 7.12 illustrates typical fringe patterns, and Table 7.2 offers useful hints about the nature of surfaces corresponding to the patterns. Once we recognize surface configurations from their fringe patterns, it is much easier to measure the configurations.

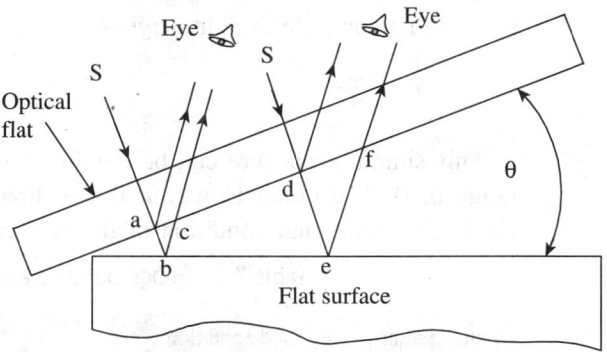

Fig. 7.10   Fringe formation in an optical flat

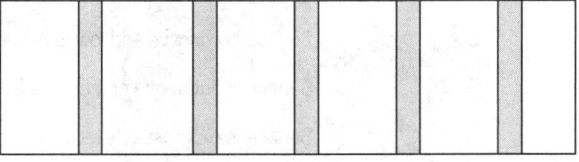

Fig. 7.11   Interference fringes

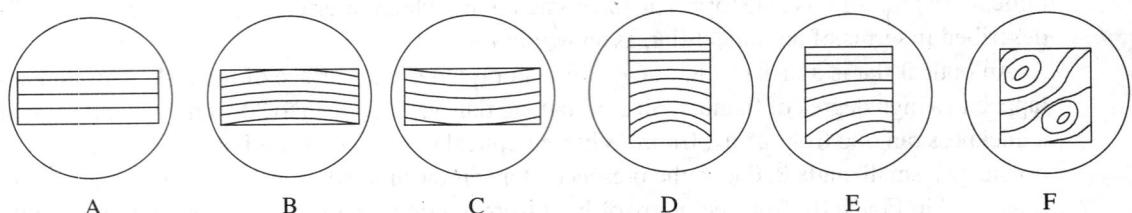

Fig. 7.12 Fringe patterns reveal surface conditions

Comparative Measurement with Optical Flats

One of the obvious uses of an optical flat is to check the heights of slip gauge blocks. The slip gauge that is to be checked is kept alongside the reference gauge on a flat table. An optical flat is then placed on top of both gauges, as shown in Fig. 7.13. Let us assume that A is the standard reference gauge block while B is the gauge block that is being inspected.

A monochromatic light source is used and the fringe patterns are observed with the help of a magnifying glass. It can be seen from the figure that the optical flat makes inclinations of θ and θ' with the top surfaces of the two slip gauges. Ideally, the two angles should be the same. However, in most cases, the angles are different by virtue of wear and tear of the surface of the slip gauge that is being inspected. This can easily be seen by looking at the fringe pattern that is formed on the two gauges, as seen from the magnified images. The fringes seen on both the gauges are parallel and same in number if both the surfaces are perfectly flat; otherwise, the number of fringes formed on the two gauges differs, based on the relationship between θ and θ'.

Now, let the number of fringes on the reference block be N over a width of l mm. If the distance between the two slip gauges is L and λ is the wavelength of the monochromatic light source, then the difference in height h is given by the following relation:

$$h = \frac{\lambda LN}{2l}$$

This simple procedure can be employed to measure very small height differences in the range of 0.01–0.1 mm. However, the accuracy of this method depends on the accuracy of the surface plate and condition of the surfaces of the specimen on which the optical flat is

Table 7.2 Fringe patterns and the resulting surface conditions

Fringe pattern	Surface condition
A	Block is nearly flat along its length.
B	Fringes curve towards the line of contact, showing that the surface is convex and high in the centre.
C	Surface is concave and low in the centre.
D	Surface is flat at one end but becomes increasingly convex.
E	Surface is progressively lower towards the bottom left-hand corner.
F	There are two points of contact, which are higher compared to other areas of the block.

resting. It is difficult to control the 'lay' of the optical flat and thus orient the fringes to the best advantage. The fringe pattern is not viewed from directly above, and the resulting obliquity can cause distortion and errors in viewing. A better way of conducting accurate measurement is to use an interferometer. While a variety of interferometers are used in metrology and physical sciences, two types are discussed in the following section: the NPL flatness interferometer and the Pitter–NPL gauge interferometer.

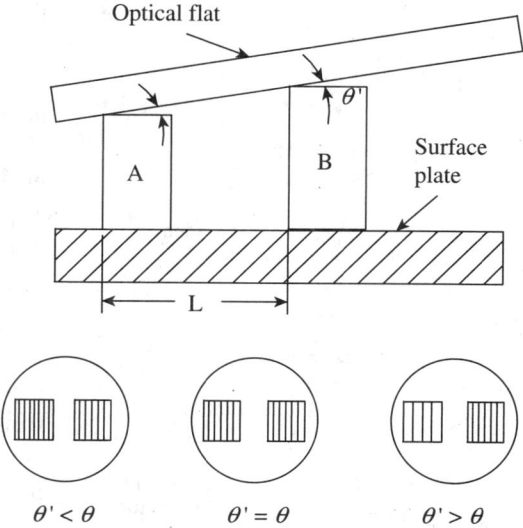

Fig. 7.13 Height measurement using an optical flat

7.5 INTERFEROMETERS

Interferometers are optical instruments that are used for very small linear measurements. They are used for verifying the accuracy of slip gauges and measuring flatness errors. Though an interferometer works on the same basic principle as that of an optical flat, it is provided with arrangements in order to control the lay and orientation of fringes. It is also provided with a viewing or recording system, which eliminates measurement errors.

7.5.1 NPL Flatness Interferometer

This interferometer was designed and developed by the National Physical Laboratory of the United Kingdom. It comprises a simple optical system, which provides a sharp image of the fringes so that it is convenient for the user to view them. The light from a mercury vapour lamp is condensed and passed through a green filter, resulting in a green monochromatic light source. The light will now pass through a pinhole, giving an intense point source of monochromatic light. The pinhole is positioned such that it is in the focal plane of a collimating lens. Therefore, the collimating lens projects a parallel beam of light onto the face of the gauge to be tested via an optical flat. This results in the formation of interference fringes. The light beam, which carries an image of the fringes, is reflected back and directed by 90° using a glass plate reflector.

The entire optical system is enclosed in a metal or fibreglass body. It is provided with adjustments to vary the angle of the optical flat, which is mounted on an adjustable tripod. In addition, the base plate is designed to be rotated so that the fringes can be oriented to the best advantage (Fig. 7.14).

Figure 7.15 illustrates the fringe pattern that is typically observed on the gauge surface as well as the base plate. In Fig. 7.15(a), the fringes are parallel and equal in number on the two surfaces. Obviously, the two surfaces are parallel, which means that the gauge surface is perfectly flat. On the other hand, in Fig. 7.15(b), the number of fringes is unequal and, since the base plate surface is ensured to be perfectly flat, the workpiece surface has a flatness error. Due to the flatness error, the optical flat makes unequal angles with the workpiece and the base plate, resulting in an unequal number of fringes. Most of the times fringes will not be parallel

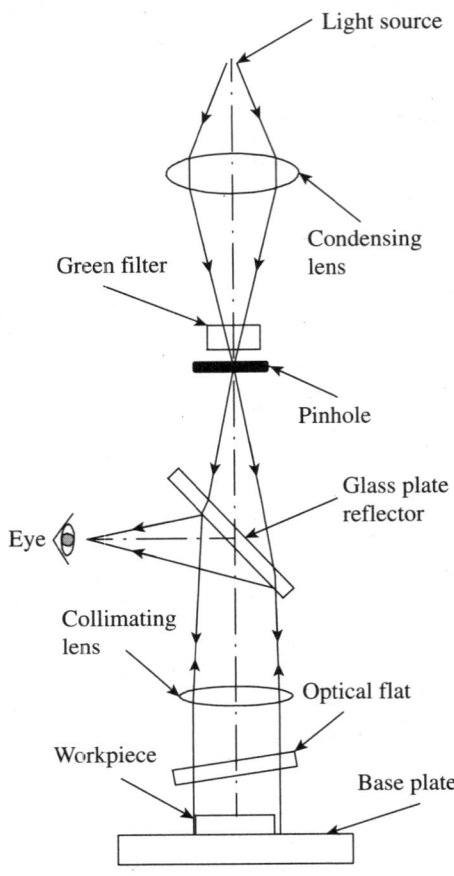

Fig. 7.14 Optical system of an NPL flatness interferometer

lines, but will curve out in a particular fashion depending on the extent of wear and tear of the upper surface of the workpiece. In such cases, the fringe pattern gives a clue about the nature and direction of wear.

Measuring Error in Parallelism

The NPL flatness interferometer is used for checking flatness between gauge surfaces. The gauge to be checked is placed on a base plate that has a high degree of flatness. If the gauge length is smaller than 25 mm, the gauge is placed on the base plate and the fringe pattern is observed. If the gauge being inspected is free from flatness error, then the fringes formed on both the gauge surface and the base plate are equally spaced. For gauges longer than 25 mm, fringe pattern on the base plate is difficult to observe. Therefore, the gauge is placed on a rotary table, as shown in Fig. 7.16. Suppose the gauge surface has flatness error, because of the angle it makes with the optical flat, a number of fringes are seen on the gauge surface. Now the table is rotated through 180°, and the surface of the gauge becomes even less parallel to the optical flat. This results in more number of fringes appearing on the gauge surface.

Let us consider a gauge that shows n_1 fringes along its length in the first position and n_2 in the second position. As seen in Fig. 7.16, the distance between the gauge and the optical flat in the first position has increased by a distance δ_1, over the length of the gauge, and in the second position by a distance δ_2. It is clear that the distance between the gauge and the optical flat changes by $\lambda/2$, between adjacent fringes.

Therefore, $\delta_1 = n_1 \times \lambda/2$ and $\delta_2 = n_2 \times \lambda/2$.
The change in angular relationship is $(\delta_2 - \delta_1)$, that is, $(\delta_2 - \delta_1) = (n_1 - n_2) \times \lambda/2$.

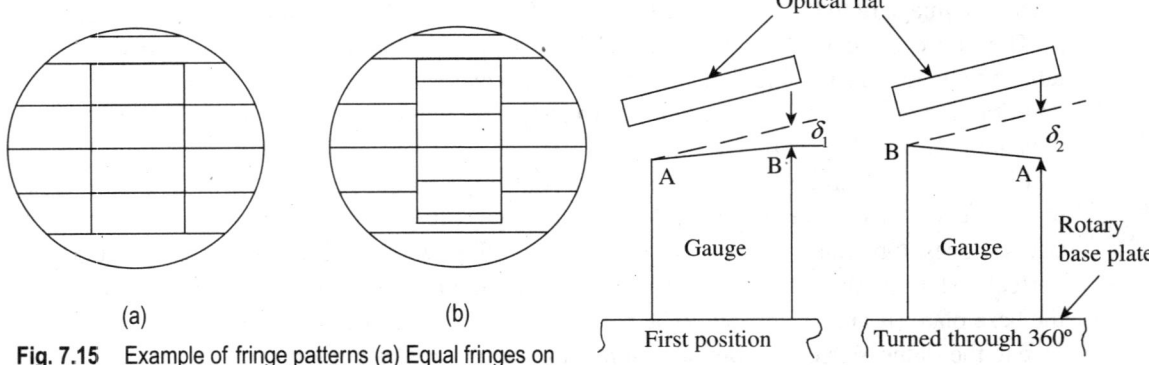

Fig. 7.15 Example of fringe patterns (a) Equal fringes on parallel surfaces (b) Unequal fringes due to flatness error

Fig. 7.16 Testing parallelism in gauges

The error in parallelism is actually $(\delta_2 - \delta_1)/2$ because of the doubling effect due to the rotation of the base plate.

Thus, $(\delta_2 - \delta_1)/2 = (n_1 - n_2)/2 \times (\lambda/2)$.

7.5.2 Pitter–NPL Gauge Interferometer

This interferometer is used for determining actual lengths of slip gauges. Since the measurement calls for a high degree of accuracy and precision, the instrument should be used under highly controlled physical conditions. It is recommended that the system be maintained at an ambient temperature of 20 °C, and a barometric pressure of 760 mmHg with a water vapour pressure of 7 mm, and contain 0.33% by volume of carbon dioxide.

The optical system of the Pitter–NPL interferometer is shown in Fig. 7.17. Light from a monochromatic source (the preferred light source is a cadmium lamp) is condensed by a condensing lens and focused onto an illuminating aperture. This provides a concentrated light source at the focal point of a collimating lens. Thus, a parallel beam of light falls on a constant deviation prism. This prism splits the incident light into light rays of different wavelengths and hence different colours. The user can select a desired colour by varying the angle of the reflecting faces of the prism relative to the plane of the base plate.

The prism turns the light by 90° and directs it onto the optical flat. The optical flat can be positioned at a desired angle by means of a simple arrangement. The slip gauge that is to be checked is kept right below the optical flat on top of the highly flat surface of the base plate. The lower portion of the optical flat is coated with a film of aluminium, which transmits and reflects equal proportions of the incident light. The light is reflected from three surfaces, namely the surface of the optical flat, the upper surface of the slip gauge, and the surface of the base plate. Light rays reflected from all the three surfaces pass through the optical system again; however, the axis is slightly deviated due to the inclination of the optical flat. This slightly shifted light is captured by another prism and turned by 90°, so that the fringe pattern can be observed and recorded by the user.

The typical fringe pattern observed is shown in Fig. 7.18. Superimposition of the fringes

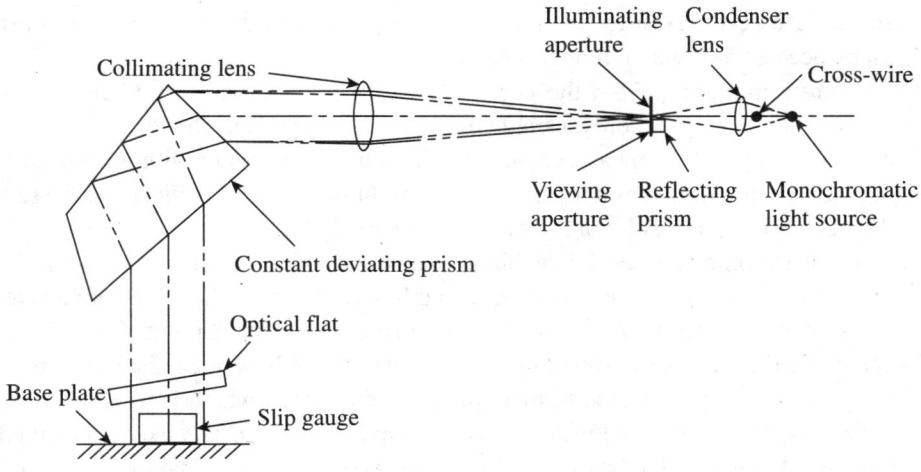

Fig. 7.17 Optical system of the Pitter–NPL gauge interferometer

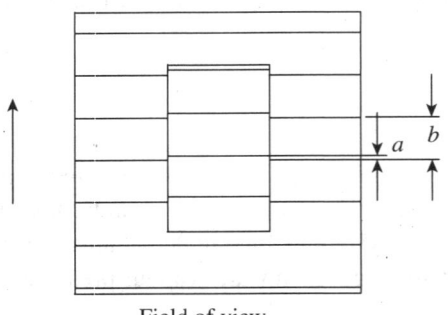

Field of view

Fig. 7.18 Field of view of fringe pattern

corresponding to the upper surface of the slip gauge upon those corresponding to the surface of the base plate is shown in Fig. 7.18. It can be seen that the two sets of fringes are displaced by an amount a with respect to each other. The value of a varies depending on the colour of the incident light. The displacement a is expressed as a fraction of the fringe spacing b, which is as follows:

$$f = a/b$$

The height of the slip gauge will be equal to a whole number of half wavelengths, n, plus the fraction a/b of the half wavelengths of the radiation in which the fringes are observed.

Therefore, the height of the slip gauge, $H = n\,(\lambda/2) + (a/b) \times (\lambda/2)$, where n is the number of fringes on the slip gauge surface, λ is the wavelength of light, and a/b is the observed fraction.

However, practitioners of industrial metrology are not happy with the values thus obtained. The fraction readings are obtained for all the three colours of cadmium, namely red, green, and violet. For each of the wavelengths, fractions a/b are recorded. Using these fractions, a series of expressions are obtained for the height of the slip gauge. These expressions are combined to get a general expression for gauge height. The Pitter–NPL gauge interferometer is provided with a slide rule, in which the wavelengths of red, green, and violet are set to scale, from a common zero. This provides a ready reckoner to speed up calculations.

7.5.3 Laser Interferometers

In recent times, laser-based interferometers are becoming increasingly popular in metrology applications. Traditionally, lasers were more used by physicists than engineers, since the frequencies of lasers were not stable enough. However now, stabilized lasers are used along with powerful electronic controls for various applications in metrology. Gas lasers, with a mixture of neon and helium, provide perfectly monochromatic red light. Interference fringes can be observed with a light intensity that is 1000 times more than any other monochromatic light source. However, even to this day, laser-based instruments are extremely costly and require many accessories, which hinder their usage.

More importantly, from the point of view of calibration of slip gauges, one limitation of laser is that it generates only a single wavelength. This means that the method of exact fractions cannot be applied for measurement. In addition, a laser beam with a small diameter and high degree of collimation has a limited spread. Additional optical devices will be required to spread the beam to cover a larger area of the workpieces being measured.

In interferometry, laser light exhibits properties similar to that of any 'normal' light. It can be represented by a sine wave whose wavelength is the same for the same colours and amplitude is a measure of the intensity of the laser light. From the measurement point of view, laser interferometry can be used for measurements of small diameters as well as large displacements. In this section, we present a simple method to measure the latter aspect, which is used for measuring machine slideways. The laser-based instrument is shown in Fig. 7.19. The fixed unit called the laser head consists of laser, a pair of semi-reflectors, and two photodiodes. The sliding unit has a corner cube

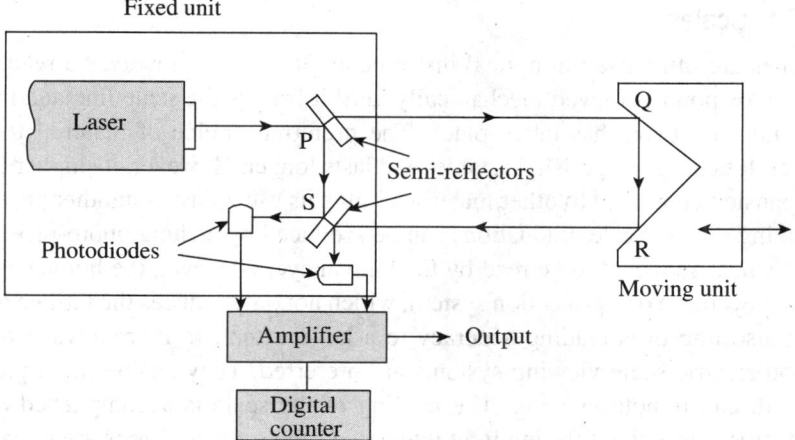

Fig. 7.19 Laser interferometer

mounted on it. The corner cube is a glass disk whose back surface has three polished faces that are mutually at right angles to each other. The corner cube will thus reflect light at an angle of 180°, regardless of the angle at which light is incident on it. The photodiodes will electronically measure the fringe intensity and provide an accurate means for measuring displacement.

Laser light first falls on the semi-reflector P, is partially reflected by 90° and falls on the other reflector S. A portion of light passes through P and strikes the corner cube. Light is turned through 180° by the corner cube and recombines at the semi-reflector S. If the difference between these two paths of light (PQRS – PS) is an odd number of half wavelengths, then interference will occur at S and the diode output will be at a minimum. On the other hand, if the path difference is an even number of half wavelengths, then the photodiodes will register maximum output.

It must have now become obvious to you that each time the moving slide is displaced by a quarter wavelength, the path difference (i.e., PQRS – PS) becomes half a wavelength and the output from the photodiode also changes from maximum to minimum or vice versa. This sinusoidal output from the photodiode is amplified and fed to a high-speed counter, which is calibrated to give the displacement in terms of millimetres. The purpose of using a second photodiode is to sense the direction of movement of the slide.

Laser interferometers are used to calibrate machine tables, slides, and axis movements of coordinate measuring machines. The equipment is portable and provides a very high degree of accuracy and precision.

7.6 SCALES, GRATINGS, AND RETICLES

The term, scale, is used when rulings are spaced relatively far apart, requiring some type of interpolating device to make accurate settings. The term, grating, is used when rulings are more closely spaced, producing a periodic pattern without blank gaps. Of course, gratings cannot be either generated or read manually. They require special readout systems, usually photoelectric. The only element that makes a microscope a measuring instrument is the reticle.

7.6.1 Scales

Scales are often used in optical instruments. It typically involves a read-out system in which an index point is moved mechanically until it frames the scale line and then reads the amount of movement that has taken place. The preferred choice of material for a scale is stainless steel. It takes good polish, is stable, and lasts longer. However, its higher thermal coefficient of expansion compared to other materials limits its use. Glass is another popular material used for making scales. Scale graduations can be produced by etching photo-resistive material.

Scales are meant to be read by the human eye. However, the human eye is invariably aided by an eyepiece or a projection system, which not only reduces the fatigue of the human operator but also improves reading accuracy to a large extent. In more advanced optical instruments, photoelectric scale viewing systems are preferred. They enable more precise settings, higher speed, and remote viewing. The reading of the scale is accomplished electronically. Photo-detectors sense the differing light intensity as the scale divisions are moved across a stationary photodetector. While the number of such light pulses indicates the distance moved, the rate of the pulses enables the measurement of speed of movement.

7.6.2 Gratings

Scales with a continuously repeating pattern of lines or groves that are closely spaced are called reticles. The line spacing may be of the order of 50–1000 per millimetre. They are invariably sensed by photo-electric read-outs. There are two types of gratings: Ronchi rulings and phase gratings. Ronchi rulings consist of strips that are alternatively opaque and transmitting, with a spacing of 300–1000 per millimetre. Phase gratings consist of triangularly shaped, contiguous grooves similar to spectroscopic diffraction gratings.

Moire Fringes

When two similar gratings are placed face to face, with their lines parallel, a series of alternating light and dark bands known as *moire fringes* will appear. When one scale moves in a direction perpendicular to the lines with respect to a stationary index grating, the fringes are seen to move at right angles to the motion. These fringes are largely free from harmonics. Two photocells in the viewing optics spaced 90 fringe-phase degrees apart are capable of generating bidirectional fringe-counting signals.

7.6.3 Reticles

As already pointed out, the main element that makes a microscope a measuring instrument is the reticle. It provides a reference in the form of cross-wires for taking measurements. The cross-wires (sometimes also called 'cross-hairs') are usually etched on glass and fitted to the eyepiece of the microscope. A variety of reticles are used with microscopes for precise setting to measure part features. Figure 7.20 illustrates the four types of reticles that are normally used.

Type A is the most common but does not provide high accuracy. The cross-wire thickness usually varies from 1 to 5 μm. This is usually used for microscopes that have a magnification of 5× for the objective lens and 10× for the eyepiece.

Better accuracy can be achieved if the lines are broken, as in reticle B. This is useful when the line on the feature is narrower than the reticle line. For precise measurement along

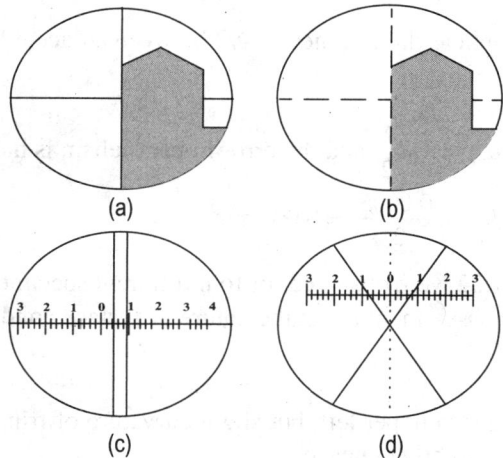

Fig. 7.20   Types of reticles (a) Type A (b) Type B
(c) Type C (d) Type D

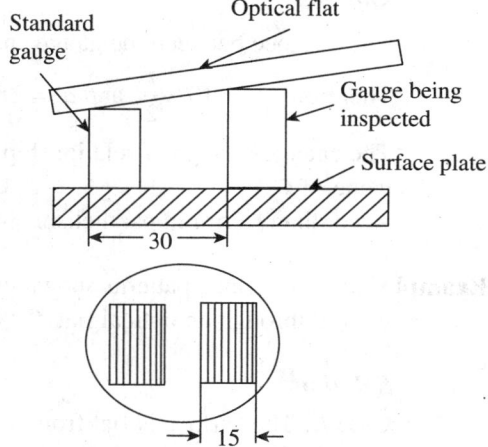

Fig. 7.21   Checking the height of the
slip gauge

a scale, reticle C is convenient. Parallel lines spaced slightly wider than the scale lines enable precise settings to be made. In this case, the eye averages any slight irregularities of the edges of the scale lines when seen in the clear spaces along each side. This is known as *bifilar reticle*.

Type D provides the highest accuracy of reading. It is preferred in measurements involving a high degree of precision like photo-etching jobs. The cross-wires are at 30° to each other. The eye has the ability to judge the symmetry of the four spaces created between the cross-wires and position the centre at the precise location for taking readings.

7.7 NUMERICAL EXAMPLES

Example 7.1   Figure 7.21 illustrates the use of an optical flat to check the height of the slip gauge against a standard gauge of 20 mm height. The wavelength of a cadmium light source is 0.509 μm. If the number of fringes on a gauge width of 15 mm is 10 and the distance between the two blocks is 30 mm, calculate the true height of the gauge being inspected.

Solution

The difference in height h is given by the following equation:

$$h = \frac{\lambda LN}{2l}$$

Therefore, $h = \dfrac{0.509 \times 30 \times 10}{1000 \times 2 \times 15} = 0.00509$ mm or 5.09 μm

Example 7.2   A slip gauge is being inspected using the NPL flatness interferometer. It is recorded that the gauge exhibits 10 fringes along its width in one position and 18 fringes in the other position. If the wavelength of the monochromatic light source is 0.5 μm, determine the error of flatness over its width.

Solution

The distance between the gauge and the optical flat changes by $\lambda/2$ between adjacent fringes.

Therefore, $\ddot{a}_1 = 10 \times \dfrac{\lambda}{2}$ and $\ddot{a}_2 = 18 \times \dfrac{\lambda}{2}$

The change in angular relationship $(\ddot{a}_1 - \ddot{a}_2) = 8 \times \dfrac{\lambda}{2}$ and the error in parallelism is half of this value.

Accordingly, error in parallelism $= 4 \times \dfrac{\lambda}{2} = 4 \times \dfrac{0.0005}{2} = 0.001\,\text{mm}.$

Example 7.3   The fringe patterns shown in Fig. 7.22 were observed for four different specimens when viewed through an optical flat. Give your assessment about the nature of surface conditions.

Solution

Case A: The surface is flat from lower right to upper left, but slight curvature of fringes away from the line of contact indicates that it is slightly concave

Case B: The surface is flat in the direction that the fringes run. However, it is higher diagonally across the centre, where the fringes are more widely spaced, than at the ends.

Case C: Circular fringes with decreasing diameters indicate the surface to be spherical. By applying a small pressure at the centre of the fringes, if the fringes are found to move towards the centre, the surface is concave. On the other hand, if the fringes move away from the centre, the surface is convex.

Case D: Parallel, straight, and uniformly spaced fringes indicate a flat surface.

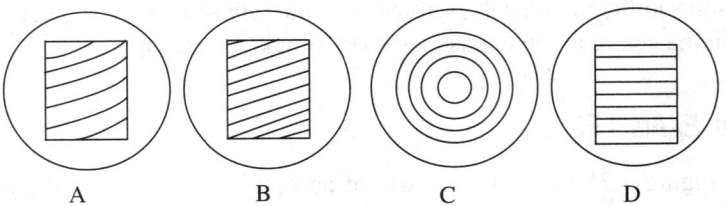

Fig. 7.22   Fringe patterns viewed through an optical flat

Example 7.4   Two flat gauges are tested for taper over a length of 25 mm on a surface plate using an optical interferometer. Determine the taper of gauge surfaces if the wavelength of light source is 0.5 μm.

(a) Gauge A: The number of fringes on the gauge surface is 15 and that on the surface plate is 5.

(b) Gauge B: The number of fringes on the gauge surface is 5 and that on the surface plate is 8.

Solution

Amount of taper for gauge A $= (15 - 5) \times \dfrac{\lambda}{2} = 10 \times \dfrac{0.5}{2} = 2.5\,\mu\text{m}.$

Amount of taper for gauge B $= (8 - 5) \times \dfrac{\lambda}{2} = 3 \times \dfrac{0.5}{2} = 0.75\,\mu\text{m}.$

A QUICK OVERVIEW

- Optical measurement provides a simple, easy, accurate, and reliable means for carrying out inspection and measurements in the industry.
- Optical magnification is one of the most widely used techniques in metrology. However, optics has three other principal applications: in *alignment*, in *interferometry*, and as an absolute *standard* of length. An optical measurement technique to check alignment employs light rays to establish references such as lines and planes. Interferometry uses a phenomenon of light to facilitate measurement at the micrometre level. Another significant application of optics is the use of light as the absolute standard of length, which has been discussed in Chapter 2.
- Tool maker's microscope is a multifunctional device that is primarily used for measurement in factory shop floors. Designed with the measurement of workpiece contours and inspection of surface features in mind, a tool maker's microscope supports a wide range of applications from shop floor inspection, and measurement of tools and machined parts, to precision measurement of test tools in a measuring room.
- An optical square is a pentagonal prism (also called pentaprism), which is useful for turning the line of sight by 90° from its original path. Regardless of the angle at which the incident beam strikes the face of the prism, it is turned through 90° by an internal reflection.
- When two light waves interact with each other, the wave effect leads to a phenomenon called *interference* of light. Instruments designed to measure interference are known as *interferometers*.
- When light from a monochromatic light source is made to fall on an optical flat, which is oriented at a very small angle with respect to a flat reflecting surface, a band of alternate light and dark patches, popularly known as fringes,

is seen by the eye. In case of a perfectly flat surface, the fringe pattern is regular, parallel, and uniformly spaced. Any deviation from this pattern is a measure of error in the flatness of the surface being measured. Fringe patterns reveal surface conditions like contour lines on a map.
- Though an interferometer works on the same basic principle as that of an optical flat, it is provided with arrangements in order to control the lay and orientation of fringes. It is also provided with a viewing or recording system, which eliminates measurement errors. Two types of interferometers have been discussed in this chapter. While the NPL flatness interferometer is used for checking flatness between gauge surfaces, the Pitter–NPL gauge interferometer is used for determining the actual lengths of slip gauges.
- In recent times, laser-based interferometers are increasingly becoming popular in metrology applications. Stabilized lasers are used along with powerful electronic controls for various applications in metrology. Gas lasers with a mixture of neon and helium provide perfectly monochromatic red light. Interference fringes can be observed with a light intensity that is 1000 times more than any other monochromatic light source. However, even to this day, laser-based instruments are extremely costly and require many accessories, which hinder their usage.
- The term 'scale' is used when rulings are spaced relatively far apart, requiring some type of interpolating device to make accurate settings. The term 'grating' is used when rulings are more closely spaced, producing a periodic pattern without blank gaps. Of course, gratings cannot be either generated or read manually. They require special read-out systems, usually photoelectric. The only element that makes a microscope a measuring instrument is the 'reticle'.

MULTIPLE-CHOICE QUESTIONS

1. What characteristic of light makes it a standard?
   (a) It is easily sensed by the human eye.
   (b) Its colour can be selected as per the user's choice.
   (c) The length of waves is known and unvarying.
   (d) It is easily refracted.

2. Which of the following typifies measurement with light waves?
   (a) Gauge blocks
   (b) Perfectly lapped surfaces
   (c) Comparison measurement
   (d) Optical flats

3. What was the contribution of Albert A. Michelson to metrology?
   (a) He used light waves to calibrate gauge blocks.
   (b) He invented laser interferometry.
   (c) He showed that only cadmium is the reliable light source for measurement.
   (d) He had nothing to do with metrology.

4. What is the significance of krypton-86 in measuring length?
   (a) It is the only wavelength that does not fluctuate.
   (b) It is the most easily reproducible wavelength.
   (c) It is less harmful to the human eye.
   (d) It produces pleasant colours.

5. The main use of a tool maker's microscope is in measuring
   (a) phase shift of monochromatic light
   (b) shape, size, and angle of small components
   (c) biological degradation of small machine components
   (d) contours of large machine parts

6. The element that makes a microscope a measuring instrument is
   (a) objective lens   (c) reticle
   (b) light beam   (d) none of these

7. The types of illumination modes available in a tool maker's microscope are
   (a) form illumination and feature illumination
   (b) form illumination and surface illumination
   (c) feature illumination and contour illumination
   (d) contour illumination and surface illumination

8. A pentaprism is used in optical squares because it
   (a) is easy to install and repair
   (b) is not affected by errors in the mounting arrangement
   (c) is sensitive to errors present in the mounting arrangements
   (d) can split light into multiple rays

9. In case of fringes, the dark areas will occur
   (a) when there is no path difference between two light rays
   (b) when the path difference of two light rays from the same source amounts to an even number of half wavelengths
   (c) when the path difference of two light rays from the same source amounts to an odd number of half wavelengths
   (d) in none of these

10. In case of fringes, the bright areas will occur
    (a) when there is no path difference between two light rays
    (b) when the path difference of two light rays from the same source amounts to an even number of half wavelengths
    (c) when the path difference of two light rays from the same source amounts to an odd number of half wavelengths
    (d) in all of these

11. An optical flat can be employed to measure height differences in the range of
    (a) 0.01–0.1 mm   (c) 10–100 mm
    (b) 1–10 mm   (d) 1–10 m

12. One major limitation of laser interferometry is that
    (a) it generates only two wavelengths
    (b) it does not have a constant wavelength
    (c) it generates only a single wavelength
    (d) its wavelength cannot be predicted

13. The term 'grating' means that
    (a) rulings are spaced relatively far apart, requiring some type of interpolating device to make accurate settings
    (b) rulings need not have any pattern
    (c) rulings follow a logarithmic scale
    (d) rulings are more closely spaced, producing a periodic pattern without blank gaps

14. When two similar gratings are placed face to face with their lines parallel, there will appear a series of bands popularly known as
    - (a) Pitter fringes
    - (b) Moire fringes
    - (c) De Morgan fringes
    - (d) Michaelson fringes
15. In a bifilar reticle, precise settings can be made because
    - (a) the reticle has cross-wires at an angle of 45°
    - (b) thick cross-wires are used
    - (c) broken lines are used for cross-wires
    - (d) parallel lines spaced slightly wider than the scale lines are used

REVIEW QUESTIONS

1. What are the essential features of an optical system?
2. Explain the principle of microscopy.
3. With the help of a neat sketch, explain the construction details of a tool maker's microscope.
4. Write a note on surface illumination modes available in a tool maker's microscope.
5. Discuss the important applications of a tool maker's microscope.
6. Explain the optical system and working principle of a profile projector.
7. What are the advantages of optical squares when used in optical measuring instruments?
8. With the help of a simple sketch, explain how optical squares are made use of in testing the squareness of machine parts.
9. Explain the phenomenon of optical interference.
10. Explain the measurement methodology involved in the use of optical flats.
11. Briefly describe the procedure involved in checking slip gauges using optical flats. What are the limitations of this method?
12. With the help of a neat sketch, explain the optical system used in an NPL flatness interferometer. Discuss the procedure of measuring error in parallelism using the same.
13. Write a note on the Pitter–NPL gauge interferometer.
14. What are the advantages and limitations of laser interferometry?
15. Explain the working principle of a laser interferometer.
16. Differentiate between scale, grating, and reticle.

PROBLEMS

1. An optical flat is being used to check the height of a slip gauge against a standard gauge of 25 mm height. The wavelength of the light source is 0.498 μm. If the number of fringes on a gauge width of 20 mm is 10 and the distance between the two blocks is 30 mm, calculate the true height of the gauge being inspected.
2. A slip gauge is being tested with the help of a reference gauge and following is the test result:
   Number of fringes on each gauge = 10
   Width of each gauge = 20 mm
   Distance between the two gauges = 50 mm
   Wavelength of light source = 0.00005 mm
   Height of the reference gauge = 25 mm
   Determine the height of the test gauge.
3. An NPL flatness interferometer is being used to determine the error of flatness of a precision component. Upon inspection, 10 fringes are observed along its width of 30 mm in one position and 20 fringes in the other position. If the wavelength of the light source is 0.503 μm, determine the error of flatness.

Answers to Multiple-choice Questions

1. (c)	2. (d)	3. (a)	4. (b)	5. (b)	6. (c)	7. (d)	8. (b)
9. (c)	10. (b)	11. (a)	12. (c)	13. (d)	14. (b)	15. (d)	

Metrology of Gears and Screw Threads

After studying this chapter, the reader will be able to

- understand the basic principles of measurement of gears and screw threads
- throw light on the geometry of spur gears and screw threads
- elucidate the measurement techniques used for the measurement of runout, pitch, profile, lead, backlash, and tooth thickness of spur gears
- analyse the various gear-measuring instruments such as gear-measuring machine, gear tooth calliper, tooth span micrometer, and Parkinson gear tester.
- explain the measurement principles of major diameter, minor diameter, effective diameter, pitch, angle, and form of screw threads.
- describe thread-measuring instruments such as bench micrometer, floating carriage micrometer, and pitch-measuring machine
- discuss the types and use of thread gauges for screw thread inspection

8.1 INTRODUCTION

Gears are the main elements in a transmission system. It is needless to say that for efficient transfer of speed and power, gears should conform perfectly to the designed profile and dimensions. Misalignments and gear runout will result in vibrations, chatter, noise, and loss of power. Therefore, one cannot understate the importance of precise measurement and inspection techniques for gears. On the other hand, threaded components should meet stringent quality requirements to satisfy the property of *interchangeability*. Geometric aspects of screw threads are quite complex and, therefore, thread gauging is an integral part of a unified thread gauging system.

The most common forms of gear teeth are involute and cycloidal. The major gear types are spur, helical, bevel, spiral, and worm gears. Coverage of the entire range of inspection methods and instrumentation is an arduous task and requires a separate volume altogether. Therefore, this chapter is confined to the major inspection methods suited for spur gears having an involute profile. We are sure that the reader will be benefited with this basic knowledge and be motivated to refer to standard books dealing with inspection of gears and screw threads. While the first part of the chapter deals with measurements of gears, the second part outlines some major techniques used for the measurement of screw threads.

8.2 GEAR TERMINOLOGY

Each gear has a unique form or geometry. The gear form is defined by various elements. An illustration of the gear highlighting the important elements is referred to as 'gear terminology'. This section explains the types of gears and their terminology.

8.2.1 Types of Gears

The common types of gears used in engineering practices are described in this section. The information provided here is very brief, and the reader is advised to read a good book on 'theory of machines' to understand the concepts better.

Spur gears These gears are the simplest of all gears. The gear teeth are cut on the periphery and are parallel to the axis of the gear. They are used to transmit power and motion between parallel shafts (Fig. 8.1).

Helical gears The gear teeth are cut along the periphery, but at an angle to the axis of the gear. Each tooth has a helical or spiral form. These gears can deliver higher torque since there are more number of teeth in a mesh at any given point of time. They can transmit motion between parallel or non-parallel shafts.

Herringbone gears These gears have two sets of helical teeth, one right-hand and the other left-hand, machined side by side (Fig. 8.2).

Worm and worm gears A worm is similar to a screw having single or multiple start threads, which form the teeth of the worm. The worm drives the worm gear or worm wheel to enable transmission of motion. The axes of worm and worm gear are at right angles to each other (Fig. 8.3).

Fig. 8.1 Spur gear

Fig. 8.2 Herringbone gear

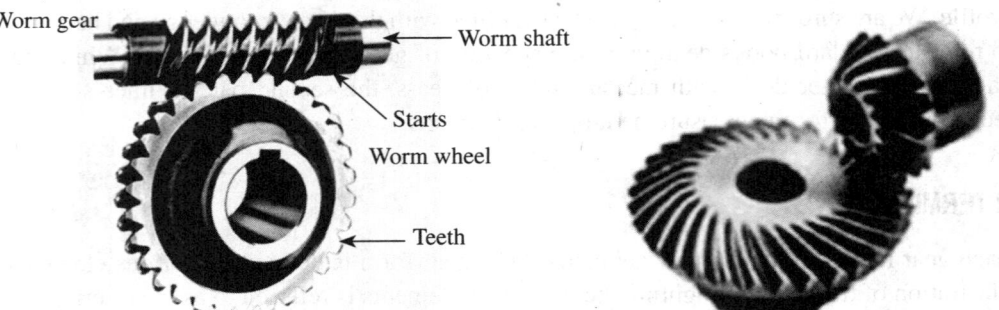

Fig. 8.3   Worm and worm gear   Fig. 8.4   Bevel gears

Bevel gears   These gears are used to connect shafts at any desired angle to each other. The shafts may lie in the same plane or in different planes (Fig. 8.4).

Hypoid gears   These gears are similar to bevel gears, but the axes of the two connecting shafts do not intersect. They carry curved teeth, are stronger than the common types of bevel gears, and are quiet-running. These gears are mainly used in automobile rear axle drives.

A gear tooth is formed by portions of a pair of opposed involutes. By far, the involute tooth profile is most preferred in gears. A clear understanding of the various terminologies associated with gears is extremely important before an attempt is made to learn about inspection and measurement of gears. The following are some of the key terminologies associated with gears, which have been illustrated in Fig. 8.5:

Base circle   It is the circle from which the involute form is generated. Only the base circle of a gear is fixed and unalterable.

Outside circle   It marks the maximum diameter of the gear up to which the involute form is extended. It is also called the addendum circle. In addition, it is the diameter of the blank from which the gear is cut out.

Pitch circle   It is the imaginary circle on which lies the centres of the pitch cylinders of two mating gears.

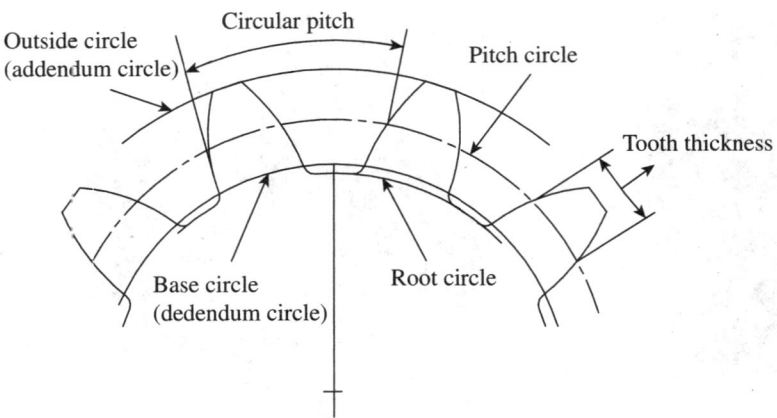

Fig. 8.5   Spur gear terminology

Root circle   It is the circle corresponding to the minimum diameter of the gear profile. However, the involute profile is limited only up to the base circle of a spur gear.

Addendum   It is the radial distance between the addendum circle and the pitch circle.

Dedendum   It is the radial distance between the pitch circle and the root circle.

Face   The portion of tooth lying between the addendum circle and the pitch circle is called the face.

Flank   The portion of tooth lying between the pitch circle and the dedendum circle is called the flank.

Circular pitch   It is the distance between corresponding points of adjacent teeth measured along the pitch circle.

Diametrical pitch   It is expressed as the number of teeth per unit diameter of the pitch circle.

Module   It is simply the metric standard for pitch. It is the linear distance (in millimetres) that each tooth of the gear would occupy if the gear teeth were spaced along the pitch diameter. Accordingly, if the pitch circle diameter of the gear is D and the number of teeth is N, then the module m is given by D/N and is expressed in millimetres.

In order to ensure interchangeability and smooth meshing of gears, standard modules are recommended. These standards are also useful for the design of gear cutting tools. The Indian Standards Institute has recommended the following modules (in mm) in order of preference:

First choice   1, 1.25, 1.5, 2, 2.5, 3, 4, 5, 6, 8, 10, 12, 16, 20

Second choice   1.125, 1.375, 1.75, 2.25, 2.75, 3.5, 4.5, 5.5, 7, 9, 11, 14, 18

Third choice   3.25, 3.75, 6.5

Tooth thickness   It is the arc distance measured along the pitch circle from its intercept with one flank to that with the other flank of the same tooth.

Base pitch   It is the distance measured around the base circle from the origin of the involute on the tooth to the origin of a similar involute on the next tooth.

Base pitch = Base circumference/Number of teeth

Table 8.1 illustrates the nomenclature of a spur gear.

Table 8.1   Spur gear nomenclature

Nomenclature	Symbol	Formula
Module	m	D/N
Diametrical pitch	DP	$N/D = \pi/p = 1/m$
Pitch	p	$P = \pi m$
Pitch circle diameter	D_p	Nm
Tooth height	h	$2.2m$
Addendum	h'	M
Dedendum	h''	$1.2m$
Outside diameter	D_0	$m(N + 2)$
Root circle diameter	D_r	$D_0 - 4.4m$
Pressure angle	α	20° or 14½°

N = Number of teeth on the gear, D = Outside diameter of the gear.

8.2.2 Line of Action and Pressure Angle

The mating teeth of two gears in the mesh make contact with each other along a common tangent to their base circle, as shown in Fig. 8.6. This line is referred to as the 'line of action'. The load or the so-called pressure between the two gears is transmitted along this line.

The angle between the line of action and the common tangent to the pitch circles is known as the *pressure angle*. Using trigonometry, it can be shown that the pressure angle is also given by the angle made between the normal drawn at the point of contact of line of action to the base circle and the line joining the centres of the gears.

8.3 ERRORS IN SPUR GEARS

A basic understanding of the errors in spur gears during manufacturing is important before we consider the possible ways of measuring the different elements of gears. A spur gear is a rotating member that constantly meshes with its mating gear. It should have the perfect geometry to maximize transmission of power and speed without any loss. From a metrological point of view, the major types of errors are as follows:

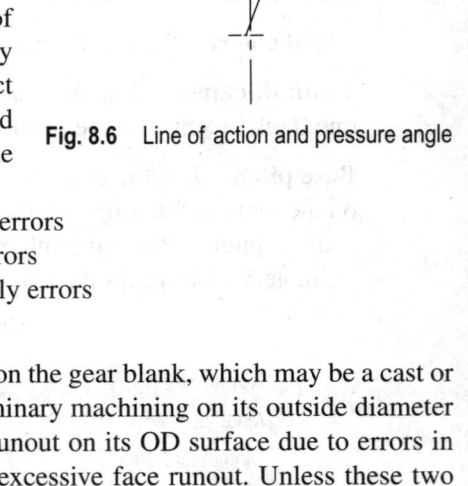

Fig. 8.6 Line of action and pressure angle

1. Gear blank runout errors
2. Gear tooth profile errors
3. Gear tooth errors
4. Pitch errors
5. Runout errors
6. Lead errors
7. Assembly errors

Gear blank runout errors   Gear machining is done on the gear blank, which may be a cast or a forged part. The blank would have undergone preliminary machining on its outside diameter (OD) and the two faces. The blank may have radial runout on its OD surface due to errors in the preliminary machining. In addition, it may have excessive face runout. Unless these two runouts are within prescribed limits, it is not possible to meet the tolerance requirements at later stages of gear manufacture.

Gear tooth profile errors   These errors are caused by the deviation of the actual tooth profile from the ideal tooth profile. Excessive profile error will result in either friction between the mating teeth or backlash, depending on whether it is on the positive or negative side.

Gear tooth errors   This type of error can take the form of either tooth thickness error or tooth alignment error. The tooth thickness measured along the pitch circle may have a large amount of error. On the other hand, the locus of a point on the machined gear teeth may not follow an ideal trace or path. This results in a loss in alignment of the gear.

Pitch errors Errors in pitch cannot be tolerated, especially when the gear transmission system is expected to provide a high degree of positional accuracy for a machine slide or axis. Pitch error can be either *single pitch error* or *accumulated pitch error*. Single pitch error is the error in actual measured pitch value between adjacent teeth. Accumulated pitch error is the difference between theoretical summation over any number of teeth intervals and summation of actual pitch measurement over the same interval.

Runout errors This type of error refers to the runout of the pitch circle. Runout causes vibrations and noise, and reduces the life of the gears and bearings. This error creeps in due to inaccuracies in the cutting arbour and tooling system.

Lead errors This type of error is caused by the deviation of the actual advance of the gear tooth profile from the ideal value or position. This error results in poor contact between the mating teeth, resulting in loss of power.

Assembly errors Errors in assembly may be due to either the centre distance error or the axes alignment error. An error in centre distance between the two engaging gears results in either backlash error or jamming of gears if the distance is too little. In addition, the axes of the two gears must be parallel to each other, failing which misalignment will be a major problem.

8.4 MEASUREMENT OF GEAR ELEMENTS

A number of standard gear inspection methods are used in the industry. The choice of the inspection procedure and methods not only depends on the magnitude of tolerance and size of the gears, but also on lot sizes, equipment available, and inspection costs. While a number of analytical methods are recommended for inspection of gears, statistical quality control is normally resorted to when large quantities of gears are manufactured. The following elements of gears are important for analytical inspection:
1. Runout
2. Pitch
3. Profile
4. Lead
5. Backlash
6. Tooth thickness

8.4.1 Measurement of Runout

Runout is caused when there is some deviation in the trajectories of the points on a section of a circular surface in relation to the axis of rotation. In case of a gear, runout is the resultant of the radial throw of the axis of a gear due to the out of roundness of the gear profile. *Runout tolerance* is the total allowable runout. In case of gear teeth, runout is measured by a specified probe such as a cylinder, ball, cone, rack, or gear teeth. The measurement is made perpendicular to the surface of revolution. On bevel and hypoid gears, both axial and radial runouts are included in one measurement.

A common method of runout inspection, called a single-probe check and shown in Fig. 8.7(a), uses an indicator with a single probe whose diameter makes contact with the flanks of adjacent teeth in the area of the pitch circle. On the other hand, in a two-probe check illustrated in Fig. 8.7(b), one fixed and one free-moving probe, are positioned on diametrically opposite sides of the gear and make contact with identically located elements of the tooth profile. The range of indications obtained with the two-probe check during a complete revolution of the gear is twice the amount resulting from the single-probe check.

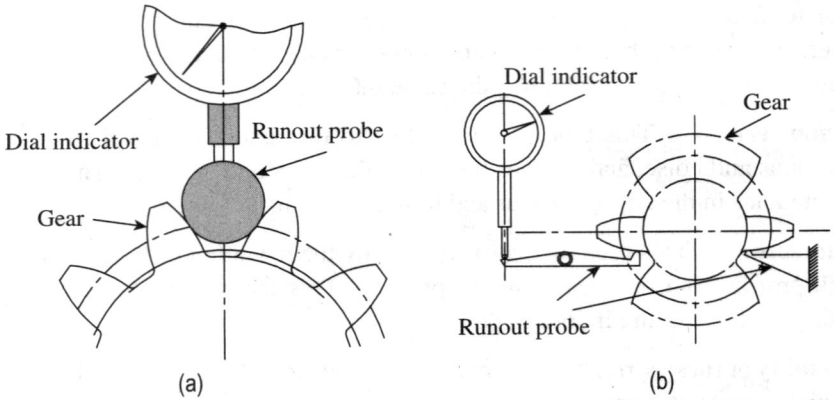

(a)   (b)

Fig. 8.7   Measurement of radial runout (a) Single-probe check (b) Two-probe check

8.4.2 Measurement of Pitch

Pitch is the distance between corresponding points on equally spaced and adjacent teeth. Pitch error is the difference in distance between equally spaced adjacent teeth and the measured distance between any two adjacent teeth. The two types of instruments that are usually employed for checking pitch are discussed in this section.

Pitch-measuring Instruments

These instruments enable the measurement of chordal pitch between successive pairs of teeth. The instrument comprises a fixed finger and a movable finger, which can be set to two identical points on adjacent teeth along the pitch circle. The pitch variation is displayed on a dial indicator attached to the instrument, as shown in Fig. 8.8. In some cases, the pitch variation is recorded on a chart recorder, which can be used for further measurements. A major limitation of this method is that readings are influenced by profile variations as well as runout of the gear.

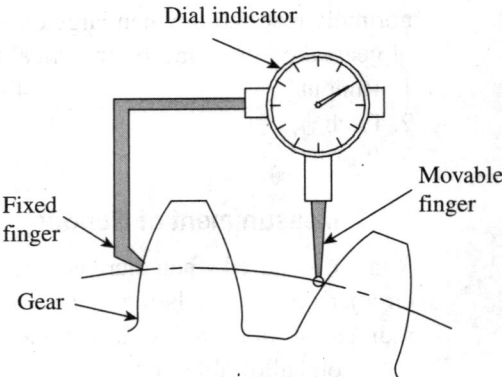

Fig. 8.8   Pitch-measuring instrument

Pitch-checking Instrument

A pitch-checking instrument is essentially a dividing head that can be used to measure pitch variations. The instrument can be used for checking small as well as large gears due to its portability. Figure 8.9 explains the measuring principle for a spur gear. It has two probes—one fixed, called the anvil, and the other movable, called the measuring feeler. The latter is connected to a dial indicator through levers.

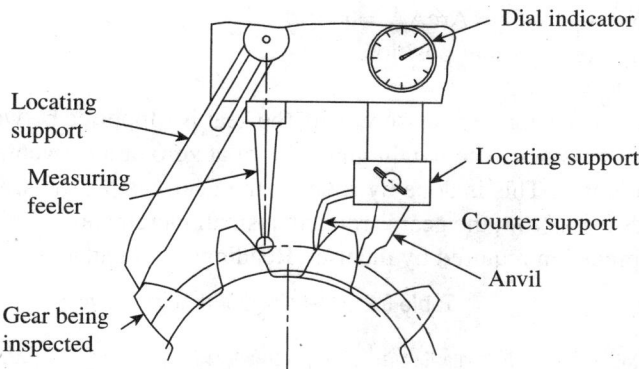

Fig. 8.9 Pitch-checking instrument

The instrument is located by two adjacent supports resting on the crests of the teeth. A tooth flank is butted against the fixed anvil and locating supports. The measuring feeler senses the corresponding next flank. The instrument is used as a comparator from which we can calculate the adjacent pitch error, actual pitch, and accumulated pitch error.

8.4.3 Measurement of Profile

The profile is the portion of the tooth flank between the specified form circle and the outside circle or start of tip chamfer. Profile tolerance is the allowable deviation of the actual tooth form from the theoretical profile in the designated reference plane of rotation. As the most commonly used profile for spur and helical gears is the involute profile, our discussions are limited to the measurement of involute profile and errors in this profile. We will now discuss two of the preferred methods of measuring a tooth profile.

Profile Measurement Using First Principle of Metrology

In this method, a dividing head and a height gauge are used to inspect the involute profile of a gear. With reference to Fig. 8.10, A is the start of the involute profile and AT is the vertical tangent to the base circle. When the involute curve is rotated through the roll angle ϕ_1, the point C_1 of the true involute profile is on the vertical tangent AT.

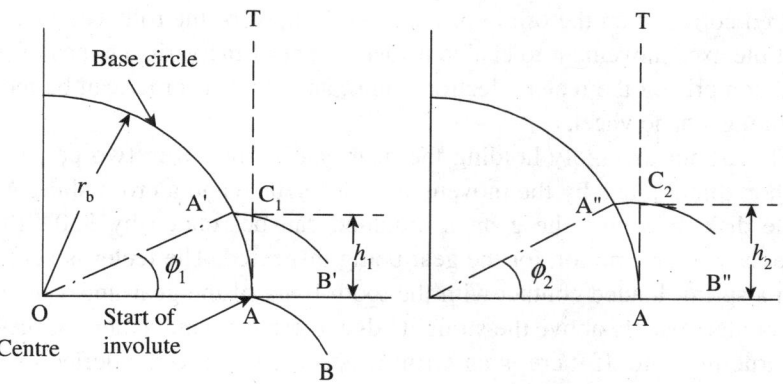

Fig. 8.10 Principle of involute measurement

Therefore, $h_1 = AC_1 = ArcAA' = r_b \times \phi_1$

Similarly, for any other position we have

$$h_n = r_b \times \phi_n$$

In order to carry out the measurement, the gear is supported between the centres of a dividing head and its tailstock. The height gauge is set at zero at a convenient position on the gear, for example, point C. This is done by rotating the gear against the stylus of the gauge so that the gauge reads zero. Now, the gear is rotated in small increments such as 1° or 1', depending on the degree of precision required by the user. Readings are tabulated as shown in Table 8.2.

Table 8.2 Readings of dividing head

Roll angle	True reading h	Actual reading	Error
0	0	0	0
1°	$h_1 = r_b \dfrac{1 \times \pi}{180}$	h_1'	$\epsilon_1 = h_1' - h_1$
2°	$h_2 = r_b \dfrac{2 \times \pi}{180}$	h_2'	$\epsilon_2 = h_2' - h_2$
.	.	.	.
.	.	.	.
.	.	.	.
n°	$h_n = r_b \dfrac{n \times \pi}{180}$	h_n'	$\epsilon_n = h_n' - h_n$

This procedure provides a reasonable and accurate means of measuring the deviation of the actual tooth profile from the true involute profile.

Profile Measurement Using Special Profile-measuring Instruments

The gear to be inspected is mounted on an arbour on the gear-measuring machine, as shown in Fig. 8.11. The probe is brought into contact with the tooth profile. To obtain the most accurate readings, it is essential that the feeler (probe) is sharp, positioned accurately, and centred correctly on the origin of the involute at 0° of the roll. The machine is provided with multiple axes movement to enable measurement of the various types of gears. The measuring head comprising the feeler, electronic unit, and chart recorder can be moved up and down by operating a handwheel.

The arbour assembly holding the gear can be moved in two perpendicular directions in the horizontal plane by the movement of a carriage and a cross-slide. Additionally, the base circle disk on which the gear is mounted can be rotated by 360°, thereby providing the necessary rotary motion for the gear being inspected. The feeler is kept in such a way that it is in a spring-loaded contact with the tooth flank of the gear under inspection. As the feeler is mounted exactly above the straight edge, there is no movement of the feeler if the involute is a true involute. If there is an error, it is sensed due to the deflection of the feeler, and is amplified by the electronic unit and recorded by the chart recorder. The movement of the

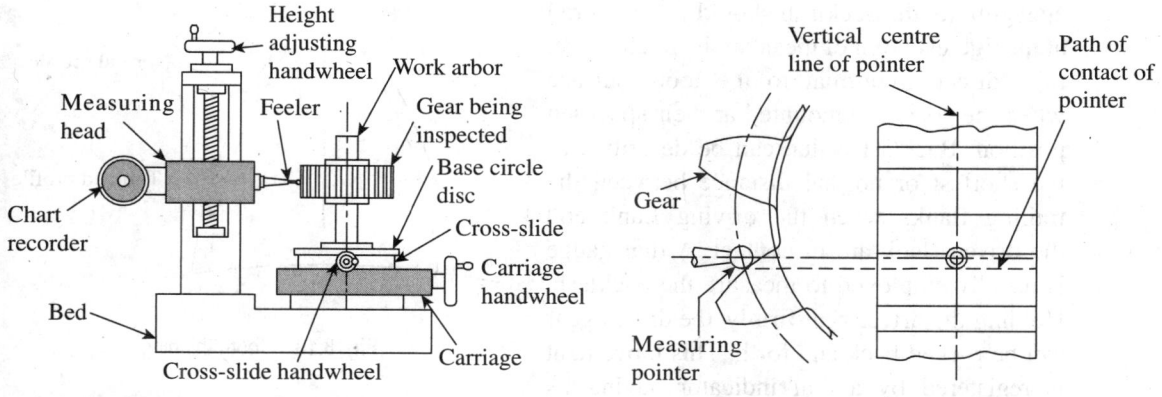

Fig. 8.11 Gear-measuring machine   **Fig. 8.12** Measurement of lead

feeler can be amplified 250, 500, or 1000 times, the amplification ratio being selected by a selector switch. When there is no error in the involute profile, the trace on the recording chart will be a straight line. Gleason gear inspection machine, a product of Gleason Metrology Systems Corporation, USA, follows the fundamental design aspect of any testing machine with the capability to handle up to 350 mm dia gears. It also integrates certain object-oriented tools to achieve faster cycle times and a better human–machine interaction.

8.4.4 Measurement of Lead

Lead is the axial advance of a helix for one complete rotation about its axis. In case of spur gears, lead tolerance is defined as the allowable deviation across the face width of a tooth surface. Control of lead is necessary in order to ensure adequate contact across the face width when gear and pinion are in mesh. Figure 8.12 illustrates the procedure adopted for checking lead tolerance of a spur gear.

A measuring pointer traces the tooth surface at the pitch circle and parallel to the axis of the gear. The measuring pointer is mounted on a slide, which travels parallel to the centre on which the gear is held. The measuring pointer is connected to a dial gauge or any other suitable comparator, which continuously indicates the deviation. The total deviation shown by the dial indicator over the distance measured indicates the amount of displacement of the gear tooth in the face width traversed.

Measurement of lead is more important in helical and worm gears. Interested readers are advised to refer to a gear handbook to learn more about the same.

8.4.5 Measurement of Backlash

If the two mating gears are produced such that tooth spaces are equal to tooth thicknesses at the reference diameter, then there will not be any clearance in between the teeth that are getting engaged with each other. This is not a practical proposition because the gears will get jammed even from the slightest mounting error or eccentricity of bore to the pitch circle diameter. Therefore, the tooth profile is kept uniformly thinned, as shown in Fig. 8.13. This results in a small play between the mating tooth surfaces, which is called a *backlash*.

We can define backlash as the amount by which a tooth space exceeds the thickness of an engaging tooth. Backlash should be measured at the tightest point of mesh on the pitch circle, in a direction normal to the tooth surface when the gears are mounted at their specified position. Backlash value can be described as the shortest or normal distance between the trailing flanks when the driving flank and the driven flank are in contact. A dial gauge is usually employed to measure the backlash. Holding the driver gear firmly, the driven gear can be rocked back and forth. This movement is registered by a dial indicator having its pointer positioned along the tangent to the pitch circle of the driven gear.

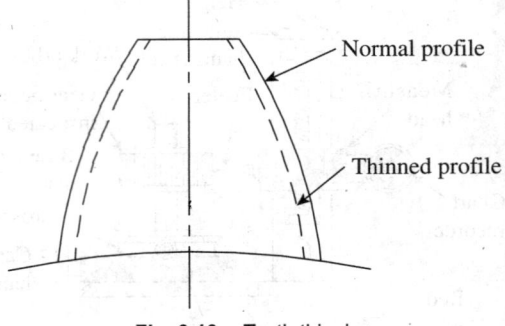

Fig. 8.13 Tooth thinning

8.4.6 Measurement of Tooth Thickness

Various methods are recommended for the measurement of gear tooth thickness. There is a choice of instruments such as the gear tooth calliper, and span gauging or tooth span micrometer. Constant chord measurement and measurement over rolls or balls are additional options. Two such methods, namely measurement with gear tooth calliper and tooth span micrometer are discussed in detail here.

Measurement with Gear Tooth Callipers

This is one of the most commonly used methods and perhaps the most accurate one. Figure 8.14 illustrates the construction details of a gear calliper. It has two vernier scales, one horizontal and the other vertical. The vertical vernier gives the position of a blade, which can slide up and down. When the surface of the blade is flush with the tips of the measuring anvils, the vertical scale will read zero. The blade position can be set to any required value by referring to the vernier scale.

From Fig. 8.15, it is clear that tooth thickness should be measured at the pitch circle (chord thickness C_1C_2 in the figure). Now, the blade position is set to a value equal to the addendum of the gear tooth and locked into position with a locking screw. The calliper is set on the gear in such a manner that the blade surface snugly fits with the top surface of a gear tooth. The two anvils are brought into close contact with the gear, and the chordal thickness is noted down on the horizontal vernier scale.

Let d = Pitch circle diameter

g_c = Chordal thickness of gear tooth along the pitch circle

h_c = Chordal height

z = Number of teeth on the gear

Chordal thickness g_c = Chord C_1C_2

$$= 2(\text{pitch circle radius}) \times \sin ç$$
$$= 2 \times \frac{d}{2} \times \sin ç$$
$$= d \sin ç$$

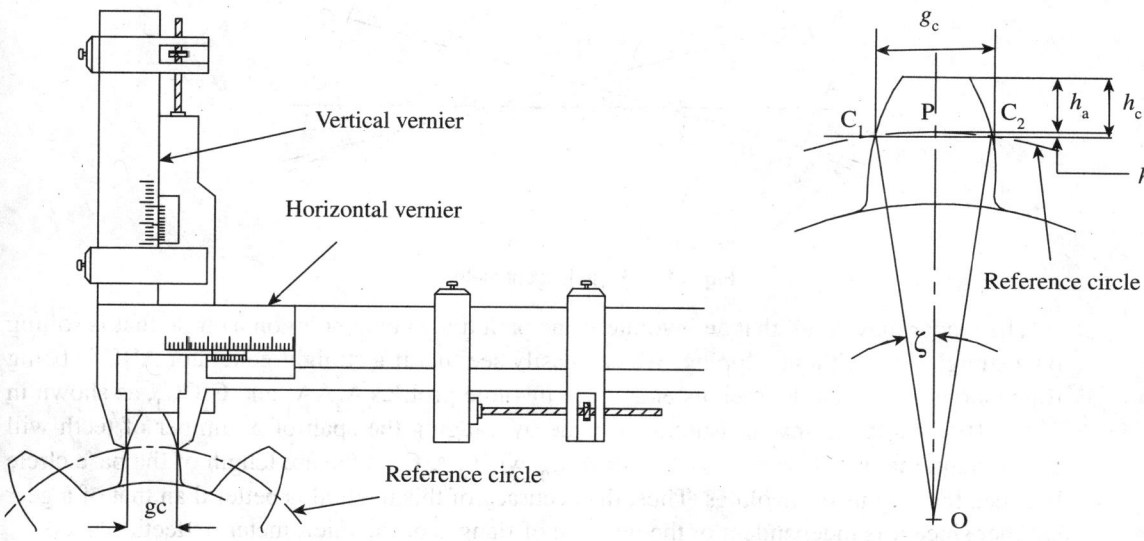

Fig. 8.14 Gear tooth calliper

Fig. 8.15 Chordal thickness and chordal height

Arc $C_1PC_2 = \dfrac{d}{2} \times 2\varsigma$ (value of ς in radians)

$$= d \times \varsigma = \frac{\pi d}{2z}$$

Therefore, $\varsigma = \dfrac{\pi}{2z}$

$g_c = d\sin(\pi/2z)$ (where $\pi/2z$ is in radians)

$g_c = d\sin(90/2z)$ (argument of sin is in degrees)

Chordal height $h_c = h_a + \Delta h = m + \Delta h$

However, $\Delta h(d - \Delta h) = \dfrac{gc}{2} \times \dfrac{gc}{2}$

and $4(\Delta h)^2 - 4\Delta h \times d + g_c^2 = 0$

$\Delta h = [d \pm \sqrt{d^2 - g_c^2}]/2$

$= [d - \sqrt{d^2 - g_c^2}]/2$; the other value is neglected because $\Delta h > d$ is not possible.

Neglecting $(\Delta h)^2$, we get $\Delta h \times d = g_c^2/4$

$\Delta h = g_c^2/4d$

Thus, $h_c = h_a + g_c^2/4d = m + g_c^2/4d$

Therefore, $g_c = d \sin (90°/z)$ (8.1)

and $h_c = m + g_c^2/4d$ (8.2)

Measurement with Tooth Span Micrometers

In this case, tooth thickness is measured by measuring the chordal distance over a number of teeth by using a *tooth span micrometer*, also called a *flange micrometer*. The inspection of gear is simple, easy, and fast. The measurement is based on the base tangent method and is illustrated in Fig. 8.16.

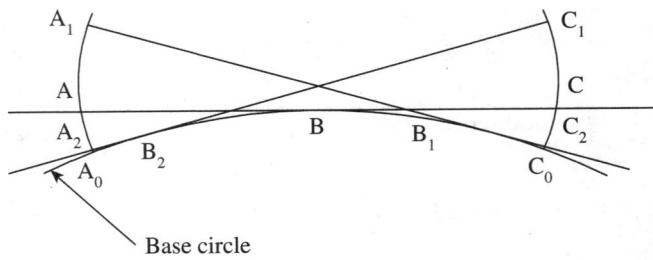

Fig. 8.16 Base tangent method

The reader may recall that an involute is the path traced by a point on a circle that is rolling on a straight line without slipping. We can easily see that if a straight generator ABC is being rolled along a base circle, then its ends trace involute profiles A_2AA_1 and C_2CC_1, as shown in Fig. 8.16. Therefore, any measurement made by gauging the span of a number of teeth will be constant, that is, $AC = A_1C_2 = A_2C_1 = A_0C_0$, where A_0C_0 is the arc length of the base circle between the origins of involutes. Thus, the accuracy of this method is better than that of a gear calliper since it is independent of the position of flanges of the micrometer on teeth flanks.

Suppose, the gear has N number of teeth and the length AC on the pitch circle corresponds to S number of teeth (called the tooth span), then $AC = (S - \frac{1}{2})$ pitches.

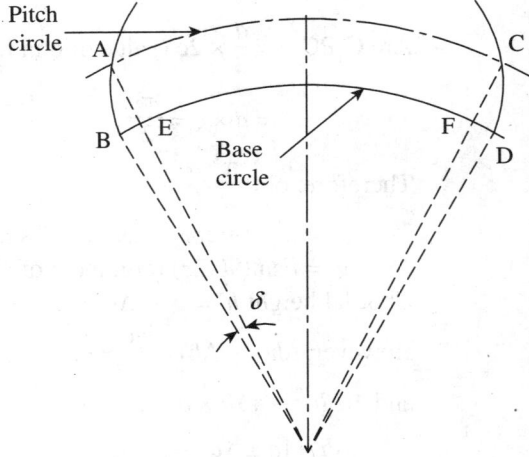

Fig. 8.17 Measurement across two opposed involutes

Therefore, angle subtended by AC = $(S - \frac{1}{2}) \times 2\pi/N$ radians

Figure 8.17 illustrates the measurement across two opposed involutes over a span of several teeth.

Involute function of pressure angle $= \delta = \tan \phi - \phi$, where ϕ is the pressure angle.

Therefore, angle of arc BD = $(S - \frac{1}{2}) \times 2\pi/N + 2(\tan \phi - \phi)$.

From Fig. 8.17, it is clear that BD = angle of arc BD $\times R_b$, where R_b is the radius of the base circle.

Thus, BD = $[(S - \frac{1}{2}) \times 2\pi/N + 2(\tan \phi - \phi)] \times R_p \cos \phi$ $\quad (R_b = R_p \cos \phi)$

$\quad = \dfrac{mN}{2} \cos \phi [(S - \frac{1}{2}) \times 2\pi/N + 2(\tan \phi - \phi)]$ $\quad (R_b = R_p \cos \phi)$

Therefore,

length of arc BD = $Nm \cos \phi \left[\dfrac{\pi S}{N} - \dfrac{\pi}{2N} + \tan \phi - \phi \right]$ $\hspace{2cm}$ (8.3)

This is the measurement W_s made across opposed involutes using the tooth span micrometer (Fig. 8.18). The following paragraphs illustrate the methodology.

Span gauging length is called the base tangent length and is denoted by W_s, where s is the number of spans. If the measurement is carried over five teeth, then s will be denoted as 5. While carrying out measurements, the micrometer reading is set to the value of W_s as determined from Eq. (8.3). If the span selected is 3, then s is taken as 3 in the equation. The micrometer reading is locked using a locking screw. Now, the micrometer is as good as an inspection gauge and can be used to check gears for accuracy of span width. The inspection is carried out with the contact of measuring flanges being made approximately at the mid-working depth of the gear teeth.

Tables that serve as ready reckoners for the span width for given values of module, number of teeth on the gear, and span width are available. Table 8.3 gives a sample of the tabulated values. The advantage of using the table is that the span width can be readily measured without having to calculate the value using relevant equations.

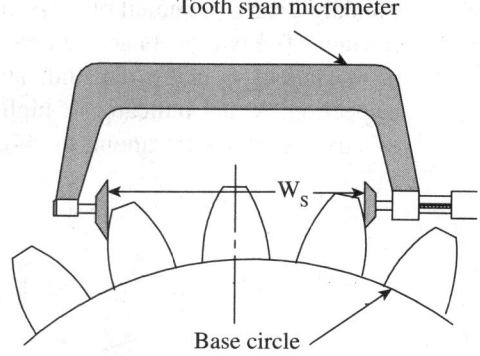

Fig. 8.18   Tooth span micrometer

Table 8.3   Values of span width

| Value of base tangent length W_s for uncorrected spur gears in mm | | | | | |
| Pressure angle = 20° | | | Module m = 1 | | |
Number of teeth on the gear (z)	Number of teeth spanned (s)	Base tangent length (W_s; mm)	Number of teeth on the gear (z)	Number of teeth spanned (s)	Base tangent length (W_s; mm)
7	1	1.5741	25	3	7.7305
8	1	1.5881	26	3	7.7445
9	2	4.5542	27	4	10.7106
10	2	4.5683	28	4	10.7246
11	2	4.5823	29	4	10.7386
–	–	–	–	–	–
–	–	–	–	–	–

8.5 COMPOSITE METHOD OF GEAR INSPECTION

Composite action refers to the variation in centre distance when a gear is rolled in tight mesh with a standard gear. It is standard practice to specify composite tolerance, which reflects gear runout, tooth-to-tooth spacing, and profile variations. Composite tolerance is defined as the allowable centre distance variation of the given gear, in tight mesh with a standard gear, for one complete revolution. The Parkinson gear testing machine is generally used to carry out composite gear inspection.

8.5.1 Parkinson Gear Tester

It is a popular gear testing machine used in metrology laboratories and tool rooms. The gear being inspected will be made to mesh with a standard gear, and a dial indicator is used to capture radial errors. The features of a Parkinson gear tester are illustrated in Fig. 8.19. The standard gear is mounted on a fixed frame, while the gear being inspected is fixed to a sliding carriage. The two gears are mounted on mandrels, which facilitate accurate mounting of gears in machines, so that a dial indicator will primarily measure irregularities in the gear under inspection. A dial indicator of high resolution is used to measure the composite error, which reflects errors due to runout, tooth-to-tooth spacing, and profile variations.

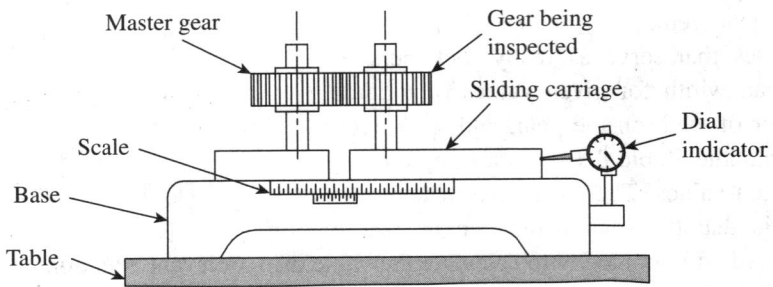

Fig. 8.19 Parkinson gear tester

To start with, the two gears are mounted on respective mandrels and the slide comprising the standard gear is fixed at a convenient position. The sliding carriage is moved along the table, the two gears are brought into mesh, and the sliding carriage base is also locked in its position. Positions of the two mandrels are adjusted in such a way that their axial distance is equal to the gear centre distance as per drawings. However, the sliding carriage is free to slide for a small distance on steel rollers under a light spring force. A vernier scale attached to the machine enables measurement of the centre distance up to 25 μm. The dial indicator is set to zero and the gear under inspection is rotated. Radial variations of the gear being inspected are indicated by the dial indicator. This variation is plotted on a chart or graph sheet, which indicates the radial variations in the gear for one complete rotation.

Many improvisations are possible to the basic machine explained in Section 8.5.1. A waxed paper recorder can be fitted to the machine so that a trace of the variations of a needle in contact with the sliding carriage is made simultaneously. The mechanism can be designed to provide a high degree of magnification.

8.6 MEASUREMENT OF SCREW THREADS

Screw thread geometry has evolved since the early 19th century, thanks to the importance of threaded fasteners in machine assemblies. The property of *interchangeability* is associated more strongly with screw threads than with any other machine part. Perhaps, the *Whitworth thread system*, proposed as early as the 1840s, was the first documented screw thread profile that came into use. A couple of decades later, the *Sellers* system of screw threads came into use in the United States. Both these systems were in practice for a long time and laid the foundation for a more comprehensive *unified screw thread system*.

Screw thread gauging plays a vital role in industrial metrology. In contrast to measurements of geometric features such as length and diameter, screw thread measurement is more complex. We need to measure inter-related geometric aspects such as pitch diameter, lead, helix, and flank angle, among others. The following sections introduce screw thread terminology and discuss the measurements of screw thread elements and thread gauging, which speeds up the inspection process.

8.7 SCREW THREAD TERMINOLOGY

Figure 8.20 illustrates the various terminologies associated with screw threads.

Screw thread   The American Society of Tool and Manufacturing Engineers (ASTME) defines a screw thread as follows: screw thread is the helical ridge produced by forming a continuous helical groove of uniform section on the external or internal surface of a cylinder or cone.

Form of thread   This is the shape of the contour of one complete thread, as seen in an axial section. Some of the popular thread forms are British Standard Whitworth, American Standard, British Association, Knuckle, Buttress, Unified, Acme, etc.

External thread   The screw thread formed on the external surface of a workpiece is called an external thread. Examples of this include bolts and studs.

Internal thread   The screw thread formed on the internal surface of a workpiece is called an internal thread. The best example for this is the thread on a nut.

Axis of thread (pitch line)   This is the imaginary line running longitudinally through the centre of the screw.

Fundamental triangle   It is the imaginary triangle that is formed when the flanks are extended till they meet each other to form an apex or a vertex.

Angle of thread   This is the angle between the flanks of a thread measured in the axial plane. It is also called an included angle.

Flank angle   It is the angle formed between a flank of the thread and the perpendicular to the axis of the thread that passes through the vertex of the fundamental triangle.

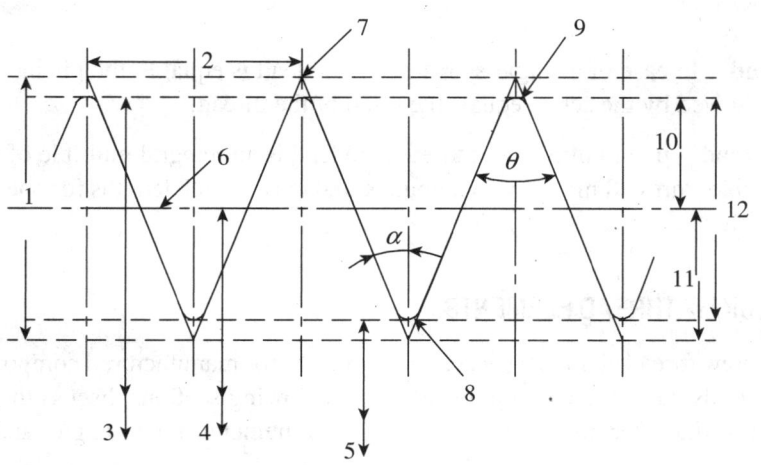

Legend
1: Angular pitch
2: Pitch
3: Major diameter
4: Pitch diameter
5: Minor diameter
6: Pitch line
7: Apex
8: Root
9: Crest
10: Addendum
11: Dedendum
12: Depth of thread
θ : Angle of thread
α : Flank angle

Fig. 8.20   Screw thread terminology

Pitch   It is the distance between two corresponding points on adjacent threads, measured parallel to the axis of the thread.

Lead   It is the axial distance moved by the screw when the crew is given one complete revolution about its axis.

Lead angle   It is the angle made by the helix of the thread at the pitch line with the plane perpendicular to the axis.

Helix angle   It is the angle made by the helix of the thread at the pitch line with the axis. This angle is measured in an axial plane.

Major diameter   In case of external threads, the major diameter is the diameter of the major cylinder (imaginary), which is coaxial with the screw and touches the crests of an external thread. For internal threads, it is the diameter of the cylinder that touches the root of the threads.

Minor diameter   In case of external threads, the minor diameter is the diameter of the minor cylinder (imaginary), which is coaxial with the screw and touches the roots of an external thread. For internal threads, it is the diameter of the cylinder that touches the crests of the threads. It is also called the root diameter.

Addendum   It is the radial distance between the major diameter and pitch line for external threads. On the other hand, it is the radial distance between the minor diameter and pitch line for internal threads.

Dedendum   It is the radial distance between the minor diameter and pitch line for external threads. On the other hand, it is the radial distance between the major diameter and pitch line for internal threads.

Effective diameter or pitch diameter   It is the diameter of the pitch cylinder, which is coaxial with the axis of the screw and intersects the flanks of the threads in such a way as to make the widths of threads and the widths of spaces between them equal. In general, each of the screw threads is specified by an effective diameter as it decides the quality of fit between the screw and a nut.

Single-start thread   In case of a single-start thread, the lead is equal to the pitch. Therefore, the axial distance moved by the screw equals the pitch of the thread.

Multiple-start thread   In a multiple-start thread, the lead is an integral multiple of the pitch. Accordingly, a double start will move by an amount equal to two pitch lengths for one complete revolution of the screw.

8.8 MEASUREMENT OF SCREW THREAD ELEMENTS

Measurement of screw thread elements is necessary not only for manufactured components, but also for threading tools, taps, threading hobs, etc. The following sections discuss the methods for measuring major diameter, minor diameter, effective diameter, pitch, angle, and form of threads.

8.8.1 Measurement of Major Diameter

The simplest way of measuring a major diameter is to measure it using a screw thread micrometer. While taking readings, only light pressure must be used, as the anvils make contact with the screw solely at points and any excess application of pressure may result in a slight deformation of anvil due to compressive force, resulting in an error in the measurement. However, for a more precise measurement, it is recommended to use a bench micrometer shown in Fig. 8.21.

A major advantage of a bench micrometer is that a fiducial indicator is a part of the measuring system. It is thus possible to apply a pressure already decided upon by referring to the fiducial indicator. However, there is no provision for holding the workpiece between the centres, unlike a floating carriage micrometer. The inspector has to hold the workpiece by hand while the readings are being taken.

The machine is essentially used as a comparator. To start with, the anvil positions are set by inserting a setting cylinder. A setting cylinder serves as a gauge and has a diameter that equals the OD of the screw thread being inspected. Now, the setting cylinder is taken out, the workpiece is inserted between the anvils, and the deviation is noted down on the micrometer head. Since the position of the fixed anvil will remain unaltered due to the setting of the fiducial arrangement, the movable anvil will shift axially depending on the variation in the value of OD of the screw being inspected. In order to sense deviations on either side of the preset value, the movable anvil will always be set to a position, which can detect small movements in either direction. The error, as measured by the micrometer head, is added to or subtracted from, as the case may be, the diameter of the setting cylinder to get the actual value of OD.

Measurement of the OD of internal threads is trickier, as it is cumbersome to take measurements using conventional instruments. An easier option is to employ some indirect measurement techniques. A cast of the thread is made, which results in a male counterpart of the internal thread. Now, the measurement can be carried out using techniques used for external threads. The cast may be made of plaster of Paris or wax.

8.8.2 Measurement of Minor Diameter

The best way of measuring a minor diameter is to measure it using a floating carriage micrometer described in Chapter 4. The carriage has a micrometer with a fixed spindle on one side and a movable spindle with a micrometer on the other side. The carriage moves on a finely ground 'V' guideway or an anti-friction guideway to facilitate movement in a direction parallel to the axis of the plug gauge mounted between centres.

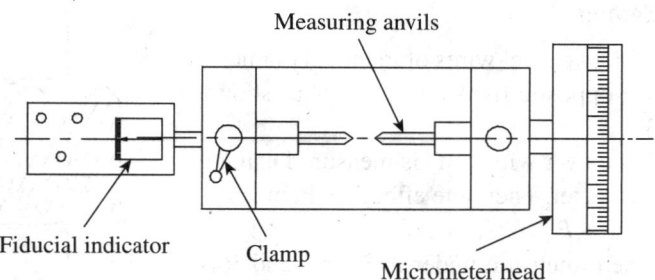

Fig. 8.21 Bench micrometer

The micrometer has a non-rotary spindle with a least count of up to 0.001 or 0.002 mm. The instrument is very useful for thread plug gauge manufacturers; in gauge calibration laboratories, established under NABL accreditation; and in standard rooms where in-house gauge calibration is carried out.

Minor diameter is measured by a comparative process, wherein small V-pieces that make contact at the root of the threads are used. The selection of V-pieces should be such that the included angle of a V-piece is less than the angle of the thread. V-pieces are placed on each side of the screw with their bases against the micrometer faces. As in the previous case, the initial reading is taken by mounting a setting cylinder corresponding to the dimension being measured. Then, the threaded workpiece is mounted between the centres and the reading is taken. The difference in the two readings directly gives the error in the minor diameter.

8.8.3 Measurement of Effective Diameter

In Section 8.7 we defined an effective diameter of a screw thread as the diameter of the pitch cylinder, which is coaxial with the axis of the screw and intersects the flanks of the threads in such a way so as to make the width of threads and widths of spaces between them equal. Since it is a notional value, it cannot be measured directly and we have to find the means of measuring it in an indirect way. Thread measurement by wire method is a simple and popular way of measuring an effective diameter. Small, hardened steel wires (best-size wire) are placed in the thread groove, and the distance over them is measured as part of the measurement process. There are three methods of using wires: one-wire, two-wire, and three-wire methods.

One-wire Method

This method is used if a standard gauge of the same dimension as the theoretical value of dimension over wire is available. First of all, the micrometer anvils are set over the standard gauge and the dimension is noted down. Thereafter, the screw to be inspected is held either in hand or in a fixture, and the micrometer anvils are set over the wire as shown in Fig. 8.22.

Micrometer readings are taken at two or three different locations and the average value is calculated. This value is compared with the value obtained with the standard gauge. The resulting difference is a reflection of error in the effective diameter of the screw. An important point to be kept in mind is that the diameter of the wire selected should be such that it makes contact with the screw along the pitch cylinder. The significance of this condition will become obvious in the two-wire method explained in the next section.

Two-wire Method

In this method, two steel wires of identical diameter are placed on opposite flanks of a screw, as shown in Fig. 8.23.

The distance over wires (M) is measured using a suitable micrometer. Then, the effective diameter,

$$D_e = T + P \qquad (8.4)$$

where T is the dimension under the wires and P is the correction factor.

And,

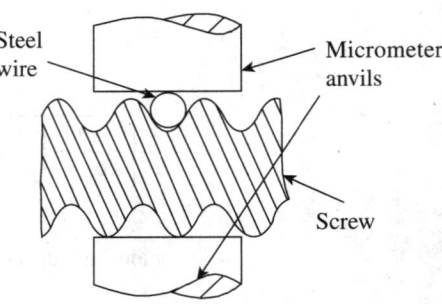

Fig. 8.22 One-wire method

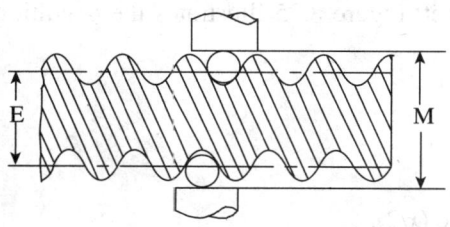

Fig. 8.23   Two-wire method

$$T = M - 2d \qquad (8.5)$$

where d is the diameter of the best-size wire.

These relationships can be easily derived by referring to Fig. 8.24.

The two wires of identical diameter are so selected that they make contact with the screw thread on the pitch line. The aforementioned equations are valid only if this condition is met.

Accordingly, from triangle OFD, $OF = \dfrac{d}{2} \operatorname{cosec} (x/2)$

$$FA = \frac{d}{2} \operatorname{cosec} (x/2) - \frac{d}{2} = \frac{d[\operatorname{cosec}(x/2) - 1]}{2}$$

$$FG = GC \cot (x/2) = \frac{p}{4} \cot (x/2) \quad \text{(because BC = pitch/2 and GC = pitch/4)}$$

Therefore, $AG = FG - FA = \dfrac{p}{4} \cot (x/2) - \dfrac{d[\operatorname{cosec}(x/2) - 1]}{2}$

Since AG accounts for the correction factor only on one side of the screw, we have to multiply this value by 2 in order to account for that on the opposite flank.

Therefore, total correction factor is as follows:

$$P = 2\,AG = \frac{p}{2} \cot (x/2) - d[\operatorname{cosec}(x/2) - 1] \qquad (8.6)$$

Although it is possible to measure the value of M, the distance over the wires, using a hand-held micrometer, this method is prone to errors. A better alternative is to use a floating carriage micrometer shown in Fig. 4.41 of Chapter 4, which helps in aligning the micrometer square to the thread, enabling more accurate readings.

Diameter of Best-size Wire

The best-size wire, of diameter d, makes contact with the thread flank along the pitch line.

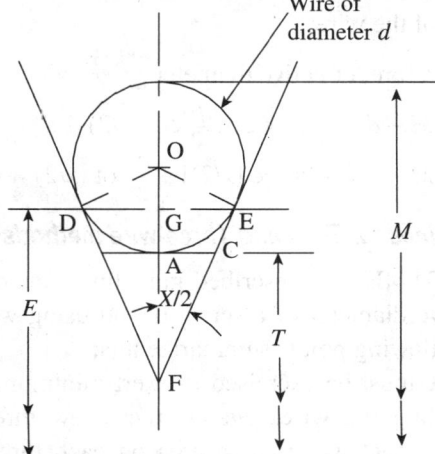

Fig. 8.24   Measurements in two-wire method

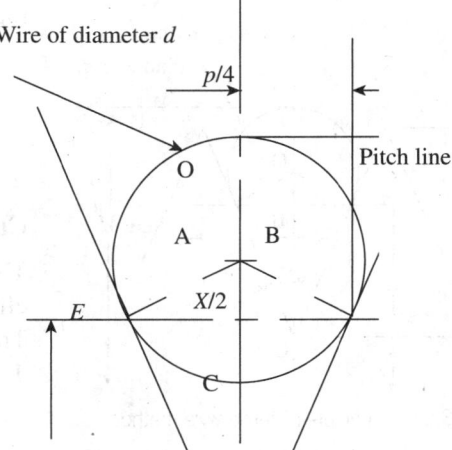

Fig. 8.25   Determination of the best-size wire

Equations (8.4)–(8.6) hold true if this condition is met. Figure 8.25 illustrates the condition achieved by the best-size wire.

In triangle OAB, sin (AOB) = AB/OB

that is, sin (90 – x/2) = AB/OB

or, OB = $\dfrac{AB}{\sin(90 - x/2)} = \dfrac{AB}{\cos(x/2)}$ = AB sec (x/2)

Diameter of the best-size wire = 2(OB) = 2(AB) sec (x/2).

However, from Fig. 8.25, AB = p/4, where p is the pitch of the thread.

Therefore, diameter of the best-size wire is

$$d = (p/2)\ \sec(x/2) \qquad (8.7)$$

Three-wire Method

The three-wire method is an extension of the principle of the two-wire method. As illustrated in Fig. 8.26, three wires are used to measure the value of M, one wire on one side and two wires on adjacent thread flanks on the other side of the screw. Measurement can be made either by holding the screw, wires, and micrometer in hand or by using a stand with an attachment to hold the screw in position. Since three wires are used, the micrometer can be positioned more accurately to measure M, the distance over the wires.

With reference to Fig. 8.27, let M be the distance over the wires, E the effective diameter of the screw, d the diameter of best-size wires, and H the height of threads.

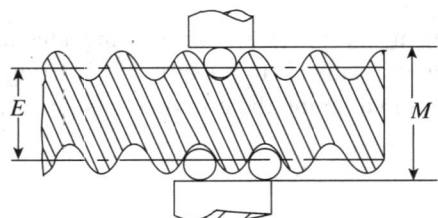

Fig. 8.26 Three-wire method

Now, OC = OA cosec (x/2) = $\dfrac{d}{2}$ cosec (x/2) $\qquad$ (8.8)

$H = \dfrac{p}{2}$ cot (x/2) and, therefore, BC = H/2

$\qquad = \dfrac{p}{4}$ cot (x/2) $\qquad$ (8.9)

If h is the height of the centre of wire from the pitch line, then h = OC – BC.

$h = \dfrac{d}{2}$ cosec (x/2) – $\dfrac{p}{4}$ cot (x/2) $\qquad$ (8.10)

Distance over wires, $M = E + 2h + 2r$, where r is the radius of the wires.

Therefore, effective diameter

$E = M - d$ cosec (x/2) $+ \dfrac{p}{2}$ cot (x/2) $- d$

$E = M - d[1 + $ cosec (x/2)$] + \dfrac{p}{2}$ cot (x/2) $\qquad$ (8.11)

Guidelines for Two- and Three-wire methods

The ASTME has prescribed guidelines for measuring the effective diameter of a screw thread using wire methods. The following points summarize this:

1. Care must be exercised to exert minimum force while holding the wires against the screw thread. Since a wire touches a minute area on each thread flank, de

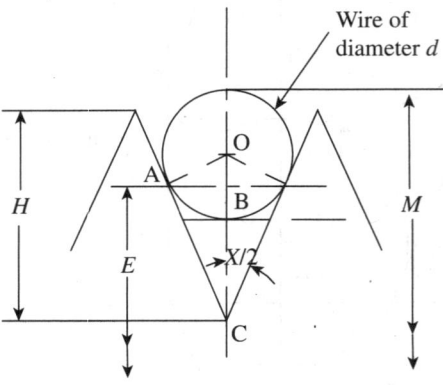

Fig. 8.27 Principle of three-wire method

formation of wire and thread will be sufficiently large to warrant some type of correction.

2. The wires should be accurately finished and hardened steel cylinders. The working surface should at least be 25 mm in length. The wires should be provided with a suitable means of suspension.

3. One set of wires should consist of three wires having the same diameter within 0.000025 mm. These wires should be measured between a flat contact and a hardened and accurately finished cylinder having a surface roughness not over 5 μm.

4. If it becomes necessary to measure the effective diameter by means of wires other than the best size, the following size limitations should be followed:

   (a) The minimum size is limited to that which permits the wire to project above the crest of the thread.

   (b) The maximum size is limited to that which permits the wire to rest on the flanks of the thread just below the crest, and not ride on the crest of the thread.

5. The wires should be free to assume their positions in the thread grooves without any restraint (the practice of holding wires in position with elastic bands can introduce errors in the measurement).

8.8.4 Measurement of Pitch

Usually, a screw thread is generated by a single-point cutting tool, with the two basic parameters being angular velocity of the workpiece and linear velocity of the tool. The tool should advance exactly by an amount equal to the pitch for one complete rotation of the workpiece. Pitch errors are bound to crop up if this condition is not satisfied. Pitch errors may be classified into the following types:

Progressive pitch error   This error occurs whenever the tool–work velocity ratio is incorrect but constant. Generally, it is caused by the pitch error in the lead screw of the machine. Figure 8.28 illustrates a progressive pitch error, in which the cumulative pitch error increases linearly along the thread numbers. Other factors contributing to the pitch error are an incorrect gear train or an approximate gear train when, for example, metric threads are being generated on basically a system suited for producing British Standard threads.

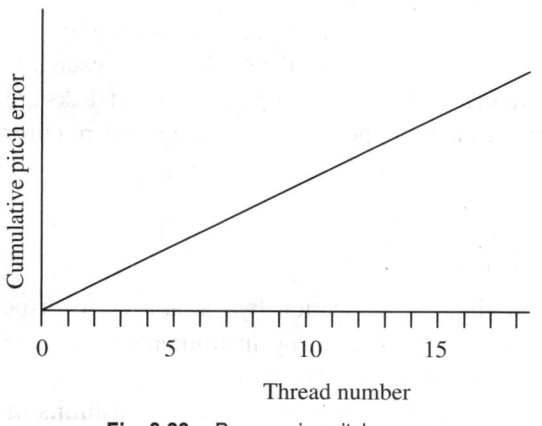

Fig. 8.28 Progressive pitch error

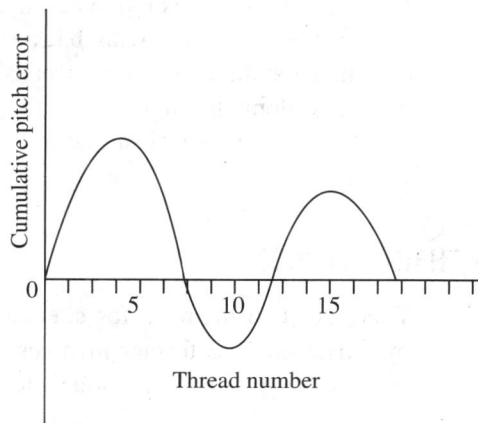

Fig. 8.29 Periodic pitch error

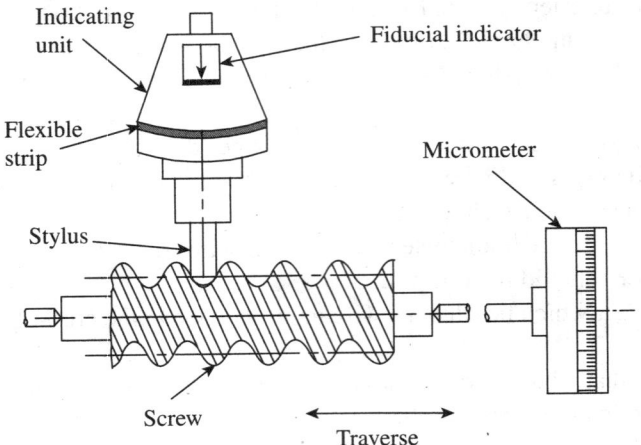

Fig. 8.30 Pitch-measuring machine

Periodic pitch error This error occurs when the tool–work velocity ratio is not constant. This results in a profile shown in Fig. 8.29. The contributing factors are pitch errors in the gear trains and/or axial movement of the lead screw due to worn out thrust bearings. A graph of cumulative pitch error versus thread number shows an approximate sinusoidal form. It can be seen that the pitch increases to a maximum value and reduces to a minimum value, and the pattern repeats throughout the length of the screw.

Drunken threads We can draw an analogy between drunken threads and a drunken (alcoholic) person, because of the fact that the thread helix will develop into a curve instead of a straight line. This results in a periodic error creeping in at intervals of one pitch. The pitch measured parallel to the thread axis will be correct, but the threads are not cut to a true helix. This error does not pose a serious problem for the function of a screw, but may result in a noticeable error for large-diameter screws.

Measurement of Pitch Error

Figure 8.30 illustrates a pitch-measuring machine.

A round-nosed stylus makes contact with the thread at approximately the pitch line. The screw, which is mounted between the centres on a slide, is given a traverse motion in a direction parallel to its axis. This makes the stylus ride over the threads. A micrometer connected to the slide enables measurement of the pitch.

A few improvisations are made in the machine for performing an accurate measurement. Since the stylus is required to make contact approximately along the pitch line, a fiducial indicator is used to ensure that it registers zero reading whenever the stylus is at the precise location. The stylus is mounted on a block supported by a thin flexible strip and a strut. This enables the stylus in moving back and forth over the threads. The micrometer reading is taken each time the fiducial indicator reads zero. These readings show the pitch error of each thread of the screw along its length and can be plotted on a graph sheet. Special graduated disks are used to fit the micrometer to suit all ordinary pitches and/or special pitches as per the requirements of the user.

8.9 THREAD GAUGES

There are two methods for checking screw threads: inspection by variables and inspection by attributes. The former involves the application of measuring instruments to measure the extent of deviation of individual elements of a screw thread. The latter involves the use of limit gauges, called *thread gauges*, to assure that the screw is within the prescribed limits of size. Thread gauges offer simpler and speedier means for checking screw threads. Gauging of screw threads is a process for assessing the extent to which screw threads conform to specified limits of size. A complete set of standards is available for thread gauging, which comprises limits of size, gauging methodology, and measurement. These standards assure interchangeability in assembly, acceptance of satisfactory threads, and rejection of those threads that are outside the prescribed limits of size.

Thread gauges are classified on the basis of the following criteria:
1. Classification of thread gauges based on the type of application
   (a) Working gauges
   (b) Inspection gauges
   (c) Master gauges
2. Classification of thread gauges based on their forms
   (a) Plug screw gauges
   (b) Ring screw gauges

Working gauges are used by production workers when the screws are being manufactured. On the other hand, inspection gauges are used by an inspector after the production process is completed. Working and inspection gauges differ in the accuracy with which they are made. Inspection gauges are manufactured to closer tolerance than working gauges and are therefore more accurate. Obviously, master gauges are even more accurate than inspection gauges. This kind of hierarchy, with respect to accuracy, ensures that the parts that are produced will have a high level of quality.

In practice, plug gauges are used to inspect external thread forms, while ring gauges are used to inspect internal thread forms. Taylor's principle of limit gauging, explained in Section 3.6.4, is also applicable for thread gauging also. It may be remembered that according to Taylor's principle, a GO gauge should check both size and geometric features, and thus be of full form. On the other hand, a NOT GO gauge should check only one dimension at a time. However, if the NOT GO gauge is made of full form, any reduction in the effective diameter due to pitch error may give misleading results. To account for this aspect, the NOT GO gauge is designed to only check for the effective diameter, which is not influenced by errors in pitch or form of the thread. The following paragraphs provide basic information on these two types of gauges.

Plug Screw Gauges

The thread form for a NOT GO gauge is truncated as shown in Fig. 8.31. This is necessary to avoid contact with the crest of the mating thread, which, while being large on the effective diameter, may have a minor diameter on the low limit. The truncation also helps in avoiding contact with the root of the nut. This modification will not have any bearing on the accuracy of measurement, because the NOT GO gauge will primarily check for only the effective diameter.

A straight thread plug gauge assures the accuracy of the basic elements of a screw thread, namely major diameter, pitch diameter, root clearance, lead, and flank angle. A taper thread plug

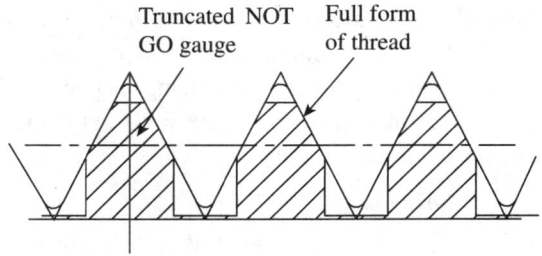

Fig. 8.31 Truncated thread form for a NOT GO gauge

gauge checks, in addition to these elements, taper of the screw as well as length from the front end to the gauge notch. The gauges are made of hardened gauge steel. The entering portion of a plug gauge is vulnerable to rapid wear and tear because of frequent contact with metal surfaces. Therefore, the ends are slightly extended with lower diameter, so that contact with the workpiece takes place gradually rather then abruptly (Fig. 8.32).

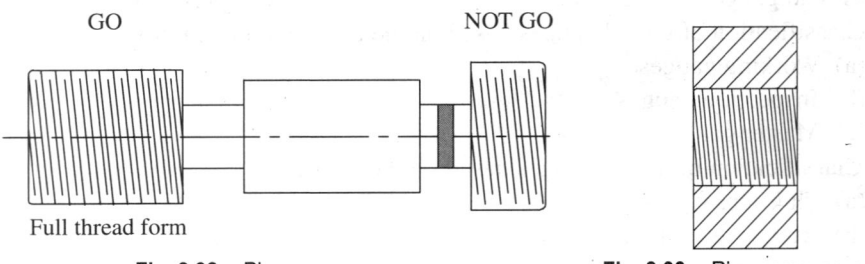

GO   NOT GO

Full thread form

Fig. 8.32 Plug screw gauge   **Fig. 8.33** Ring screw gauge

Ring Screw Gauges

Ring gauges are used to check external thread forms like bolts. Similar to plug gauges, a system of limit gauges is provided by the full-form GO gauge and NOT GO gauge to check the effective diameter (Fig. 8.33).

8.10 NUMERICAL EXAMPLES

Example 8.1   The following data is available for the measurement of chordal thickness of a gear having an involute profile: the number of teeth = 32, addendum circle diameter = 136 mm, and pressure angle = 20°. Determine the chordal height to which the gear tooth calliper should be set during measurement.

Solution

Pitch circle diameter, $d = mz$, where m is the module and z is the number of teeth on the gear.
However, m = (addendum circle diameter)/$(z + 2)$
Therefore, $m = 136/(32 + 2) = 4$
$$d = 4(32) = 128 \text{ mm}$$
Theoretical value of chordal thickness $g_c = d \sin (90°/z) = 128 \sin (90/32) = 6.28$ mm
Chordal height $h_c = m + g_c^2/4d = 4 + (6.28)^2/4(128) = 4.077$ mm
Thus, the gear tooth calliper should be set to a chordal height of 4.077 mm

Example 8.2   A tooth span micrometer is being used to measure the span across three teeth. Three trials are conducted and the average value of span width is 31.120 mm. The following data is available for the gear being inspected: the number of teeth = 32, addendum circle diameter = 136 mm, pressure angle = 20°, and span of measurement = 3. Determine the percentage error of measurement.

Solution

Theoretical span width $W_{3T} = Nm \cos \phi \left[\dfrac{\pi S}{N} - \dfrac{S}{2N} + \tan \phi - \phi \right]$

Here, N is the number of teeth on the gear, m is the module, s is the span, and ϕ is the pressure angle.

However, m = (addendum circle diameter)/$(z + 2)$
Therefore, $m = 136/(32 + 2) = 4$
Therefore, $W_{3T} = (32)\,(4)\cos 20 \left[\dfrac{\pi\,3}{32} - \dfrac{\pi}{2(32)} + \tan 20 - \pi/9 \right]$
$\quad\quad\quad\quad = 31.314\,\text{mm}$
Error $\quad\quad\quad = 31.314 - 31.120 = 0.194\,\text{mm}$
Percentage error $= 0.194/31.314 = 0.006\%$

Example 8.3 It is required to measure the effective diameter of a screw using the two-wire method. The distance across 10 threads measured using a scale is 12.5 mm. Determine the size of the best wire for metric threads.

Solution

Pitch $p = 12.5/10 = 1.25\,\text{mm}$
$\quad\quad d = (p/2)\sec(x/2)$, where $x = 60°$ for metric threads.

Therefore, diameter of the best-size wire $d = \dfrac{1.25}{2}\sec(60/2) = 0.722\,\text{mm}$.

Example 8.4 A metric screw thread is being inspected using the two-wire method in order to measure its effective diameter and the following data is generated: Pitch = 1.25 mm, diameter of the best-size wire = 0.722 mm, and distance over the wires = 25.08 mm. Determine the effective diameter of the screw thread.

Solution

Effective diameter, $D_e = T + P$
where T is the dimension under the wires and P is the correction factor.
$\quad\quad T = M - 2d$
$\quad\quad P = \dfrac{p}{2}\cot(x/2) - d[\operatorname{cosec}(x/2) - 1]$
Therefore, $P = \dfrac{1.25}{2}\cot(60/2) - 0.722\left[\operatorname{cosec}\left(\dfrac{60}{2}\right) - 1\right] = 0.3605\,\text{mm}$
$\quad\quad T = 25.08 - 2(0.722) = 23.636\,\text{mm}$
$\quad\quad D_e = 23.636 + 0.3605 = 23.9965\,\text{mm}$

A QUICK OVERVIEW

- Gears are the main elements in a transmission system. It is needless to say that for efficient transfer of speed and power, gears should perfectly conform to the designed profile and dimensions. Misalignments and gear runout will result in vibrations, chatter, noise, and loss of power. On the other hand, threaded components should meet stringent quality requirements to satisfy the property of *interchangeability*.
- The following elements of gears are important for an analytical inspection: runout, pitch, profile, lead, backlash, and tooth thickness.
- A profile-measuring machine is one of the most versatile machines used in metrology. The machine is provided with multiple axes movement to enable measurement of various types of gears. The measuring head comprising the feeler, electronic unit, and chart recorder can be moved up and down by operating a handwheel. The movement of the feeler can be amplified

250, 500, or 1000 times, the amplification ratio being selected by a selector switch.

- A gear calliper has two vernier scales, one horizontal and the other vertical. The vertical vernier gives the position of a blade, which can slide up and down. The blade position is set to a value equal to the addendum of the gear tooth and the two anvils are brought into close contact with the gear; the chordal thickness can be measured on the horizontal vernier scale.
- Gear tooth thickness can be measured by measuring the chordal distance over a number of teeth by using a *tooth span micrometer*, also called a *flange micrometer*. The inspection of gear is simple, easy, and fast. The measurement is based on the base tangent method.
- The Parkinson gear tester is a composite gear testing machine. The gear being inspected will be made to mesh with a standard gear. A dial indicator of high resolution is used to measure the composite error, which reflects errors due to runout, tooth-to-tooth spacing, and profile variations.
- Screw thread gauging plays a vital role in industrial metrology. In contrast to measurements of geometric features such as length and diameter, screw thread measurement is more complex. We need to measure inter-related geometric aspects such as pitch diameter, lead, helix, and flank angle, to name a few.
- A major advantage of a bench micrometer is that a fiducial indicator is a part of the measuring system. It is thus possible to apply pressure that has ben decided upon by referring to the fiducial indicator. This prevents damage of threads and provides longer life to the measuring anvils.
- Thread measurement by wire method is a simple and popular way of measuring effective diameter. Small, hardened steel wires (best-size wire) are placed in the thread groove, and the distance over them is measured as part of the measurement process. There are three methods of using wires: one-wire, two-wire, and three-wire methods.
- The best-size wire of diameter d makes contact with the thread flank along the pitch line.
- We can draw an analogy between drunken threads and a drunken (alcoholic) person, because of the fact that the thread helix will develop into a curve instead of a straight line. This results in periodic error creeping in at intervals of one pitch. The pitch measured parallel to the thread axis will be correct, but the threads are not cut to a true helix.
- In practice, plug gauges are used to inspect external thread forms, while ring gauges are used to inspect internal thread forms. In accordance with Taylor's principle, a GO gauge should check both size and geometric features, and thus be of full form. On the other hand, a NOT GO gauge should check only one dimension at a time. The NOT GO gauge is designed to check for only effective diameter, which is not influenced by errors in pitch or form of the thread.

MULTIPLE-CHOICE QUESTIONS

1. The involute profile of a spur gear is limited to only the
   - (a) root circle
   - (b) base circle
   - (c) pitch circle
   - (d) addendum circle
2. The angle between the line of action and the common tangent to the pitch circles is known as
   - (a) flank angle
   - (b) tooth angle
   - (c) included angle
   - (d) pressure angle
3. The path traced by a point on a circle that is rolling on a straight line without slipping is
   - (a) involute
   - (b) cycloid
   - (c) epicycloid
   - (d) hypocycloid
4. The tooth profile of mating gears is kept uniformly thinned, which results in a small play between mating tooth surfaces. This is called
   - (a) backlash
   - (b) pitch correction
   - (c) lead correction
   - (d) none of these
5. In order to measure the chordal thickness of a

gear using a gear calliper, the position of the blade is set to

(a) the entire depth of the gear tooth
(b) addendum of the gear tooth
(c) dedendum of the gear tooth
(d) top surface of the gear tooth

6. Which of the following is the tester in which the gear being inspected is made to mesh with a standard gear and a dial indicator is used to capture the radial errors?
(a) Pitch-checking instrument
(b) Johnson gear tester
(c) Parkinson gear tester
(d) McMillan gear tester

7. The angle formed between a flank of the thread and the perpendicular to the axis of the thread, which passes through the vertex of the fundamental triangle, is called
(a) a helix angle   (c) a lead angle
(b) a flank angle   (d) an included angle

8. The indicator that enables the application of a pressure already decided upon on the screw thread in a bench micrometer is called
(a) a fiducial indicator
(b) a pressure indicator
(c) a span indicator
(d) none of these

9. In wire methods, the diameter of the wire selected should be such that it makes contact with the screw along the
(a) outer diameter   (c) root diameter
(b) pitch cylinder   (d) axis of the screw

10. In a two-wire method, diameter of the best-size wire is given by

(a) $d = (p/2) \sec (x/2)$
(b) $d = (p/4) \sec (x/2)$
(c) $d = (p/2) \csc (x/2)$
(d) $d = (p/2) \cot (x/2)$

11. The pitch error that occurs whenever the tool–work velocity ratio is incorrect but constant is referred to as a
(a) cyclic error
(b) velocity error
(c) progressive error
(d) non-progressive error

12. In which of the following cases is the pitch measured parallel to the thread axis correct, but the threads are not cut to a true helix?
(a) Drunken threads   (c) Whitworth threads
(b) Sunken threads   (d) Metric threads

13. The two methods of inspecting screw threads are
(a) inspection by constants and variables
(b) inspection by variables and attributes
(c) inspection by quality and cost
(d) inspection of attributes and constants

14. Thread gauges that are used to inspect external thread forms are called
(a) plug screw gauges
(b) ring screw gauges
(c) external screw gauges
(d) all of these

15. A NOT GO screw gauge will primarily check for
(a) outer diameter and nothing else
(b) inside diameter and nothing else
(c) effective diameter and nothing else
(d) all of these

REVIEW QUESTIONS

1. Which elements of a spur gear require inspection? Name at least one instrument that is used for measuring each of these elements.
2. What is the radial runout of a gear? How is it measured?
3. With the help of a neat sketch, describe the construction and working of a pitch-measuring instrument.
4. What is the principle of involute measurement?
5. Define profile tolerance. How is the profile of a spur gear traced using a profile-measuring instrument?
6. Is backlash inevitable in a gear pair? Discuss.
7. Explain how a gear calliper enables an accurate measurement of chordal thickness of a spur gear.
8. 'The accuracy of the base tangent method is better than the system of using a gear calliper, since it is independent of the position of the

flanges of the micrometer on teeth flanks.' Justify this statement.
9. Write a note on the Parkinson gear tester.
10. With the help of a sketch, discuss screw thread terminologies.
11. Define the effective diameter of a screw thread.
12. With the help of a sketch, explain the construction and working of a bench micrometer.
13. Differentiate between two- and three-wire methods.
14. Derive the expression for the best-size wire in a two-wire method.
15. List the guidelines for the proper application of two-wire/three-wire methods.
16. Distinguish between progressive pitch error and periodic pitch error.
17. What is drunken thread?
18. Give a classification of thread gauges.
19. How is Taylor's principle applicable to thread gauging?
20. What is the need for using the truncated thread form in a NOT GO screw gauge?

PROBLEMS

1. The chordal thickness of an involute gear is being measured using a gear calliper. Determine the chordal height to which the gear tooth calliper should be set during measurement, if the number of teeth = 28, addendum circle diameter = 120 mm, pressure angle = 20°, and least count of the calliper is 0.02 mm.
2. Determine the effective diameter of a metric screw using the three-wire method. The following data is available: diameter of the best-size wire = 0.740 mm, distance over the wires = 25.58 mm, and pitch = 1.25 mm.
3. A metric screw thread is being inspected using the two-wire method in order to measure its effective diameter and the following data is generated: pitch = 1.5 mm, diameter of the best-size wire = 0.866 mm, distance over the wires = 26.58 mm, and thread angle = 60°. Determine the effective diameter of the screw thread.

Metrology of Surface Finish

After studying this chapter, the reader will be able to

- appreciate the importance of surface texture measurement and its significance
- understand the basic reasons for surface irregularities
- explain the terminology associated with the quantification and measurement of surface irregularities
- describe the surface texture characteristics and their symbolic representations
- elucidate the various methods of measurement of surface roughness
- explain the relationship between wavelength of surface roughness, frequency, and cut-off

9.1 INTRODUCTION

In contrast to the concepts we have studied hitherto, surface metrology is basically concerned with deviations between points on the same surface. On the other hand, in all other topics, the fundamental concern has been the relationship between a feature of a part or assembly and some other feature. Even though surface texture is important in many fields of interest such as aesthetics and cosmetics, among others, the primary concern in this chapter pertains to manufactured items that are subject to stress, move in relation to one another, and have close fits joining them. Surface roughness (a term used in a general way here, since it has specific connotations that will be explained shortly) or surface texture depends, to a large extent, on the type of the manufacturing operation. If rough surface for a part is acceptable, one may choose a casting, forging, or rolling operation. In many cases, the surfaces that need to contact each other for some functional requirement have to be machined, possibly followed by a finishing operation like grinding.

The reasons for pursuing surface metrology as a specialized subject are manifold. We would like to make our products operate better, cost less, and look better. In order to achieve these objectives, we need to examine the surfaces of the parts or components more closely, at the microscopic level. It would be naive to assume that two apparently flat contacting surfaces are in perfect contact throughout the apparent area of contact. Most of the earlier laws of friction were based on this assumption (perhaps until 1950). In reality, surfaces have *asperities*, which refer to the peaks and valleys of surface irregularities. Contact between the mating parts is believed to take place at the peaks. When the parts are forced against each other, they deform either elastically or plastically. In case of elastic behaviour, they return to the full height after deformation by the mating surface. If they behave plastically, some of the deformation is permanent. These aspects have a bearing on the friction characteristics of the parts in contact. As mechanical engineering is primarily concerned with machines and moving parts that are designed to precisely fit with each other, surface metrology has become an important topic in engineering metrology.

9.2 SURFACE METROLOGY CONCEPTS

If one takes a look at the topology of a surface, one can notice that surface irregularities are superimposed on a widely spaced component of surface texture called waviness. Surface irregularities generally have a pattern and are oriented in a particular direction depending on the factors that cause these irregularities in the first place. Figure 9.1 illustrates some of these features.

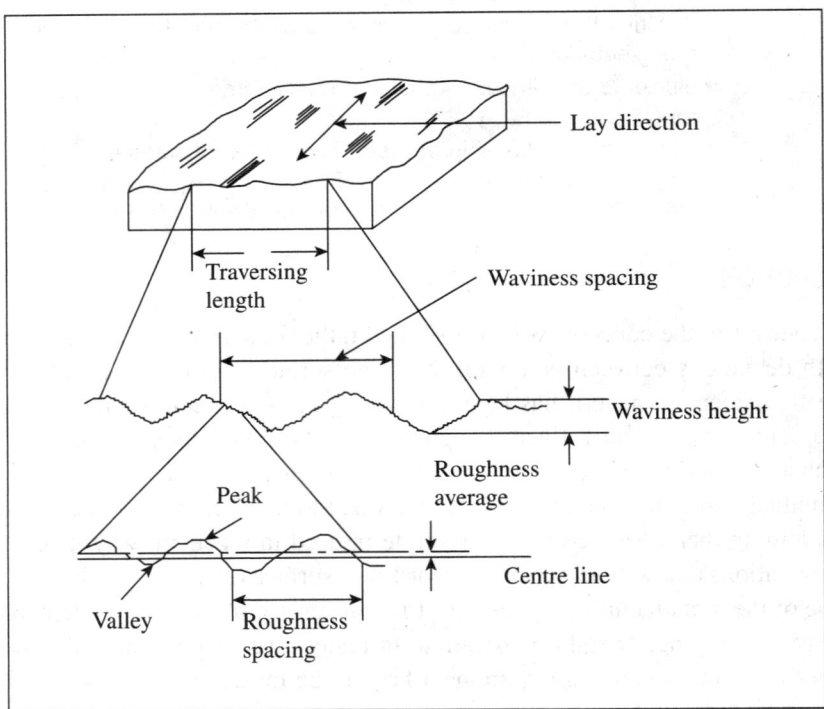

Fig. 9.1 Waviness and roughness

Surface irregularities primarily arise due to the following factors:
1. Feed marks of cutting tools
2. Chatter marks on the workpiece due to vibrations caused during the manufacturing operation
3. Irregularities on the surface due to rupture of workpiece material during the metal cutting operation

4. Surface variations caused by the deformation of workpiece under the action of cutting forces
5. Irregularities in the machine tool itself like lack of straightness of guideways

Thus, it is obvious that it is practically impossible to produce a component that is free from surface irregularities. Imperfections on a surface are in the form of succession of hills and valleys varying in both height and spacing. In order to distinguish one surface from another, we need to quantify surface roughness; for this purpose, parameters such as height and spacing of surface irregularities can be considered. In mechanical engineering applications, we are primarily concerned with the roughness of the surface influenced by a machining process. For example, a surface machined by a single-point cutting tool will have a roughness that is uniformly spaced and directional. In the case of a finish machining, the roughness is irregular and non-directional. In general, if the hills and valleys on a surface are closely packed, the wavelength of the waviness is small and the surface appears rough. On the other hand, if the hills and valleys are relatively far apart, waviness is the predominant parameter of interest and is most likely caused by imperfections in the machine tool. If the hills and valleys are closely packed, the surface is said to have a primary texture, whereas surfaces with pronounced waviness are said to have a secondary texture.

9.3 TERMINOLOGY

Roughness   The American Society of Tool and Manufacturing Engineers (ASTME) defines roughness as the finer irregularities in the surface texture, including those irregularities that result from an inherent action of the production process. Roughness spacing is the distance between successive peaks or ridges that constitute the predominant pattern of roughness. Roughness height is the arithmetic average deviation expressed in micrometres and measured perpendicular to the centre line.

Waviness   It is the more widely spaced component of surface texture. Roughness may be considered to be superimposed on a wavy surface. Waviness is an error in form due to incorrect geometry of the tool producing the surface. On the other hand, roughness may be caused by problems such as tool chatter or traverse feed marks in a supposedly geometrically perfect machine. The spacing of waviness is the width between successive wave peaks or valleys. Waviness height is the distance from a peak to a valley.

Lay   It is the direction of the predominant surface pattern, ordinarily determined by the production process used for manufacturing the component. Symbols are used to represent lays of surface pattern, which will be discussed in Section 9.5.

Flaws   These are the irregularities that occur in isolation or infrequently because of specific causes such as scratches, cracks, and blemishes.

Surface texture   It is generally understood as the repetitive or random deviations from the nominal surface that form the pattern of the surface. Surface texture encompasses roughness, waviness, lay, and flaws.

Errors of form   These are the widely spaced repetitive irregularities occurring over the full length of the work surface. Common types of errors of form include bow, snaking, and lobbing.

9.4 ANALYSIS OF SURFACE TRACES

It is required to assign a numerical value to surface roughness in order to measure its degree. This will enable the analyst to assess whether the surface quality meets the functional requirements of a component. Various methodologies are employed to arrive at a representative parameter of surface roughness. Some of these are 10-point height average (Rz), root mean square (RMS) value, and the centre line average height (Ra), which are explained in the following paragraphs.

9.4.1 Ten-point Height Average Value

It is also referred to as the *peak-to-valley height*. In this case, we basically consider the average height encompassing a number of successive peaks and valleys of the asperities. As can be seen in Fig. 9.2, a line AA parallel to the general lay of the trace is drawn. The heights of five consecutive peaks and valleys from the line AA are noted down.

The average peak-to-valley height Rz is given by the following expression:

$$Rz = \frac{(h_1 + h_3 + h_5 + h_7 + h_9) - (h_2 + h_4 + h_6 + h_8 + h_{10})}{5} \times \frac{1000}{\text{Vertical magnification}}\ \mu m$$

9.4.2 Root Mean Square Value

Until recently, RMS value was a popular choice for quantifying surface roughness; however, this has been superseded by the centre line average value. The RMS value is defined as the square root of the mean of squares of the ordinates of the surface measured from a mean line. Figure 9.3 illustrates the graphical procedure for arriving at an RMS value.

With reference to this figure, if h_1, h_2, ..., h_n are equally spaced ordinates at points 1, 2, ..., n, then

$$h_{RMS} = \frac{\sqrt{(h_1^2 + h_2^2 + \ldots + h_n^2)}}{n}$$

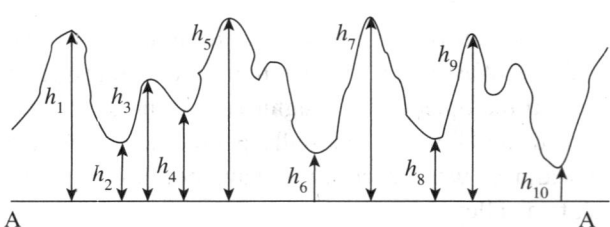

Fig. 9.2 Measurement to calculate the 10-point height average

9.4.3 Centre Line Average Value

The Ra value is the prevalent standard for measuring surface roughness. It is defined as the average height from a mean line of all ordinates of the surface, regardless of sign. With reference to Fig. 9.4, it can be shown that

$$Ra = \frac{A_1 + A_2 + \ldots + A_N}{L}$$

$$= \Sigma A/L$$

Interestingly, four countries (USA, Canada, Switzerland, and Netherlands) have exclusively adopted Ra value as the standard for measuring surface roughness. All other countries have included other assessment methods in addition to the Ra method. For

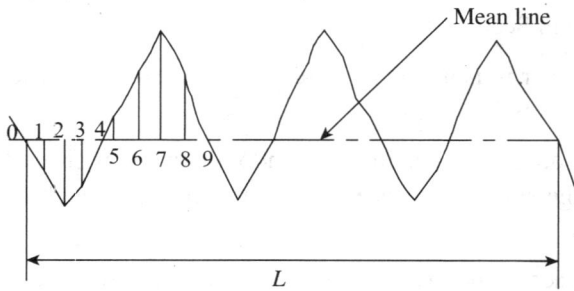

Fig. 9.3 Representation of an RMS value

instance, France has seven additional standards.

It should be mentioned here that the Ra value is an index for surface texture comparison and not a dimension. This value is always much less than the peak-to-valley height. It is generally a popular choice as it is easily understood and applied for the purpose of measurement. The bar chart shown in Fig. 9.5 illustrates the typical Ra values obtained in basic manufacturing operations.

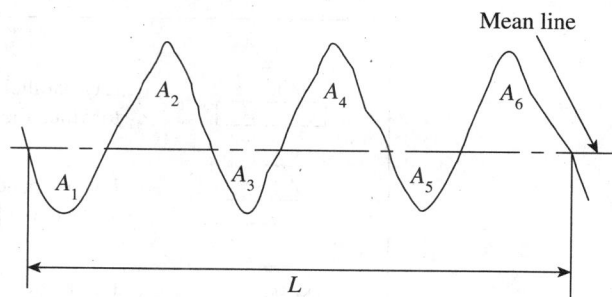

Fig. 9.4 Representation of Ra value

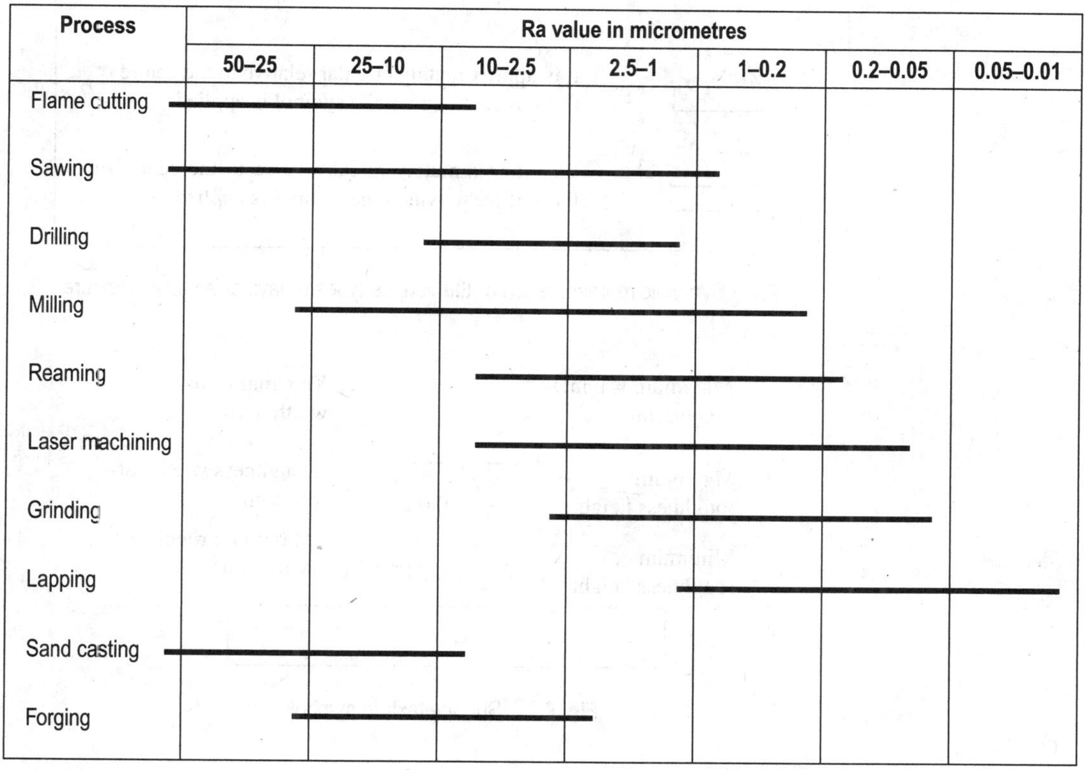

Process	Ra value in micrometres						
	50–25	25–10	10–2.5	2.5–1	1–0.2	0.2–0.05	0.05–0.01
Flame cutting							
Sawing							
Drilling							
Milling							
Reaming							
Laser machining							
Grinding							
Lapping							
Sand casting							
Forging							

Fig. 9.5 Bar chart indicating the range of Ra values for various manufacturing operations

Note: The bars indicate the entire range. In most cases, the Ra value is restricted to the mid 50% portion of the bars.

9.5 SPECIFICATION OF SURFACE TEXTURE CHARACTERISTICS

Design and production engineers should be familiar with the standards adopted for specification of the characteristics of surface texture. Symbols are used to designate surface irregularities such as the lay of surface pattern and the roughness value. Figure 9.6 illustrates the symbolic

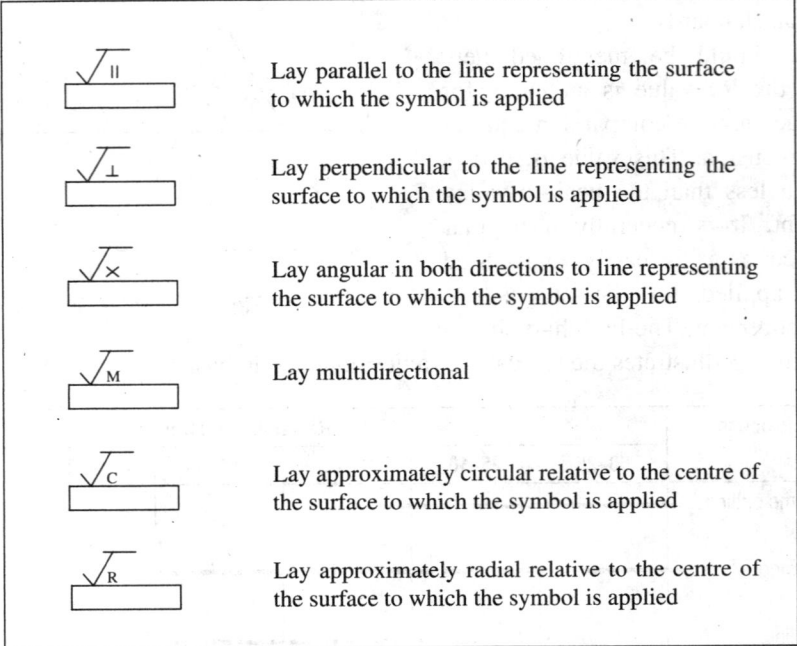

Fig. 9.6 Symbolic representation of the various types of lays of a surface texture

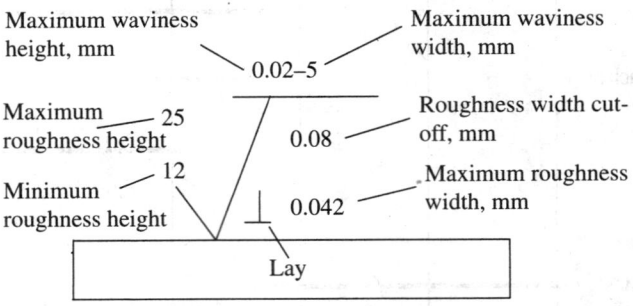

Fig. 9.7 Surface texture symbols

representation of the various types of lays and Fig. 9.7 highlights surface texture symbols with specifications.

9.6 METHODS OF MEASURING SURFACE FINISH

There are basically two approaches for measuring surface finish: comparison and direct measurement. The former is the simpler of the two but is more subjective in nature. The comparative method advocates assessment of surface texture by observation or feel of the surface. Microscopic examination is an obvious improvisation of this method. However, it still has two major drawbacks. First, the view of a surface may be deceptive; two surfaces that appear identical may be quite different. Second, the height of the asperities cannot be readily determined. Touch is perhaps a better method than visual observation. However, this method is also subjective in nature and depends, to a large extent, on the judgement of a person, and therefore not reliable.

These limitations have driven metrology experts to devise ways and means of directly measuring surface texture by employing direct methods. Direct measurement enables a numerical value to be assigned to the surface finish. The following sections explain the popular methods for the determination of surface texture.

9.7 STYLUS SYSTEM OF MEASUREMENT

The stylus system of measurement is the most popular method to measure surface finish. The operation of stylus instruments is quite similar to a phonograph pickup. A stylus drawn across the surface of the workpiece generates electrical signals that are proportional to the dimensions of the asperities. The output can be generated on a hard copy unit or stored on some magnetizable media. This enables extraction of measurable parameters from the data, which can quantify the degree of surface roughness. The following are the features of a stylus system:

1. A skid or shoe drawn over the workpiece surface such that it follows the general contours of the surface as accurately as possible (the skid also provides the datum for the stylus)
2. A stylus that moves over the surface along with the skid such that its motion is vertical relative to the skid, a property that enables the stylus to capture the contours of surface roughness independent of surface waviness
3. An amplifying device for magnifying the stylus movements
4. A recording device to produce a trace or record of the surface profile
5. A means for analysing the profile thus obtained

9.7.1 Stylus and Datum

There are two types of stylus instruments: true datum and surface datum, which are also known as *skidless* and *skid* type, respectively. In the skidless instrument, the stylus is drawn across the surface by a mechanical movement that results in a precise path. The path is the datum from which the assessment is made. In the skid-type instrument, the stylus pickup unit is supported by a member that rests on the surface and slides along with it. This additional member is the skid or the shoe. Figure 9.8 illustrates the relationship between the stylus and the skid.

Skids are rounded at the bottom and fixed to the pickup unit. They may be located in front of or behind the stylus. Some instruments use a shoe as a supporting slide instead of a skid. Shoes are flat pads with swivel mountings in the head. The datum created by a skid or a shoe is the locus of its centre of curvature as it slides along the surface.

The stylus is typically a diamond having a

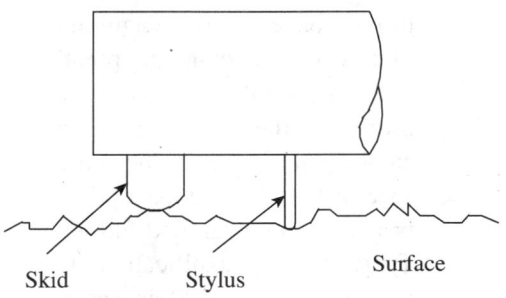

Fig. 9.8 Skid and stylus type

Skid   Stylus   Surface

cone angle of 90° and a spherical tip radius of 1–5 µm or even less. The stylus tip radius should be small enough to follow the details of the surface irregularities, but should also have the strength to resist wear and shocks. Stylus load should also be controlled so that it does not leave additional scratch marks on the component being inspected.

In order to capture the complete picture of surface irregularities, it is necessary to investigate waviness (secondary texture) in addition to roughness (primary texture). Waviness may occur with the same lay as the primary texture. While a pointed stylus is used to measure roughness, a blunt stylus is required to plot the waviness.

9.8 STYLUS PROBE INSTRUMENTS

In most stylus-based instruments, a stylus drawn across the surface of a component being inspected generates electrical signals that are proportional to the changes in the surface asperities. An electrical means of amplifying signals, rather than a purely mechanical one, minimizes the pressure of the stylus on the component. Changes in the height of asperities may be directly read by a meter or a chart. Most instruments provide a graph of the stylus path along the surface. The following paragraphs explain some of the popular stylus probe instruments used for measuring surface roughness.

9.8.1 Tomlinson Surface Meter

This is a mechanical–optical instrument designed by Dr Tomlinson of the National Physical laboratory of the UK. Figure 9.9 illustrates the construction details of the Tomlinson surface meter. The sensing element is the stylus, which moves up and down depending on the irregularities of the workpiece surface. The stylus is constrained to move only in the vertical direction because of a leaf spring and a coil spring. The tension in the coil spring P causes a similar tension in the leaf spring. These two combined forces hold a cross-roller in position between the stylus and a pair of parallel fixed rollers. A shoe is attached to the body of the instrument to provide the required datum for the measurement of surface roughness.

A light spring steel arm is attached to the cross-roller and carries a diamond tip. The translatory motion of the stylus causes rotation of the cross-roller about the point A, which in turn is converted to a magnified motion of the diamond point. The diamond tip traces the profile of the workpiece on a smoked glass sheet. The glass sheet is transferred to an optical projector and magnified further. Typically, a magnification of the order of 50–100 is easily achieved in this instrument.

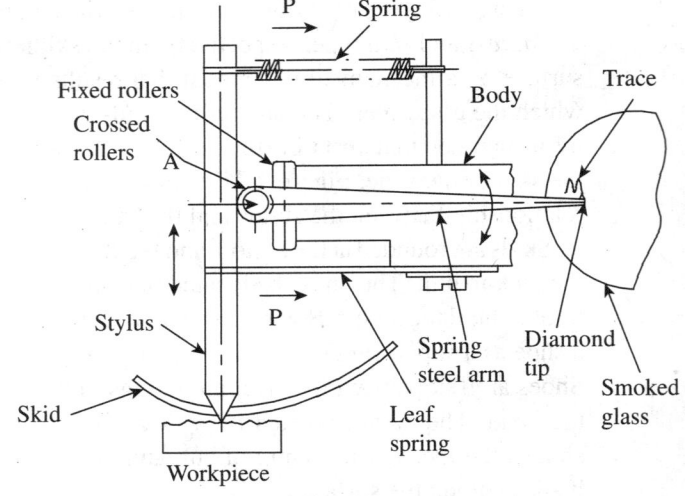

Fig. 9.9 Tomlinson surface meter

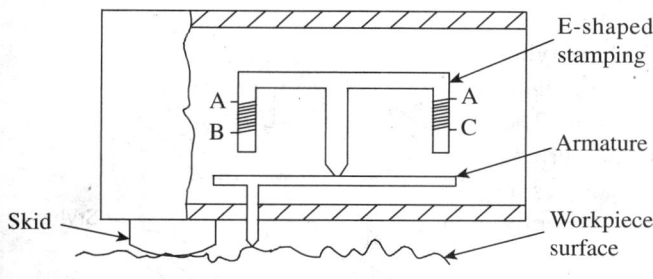

Fig. 9.10   Taylor–Hobson talysurf

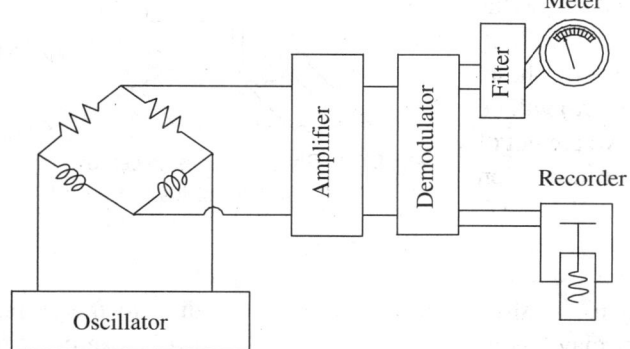

Fig. 9.11   Bridge circuit and electronics

In order to get a trace of the surface irregularities, a relative motion needs to be generated between the stylus and the workpiece surface. Usually, this requirement is met by moving the body of the instrument slowly with a screw driven by an electric motor at a very slow speed. Anti-friction guide-ways are used to provide friction-free movement in a straight path.

9.8.2 Taylor–Hobson Talysurf

The Taylor–Hobson talysurf works on the same principle as that of the Tomlinson surface meter. However, unlike the surface meter, which is purely a mechanical instrument, the talysurf is an electronic instrument. This factor makes the talysurf a more versatile instrument and can be used in any condition, be it a metrology laboratory or the factory shop floor.

Figure 9.10 illustrates the cross section of the measuring head. The stylus is attached to an armature, which pivots about the centre of piece of an E-shaped stamping. The outer legs of the E-shaped stamping are wound with electrical coils. A predetermined value of alternating current (excitation current) is supplied to the coils. The coils form part of a bridge circuit. A skid or shoe provides the datum to plot surface roughness. The measuring head can be traversed in a linear path by an electric motor. The motor, which may be of a variable speed type or provided with a gear box, provides the required speed for the movement of the measuring head.

As the stylus moves up and down due to surface irregularities, the armature is also displaced. This causes variation in the air gap, leading to an imbalance in the bridge circuit. The resulting bridge circuit output consists of only modulation. This is fed to an amplifier and a pen recorder is used to make a permanent record (Fig. 9.11). The instrument has the capability to calculate and display the roughness value according to a standard formula.

9.8.3 Profilometer

A profilometer is a compact device that can be used for the direct measurement of surface texture. A finely pointed stylus will be in contact with the workpiece surface. An electrical pickup attached to the stylus amplifies the signal and feeds it to either an indicating unit or a recording unit. The stylus may be moved either by hand or by a motorized mechanism.

The profilometer is capable of measuring roughness together with waviness and any other surface flaws. It provides a quick-fix means of conducting an initial investigation before attempting a major investigation of surface quality.

9.9 WAVELENGTH, FREQUENCY, AND CUT-OFF

The complete traverse length of the stylus instrument is called the *measuring traverse length*. It is divided into several sampling lengths. The sampling length is chosen based on the surface under test. Generally, results of all the samples in the measuring traverse length are averaged out by the instrument to give the final result.

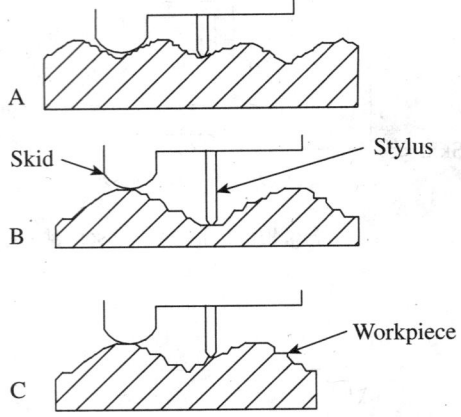

Fig. 9.12 Phase relationship between the skid and stylus

Skids simplify surface assessment while using stylus instruments. However, there is a distortion because of phase relationship between the stylus and the skid. This aspect is illustrated in Fig. 9.12. In case A, the stylus and the skid are in phase. Therefore, roughness (the primary texture) will be relatively undistorted. In case B, the two are out of phase. In this situation, waviness superimposes on the roughness reading and is misleading. In case C also, the stylus and skid are out of phase, resulting in an unrealistic interpretation of roughness value.

Thus, since the skid, like the stylus, is also rising and falling according to the surface asperities, stylus height measurement may be distorted. Therefore, care must be exercised for the selection of sampling length.

9.9.1 Cut-off Wavelength

The frequency of the stylus movement as it rises up and down the workpiece surface is determined by the traversing speed. Assuming that f is the frequency of the stylus movement, λ is the surface wavelength, and v is the traverse speed, one gets the following equation:

$$f = v/\lambda$$

Therefore, $f \propto 1/\lambda$, if v remains constant.

For surfaces produced by single-point cutting tools, a simple guideline for selecting cut-off wavelength is that it should not exceed one feed spacing. However, for many fine irregular surfaces, a cut-off length of 0.8 mm is recommended. Table 9.1 illustrates the recommended cut-off wavelengths for machining processes.

Table 9.1 Recommended wavelengths for machining processes

Finishing processes	Cut-off length (mm)
Superfinishing, lapping, honing, diamond boring, polishing, and buffing	0.25–0.8
Grinding	0.25–0.8
Turning, reaming, and broaching	0.8–2.5
Boring, milling, and shaping	0.8–8.0
Planning	2.5–25

9.10 OTHER METHODS FOR MEASURING SURFACE ROUGHNESS

In addition to the stylus-based methods of surface roughness measurement explained in Section 9.8, this section presents in brief some of the alternative methods used in the industry.

9.10.1 Pneumatic Method

The *air leakage method* is often used for assessing surface texture. A pneumatic comparator is used for conducting mass inspection of parts. Compressed air is discharged from a self-aligning nozzle held close to the surface being inspected. Depending on height variations in the surface irregularities, the gap between the nozzle tip and the workpiece surface varies. This results in the variation of flow rate of air, which in turn varies the rotation speed of a rotameter. Rotation of the rotameter is an indication of surface irregularities. Alternatively, a float can also be used to measure surface deviations. The comparator is initially set using reference gauges.

9.10.2 Light Interference Microscopes

The light interference technique offers a non-contact method of assessing surface texture. Advantages of this method are that it allows an area of the workpiece surface to be examined, a wide range of magnifications to be used, and the opportunity for making a permanent record of the fringe pattern using a camera. Good magnification capability allows good resolution up to a scratch spacing of 0.5 µm.

A monochromatic light passing through an optical flat and falling on the workpiece surface generates the fringe pattern. The technique of measurement using interference fringes has already been explained in Chapter 7. However, assessment of surface irregularities cannot be directly related to the Ra value. Master specimens are used to generate a reference fringe pattern, which is compared with the fringe pattern of the workpiece in order to arrive at a conclusion regarding surface quality. This method provides a viable alternative for inspecting soft or thin surfaces, which normally cannot be examined using stylus instruments.

9.10.3 Mecrin Instrument

The Mecrin instrument assesses surface irregularities through frictional properties and the average slope of the irregularities. This gauge is suited for surfaces manufactured by processes such as grinding, honing, and lapping, which have low Ra values in the range 3–5 µm. Figure 9.13 illustrates the working principle of this instrument.

A thin metallic blade is pushed against the workpiece surface at a certain angle. The blade may slide or buckle, depending on the surface roughness and the angle of attack. At lower angles of attack, the blade tip will slide over the surface of the workpiece. As the angle of attack is increased, a critical

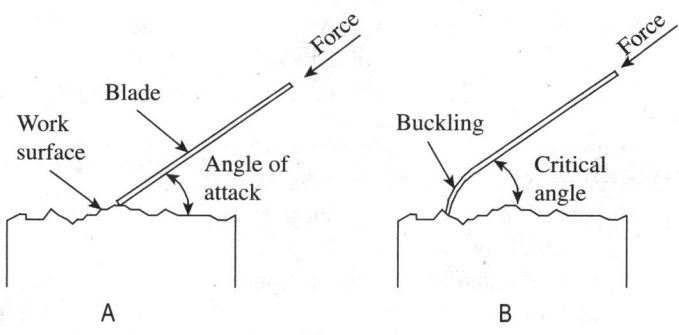

Fig. 9.13 Principle of the Mecrin instrument

value is reached at which the blade starts to buckle. This critical angle is a measure of the degree of roughness of the surface. The instrument is provided with additional features for easier handling. A graduated dial will directly give the reading of roughness value.

A QUICK OVERVIEW

- Workpiece surfaces have *asperities*, which are the peaks and valleys of surface irregularities. Contact between mating parts is believed to take place at the peaks. When the parts are forced against each other, they deform either elastically or plastically. In case of elastic behaviour, they return to the full height after deformation by the mating surface. If they behave plastically, some of the deformation is permanent. As these aspects have a bearing on the friction characteristics of the parts in contact, the study of surface texture has become an important part of metrology.

- To measure the degree of surface roughness, it is required to assign a numerical value to it. This will enable the analyst to assess whether the surface quality meets the functional requirements of a component. Various methodologies are employed to arrive at a representative parameter of surface roughness. Some of these are 10-point height average, RMS value, and the centre line average height.

- There are two approaches for measuring surface finish: comparison and direct measurement. The former is the simpler of the two, but is more subjective in nature. The comparative method advocates assessment of surface texture by observation or feel of the surface. On the other hand, direct measurement is more reliable since it enables a numerical value to be assigned to the surface finish.

- The stylus system of measurement is the most popular method to measure surface finish. Operation of stylus instruments is quite similar to a phonograph pickup. A stylus drawn across the surface of the workpiece generates electrical signals that are proportional to the dimensions of the asperities. The output can be generated on a hard copy unit or stored on some magnetizable media. This enables extraction of measurable parameters from the data, which can quantify the degree of surface roughness.

- Among the stylus-based measurement systems, the Tomlinson surface meter and Taylor–Hobson talysurf are popular.

- The complete traverse length of the stylus instrument is called the measuring traverse length. It is divided into several sampling lengths. The sampling length is chosen based on the surface under test. Generally, the results of all the samples in the measuring traverse length are averaged out by the instrument to give the final result.

- The frequency of the stylus movement as it rises up and down the workpiece surface is determined by the traversing speed. If f is the frequency of the stylus movement, λ is the surface wavelength, and v is the traverse speed, then $f = v/\lambda$.

MULTIPLE-CHOICE QUESTIONS

1. Surface texture depends to a large extent on
   (a) material composition
   (b) type of manufacturing operation
   (c) skill of the operator
   (d) accuracy of measurement

2. Peaks and valleys of surface irregularities are called
   (a) waves   (c) asperities
   (b) manifolds   (d) perspectives

3. While roughness is referred to as a primary

texture, _____ is called a secondary texture.

(a) waviness   (c) error of form
(b) lay   (d) error of geometry

4. The direction of the predominant surface pattern, ordinarily determined by the production process used for manufacturing the component, is referred to as
   (a) lay   (c) waviness
   (b) flaw   (d) none of these

5. Irregularities that occur in isolation or infrequently because of specific causes such as scratches, cracks, and blemishes are called
   (a) surface texture (c) waviness
   (b) lay   (d) flaws

6. The direction of a lay is
   (a) the direction that the stylus trace is made
   (b) the direction of the asperities
   (c) perpendicular to the asperities
   (d) any selected straight line taken as reference

7. The average height from a mean line of all ordinates of the surface, regardless of sign, is the
   (a) RMS value   (c) Rz value
   (b) Ra value   (d) Rm value

8. The datum created by a skid or shoe is the locus of its _____ as it slides along the workpiece surface.
   (a) centre of curvature
   (b) centre of its bottom
   (c) centre line
   (d) all of these

9. A _____ provides a quick-fix means of conducting an initial investigation before attempting a major investigation of surface quality.
   (a) Tomlinson surface meter
   (b) Taylor–Hobson talysurf

   (c) light interference microscope
   (d) profilometer

10. Which of the following is the best analogy for the trace of a stylus instrument?
    (a) A topographical map
    (b) A rolling ball
    (c) A pin-ball machine
    (d) A phonograph

11. What characteristics of asperities are quantified by the stylus instruments?
    (a) Peakedness   (c) Heights
    (b) Percentages   (d) Volumes

12. The frequency of the stylus movement as it rises up and down the workpiece surface is determined by
    (a) the traversing speed
    (b) the length of stylus
    (c) the curvature of skid
    (d) all of these

13. The measurement of roughness is relatively undistorted
    (a) irrespective of the phase difference between the stylus and the skid
    (b) if the stylus and the skid are in phase
    (c) if the stylus and the skid have a large phase difference
    (d) in none of these cases

14. In graphs, heights of asperities are exaggerated compared to their spacings. This is known as
    (a) aspect ratio   (c) articulated ratio
    (b) asperities ratio (d) distortion ratio

15. The Mecrin instrument assesses the surface irregularities through
    (a) fringe pattern
    (b) air-leakage method
    (c) frictional properties
    (d) thermal properties

REVIEW QUESTIONS

1. What is the justification for studying surface metrology as a specialized subject?

2. With the help of an illustration, explain the following terms: roughness, waviness, lay, and flaws.

3. What are the primary reasons for surface irregularities?

4. Explain the following methods of quantifying surface roughness: (a) Rz value, (b) RMS value, and (c) Ra value.

5. Identify the prominent types of 'lay' by means of symbols only.

6. Distinguish between comparison and direct measurement of surface roughness.
7. List the major features of the stylus system of measurement.
8. Distinguish between skidless and skid-type instruments.
9. With the help of a neat sketch, explain the working principle of the Tomlinson surface meter.
10. With the help of a neat sketch, explain the Taylor–Hobson talysurf.
11. What is a profilometer?
12. Discuss the effect of phase relationship between the stylus and the skid on measurement accuracy.
13. What is a cut-off wavelength? List the recommended cut-off wavelengths for some of the typical manufacturing operations.
14. How is the interferometry technique useful for measurement of surface irregularities?
15. What is the measurement concept in the 'Mecrin' instrument?

Answers to Multiple-choice Questions

1. (b)	2. (c)	3. (a)	4. (a)	5. (d)	6. (b)	7. (b)	8. (a)
9. (d)	10. (d)	11. (c)	12. (a)	13. (b)	14. (d)	15. (c)	

Miscellaneous Metrology

After studying this chapter, the reader will be able to

- apply laser-based technology for precise measurements
- understand the construction and working principle of coordinate measuring machines
- explain the ways and means of testing alignments of machine tools
- elucidate the technology of automated inspection and machine vision

10.1 INTRODUCTION

Hitherto, we have discussed various instruments and measurement techniques under specific headings such as linear measurement and angular measurement. However, there are certain instruments, measuring machines, and techniques, which are difficult to label under these headings, but are of utmost importance in the field of metrology. This chapter discusses some of these instruments and measurement techniques, without which our knowledge of metrology will be incomplete.

We dealt with laser interferometry in Chapter 7. Apart from this technique, precision instrumentation based on laser principles is increasingly being used in applications like machine tool assembly to ensure perfect alignment of machine parts. Concurrently, manual handling of work parts or machines is also being minimized or eliminated in order to ensure positional accuracy and reliability. For instance, the principle of flexible manufacturing system (FMS) advocates complete automation of a work cell comprising several machines, transfer mechanisms, and inspection stations. This calls for fully automated inspection machines that have the required on-board electronics to seamlessly integrate in such a manufacturing environment. *Coordinate measuring machines* (CMMs), which can provide such a capability, are nowadays an integral part of a modern factory. Here, we present the construction, working, and applications of CMMs, which will be of interest to the student.

Machine tools are the drivers of the modern factory systems. The more the accuracy and precision ensured in their manufacture, the more precise and accurate are the components machined out of them. This chapter deals with the standard procedure for ensuring accuracy and precision of manufacture by carrying out *acceptance tests* on machine tools. As the name itself implies, a machine tool can be accepted as fit for production only if it passes all the acceptance test criteria.

We have added special sections on *automated inspection* and *machine vision*. While the former facilitates 100% inspection of work parts, the latter provides human-like capability to an inspection machine to carry out on-line, visual inspection of work parts. Thus, this chapter is an interesting mix of various topics that are essential in a modern factory system.

10.2 PRECISION INSTRUMENTATION BASED ON LASER PRINCIPLES

Light amplification by stimulated emission of radiation (laser) produces an intense emergent beam of light that can be parallel to a high degree or can be focused onto a very small area. Although a number of materials may be used to produce lasers, the helium–neon gas laser is the most popular for applications in metrology.

For the purpose of measurement, laser has properties similar to 'normal' light. It can be represented as a sine wave whose wavelength remains the same for a given colour. The amplitude is a measure of the intensity of laser light. More importantly, laser has certain additional properties that are not possessed by ordinary light. Some of these are described here:

1. Laser light is *monochromatic*. It has a bandwidth in the range of 0.4–0.5 µm. Stabilized lasers have still narrower bandwidths, with the result that very high resolution can be achieved during measurement.
2. Laser light is *coherent*. In normal light, the rays are randomly phased, resulting in partial interference within the beam. In contrast, laser rays are all in phase, producing a coherent beam of light.
3. Laser light is naturally *collimated*. The rays in a laser beam are perfectly parallel with little divergence and scatter.

These factors combine to produce a beam that is perfectly parallel and very small in size. The light is extremely bright and can produce images or fringes that are quite sharp when employed in an optical system. Therefore, laser-based instrumentation is the best choice for precise measurement. Since practical realization of metre is closely related to the radiation of stabilized frequency lasers, laser interferometers are used for precise and traceable length measurements. The simplest form of laser measurement consists of a laser, an interferometer, a reflector, and a receiver, as shown in Fig. 10.1. The laser, interferometer, and receiver remain stationary while the retroreflector senses the variables to be measured. A laser transducer is basically a comparator that measures only the relative change of position between the interferometer and the retroreflector. In other words, it does not provide an absolute measurement of length.

The laser source is typically a double-frequency He–Ne laser. A double-frequency radiation source is required since the interfering measuring and reference beams must have slightly different frequencies and photo detectors to detect the phase shift between these two beams. The two frequencies are separated by their polarization state, so that a polarization beam splitter can generate a measurement beam with the frequency f_1 and a reference beam with the frequency f_2. Movement of the measurement reflector with a velocity v causes a frequency shift δf_1 in the measurement beam due to Doppler effect. This shift will increase or decrease depending on the direction of movement of the measurement reflector. When counting the periods of reference and measurement signals simultaneously, their difference is proportional to displacement. The direction of movement can be directly determined from the sign of this difference. The comparison of reference and measurement signals is electronically processed to

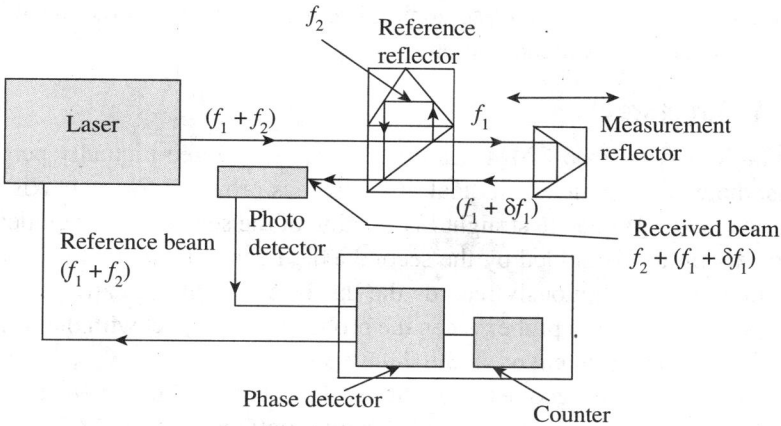

Fig. 10.1 Laser transducer system

generate displacement information. Laser transducers can measure up to six independent axes of displacement using one transducer head.

Section 7.5.3 in Chapter 7 deals with the application of laser interferometry for measuring large displacements like machine slideways.

10.3 COORDINATE MEASURING MACHINES

The term *measuring machine* generally refers to a single-axis measuring instrument. Such an instrument is capable of measuring one linear dimension at a time. The term *coordinate measuring machine* refers to the instrument/machine that is capable of measuring in all three orthogonal axes. Such a machine is popularly abbreviated as CMM. A CMM enables the location of point coordinates in a three-dimensional (3D) space. It simultaneously captures both dimensions and orthogonal relationships. Another remarkable feature of a CMM is its integration with a computer. The computer provides additional power to generate 3D objects as well as to carry out complex mathematical calculations. Complex objects can be dimensionally evaluated with precision and speed.

The first batch of CMM prototypes appeared in the United States in the early 1960s. However, the modern version of CMM began appearing in the 1980s, thanks to the rapid developments in computer technology. The primary application of CMM is for inspection. Since its functions are driven by an on-board computer, it can easily be integrated into a computer-integrated manufacturing (CIM) environment. Its potential as a sophisticated measuring machine can be exploited under the following conditions:

Multiple features The more the number of features (both dimensional and geometric) that are to be controlled, the greater the value of CMM.

Flexibility It offers flexibility in measurement, without the necessity to use accessories such as jigs and fixtures.

Automated inspection Whenever inspection needs to be carried out in a fully automated environment, CMM can meet the requirements quite easily.

High unit cost If rework or scrapping is costly, the reduced risk resulting from the use of a CMM becomes a significant factor.

10.3.1 Structure

The basic version of a CMM has three axes, along three mutually perpendicular directions. Thus, the work volume is cuboidal. A carriage is provided for each axis, which is driven by a separate motor. While the straight line motion of the second axis is guided by the first axis, the third axis in turn is guided by the second axis. Each axis is fitted with a precision measuring system, which continuously records the displacement of the carriage from a fixed reference. The third axis carries a probe. When the probe makes contact with the workpiece, the computer captures the displacement of all the three axes.

Depending on the geometry of the workpiece being measured, the user can choose any one among the five popular physical configurations. Figure 10.2 illustrates the five basic configuration types: cantilever (Fig. 10.2a), bridge (Fig. 10.2b), column (Fig. 10.2c), horizontal arm (Fig. 10.2d), and gantry (Fig. 10.2e).

Cantilever The vertically positioned probe is carried by a cantilevered arm. The probe moves up and down along the Z-axis, whereas the cantilever arm moves in and out along the Y-axis (lateral movement). The longitudinal movement is provided by the X-axis, which is basically the work table. This configuration provides easy access to the workpiece and a relatively large work volume for a small floor space.

Bridge A bridge-type configuration is a good choice if better rigidity in the structure is required. The probe unit is mounted on a horizontal moving bridge, whose supports rest on the machine table.

Column This configuration provides exceptional rigidity and accuracy. It is quite similar in construction to a jig boring machine. Machines with such a configuration are often referred to as *universal measuring machines*.

Horizontal arm In this type of configuration, the probe is carried by the horizontal axis. The probe assembly can also move up and down along a vertical axis. It can be used for gauging larger workpieces since it has a large work volume. It is often referred to as a layout.

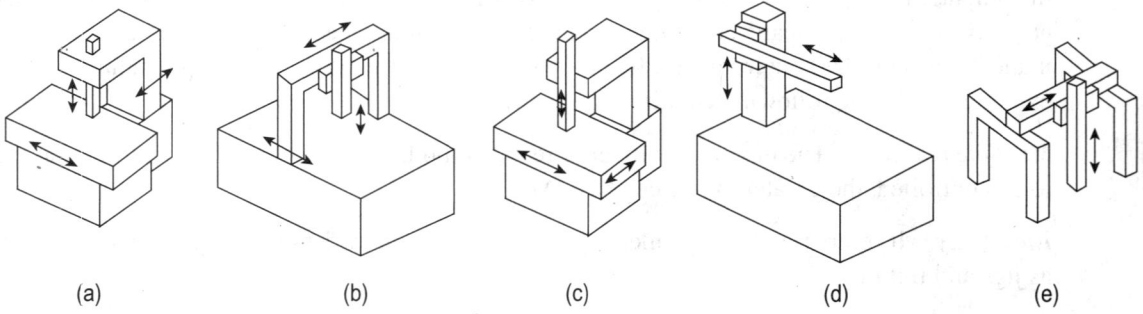

(a) (b) (c) (d) (e)

Fig. 10.2 Basic configuration of a CMM (a) Moving lever cantilever arm type
(b) Moving bridge type (c) Column type (d) Moving RAM horizontal arm type (e) Gantry type

Gantry In this configuration, the support of the workpiece is independent of the *X*- and *Y*-axis. Both these axes are overhead and supported by four vertical columns from the floor. The operator can walk along with the probe, which is desirable for large workpieces.

Some of the machines may have rotary tables or probe spindles, which will enhance the versatility of the machines. The work space that is bounded by the limits of travel in all the axes is known as the *work envelop*. Laser interferometers are provided for each of the axes if a very precise measurement is necessary.

10.3.2 Modes of Operation

Modes of operation are quite varied in terms of type of construction and degree of automation. Accordingly, CMMs can be classified into the following three types based on their modes of operation:
1. Manual
2. Semi-automated
3. Computer controlled

The manual CMM has a free-floating probe that the operator moves along the machine's three axes to establish contact with part features. The differences in the contact positions are the measurements. A semi-automatic machine is provided with an electronic digital display for measurement. Many functions such as setting the datum, change of sign, and conversion of dimensions from one unit to another are done electronically.

A computer-controlled CMM has an on-board computer, which increases versatility, convenience, and reliability. Such machines are quite similar to CNC machines in their control and operation. Computer assistance is utilized for three major functions. Firstly, a programming software directs the probe to the data collection points. Secondly, measurement commands enable comparison of the distance traversed to the standard built into the machine for that axis. Thirdly, computational capability enables processing of the data and generation of the required results.

10.3.3 Probe

The probe is the main sensing element in a CMM. Generally, the probe is of 'contact' type, that is, it is in physical contact with the workpiece when the measurements are taken. Contact probes may be either 'hard' probes or 'soft' probes. However, some CMMs also use a non-contact-type.

Figure 10.3 illustrates the main components of a probe assembly. A probe assembly comprises the probe head, probe, and stylus. The probe is attached to the machine quill by means of the probe head and may carry one or more styli. Some of the probes are motorized and provide additional flexibility in recording coordinates.

The stylus is integral with hard probes and comes in various shapes such as pointed, conical, and ball end. As a power feed is used to move the probe along different axes, care should be exercised when contact is made with the workpiece to ensure that excessive force is not applied on the probe. Excessive contact force may distort either the probe itself or the workpiece, resulting in inaccuracy in measurement. Use of soft probes mitigates this problem to a large

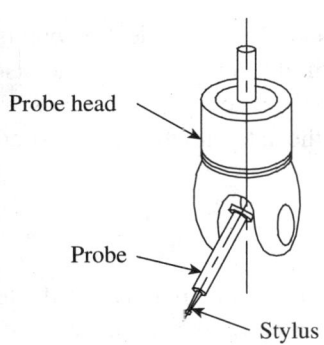

Probe head

Probe

Stylus

Fig. 10.3 Probe assembly

extent. Soft probes make use of electronic technology to ensure application of optimum contact pressure between the probe and the workpiece. Linear voltage differential transformer heads are generally used in electronic probes. However, 'touch trigger' probes, which use differences in contact resistance to indicate deflection of the probe, are also popular.

Some measurement situations, for example, the inspection of printed circuit boards, require non-contact-type probes. Measurement of highly delicate objects such as clay or wax models may also require this type of probe. Most non-contact probes employ a *light beam stylus*. This stylus is used in a manner similar to a soft probe. The distance from the point of measurement is known as *standoff* and is normally 50 mm. The system provides 200 readings per second for surfaces with good contrast. The system has high resolution of the order of 0.00005 mm. However, illumination of the workpiece is an important aspect that must be taken into consideration to ensure accurate measurement.

Probe Calibration

A remarkable advantage of a CMM is its ability to achieve a high level of accuracy even with reversal in the direction of measurement. It does not have the usual problems such as backlash and hysteresis associated with measuring instruments. However, the probe may mainly pose a problem due to deflection. Therefore, it needs to be calibrated against a master standard. Figure 10.4 illustrates the use of a slip gauge for calibration of the probe.

Calibration is carried out by touching the probe on either side of the slip gauge surface. The nominal size of the slip gauge is subtracted from the measured value. The difference is the 'effective' probe diameter. It differs from the measured probe diameter because it contains the deflection and backlash encountered during measurement. These should nearly remain constant for subsequent measurements.

10.3.4 Operation

This section explains the operation or the measurement process using a CMM. Most modern CMMs invariably employ computer control. A computer offers a high degree of versatility, convenience, and reliability. A modern CMM is very similar in operation to a computer numerical control (CNC) machine, because both control and measurement cycles are under the control of the computer. A user-friendly software provides the required functional features. The software comprises the following three components:

1. *Move* commands, which direct the probe to the data collection points
2. *Measurement* commands, which result in the comparison of the distance traversed to the standard built into the machine for that axis
3. *Formatting* commands, which translate the data into the form desired for display or printout

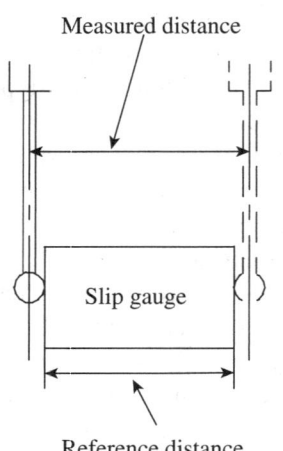

Measured distance

Slip gauge

Reference distance

Fig. 10.4 Probe calibration

Machine Programming

Most measurement tasks can be carried out using readily available *subroutines*. The subroutines are designed based on the frequency with which certain measurement tasks recur in practice. An operator only needs to find the subroutine in a menu displayed by the computer. The operator then inputs the data collection points, and using simple keyboard commands the desired results can be obtained. The subroutines are stored in the memory and can be recalled whenever the need arises. Figure 10.5 illustrates a few typical subroutines that are used in CMMs.

A circle can be defined by specifying three points lying on it. This is shown in Fig. 10.5(a). The program automatically calculates the centre point and the diameter of the best-fit circle. A cylinder is slightly more complex, requiring five points. The program determines the best-fit cylinder and calculates the diameter, a point on the axis, and a best-fit axis (Fig. 10.5b).

Situations concerning the relationship between planes are common. Very often, we come across planes that need to be perfectly parallel or perpendicular to each other. Figure 10.5(c) illustrates a situation where the perpendicularity between two planes is being inspected. Using a minimum of two points on each line, the program calculates the angle between the two lines. Perpendicularity is defined as the tangent of this angle. In order to assess the parallelism between two planes (Fig. 10.5d), the program calculates the angle between the two planes. Parallelism is defined as the tangent of this angle.

In addition to subroutines, a CMM needs to offer a number of utilities to the user, especially mathematical operations. Most CMMs have a measurement function library. The following are some typical library programs:

1. Conversion from SI (or metric) to British system
2. Switching of coordinate systems, from Cartesian to polar and vice versa
3. Axis scaling
4. Datum selection and resetting
5. Nominal and tolerance entry

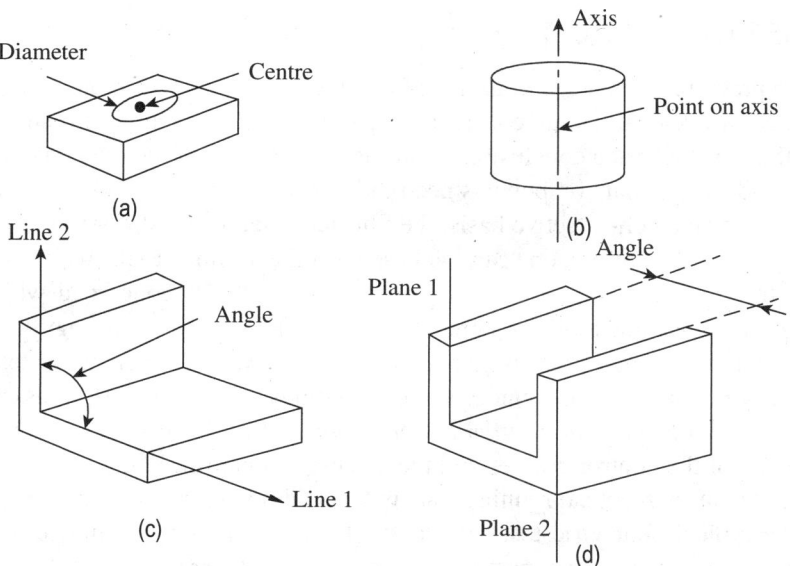

Fig. 10.5 Typical subroutines used in a CMM (a) Circle (b) Cylinder
(c) Perpendicularity between two planes (d) Parallelism between two planes

6. Bolt-circle centre and diameter
7. Statistical tools

10.3.5 Major Applications

The CMM is a sophisticated equipment, which offers tremendous versatility and flexibility in modern manufacturing applications. It uses the fundamental principles of metrology to an extent that is not matched by any other measurement instrument. However, its use is limited to situations where production is done in small batches but products are of high value. It is especially useful for components of varied features and complex geometry. In addition to these factors, a CMM is a good choice in the following situations:

1. A CMM can easily be integrated into an automated inspection system. The computer controls easy integration in an automated environment such as an FMS or a CIM. The major economic benefit is the reduction in downtime for machining while waiting for inspection to be completed.

2. A CMM may be interfaced with a CNC machine so that machining is corrected as the workpiece is inspected. A further extension of this principle may include computer-assisted design and drafting (CADD).

3. Another major use (or abuse?) of CMMs is in reverse engineering. A complete 3D geometric model with all critical dimensions can be built where such models do not exist. Once the geometric model is built, it becomes easier to design dies or moulds for manufacturing operations. Quite often, companies create 3D models of existing critical dies or moulds of their competitors or foreign companies. Subsequently, they manufacture the dies, moulds, or components, which create a *grey market* for such items in the industry.

10.4 MACHINE TOOL METROLOGY

In the previous chapters, we discussed the need to produce accurate and precise components. We also saw several instruments and comparators that can help us measure various dimensions and thereby arrive at a conclusion regarding the accuracy of the manufactured components. We have understood that components need to be manufactured to such accuracy that they may be assembled on a non-selective basis, the finished assemblies conforming to stringent functional requirements. Therefore, one can easily see that the machine tools, which produce components, have to be extremely accurate. The metrological aspects concerned with the assessment of accuracy of machine tools are popularly called machine tool metrology.

Machine tool metrology is primarily concerned with the geometric tests of the alignment accuracy of machine tools under static conditions. It is important to assess the alignment of various machine parts *in relation to one another*. It is also necessary to assess the quality and accuracy of the control devices and the driving mechanism in the machine tool. In addition to geometric tests, practical running tests will also throw light on the accuracy of a machine tool.

The typical geometric tests conducted for machine tools comprise tests for straightness, flatness, squareness, and parallelism. Running tests are conducted to evaluate geometric tolerances such as roundness and cylindricity. The next section gives a general overview of these tests.

10.4.1 Straightness, Flatness, Parallelism, Squareness, Roundness, Cylindricity, and Runout

Straightness, flatness, parallelism, squareness, roundness, and cylindricity are measures of geometric accuracy of machine parts. Geometric accuracy is of utmost importance, especially to ensure the accuracy in relative engagement or motion of various machine parts. The following paragraphs outline, in brief, the meaning and measurement methods of these measures of accuracy.

Straightness

A line is said to be straight over a given length if the deviation of various points on the line from two mutually perpendicular reference planes remains within stipulated limits. The reference planes are so chosen that their intersection is parallel to the straight line lying between the two specific end points. The tolerance on the straightness of a line is defined as the maximum deviation of the spread of points on either side of the reference line, as shown in Fig. 10.6.

The maximum spread of deviation with respect to the reference line is a measure of straightness accuracy. The lesser the deviation or spread, the better the straightness accuracy of a machine part. Straightness can be measured in various ways, depending on the need for accuracy in measurement, right from using a spirit level to sophisticated laser-based measurement devices. Section 5.6.2 in Chapter 5 explains the use of an autocollimator to measure straightness of machine guideways.

Flatness

Machine tool tables, which hold workpieces during machining, should have a high degree of flatness. Many metrological devices like the sine bar invariably need a perfectly flat surface plate. Flatness error may be defined as the minimum separation of a pair of parallel planes that will just contain all the points on the surface. Figure 10.7 illustrates the measure of flatness error a.

It is possible, by using simple geometrical approaches, to fit a best-fit plane for the macro-surface topography. Flatness is the deviation of the surface from the best-fit plane. According to IS: 2063-1962, a surface is deemed to be flat within a range of measurement when the variation of the perpendicular distance of its points from a geometrical plane (this plane should be exterior to the surface to be tested) parallel to the general trajectory of the plane to be tested remains below a given value. The geometrical plane may be represented either by means of a surface plane or by a family of straight lines obtained by the displacement of a straight edge, a spirit level, or a light beam. While there are quite a few methods for measuring flatness, such as the beam comparator method, interferometry technique, and laser beam measurement,

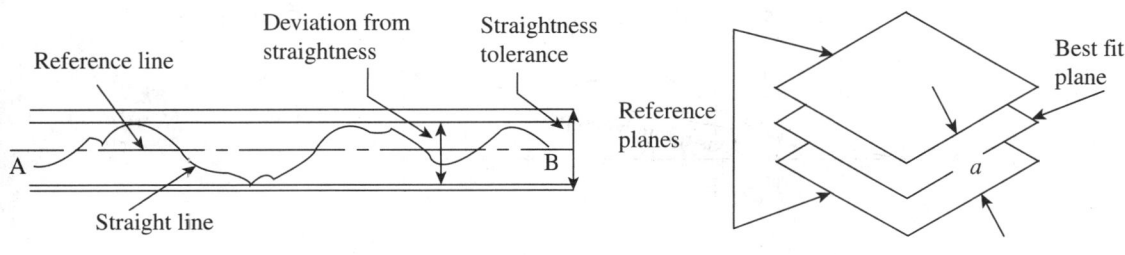

Fig. 10.6 Straightness of a line

Fig. 10.7 Measure of flatness error

the following paragraphs explain the simplest and most popular method of measuring flatness using a spirit level or a clinometer.

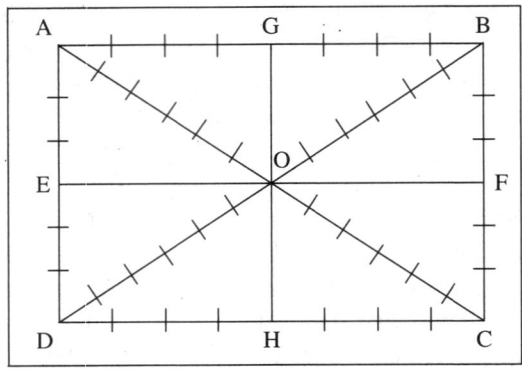

Fig. 10.8 Grid lines for flatness test

Measurement of flatness error Assuming that a clinometer is used for measuring angular deviations, a grid of straight lines, as shown in Fig. 10.8, is formulated. Care is taken to ensure that the maximum area of the flat table or surface plate being tested is covered by the grid. Lines AB, DC, AD, and BC are drawn parallel to the edges of the flat surface; the two diagonal lines DB and AC intersect at the centre point O. Markings are made on each line at distances corresponding to the base length of the clinometer.

The following is a step-by-step procedure to measure flatness error:

1. Carry out the straightness test, as per the procedure described in Chapter 5, on all the lines and tabulate the readings up to the cumulative error column. Figure 10.9 gives an example of line AB.

2. We know that a plane is defined as a 2D entity passing through a minimum of three points not lying on the same straight line. Accordingly, a plane passing through the points A, B, and D is assumed to be an arbitrary plane, relative to which the heights of all other points are determined. Therefore, the ends of lines AB, AD, and BD are corrected to zero and the heights of points A, B, and D are forced to zero.

3. The height of the centre 'O' is determined relative to the arbitrary plane ABD. Since O is also the mid-point of line AC, all the points on AC can be fixed relative to the arbitrary plane ABD. Assume A = 0 and reassign the value of O on AC to the value of O on BD. This will readjust all the values on AC in relation to the arbitrary plane ABD.

4. Next, point C is fixed relative to the plane ABD; points B and D are set to zero. All intermediate points on BC and DC are also adjusted accordingly.

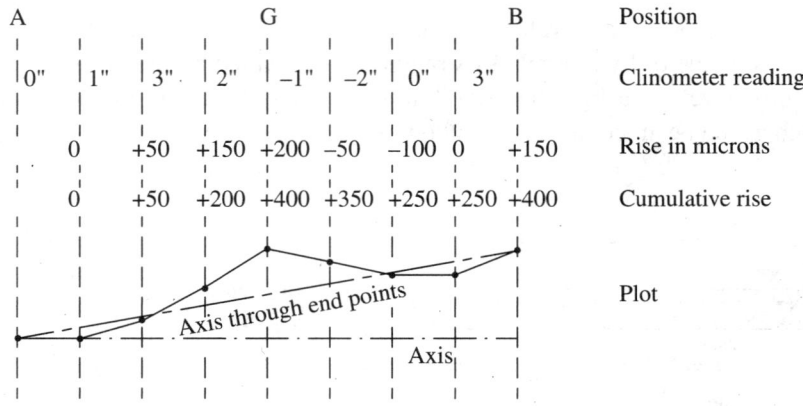

Fig. 10.9 Straightness plot for line AB

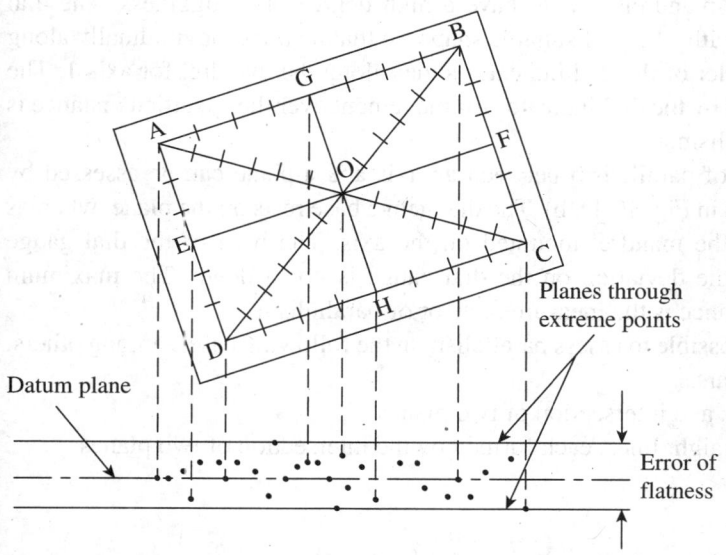

Fig. 10.10   Plot of heights of all points with reference to the datum plane ABD

5. The same procedure applies to lines EF and GH. The mid-points of these lines should also coincide with the known mid-point value of O.

6. Now, the heights of all the points, above and below the reference plane ABD, are plotted as shown in Fig. 10.10. Two lines are drawn parallel to and on either side of the datum plane, such that they enclose the outermost points. The distance between these two outer lines is the flatness error.

Some authors argue that the reference plane ABD that is chosen in this case may not be the best datum plane. They recommend further correction to determine the minimum separation between a pair of parallels that just contains all the points on the surface. However, for all practical purposes, this method provides a reliable value of flatness error, up to an accuracy of 10 µm.

Parallelism

In geometry, parallelism is a term that refers to a property in Euclidean space of two or more lines or planes, or a combination of these. The assumed existence and properties of parallel lines are the basis of Euclid's parallel postulate. Two lines in a plane that do not intersect or touch at a point are called parallel lines. Likewise, a line and a plane, or two planes, in 3D Euclidean space that do not share a point are said to be parallel. We come across various situations in machine tool metrology where two axes or an axis and a plane are required to be perfectly parallel to meet functional requirements. Figure 10.11 illustrates two typical cases of parallelism.

In Fig. 10.11(a), the parallelism between two axes is illustrated. Since the axis of a component or part is notional rather than physical, we need to use mandrels fitted along the axes under

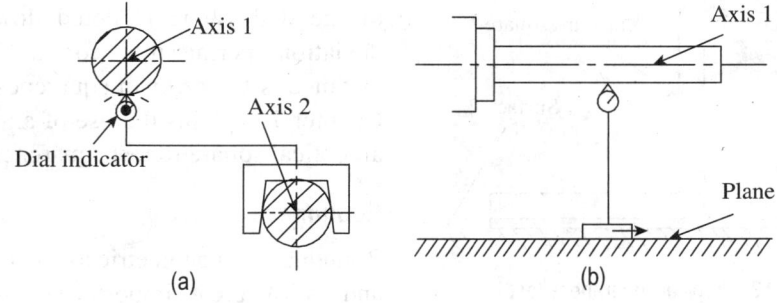

Fig. 10.11   Parallelism (a) Between two axes (b) Between an axis and a plane

investigation. The surfaces of mandrels should have a high degree of straightness. The dial indicator is held on a support with a base of suitable shape, so that it slides longitudinally along the mandrel of axis 2. The feeler of the dial indicator slides along the mandrel for axis 1. The maximum deviation registered by the dial indicator for movement over the specified distance is the measure of error of parallelism.

In a similar manner, error of parallelism between an axis and a plane can be assessed by having the arrangement shown in Fig. 10.11(b). The dial gauge base rests on the plane, whereas the feeler is in contact with the mandrel mounted on the axis. The base of the dial gauge is moved longitudinally and the deviation on the dial gauge is noted down. The maximum deviation over a specified distance is the measure of error of parallelism.

On similar lines, it is also possible to assess parallelism in the following cases, among others:
1. Parallelism between two planes
2. Parallelism between an axis and intersection of two planes
3. Parallelism between two straight lines, each formed by the intersection of two planes
4. Parallel motion

Squareness

Very often, two related parts of a machine need to meet perfect squareness with each other. In fact, the angle 90° between two lines or surfaces or their combinations, is one of the most important requirements in engineering specifications. For instance, the cross-slide of a lathe must move at exactly 90° to the spindle axis in order to produce a flat surface during facing operation. Similarly, the spindle axis of a drilling machine and a vertical milling machine should be perfectly square with the machine table. From a measurement perspective, two planes, two straight lines, or a straight line and a plane are said to be square with each other when error of parallelism in relation to a *standard square* does not exceed a limiting value. The standard square is an important accessory for conducting the squareness test. It has two highly finished surfaces that are perpendicular to each other to a high degree of accuracy. Figure 10.12 illustrates the use of the standard square for conducting this test.

Two surfaces need to have a high degree of squareness. The base of a dial gauge is mounted on one of the surfaces, and the plunger is held against the surface of the standard square and set to zero. Now, the dial gauge base is given a traversing motion in the direction shown in the figure, and deviation of the dial gauge is noted down. The maximum deviation permissible for a specific traversing distance is the error in squareness. Section 7.2.3 in Chapter 7 explains the use of an autocollimator and an optical square to test squareness.

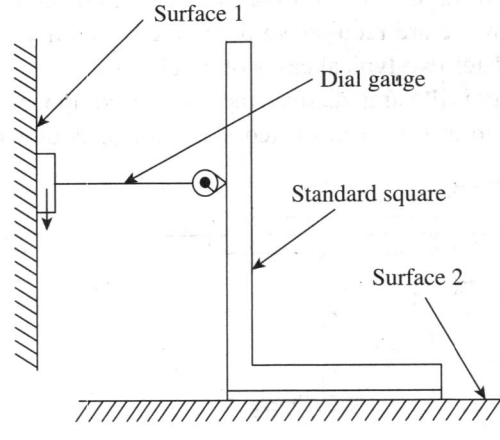

Fig. 10.12 Typical squareness test

Roundness

Roundness is a geometric aspect of surface metrology and is of great importance because the number

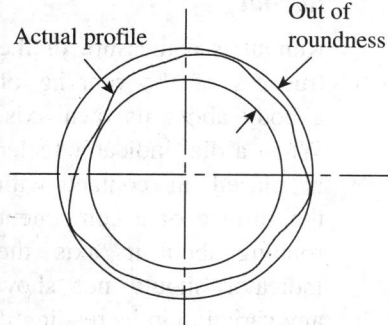

Fig. 10.13   Out of roundness

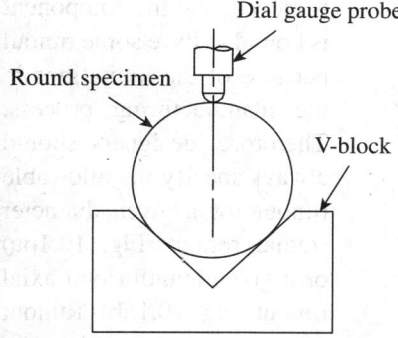

Fig. 10.14   Use of a V-block for measuring out of roundness

of rotational bearings in use is much more than that of linear bearings. Many machine parts, such as a machine spindle or the hub of a gear, have circular cross sections; these parts should have roundness with a very high degree of accuracy.

Roundness is defined as a condition of surface of revolution where all points of the surface intersected by any plane perpendicular to a common axis are equidistant from the axis. It is obvious that any manufactured part cannot have perfect roundness because of limitations in the manufacturing process or tools; we need to determine how much deviation from perfect roundness can be tolerated so that the functional aspects of the machine part are not impaired. This leads to the definition of *out of roundness* as a measure of roundness error of a part. Figure 10.13 explains out of roundness or *roundness error*. It is the radial distance between the minimum circumscribing circle and the maximum inscribing circle, which contain the profile of the surface at a section perpendicular to the axis of rotation.

Roundness error can be measured in various ways. Figure 10.14 illustrates the use of a V-block to measure it. Accessories required for the measurement comprise a surface plate, a V-block, and a dial gauge with a stand. The V-block is kept on the surface plate, and the cylindrical work part is positioned on the V-block. Care should be taken to ensure that the axis of the work part is parallel to the edges of the 'V' of the V-block. The dial gauge is mounted on its stand and the plunger is made to contact the surface of the work part. A light contact pressure is applied on the plunger so that it can register deviations on both the plus and minus sides. The dial gauge reading is set to zero. Now the work part is slowly rotated and the deviations of the dial indicator needle on both the plus and minus sides are noted down. The difference in reading for one complete rotation of the work part gives the value of out of roundness.

Cylindricity

A cylinder is an envelope of a rectangle rotated about an axis. It is bound between two circular and parallel planes. Quite often, we come across precision components such as the bore of an automotive cylinder or the bore of a hydraulic cylinder, which should have a high degree of cylindricity to give the best performance. Cylindricity is a measure of the degree of conformance of a component to the ideal cylinder for which it is designed.

The general way of describing cylindricity of a component is by the *minimum-zone method*. This is described as the radial separation of two coaxial cylinders fitted to the total measured surface under test in such a way that their radial difference is minimum. Figure 10.15(a) illustrates the definition of cylindricity, whereas Fig. 10.15(b) illustrates the specification of geometric/dimensional tolerance of cylindricity.

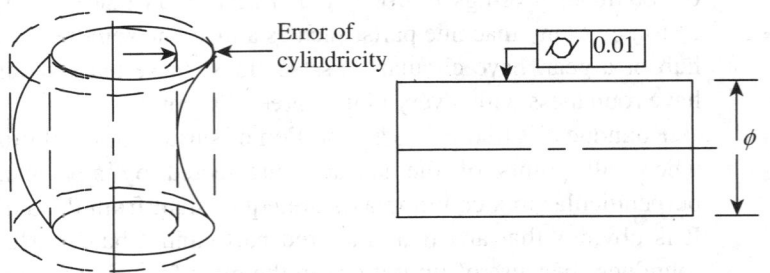

Fig. 10.15 Cylindricity (a) Definition (b) Representation

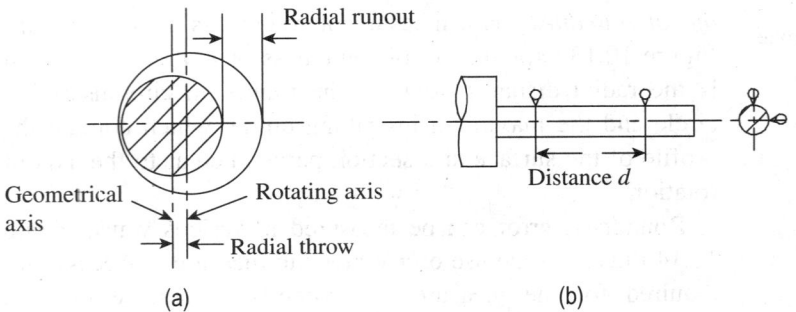

Fig. 10.16 Runout (a) Radial runout (b) Axial runout

Runout

Runout is a measure of the trueness of the running of a body about its own axis. When a dial indicator feeler is placed in contact with the surface of a component rotating about its axis, the indicator should not show any variation in its reading if the runout is zero. Obviously, this condition is rarely achieved, and the component is bound to have some runout because of the limitations in the manufacturing process. Therefore, designers should always specify the allowable runout for a given diameter (radial runout, Fig. 10.16a) or a given length (total axial runout, Fig. 10.16b). Runout tolerance is very important for components such as bearings, machine spindles, and axles.

The radial runout of a component with a circular cross-section is twice the radial throw of the axis in a given section, as illustrated in Fig. 10.16(a). In case of components having considerable thickness, runout test is carried out at three sections in order to generate a more realistic picture of the runout. On the other hand, the total axial runout is twice the difference in radial throw at two sections separated by a distance d. As can be seen in Fig. 10.16(b), the runout test is carried out along two planes, one horizontal and the other vertical.

10.4.2 Acceptance Tests for Machine Tools

The basic objective of conducting acceptance tests is to ensure that all relative movements of the machine tool conform well within the accepted limits of deviations from designed values. This is important since the dimensions of the work part depend, to a large extent, on the accuracy of the mating parts of the machine tool. The phrase *acceptance test* is coined to signify the fact that the machine buyer will inspect the alignment of various parts of the machine tool in detail and will 'accept' the machine tool from the vendor's factory only after it conforms to accepted norms. We can broadly classify the various tests under the following groups:

1. Tests for ensuring levelling of the machine tool in both horizontal and vertical planes
2. Tests for flatness of machine bed and straightness as well as parallelism of bed ways and bearing surfaces
3. Tests for perpendicularity of guideways with respect to other guideways or bearing surfaces
4. Tests for true running of the machine spindle and its axial movements

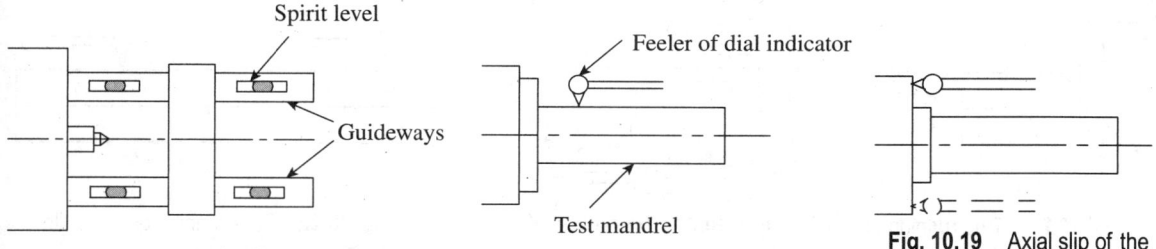

Fig. 10.17 Levelling of the machine

Fig. 10.18 True running of the main spindle

Fig. 10.19 Axial slip of the main spindle

5. Tests for assessing parallelism of spindle axis to guideways or bearing surfaces
6. Tests for line of movements for slides, cross-slides, and carriages
7. Practical tests for assessing dimensional and geometric accuracy in machining

Acceptance Tests for Lathes

The following are the important tests carried out on a lathe:

Levelling of machine First and foremost, the machine should be checked for accuracy of levelling. The machine should be installed such that the lathe bed is truly horizontal. A sensitive spirit level or a clinometer can be used to verify the levelling of the machine. The spirit level is moved over the designated distance specified in the test chart, and the deviation is noted down. The test is carried out in both longitudinal and transverse directions. The positioning of the spirit level on a guideway is illustrated in Fig. 10.17.

True running of main spindle The main spindle, while running, should not have any play or radial deviations from its axis. This is tested using a test mandrel of acceptable quality. The mandrel is fitted to the spindle bore and the dial indicator feeler is made to contact the mandrel surface as shown in Fig. 10.18. The spindle is gently rotated by hand, and the deviations of the dial indicator are noted down. The dial indicator base is mounted on the carriage. The deviation should be within acceptable limits.

Axial slip of main spindle The spindle should have true running in a direction parallel to its axis. This is easily checked by placing the dial indicator feeler against the spindle face and giving slight rotation to the spindle. The deviations should be within acceptable limits. The test is repeated at a diametrically opposite location to ensure that the spindle does not have an axial slip (Fig. 10.19).

True running of headstock centre The headstock centre is the live centre of the machine; if it is not true, accuracy of the workpiece will suffer. The workpiece will develop eccentricity if the error is too much. The feeler of the dial indicator is pressed perpendicular to the taper surface of the centre, and the spindle is rotated. The deviation indicates the trueness of the headstock centre. The test procedure is illustrated in Fig. 10.20.

Parallelism of main spindle Parallelism of the spindle is crucial for generating accurate dimensions. Any error in parallelism will result in a tapered surface after machining. In order to test parallelism, a test mandrel is made use of. It is fitted into the spindle bore and dial gauge readings are taken over a specified length, as shown in Fig. 10.21. Two readings are taken, one

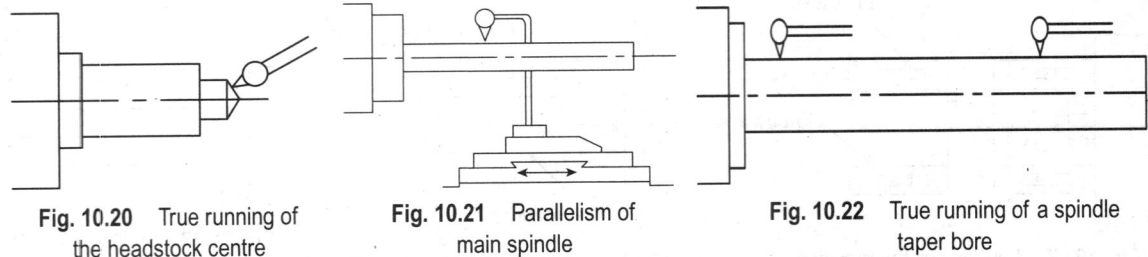

Fig. 10.20 True running of the headstock centre   **Fig. 10.21** Parallelism of main spindle   **Fig. 10.22** True running of a spindle taper bore

in a horizontal plane and the other in a vertical plane, on one of the sides of the mandrel. It is important to see that excess overhang of the mandrel does not result in a sag due to its own weight.

True running of taper bore of main spindle   The lathe spindle bore has a standard taper. Unless this taper is concentric with the spindle axis, workpieces will have undesired taper or eccentricity. A test mandrel is fitted to the tapered bore of the spindle, and dial gauge readings are taken at the two extreme ends of the mandrel. This value should be well within the allowable limits (Fig. 10.22).

Parallelism of tailstock sleeve to carriage movement   A tailstock is generally used to hold long workpieces. The dead centre of the tailstock is mounted on the end of the sleeve, which moves in an axial direction. The sleeve surface should be perfectly parallel to the movement of the carriage. A mandrel is put in the sleeve socket and dialling is done by mounting the dial gauge base on the tool post, as shown in Fig. 10.23. The test should be carried out in both the horizontal and vertical planes.

Parallelism of tailstock guideways to carriage movement   The tailstock guideways should be parallel to the movement of the carriage. Whenever long jobs are being turned, it becomes necessary to shift the tailstock along its guideways. While doing so, the job axis should perfectly coincide with the tailstock centre. If this condition is not satisfied, the workpiece will develop an undesirable taper.

A test block, specially designed for the purpose, is placed on the guideways, as shown in Fig. 10.24. The dial indicator base is mounted on the carriage and the feeler is made to contact the workpiece. The carriage is moved for the specified distance and the deviation is noted down. The test is carried out in both the horizontal and vertical planes.

Practical tests   Actual machining is carried out in order to ascertain not only the accuracy of alignment but also the rigidity of the machine tool. At least three turning operations are

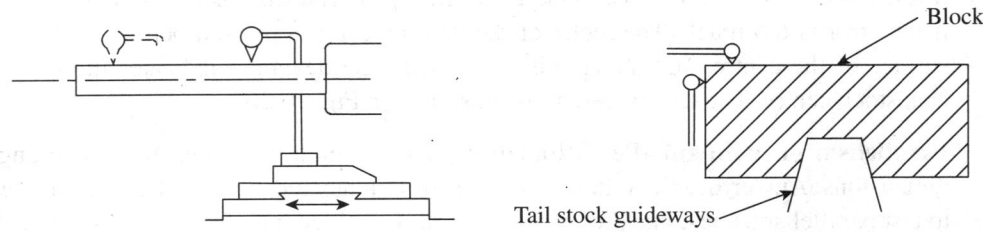

Fig. 10.23 Parallelism of a tailstock sleeve   **Fig. 10.24** Parallelism of tailstock guideways

Table 10.1 Recommended practical tests

Operation	Workpiece diameter (mm)	Permissible error (mm)
Chucking	50–100	0.01
Turning between centres	100–200	0.02
Facing	100–200	0.02

mandatory, namely chucking (using the chuck only), turning between centres, and facing. The operations are performed with prescribed values of cutting speed, feed rate, and depth of cut. Table 10.1 illustrates the recommended ways of conducting practical tests on lathes.

Acceptance Tests for Milling Machines

It is presumed that the student is aware of the various types of milling machines. However, we will deal with some of the important tests conducted on a horizontal milling machine in order to provide the student with a fair knowledge of the methodology.

Axial slip of spindle   A spindle may have an axial slip, which is the axial movement of the spindle during its rotation. Axial slip may occur due to the following reasons: errors due to worn-out spindle bearings, face of the locating shoulder of the spindle not being in a plane perpendicular to the axis of the spindle, and irregularities in the front face of the spindle. Figure 10.25 illustrates the test for measuring the axial slip of the spindle. The feeler of the dial gauge is held against the front face of the spindle and the base is mounted on the table. The position of the dial gauge (in order to measure axial slip) is denoted by A in the figure. The spindle is gently rotated by hand and the dial gauge reading is noted down. The test is repeated at a diametrically opposite spot. The maximum deflection should be well within the prescribed limits.

Eccentricity of external diameter of spindle   Position B of the dial gauge shown in Fig. 10.25 is used to determine the eccentricity of the external diameter of the spindle. The feeler is made to contact the spindle face radially, and the dial gauge base is mounted on the machine table. The spindle is gently rotated by hand and the dial gauge deviation is noted down. The maximum deviation gives the eccentricity of the external diameter of the spindle, and it should be well within specified limits.

True running of inner taper of spindle   The spindle of a milling machine is provided with a standard taper, which matches with the tooling used on the machine. The axis of the taper should be perfectly coincident with the axis of the spindle. Otherwise, the workpieces will develop undesired taper or eccentricity after machining. This condition is checked by using a test mandrel that fits into the spindle taper. The dial gauge is mounted on the machine table, and the feeler is made to contact the mandrel at one end of the mandrel, as shown in Fig. 10.26. The maximum deviation of the dial gauge is noted by gently rotating the spindle by hand. The test is repeated at the other end of the mandrel.

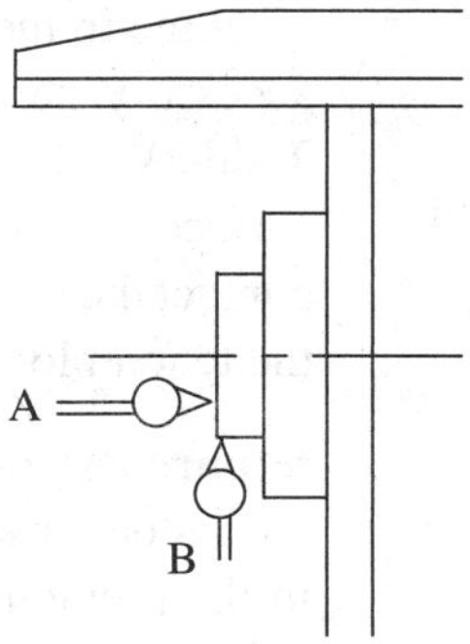

Fig. 10.25 Measuring axial slip of a spindle

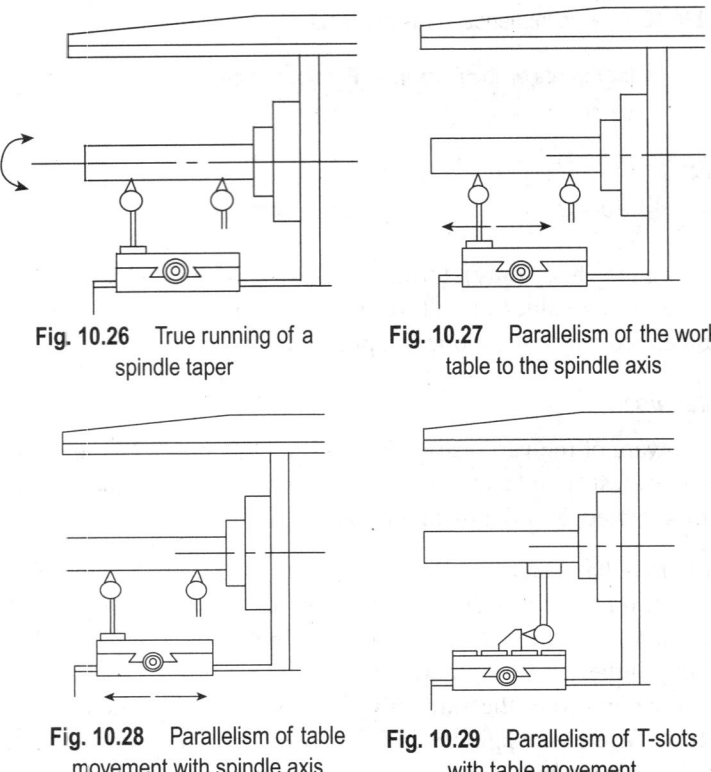

Fig. 10.26 True running of a spindle taper

Fig. 10.27 Parallelism of the work table to the spindle axis

Fig. 10.28 Parallelism of table movement with spindle axis

Fig. 10.29 Parallelism of T-slots with table movement

Parallelism of work table surface to spindle axis Workpieces are clamped on the top surface of the work table of the machine. We should ensure that the work table surface is parallel to the spindle axis; otherwise, milled surfaces will not meet quality requirements. A test mandrel is fitted to the spindle. If the machine has an arbour, the arbour surface is checked for parallelism. The feeler of the dial gauge is made to contact the mandrel or arbour, as the case may be, and the base of the dial gauge is kept on the table surface. Now the dial gauge base is moved on the table surface in a direction parallel to the spindle axis till the extreme end, as shown in Fig. 10.27. The dial gauge deflection is noted down, which should be well within the permissible limits.

Parallelism of transverse movement of table to spindle axis The machine table is set in its mean position and the dial gauge base is mounted on the table. The feeler is made to contact the mandrel and the table is given motion in the transverse direction. The test is carried out both in the horizontal and vertical planes of the mandrel. The deviation on the dial indicator should be within permissible limits (Fig. 10.28).

Parallelism of T-slots to table movement A number of T-slots are provided on the machine table to enable accurate clamping of work holding devices, which in turn hold the workpieces. The vertical surfaces of the T-slots should be perfectly parallel to the longitudinal axis of the machine in the horizontal plane. Generally, the test is carried out for the central T-slot, and an accessory, namely a tenon, is used. A tenon is a 150 mm long simple bracket, which fits into the T-slot. While sitting in a T-slot, a butting surface projects out, which is parallel to the vertical surface of a T-slot. The dial gauge base is mounted on the spindle and the feeler is made to contact the tenon, as shown in Fig. 10.29. Now, the machine table is moved longitudinally while the tenon block is held stationary. Deviations from parallelism are noted from the dial gauge.

Squareness of T-slots with machine spindle Unless the T-slots are parallel to the spindle axis, slots or key ways cut on the workpiece will not be aligned in the required direction. As in the previous case, this test is also generally carried out for the central T-slot. The table is approximately set in a mid-position, and a tenon block is inserted in the T-slot. The dial gauge base is fixed to the mandrel and the feeler is brought in contact with the vertical surface of the

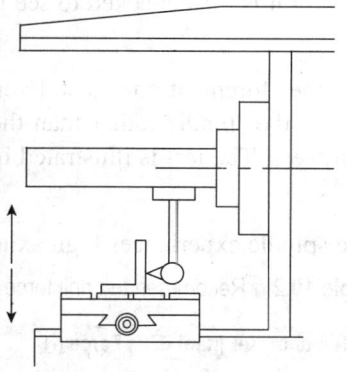

Fig. 10.30 Squareness of T-slots with the spindle

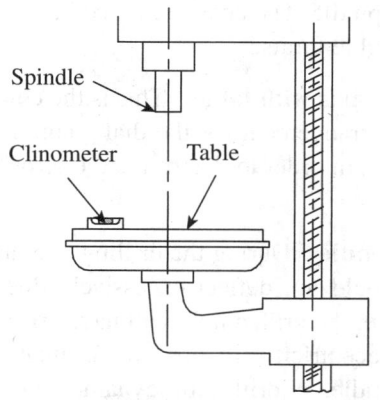

Spindle

Clinometer   Table

Fig. 10.31 Flatness of table

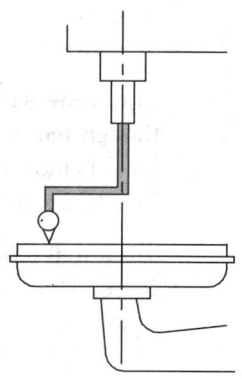

Fig. 10.32 Squareness of table with the spindle axis

tenon. Now, the tenon is moved towards one of the ends and the dial gauge swung to establish contact with the tenon. The reading on the dial gauge is noted. The tenon block is moved towards the other end and the dial gauge swung the other way in order to establish contact again. The difference in reading, which is a measure of the squareness error, is noted down. Figure 10.30 illustrates the test set-up.

The preceding paragraphs provided some insight into the important acceptance tests carried out on a horizontal milling machine. It is recommended that 25–30 different tests be carried out on a milling machine before it can be accepted for production work. Only a few tests are highlighted here. The student is advised to visit the metrology laboratory in the institute and go through the detailed acceptance charts provided by the machine suppliers.

Subsequent to the aforementioned tests, it is recommended to carry out practical tests by milling test pieces. One has to check for accuracy in surface milling, end milling, and slot milling before certifying the machine to be fit for use on a production shop floor.

Acceptance Tests for Drilling Machines

There are various types of drilling machines such as sensitive drilling machine, pillar-type drilling machine, and radial drilling machine. This section deals with the acceptance tests recommended for a pillar-type drilling machine.

Flatness of clamping surface of table   This test is performed for the table of the milling machine. The table surface provides support to either clamp the workpiece directly or to clamp a fixture that holds the workpiece. The test is performed using a sensitive spirit level or a clinometer with good resolution. Error of flatness is determined as explained in Section 10.4.1 of this chapter. Figure 10.31 illustrates the use of a clinometer to determine flatness.

Squareness of clamping surface of table with spindle axis   It is important to ensure that the clamping surface of the machine table is perfectly square with the axis of the spindle. Unless this condition is satisfied, the drilled hole will not be parallel to the spindle axis. The arrangement for the test is shown in Fig. 10.32.

The dial gauge base is mounted on the spindle, and the feeler is made to touch the machine table and set to zero. Now, the table is rotated slowly by 180°, without disturbing the dial gauge base, which is in the spindle. The change in reading is noted down; it is then checked to see if it is within the permissible limits.

Squareness of spindle axis with table   This is the corollary to the aforementioned test. Even though the mounting arrangement for the dial gauge is the same, the spindle rather than the table is moved by 180°, in order to determine the error in squareness. The test is illustrated in Fig. 10.33.

Total deflection of spindle   During the drilling operation, the spindle experiences high axial force. The spindle should not deflect excessively due to this force. Otherwise, the drilled hole will have error of straightness and eccentricity. In order to evaluate deflection of the spindle, a drill tool dynamometer (DTD) is used. The DTD provides a means of applying a known amount of load on the spindle.

The drill spindle is loaded by moving the drill head downwards and recording the value of force on the DTD display screen. The base of the dial indicator is placed on the machine table, as shown in Fig. 10.34. The feeler is held against the spindle face. The recommended pressure is applied on the spindle and the dial gauge deflection is noted down. While Table 10.2 lists the recommended value of force for various drill diameters, Table 10.3 gives the permissible errors for various centre distances.

Table 10.2 Recommended drill force

Drill diameter (mm)	Force (N)
6	1000
10	2000
16	3500
20	5500
25	7500
32	9500
40	12,000
50	15,000

Table 10.3 Permissible errors

Centre distance (mm)	Permissible error (mm)
Up to 200	0.4
200–300	0.6
300–400	0.8
400 and above	1.0

True running of spindle taper   The true running of a spindle taper is tested using a test mandrel. The test mandrel is loaded in the spindle and the dial gauge base is fixed on the machine table, as shown in Fig. 10.35. The feeler is made to contact the mandrel surface and the spindle is gently rotated by hand. The dial indicator

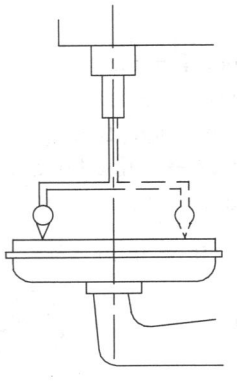

Fig. 10.33   Squareness of the spindle axis with table

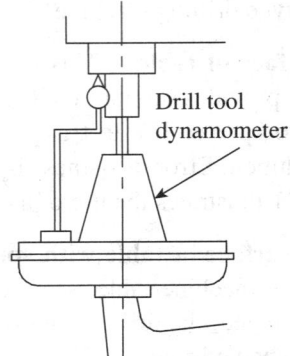

Fig. 10.34   Deflection of a spindle

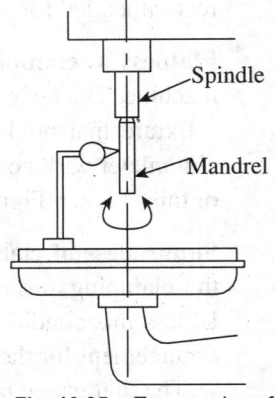

Fig. 10.35   True running of a spindle taper

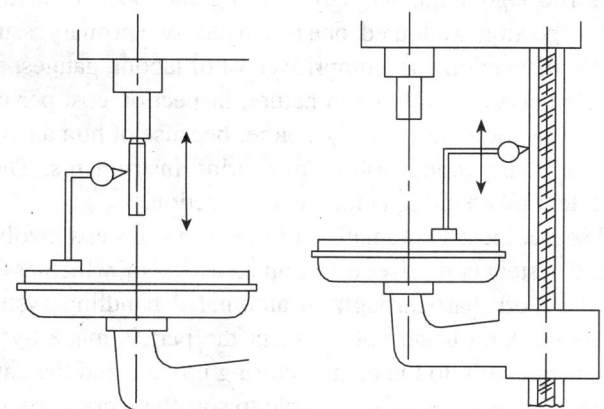

Fig. 10.36 Parallelism of a spindle axis with its vertical movement

Fig. 10.37 Squareness of guideways with table

reading is noted down and it is ascertained if the reading is within permissible limits. The test is repeated at three different locations to ensure its validity.

Parallelism of spindle axis to its vertical movement While drilling holes, the spindle is given feed in the vertical direction, and therefore, it is necessary to ensure that the vertical movement of the spindle is parallel to its axis. The test set-up is illustrated in Fig. 10.36. The mandrel is fixed in the spindle, and the dial gauge is mounted with the feeler making contact with the mandrel at the bottom. The spindle is moved vertically by hand feed and the dial gauge reading is noted down. The test is repeated in another plane by rotating the spindle through 90°. The total deflection shown by the dial gauge should be well within the allowable limits.

Squareness of drill head guideways with table The drill head in a pillar-type machine moves up and down in order to accommodate various workpiece sizes. This up and down movement is facilitated by guideways (Fig. 10.37). Therefore, it is important to ensure that this motion of the drill head is perfectly square with the clamping surface of the work table. If this condition is not met, the drilled holes will be skewed from the vertical axis and the workpiece will fail to pass inspection.

10.5 AUTOMATED INSPECTION

The modern industry is adopting high technology manufacturing solutions to achieve a high production rate as well as to improve productivity. Manufacturing systems such as FMS and CIM call for integration of production and inspection equipment into fully automated systems. An inspection equipment should have electronic control, which facilitates pre-programmed tasks and provides an easy interface with robotic or work-cell controller. The inspection equipment should have the ability to carry out on-line inspection and segregate good and bad parts for further processing.

Inspection constitutes the operational part of quality control. We are basically assessing dimensional accuracy, surface finish, and overall appearance of various components. In a typical manufacturing scenario, components are manufactured in large quantities and it becomes a herculean task to carry out 100% inspection. In such cases, we resort to statistical means by drawing samples out of the population. This method has one serious flaw—there is always a likelihood of defective parts getting through a lot that has passed the sampling test. In other words, nothing less than 100% good quality must be tolerated. Another important factor is the human factor. In a manual inspection method, inspectors carry out the inspection procedure. This can always lead to human errors, where bad parts may slip through and good parts may be rejected.

In principle, the only way to achieve 100% good quality is by carrying out 100% inspection. This is easier said than done. If manual inspection is adopted, one has to face two primary issues. Firstly, the cost of inspection may go up. Inspection cost comprises cost of labour, gauges, and accessories. Since most of these cost elements are variable in nature, inspection cost per unit may be substantial. Secondly, errors in inspection may also rise, either because of human error due to fatigue or because of wrong usage and application of measuring instruments. These problems can be addressed to a large extent by adopting automated inspection.

Automated inspection, in a general sense, means automation of one or more steps involved in the inspection procedure. A lower-end system comprises an automated system, which assists the operator/inspector by presenting the work part through an automated handling system. However, the actual inspection and the decision to accept or reject the part is made by the operator. A higher-end system is fully integrated into the manufacturing process and the entire inspection process is automated. The inspection system is also able to sort the work parts into good and bad pieces (rejected). The inspection of work parts can be done either off-line or on-line. Off-line inspection is done after a particular manufacturing operation, and invariably there is a delay between processing and inspection. This type of inspection is preferred when the process capability is well within the specified design tolerance and the production is done in large batch sizes with short cycle time.

On-line inspection is becoming increasingly popular, especially in an FMS or a CIM environment. It has two variants: *on-line/in-process inspection* and *on-line/post-process inspection*. The former is a more sophisticated one, wherein a feedback system comprising sensors and inspection devices perform inspection in an ongoing manufacturing process. The dimension of interest is continuously monitored, and the process is stopped as soon as the desired dimension is achieved. Such a system ensures 100% assurance of quality but is rather expensive to implement. As an example, Fig. 10.38 illustrates the configuration of such a system on a CNC machine. The CNC program is keyed in and simulated, and a dry run is performed to remove the bugs in the program. Trial production is performed, which not only ensures proper machining but also helps in setting the in-process gauges. When the actual production commences, in-process gauges will monitor the dimensions of interest, and withdraw the slide and stop the machine once the desired dimension is achieved. There is a provision to incorporate off-set corrections for change or wear of tools.

A slightly modified version, which is more popular, is the on-line/post-process inspection. The gauging and inspection of work parts is carried out following the production process. Even though it is performed subsequent to the production process, it is still considered on-line inspection since the inspection station is integrated into the manufacturing sequence. A transfer system automatically transfers the work parts from the machine to the inspection station and sorts the parts into good and defective ones.

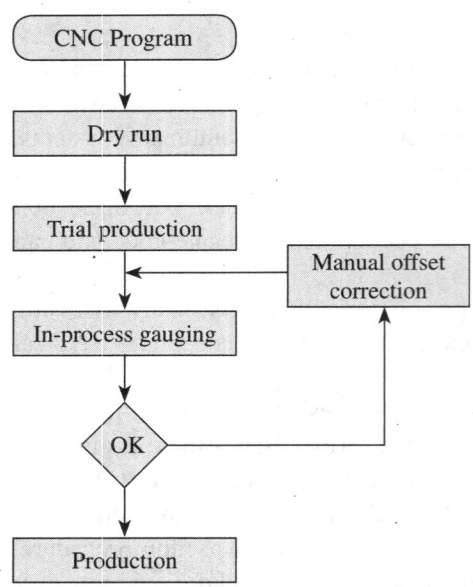

Fig. 10.38 On-line/In-process inspection

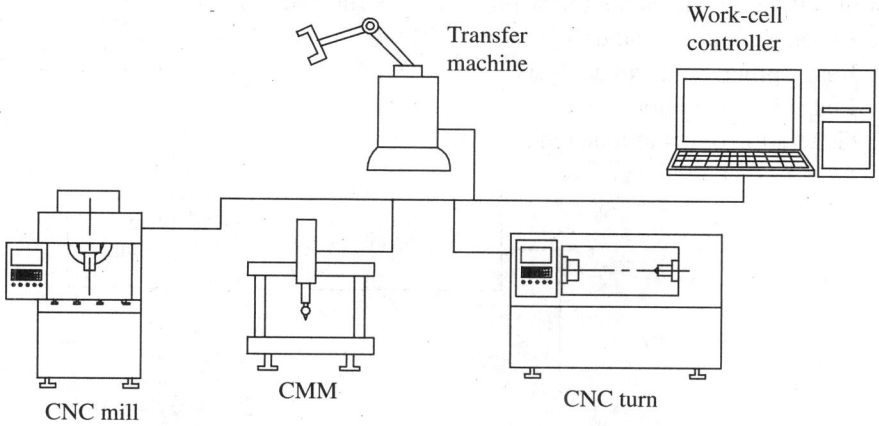

Fig. 10.39 Flexible inspection system

10.5.1 Flexible Inspection System

In recent times, the flexible inspection system (FIS) has become quite popular, specifically in an FMS environment. This system is illustrated in Fig. 10.39. A computer serves as a real-time controller to integrate the functions of several machines with the inspection station. It also ensures smooth performance of tasks in a required sequence. It can be seen in Fig. 10.39 that a typical FIS comprises one or more CNC machines, a transfer machine (alternatively a robotic handling system), and a CMM. The controller continuously monitors the function of the CNC machine and, as per the process plan, directs the transfer machine to shift the work part to the CMM to carry out on-line/post-process inspection. The data from the CMM is fed into the controller for analysis and feedback control.

A high-tech factory may have a number of FIS systems. The quality control data and inspection data from each station are fed through the terminals to a main computer. This enables performance of machine capability studies for individual machines. Another important feature of an FIS is dynamic statistical sampling of parts, which provides vital clues to modify the number of parts and part features to be inspected based on the inspection history.

10.6 MACHINE VISION

A machine vision system enables the identification and orientation of a work part within the field of vision, and has far-reaching applications. It can not only facilitate automated inspection, but also has wide ranging applications in robotic systems. Machine vision can be defined as the acquisition of image data of an object of interest, followed by processing and interpretation of data by a computer program, for useful applications.

10.6.1 Stages of Machine Vision

The principal applications in inspection include dimensional gauging, measurement, and verification of the presence of components. The operation of a machine vision system, illustrated in Fig. 10.40, involves the following four important stages:

1. Image generation and digitization
2. Image processing and analysis
3. Image interpretation
4. Generation of actuation signals

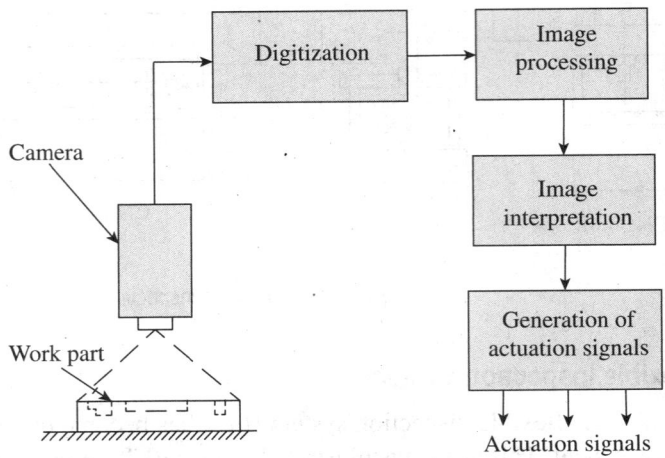

Fig. 10.40   Configuration of a machine vision system

Image Generation and Digitization

The primary task in a vision system is to capture a 2D or 3D image of the work part. A 2D image captures either the top view or a side elevation of the work part, which would be adequate to carry out simple inspection tasks. While the 2D image is captured using a single camera, the 3D image requires at least two cameras positioned at different locations. The work part is placed on a flat surface and illuminated by suitable lighting, which provides good contrast between the object and the background. The camera is focused on the work part and a sharp image is obtained. The image comprises a matrix of discrete picture elements popularly referred to as pixels. Each pixel has a value that is proportional to the light intensity of that portion of the scene. The intensity value for each pixel is converted to its equivalent digital value by an *analog-to-digital converter* (ADC).

This digitized frame of the image is referred to as the *frame buffer*. While Fig. 10.41(a) illustrates the object kept in the scene of vision against a background, Fig. 10.41(b) shows the division of the scene into a number of discrete spaces called pixels. The choice of camera and proper lighting of the scene are important to obtain a sharp image, having a good contrast with the background.

Two types of cameras are used in machine vision applications, namely *vidicon cameras* and *solid-state cameras*. Vidicon cameras are analog cameras, quite similar to the ones used in conventional television pictures.

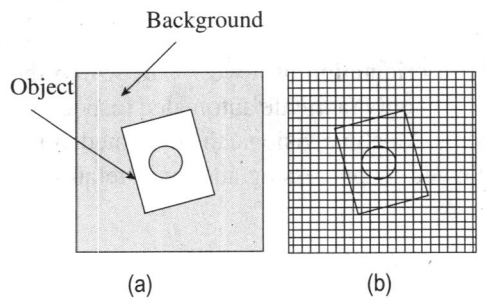

Fig. 10.41   Vision system (a) Object and background (b) Matrix of pixels

The image of the work part is focused onto a photoconductive surface, which is scanned at a frequency of 25–30 scans per second by an electron beam. The scanning is done in a systematic manner, covering the entire area of the screen in a single scan. Different locations on the photoconductive surface, called pixels, have different voltage levels corresponding to the light intensity striking those areas. The electron beam reads the status of each pixel and stores it in the memory.

Solid-state cameras are more advanced and function in digital mode. The image is focused onto a matrix of equally spaced photosensitive elements called pixels. An electrical charge is generated in each element depending on the intensity of light striking the element. The charge is accumulated in a storage device. The status of every pixel, comprising either the grey scale or the colour code, is thus stored in the frame buffer. Solid-state cameras have become more popular because they adopt more rugged and sophisticated technology and generate much sharper images. Charge-coupled-device (CCD) cameras have become the standard accessories in modern vision systems.

Image Processing and Analysis

The frame buffer stores the status of each and every pixel. A number of techniques are available to analyse the image data. However, the information available in the frame buffer needs to be refined and processed to facilitate further analysis. The most popular technique for image processing is called *segmentation*. Segmentation involves two stages: *thresholding* and *edge detection.*

Thresholding converts each pixel value into either of the two values, white or black, depending on whether the intensity of light exceeds a given threshold value. This type of vision system is called a *binary vision* system. If necessary, it is possible to store different shades of grey in an image, popularly called the *grey-scale system.* If the computer has a higher main memory and a faster processor, an individual pixel can also store colour information. For the sake of simplicity, let us assume that we will be content with a binary vision system. Now the entire frame of the image will comprise a large number of pixels, each having a binary state, either 0 or 1. Typical pixel arrays are 128×128, 256×256, 512×512, etc.

Edge detection is performed to distinguish the image of the object from its surroundings. Computer programs are used, which identify the contrast in light intensity between pixels bordering the image of the object and resolve the boundary of the object.

In order to identify the work part, the pattern in the pixel matrix needs to be compared with the templates of known objects. Since the pixel density is quite high, one-to-one matching at the pixel level within a short time duration demands high computing power and memory. An easier solution to this problem is to resort to a technique known as *feature extraction*. In this technique, an object is defined by means of its features such as length, width, diameter, perimeter, and aspect ratio. The aforementioned techniques—thresholding and edge detection—enable the determination of an object's area and boundaries.

Image Interpretation

Once the features have been extracted, the task of identifying the object becomes simpler, since the computer program has to match the extracted features with the features of templates already stored in the memory. This matching task is popularly referred to as *template matching*. Whenever a match occurs, an object can be identified and further analysis can be carried out. This interpretation function that is used to recognize the object is known as *pattern recognition*. It is needless to say that in order to facilitate pattern recognition, we need to create templates or a database containing features of the known objects. Many computer algorithms have been developed for template matching and pattern recognition. In order to eliminate the possibility of wrong identification when two objects have closely resembling features, *feature weighting* is resorted to. In this technique, several features are combined into a single measure by assigning a weight to each feature according to its relative importance in identifying the object. This adds an additional dimension in the process of assigning scores to features and eliminates wrong identification of an object.

Generation of Actuation Signals

Once the object is identified, the vision system should direct the inspection station to carry out the necessary action. In a flexible inspection environment, a discussed in Section 10.5.1, the work-cell controller should generate the actuation signals to the transfer machine to transfer the work part from machining stations to the inspection station and vice versa. Clamping, de-clamping, gripping, etc., of the work parts are done through actuation signals generated by the work-cell controller.

10.6.2 Applications of Machine Vision in Inspection

Machine vision systems are used for various applications such as part identification, safety monitoring, and visual guidance and navigation. However, by far, their biggest application is in automated inspection. It is best suited for mass production, where 100% inspection of components is sought. The inspection task can either be in on-line or off-line mode. The following are some of the important applications of machine vision system in inspection:

Dimensional gauging and measurement   Work parts, either stationary or moving on a conveyor system, are inspected for dimensional accuracy. A simpler task is to employ gauges that are fitted as end effectors of a transfer machine or robot, in order to carry out gauging, quite similar to a human operator. A more complicated task is the measurement of actual dimensions to ascertain the dimensional accuracy. This calls for systems with high resolution and good lighting of the scene, which provides a shadow-free image.

Identification of surface defects   Defects on the surface such as scratch marks, tool marks, pores, and blow holes can be easily identified. These defects reveal themselves as changes in reflected light and the system can be programmed to identify such defects.

Verification of holes   This involves two aspects. Firstly, the count of number of holes can be easily ascertained. Secondly, the location of holes with respect to a datum can be inspected for accuracy.

Identification of flaws in a printed label   Printed labels are used in large quantities on machines or packing materials. Defects in such labels such as text errors, numbering errors, and graphical errors can be easily spotted and corrective action taken before they are dispatched to the customer.

A QUICK OVERVIEW

- Laser-based instrumentation is the best choice for precise measurement, because practical realization of the metre is closely related with the radiation of stabilized frequency lasers.
- A laser transducer is basically a comparator that measures only the relative change of position between the interferometer and the retroreflector. In other words, it does not provide an absolute measurement of length.
- The CMM enables the location of point co-ordinates in a 3D space. It simultaneously captures both dimensions and orthogonal relationships. Another remarkable feature of CMM is its integration with a computer. The computer provides the additional power to generate 3D objects as well as to carry out complex mathematical calculations. Thus, complex objects can be dimensionally evaluated with precision and speed.
- The five basic configuration types of CMM are (a) cantilever, (b) bridge, (c) column, (d) horizontal arm, and (e) gantry.
- A computer-controlled CMM has an on-board computer, which increases versatility, convenience, and reliability. Computer assistance is utilized for three major functions. Firstly, a programming software directs the probe to the data collection points. Secondly, measurement commands enable comparison of the distance traversed to the standard built into the machine for that axis. Thirdly, computational capability enables processing of the data and generation of the required results.
- Machine tool metrology is primarily concerned with the geometric tests of the alignment accuracy of machine tools under static conditions. Standard tests are done to assess the alignment of various machine parts in relation to one another. The phrase *acceptance test* is coined to signify the fact that the machine buyer will inspect the alignment of various parts of the machine tool in detail and will 'accept' the machine tool from the vendor's factory only after it conforms to accepted norms.
- Automated inspection, in a general sense, means automation of one or more steps involved in the inspection procedure. A lower-end system comprises an automated system, which assists the operator/inspector by presenting the work part through an automated handling system. However, the actual inspection and the decision to accept or reject the part is made by the operator. A higher-end system is fully integrated into the manufacturing process and the entire inspection process is automated. The inspection system is also able to sort the work parts into good and bad pieces (rejected).
- In recent times, the FIS has become quite popular, specifically in an FMS environment. A computer serves as a real-time controller to integrate the functions of several machines with the inspection station. It also ensures smooth performance of inspection tasks in the required sequence.
- A machine vision system enables the identification and orientation of a work part within the field of vision, and has far-reaching applications. It can not only facilitate automated inspection but also has wide ranging applications in robotic systems. Machine vision can be defined as the acquisition of image data of an object of interest, followed by processing and interpretation of data by a computer program, for useful applications.

MULTIPLE-CHOICE QUESTIONS

1. Laser can be represented as a sine wave whose wavelength remains constant for
   (a) a given temperature
   (b) a given frequency
   (c) a given colour
   (d) none of these

2. In metrology applications, lasers are employed in the bandwidth of
   (a) 0.04–0.05 µm
   (b) 0.4–0.5 µm
   (c) 4–5 µm
   (d) 40–50 µm

3. In a laser transducer, the movement of the measurement reflector with a velocity v causes a frequency shift δf_1 in the measurement beam and is known as
   (a) Doppler effect
   (b) Seibek effect
   (c) Bipolar effect
   (d) Johnson effect

4. The CMM enables the location of point coordinates in a
   (a) 3D space
   (b) 2D space
   (c) horizontal plane only
   (d) vertical plane only

5. The _____ type configuration is a good choice for CMM if better rigidity in the structure is required.
   (a) cantilever
   (b) moving ram
   (c) bridge
   (d) gantry

6. In a CMM, the probes that make use of electronic technology to ensure application of optimum contact pressure are called
   (a) soft probes
   (b) hard probes
   (c) powered probes
   (d) mild probes

7. Which of the following commands does not belong to the 'measurement function library' of a CMM?
   (a) Conversion from SI units to British
   (b) Switching of coordinate system
   (c) Formatting
   (d) Datum selection

8. The minimum separation of a pair of parallel planes that will just contain all the points on the surface is a measure of
   (a) straightness
   (b) parallelism
   (c) squareness
   (d) flatness

9. ___ has two highly finished surfaces, which are perpendicular to each other with a high degree of accuracy and are used as an accessory for testing squareness.
   (a) An optical square
   (b) A standard square
   (c) A set square
   (d) All of these

10. The general way of describing cylindricity of a component is by the
    (a) minimum-zone method
    (b) maximum-zone method
    (c) limited-zone method
    (d) cylinder-zone method

11. In the acceptance test to evaluate deflection of the spindle, the main accessory is
    (a) a clinometer
    (b) a spirit level
    (c) an LTD
    (d) a DTD

12. In case of a machine vision system, the discrete addressable picture elements are popularly referred to as
    (a) photo detectors
    (b) pixels
    (c) pixars
    (d) data points

13. The digitized frame of the image in a machine vision system is referred to as the
    (a) ADC
    (b) frame buffer
    (c) vision buffer
    (d) DAC

14. Which of the following is not related to feature extraction?
    (a) Length
    (b) Width
    (c) Perimeter
    (d) Density

15. In the process of feature extraction, ____ is done in order to eliminate the possibility of wrong identification when two objects have closely resembling features.
    (a) feature weighting
    (b) feature building
    (c) feature indexing
    (d) feature elimination

REVIEW QUESTIONS

1. What are the special properties possessed by lasers that make them suitable for metrology applications?

2. With the help of an illustration, explain the working principle of a laser transducer system.

3. What is the need for using a double-frequency radiation source in a laser transducer?

4. What is the significance of the word 'coordinate'

in a CMM? What conditions warrant the use of a CMM in the industry?

5. Compare the pros and cons of the five different configurations of a CMM.

6. How does a semi-automated CMM differ from a computer-controlled CMM?

7. Distinguish between the following with respect to CMM probes:
   (a) Contact and non-contact probes
   (b) Hard and soft probes

8. Briefly discuss the three basic types of commands in programming of CMMs.

9. Discuss the major applications of CMMs.

10. What is the need for acceptance tests?

11. Define straightness, flatness, parallelism, squareness, and roundness.

12. Clarify the meaning of the following tests and their significance on a machine tool:
   (a) True running of a spindle
   (b) Squareness of guideways with table
   (c) Total deflection of the spindle
   (d) Parallelism of table movement with spindle axis

13. How does on-line/in-process inspection differ from on-line/post-process inspection?

14. Explain the configuration of a flexible inspection system.

15. Briefly explain the following in relation to a machine vision system:
   (a) Segmentation
   (b) Thresholding
   (c) Edge detection
   (d) Feature extraction

Inspection and Quality Control

After studying this chapter, the reader will be able to

- appreciate the importance of inspection and quality control practices in the industry
- understand the basic concepts of statistical quality control (SQC) and important SQC tools like control charts
- work out solutions for typical problems involving determination of control limits and assess whether a manufacturing process is in control or not
- explain the management perspective of quality control such as total quality management, six sigma, and related quality tools
- understand the objectives, scope, and implementation of ISO 9000 series certification

11.1 INTRODUCTION

'Quality' is a buzzword in modern economy. Customers in the modern world expect a high level of quality in products and services. The *International Organization for Standardization* (ISO) defines quality as 'the degree to which a set of inherent characteristics (distinguishing features) fulfil requirements, i.e., needs or expectations that are stated, generally implied or obligatory'. Implicit in this definition is the fact that quality of a product is related to the fitness of purpose. For example, a Mercedes Benz car designed for use in a modern metropolitan city will be utterly useless for a buyer in a hilly area who wishes to use it to move around in his coffee estate. Even though the product is certified as world class, it may be unfit for use by such a customer. In other words, today's concept of quality is driven by customer needs and aspirations. This chapter starts with the basic functions of inspection and statistical quality control (SQC), and deals with concepts such as *total quality management* (TQM) and *six sigma*, the customer centric approaches for achieving high quality of products, processes, and delivery.

In earlier times, *quality control* (QC) was a simple matter and the term *quality assurance* (QA) had not yet been invented. After the initial machining or some kind of processing, the parts ended up in the assembly shop. If a part did not fit properly, the concerned shop was advised to reset the machine or advance the cutting tool slightly. In today's industry, because of increased complexity of products and processes, and mass production, this approach is no longer feasible. It is costly and now unnecessary, thanks to SQC.

Although statistics as a branch of mathematics has been well understood for a long time, it was the efforts of Dr Walter A. Shewhart of Bell Telephone Laboratories (around 1924) that established the technique for applying statistics to the control of industrial processes. Control charts, which will be explained later in this chapter, are fundamental to SQC and still bear the name of their inventor. While inspection tells us whether a particular manufactured part is within tolerance limits or not, SQC goes a step further and gives a clear indication of whether a manufacturing process is under control and capable of meeting dimensional accuracy.

While SQC provides a proven approach to assess quality, it is a subset of a broader concept of quality management called 'total quality management'. This management approach is customer-centric but encompasses all the players in a product's value chain, from suppliers and producers to customers and the market. It deals with the best management practices for improvement of organizational performance and is aimed at achieving excellence. This chapter gives a detailed account of TQM philosophy and procedures, followed by a discussion on six sigma, a quality management tool developed by the Japanese management gurus. We end the chapter with a brief outline of the ISO 9000 series quality certification and its importance.

11.2 INSPECTION

It is presumed that the student has a fair knowledge of the various manufacturing processes such as machining, forging, casting, sheet metal work, and so on. Every manufacturing operation calls for inspection of a part before it is shunted out to the next operation or assembly. In a product life cycle, the design engineer creates process sheets comprising part drawings, which clearly indicate the various dimensions and tolerances to be achieved for a component before it goes into assembly. These drawings are passed on to the process planning engineer, who releases process sheets to the manufacturing departments. If the final product involves 1000 parts, then at least 1000 process sheets need to be prepared for release to the manufacturing shops. A process sheet provides the required guidelines to machine operators regarding the use of appropriate tools, process parameters, and more importantly, inspection gauges.

Inspection is the scientific examination of work parts to ensure adherence to dimensional accuracy, surface texture, and other related attributes. It is an integral part of the QA mechanism, which ensures strict adherence to the stated design intent. The American Society of Tool and Manufacturing Engineers (ASTME) defines inspection as 'the art of critically examining parts in process, assembled sub-systems, or complete end products with the aid of suitable standards and measuring devices which confirm or deny to the observer that the particular item under examination is within the specified limits of variability'. Inspection is the responsibility of operators as well as inspectors, who are specially trained to carry out inspection in a proper manner. In simple terms, the inspector inspects the parts after a particular manufacturing process and certifies whether the parts are of an acceptable standard, or whether they have to be rejected and denied further movement towards final assembly. Inspection also generates data regarding percentage rejection, number of parts sent for rework, etc., which are useful for removing process deficiencies and improve the throughput.

Inspection is carried out at three stages: receiving inspection, in-process inspection, and final inspection. A manufacturing organization sources raw materials and semi-finished components from suppliers and sub-contractors. Therefore, it is important to ensure that all such materials and components meet the quality requirements. Receiving inspection is a critical activity that ensures adequate quality of all incoming goods. In-process inspection comprises all inspection tests and procedures implemented within the confines of the factory. The scope of in-process inspection can be decided by answering the following questions:

1. What to inspect?
2. Where to inspect?
3. How much to inspect?

The answer to the first question is provided by the detailed examination of key characteristics that are related to quality or cost. One should examine the drawings released by the product designer and accordingly plan the various dimensions and attributes that need to be checked for design compliance at various stages of manufacturing. The second question is more related to the mechanism of conducting such tests. Some components can be inspected on the shop floor while some may need to be shifted to a controlled environment or a special-purpose measuring machine in order to carry out the inspection. By far, the third question is always the trickiest! Given a choice, the production engineer would like to inspect each component after every operation on the shop floor. This would be possible if the number of components and processes are limited. However, in a mass production industry, for example, an automobile industry, thousands of components need to be manufactured for hundreds of vehicles. In such cases, 100% inspection will be time consuming as well as prohibitively expensive. On the other hand, we can entirely do away with inspection. The logical step, therefore, is to adopt selective inspection by drawing representative samples from the entire lot, based on certain statistically valid procedures. This method is popularly referred to as *acceptance sampling*. The second part of this chapter addresses this method in more detail.

The decision to do away with inspection may also make economic sense in some cases. Management experts recommend the following economic model to decide if inspection should be carried out or not.

Let C_1 be the cost of inspection and removal of non-conforming item, C_2 the cost of repair, and p the true fraction non-conforming.

Then, the break-even point is given by $p \times C_2 = C_1$.

If $p > C_1/C_2$, use 100% inspection.

If $p < C_1/C_2$, avoid inspection.

The final inspection is carried out after the product is completely assembled or manufactured, and is ready for delivery to the customer. In a few cases like the sales of machine tools, the customer himself/herself would prefer to conduct acceptance tests before accepting a machine tool from the manufacturer.

11.3 SPECIFYING LIMITS OF VARIABILITY

The whole question of inspection arises because of variability in production processes. No production process, irrespective of whether it is machining, forging, or casting, can assure 100% conformance to intended dimensions and surface quality. For this precise reason, dimensional tolerance is provided for all manufactured components. The ISO has recommended a tolerance value for every manufacturing process. This is a very comprehensive bunch of standards that covers metallurgical requirements, bearings, gears, shafts, oil groves, and so on. However, the design engineer has to exercise good judgement while specifying fits and tolerances. Too close a fit or a narrow band of tolerance will demand a more precise manufacturing process. In order to ensure adherence to close fits and tolerances, the inspection cost will also go up.

Inspection cost comprises the following components:

1. Engineering cost comprising cost of design and manufacture of inspection gauges and tools
2. Cost of gauges, measuring instruments, and utilities (e.g., air-conditioned or temperature-controlled environment for conducting inspection)
3. Cost of labour (which is employed to conduct inspection)

Most companies specify the limits of variability or tolerance based on several considerations apart from the purely engineering considerations proposed by ISO. Some of the other considerations are as follows:

Customer and market needs   An industrial customer is more demanding than an ordinary consumer of household goods. The quality demands of an industrial customer are more stringent and therefore the inspection cost will go up.

Manufacturing facility   A modern factory will enable imposition of close tolerances since the process variability is within narrower limits. Otherwise, a wide process variability of machines and equipment will pose a major hindrance for specifying close tolerances. This is also coupled with a higher inspection cost as more number of parts need to be inspected to assure that only good parts go into the final assembly.

Manpower   This is an important dimension in QC and has important implications on inspection cost. In developing countries, cheap manpower is abundantly available and can be put to good use in a predominantly manual mode of inspection. However, a manual inspection method is more prone to mistakes. Even though, on the face of it, inspection cost may appear to be lesser because of cheap labour, inspection errors may prove to be costlier in the final outcome. In a modern economy, paucity of labour would necessitate increased automation in inspection. Although high accuracy and less inspection time are guaranteed, the initial investment and maintenance of advanced inspection equipment may be prohibitively high.

Management   The vision, goals, and plans of the management of a company also influences the importance attached to the production of quality products. The quality-driven approach of the management ensures investment on high-quality manufacturing machines and equipment, which in turn facilitates choice of close fits and tolerances. There will be a conscious effort to reduce variability in a manufacturing process, which in turn demands stringent inspection methods and tools.

Financial strength   A company with good financial prowess would be willing to spend more financial resources on top of class machines, equipment, and tools. This will naturally lead to the choice of high-quality measuring and inspection tools and instruments. The company would also be willing to adopt the best inspection methods to achieve zero-defect products.

The following sections deal with the requirements of inspection gauges and the management-related issues in inspection. In order to appreciate the design considerations for inspection gauges, we need to understand the practice of assigning dimensions and tolerances for work parts.

11.4 DIMENSIONS AND TOLERANCES

A dimension can be defined as the perfect separation between two definable points, called *features*, either on one part or between parts. In other words, a dimension is the statement of the intended size of a particular feature, while measurement is the statement of its actual size. The features of a part are bounded by lines and areas. Most lines involved in measurement are actually edges formed by the intersection of two planes. Various types of edges present different measurement problems. There are also angular dimensions, surface finish dimensions, among others, which have to be specified. A dimension is looked at differently by different people associated with it. The designer's concept for a perfect part determines the *dimension*. The machine operator's machining results in the *feature* of the part. The inspector's measurement verifies the machine operator's work to the designer's concept of dimension. Figure 11.1 illustrates the three aspects of a dimension. While Fig.11.1(a) represents the designer's concept of dimension, Fig. 11.1(b) is the result of the generation of features after production. Figure 11.1(c) shows the inspector's concept of dimension, which is measured to an acceptable level of accuracy.

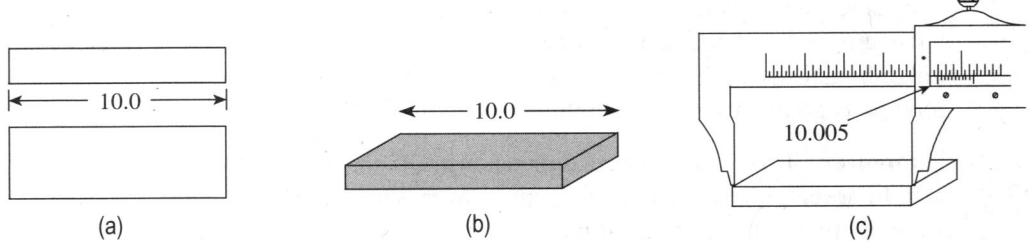

| (a) | (b) | (c) |

Fig. 11.1   Three aspects of a dimension (a) Designer's concept (b) Production concept (c) Inspector's concept

Dimensions and tolerances are specified on each and every part drawing. The dimensions given on the part need to be achieved in order to ensure proper functioning of the part either independently or in an assembly with other parts. No other dimensions are given except those needed to produce or inspect the part. The dimensions are given with absolute clarity so that there is no ambiguity on the part of machine operators or inspectors of parts. The designer should not burden them with additional calculations, which widen the scope of committing mistakes.

It is humanly impossible to manufacture components to an exact dimension because of variations in production processes, tooling, workmanship, etc. Moreover, an assembly will function satisfactorily with some variations in component sizes. In addition, the exact size will be too expensive to produce. In order to inform the production people how much variation from the exact size is permissible, the designer specifies tolerance for most of the dimensions. Tolerance can be defined as the total permissible variation of a given dimension. Thus, tolerance in a way shifts the responsibility of manufacturing quality components and products to the production engineer.

Selection of inspection gauges and instruments is directly linked to the level of tolerance specified by the designer. Very high tolerance leads to the production of bad parts, which ultimately results in products of a bad quality. On the other hand, very close tolerances demand excessively precise gauges and instruments. The additional costs of such measurements are real but obscured. Therefore, while assigning tolerances, the decision is not entirely metrological, but rather managerial in nature.

Engineering tolerances may be broadly classified into three groups:
1. Size tolerances
2. Geometric tolerances
3. Positional tolerances

Size tolerances are allowable deviations of dimensions for length, diameter, angle, etc. *Geometric tolerances* are specified for a specific geometric feature such as straightness, flatness, and squareness. Geometric tolerances are extremely important for ensuring perfect alignment and functional accuracy of the various parts in a machine. *Positional tolerance* provides an effective way of controlling the relative positions of mating features where interchangeability is the primary requirement.

Gauging is a form of inspection, which has been explained in the next two sections of this chapter. In general, inspection means open set-up inspection and gauging means *attribute gauging*. Gauging inspects one or few attributes at a time and speeds up the inspection process. The most popular are the GO and NO GO gauges, which accept (GO) or reject (NO GO) the features being checked.

11.5 SELECTION OF GAUGING EQUIPMENT

Generally, inspection gauges are designed by tool engineers in the 'tool design' department. It calls for a sound understanding of the manufacturing processes, tolerancing, tool design, and engineering materials. Gauges can be broadly classified into two types: attribute gauges and variable-type gauges. Attribute gauges such as ring and plug gauges are simple and convenient to use. They provide the operator with a simple 'yes' or 'no' answer, indicating if the part should be accepted or rejected. On the other hand, variable-type gauges such as dial indicators, callipers, and pneumatic gauges are basically measuring devices that can also function as gauges. While attribute gauges are capable of checking only one particular dimension, variable gauges can be set to the required value by the operator. Table 11.1 provides general guidelines for the selection of an appropriate gauge based on the tolerance provided for the work parts.

Table 11.1 Recommended gauges based on tolerance level

Tolerance value (mm)	Recommended gauges/measuring devices
0.100–1	Steel rule
0.050–0.1	Plug and ring gauges, vernier calliper, micrometer, dial indicator, etc.
0.010–0.05	Vernier micrometer, autocollimator, gauge blocks, optical devices, pneumatic gauges, etc.
0.001–0.01	Electronic gauges, interferometers, laser-based devices, etc.

It is general practice to assign a tolerance band that is 1/10th of the work tolerance for the manufacture of inspection gauges such as plug and ring gauges. Obviously, this calls for a highly precise process to fabricate the gauges. Every major manufacturing organization will have a tool room, where the gauges are fabricated. The tool room will have the most precise machines and highly skilled workmen who can fabricate the gauges to the required accuracy.

Whenever the tolerance level is less than 0.01 mm, it becomes necessary to conduct inspection in a controlled environment. For instance, in an automobile plant, the piston and the cylinder bore need to be matched to provide a clearance of up to 5 μm accuracy. In such circumstances, the inspection process also calls for grading the cylinder bores and ensuring a perfect match with the pistons, which are usually sourced from a supplier. The following are the best practices to ensure proper inspection:

1. A separate gauge laboratory must be installed to carry out inspection.
2. The gauge laboratory should have temperature and humidity control options and should be free from dust and smoke.
3. The laboratory should have precise measuring equipment capable of measuring up to a fraction of a micrometre.
4. It should have a good stock of master gauges under close surveillance.
5. All master gauges, in turn, should have undergone regular inspections and be traceable to the National Bureau of Standards.

11.6 GAUGE CONTROL

Gauging work parts is one of the most critical activities in a manufacturing organization. It ensures that only good parts will go into the final assembly, thereby ensuring the roll-out of high-quality products. Imagine a scenario where the piston in an automotive engine is wrongly matched with the cylinder bore. The vehicle will come back to the showroom with an extremely unhappy customer demanding immediate action. This creates bad publicity for the company, which it cannot afford in a highly competitive industry. Therefore, ensuring that only good parts, which meet the dimensional and tolerance requirements, are cleared for final assembly is extremely important.

Thousands of components need to be inspected every day in a typical engineering industry. It is important to ensure that the right gauges are available at the right time and at the right place. While design and fabrication of gauges are the job of the *tool design department*, issue and maintenance of gauges are taken care of by a *gauge control section* in the *quality control department* (*QCD*). The personnel in the gauge control section should only report to the QCD head, and the line of authority should be such that the production people not interfere with their decisions (to accept or reject parts). Their major responsibilities are to monitor the condition of gauges and other inspection equipment, carry out their periodic calibration, and ensure that they are immediately replaced if found to be unfit for use. The personnel should follow standard procedures and protocols, and diligently maintain the inspection records. The following are some of the major functions of the gauge control department:

1. Assign a unique code to each and every gauge and inspection equipment and maintain historical records from inception to the scrapping stage.
2. Maintain a clean and controlled (for temperature and humidity) environment for storing all

the gauges. Suitable storage racks and enclosures need to be used with proper security.

3. A system of recording issue and receipt of gauges either to the workers or QC inspectors should be in place. Immediate action should be initiated for the non-receipt of gauges. Obviously, a computer-based gauge management system is essential to carry out this function effectively. Information about the current deployment of a gauge should be available at the click of a button.

4. Provision should be made for carrying costly gauges or inspection equipment in protective boxes during transit from the gauge control section to the manufacturing areas.

5. All new gauges must be thoroughly checked before deploying them for inspection.

6. Schedules should be drawn for periodic calibration of gauges and adhered to diligently. Trained manpower should be deployed to repair the gauges whenever necessary.

7. The gauge control section should provide useful feedback to the company management regarding budgeting, reliable suppliers of gauges and inspection equipment, possible improvements in the design of gauges, avoidance of redundancy, opportunities for cost reduction, and so on.

11.7 QUALITY CONTROL AND QUALITY ASSURANCE

The famous management guru Dr Juran stated that *QC = defect prevention*. The ASTME define QC as the directed effort of all elements of a company towards competitive standards with minimum losses to the company.

Control is the activity of ensuring conformance to the stated specifications and taking corrective actions when necessary to rectify problems. The primary objective of QC is to ensure conformance to design specifications in terms of dimensions, tolerances, surface texture, etc., at each and every stage of a manufacturing process. QC includes the activities from the suppliers, through production, and to the customers. Incoming materials are examined to make sure that they meet the required specifications. The quality of partially completed products is analysed to determine if production processes are functioning properly. Finished goods and services are studied to determine if they meet customer expectations.

Figure 11.2 illustrates the role of QC in a production cycle. The production cycle begins with the receipt of raw materials and semi-finished components (inputs). Incoming material is

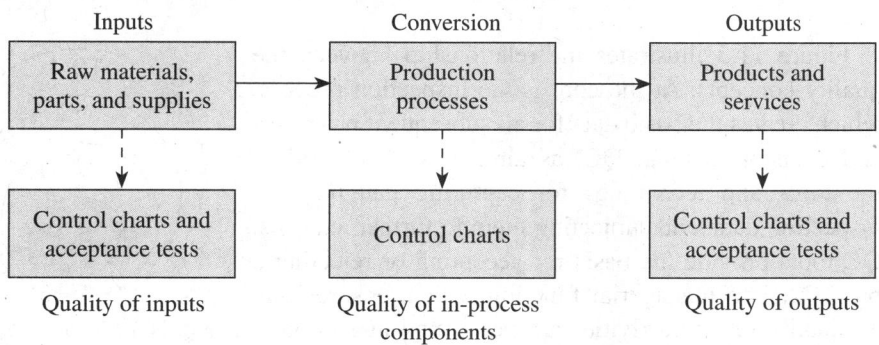

Fig. 11.2   Role of quality control in a production cycle

examined to make sure that it meets the appropriate specifications. The raw material undergoes various production processes, and the finished product is rolled out to the customer after the final inspection. The production and QC engineers should ensure strict adherence to quality at each stage of the production cycle. QC strategy involves deployment of various QC tools. Statistical techniques such as control charts and acceptance plans provide reliable and quicker procedures for inspection in an industry of mass production.

Quite often we come across QA, another popular concept. It sounds quite similar to QC and you may wonder whether both are the same. There is a clear distinction between QC and QA. While the former comprises a number of analytical measurements and tools used to assess the quality of products, the latter pertains to the overall management plan to guarantee the integrity of data. It is a comprehensive approach to build a scientific and reliable system to assure quality.

According to the ISO, QA comprises all planned and systematic activities, which provide adequate confidence that an entity will fulfil requirements for quality. The following are the major goals of QA:

1. Plan and implement all activities to satisfy quality standards for a product.
2. Ensure continuous quality improvement
3. Evolve benchmarks to generate ideas for quality improvement.
4. Ensure compliance with relevant statutory and safety requirements.
5. Provide assurance to all customers and authorities regarding
   (a) total traceability of products
   (b) suitability of products to customer requirements

Table 11.2 provides the comparison between QC and QA.

Table 11.2 Comparison between quality control and quality assurance

Quality control	Quality assurance
Finding deviations from stated quality standards and fixing them	Improving the QC process rather than fixing errors in manufacturing
Putting major responsibility on QC inspectors	Shifting responsibility to middle- and top-level managers
Basically a line function	Basically a staff function
Activities: testing, inspection, drawing control charts, etc.	Major activities: defining quality process, quality audit, selection of tools, etc.

Figure 11.3 illustrates the relationship between the quality concepts. At the core is the inspection process, which creates the basic data for all subsequent processes and decision-making. QC, as already said, comprises the tools and techniques for capturing gauging and inspection data, and subjecting them to further analysis. QC tools provide the basis for accepting or rejecting a part. QA is a managerial function, which ensures that all quality-related activities are performed in a diligent and optimum manner. However, all the aforementioned

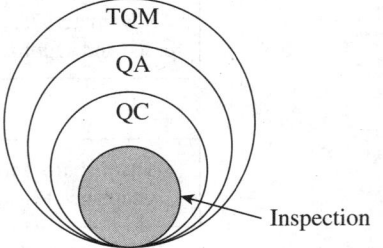

Fig. 11.3 Relationship between various quality concepts

activities are subsets of a corporate-level management strategy, popularly referred to as *total quality management*. While the first three activities pertain to operational levels and the middle management level, TQM involves corporate strategies and commitment of the top management to ensure high level of quality, not only in terms of product quality, but also in each and every sphere of activity.

11.8 STATISTICAL QUALITY CONTROL

Ideally, it is necessary to inspect each type of raw material, and semi-finished and finished components at every stage of production. This is easier said than done, because this is a Herculean task in a large industry. Statistical techniques provide a reliable and economical method for ensuring QC, which is known as *statistical quality control* (*SQC*). This section introduces the various statistical tools and methodologies that are implemented in the industry.

Although statistics as a branch of mathematics has been well understood for several centuries, its importance in engineering applications came to be realized in the 19th century. Dr Walter A. Shewhart of Bell Telephone Laboratories of USA first established the technique of applying statistics to the control of industrial production. He invented the control charts to assess whether a production process was under QC or not. Despite the proven advantages of SQC, it did not come into extensive use until the unprecedented demands of World War II. There was a heavy demand for aircrafts, battle tanks, arms, and ammunitions at the war front. In USA, new factories were built in the early 1940s to supply war equipment. However, the war effort demanded high-quality products and the resources to carry out 100% inspection were just not available. Shewhart's techniques provided timely aid to control quality and deliver supplies to the front lines on time. Shewhart created a system for tracking variation and identifying its causes. His system of *statistical process control* (SPC) was further developed and popularized by his one-time colleague, Edward Deming. Deming showed how SPC could be achieved by understanding process variability.

11.8.1 Process Variability

Consider the example of shafts made in large quantities on a lathe. If the diameter of each shaft is measured after the turning operation, we would expect to see some variability of the measurements around a mean value. These observed random variations in measurements might be due to variations in the hardness of material, power surges affecting the machine tool, or even errors in making measurements on the finished shaft. These errors are referred to as *random errors* and are inherent in any manufacturing processes.

On the other hand, consider the case of the cutting tool beginning to dull. With continuous dulling of the cutting edge, the diameter of the component will increase gradually. As the spindle bearings wear out, the dimensional variations will keep increasing. Such variations are caused due to assignable causes and are not random; these are referred to as *systematic errors*.

These two types of variations call for different managerial responses. Although one of the goals of quality management is *constant improvement* by the reduction of inherent variation, this cannot ordinarily be accomplished without changing the process. It would not make sense to change the process unless one is sure that all assignable causes for variation have been identified and brought under control. The lesson here is that if the process is out of control because some special cause for variation is still present, identify and correct the cause of that variation. Then, when the process has been brought under control, the quality can be improved by designing the process to reduce its inherent variability.

There are two major categories in SQC: SQC by attributes and SQC by variables. An attribute is a characteristic that is of a yes or no type. The GO and NO GO inspection gauges are examples of attribute inspection. On the other hand, a variable is a characteristic that can be stated in scalar values along a continuous scale. Note that the results of variable inspection can always be stated in terms of attributes, but not vice versa. We shall discuss SQC by attributes, followed by SQC by variables.

11.8.2 Importance of Sampling

SQC is based on *sampling*. A sample is defined as a representative portion of a group taken as evidence of the quality level or character of the entire group. However, there is always the risk that a sample that is drawn may not be representative of the group. This aspect has led to the development of a number of sampling theories, with the objective of ensuring selection of representative samples. The QC engineer has to be judicious in selecting the appropriate sampling method as well as a suitable sample size.

For example, consider a box of 1000 parts among which 100 are defective (that is 10% of the parts are defective), and let us assume that the company policy is to reject the entire lot if 10% of the items drawn are defective. If a sample of 10 is drawn, there is no guarantee that one of them would be defective; it can contain none, one, or several defective items. If the sample size is increased to 100, the odds of it containing the representative 10% would be much greater. At a sample size of 500, it is almost guaranteed that 10% of the drawn parts would be defective. Thus, the larger the sample, the closer it comes to being representative of the entire lot. It is important that a representative selection is made in sampling. This is called a *random sample*. This is achieved only when the parts are drawn at random from the lot. In many cases, this is done by digging the individual samples out of various places in their bulk container. When the parts can be identified by individual numbers, a random number table can be generated and used for picking the sample.

We can thus infer the following: the larger the sample size, the more representative is the sample of the lot from where it has been drawn. However, this negates the very purpose of sampling, which is to avoid 100% inspection and reduce inspection time and cost to a great degree. Statisticians recommend an *economic sampling plan* to balance the two basic requirements.

Economic Sampling Plan

While planning a sample, we need to consider two factors:
1. What is the least batch size acceptable? This is known as *lot tolerance per cent defective* (LTPD).
2. What is the risk involved in accepting a batch that is lesser than the LTPD in an optimistic sample? This is known as *consumer's risk*.

The most economic plan is the one that gives the required degree of protection to the producer for the least total amount of inspection per batch. In the long run, a manufacturing process will produce, when running normally, average per cent defectives, known as the process average.

The producer takes a risk that, when running at the process average, a pessimistic sample will reject an acceptable batch.

Let N be the batch size (or lot size), n the sample size, and R the producer's risk.

Then, the total inspection per batch I in the long run, for a given sampling plan, is given by $I = n + (N - n)R$.

It is seen that I is a function of batch size and R is a function of sample size. If I is plotted against sample size for a given batch size, a curve of the type shown in Fig. 11.4 is obtained. This gives the most economical sample size for a given batch size.

Sampling Methods

There are four popular sampling methods:
1. Simple random sampling
2. Constant interval sampling
3. Stratified sampling
4. Cluster sampling

Simple random sampling   Simple random sampling selects samples by methods that allow each possible sample to have an equal probability of being picked and each item in the entire population to have an equal chance of being included in the sample. The easiest way to randomly select a sample is to use random numbers. These numbers can be generated either by a computer programmed to scramble numbers or by a table of random numbers.

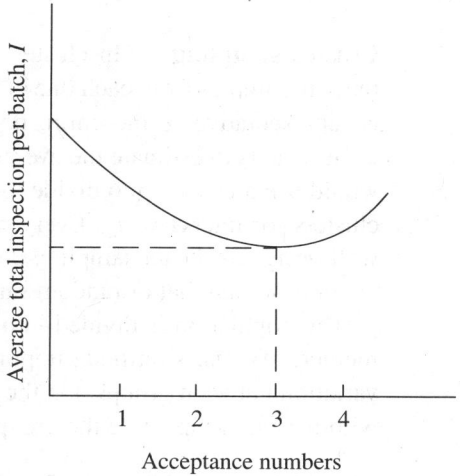

Fig. 11.4   Inspection per batch vs batch size

Constant interval sampling   This sampling is done based on the sequence at which parts leave the production operation. This may be every third part or perhaps every 10th part. The sampling plan is done based on the experience of an SQC professional. For example, if the lot size is 200 and 10 samples are needed, the sampling interval is 200/10 = 20. The point in production at which the first sample is drawn may be determined by a random number in the interval 1–19. Following that, all other samples are drawn according to the selected interval, in this case, 20 parts. Suppose, the first number turns out to be 7, then part numbers to be drawn are 7, 27, 47, 67, 87, 107, 127, 147, 167, and 187.

Stratified sampling   In certain situations, one continuous sample may not provide the desired information. For example, consider the production run of one part divided among several different machines or different operators. In such cases, the sub-lot sizes may differ. Therefore, the sample size for each sub-lot should be determined separately. Thus, stratified sampling is appropriate when the population is already divided into groups of different sizes and we wish to acknowledge this fact. Let us assume that a part can be machined on any one of four machines of a similar type, but of different age. In order to carry out inspection for dimensional accuracy, random samples can be drawn from each of the four machines by giving weight to the age of the machines. A probable sampling plan for this case is shown in Table 11.3.

Table 11.3 Stratified sampling plan

Machine number	Machine age	Percentage of total
1	Less than 1 year	08
2	Between 1 and 3 years	12
3	Between 3 and 5 years	18
4	Between 5 and 8 years	27
5	More than 8 years	35

Cluster sampling   In cluster sampling, we divide the population into clusters and select random samples from each one of these clusters. The underlying assumption is that each cluster is representative of the entire population. For instance, if a company manufacturing motor cycles wants to estimate the average number of young males per household in a large city, they would use a city map to divide the territory into clusters and then choose a certain number of clusters for interviewing. Every household in each of these clusters would be interviewed. A well-designed cluster sampling procedure can produce a more precise sample at a considerably lesser cost than that of random sampling.

The population is divided into well-defined groups in both stratified and cluster sampling methods. We use stratified sampling when each group has small variation within itself but wide variation between groups. On the other hand, cluster sampling involves considerable variation within each group while the groups are essentially similar.

11.8.3 Statistical Quality Control by Attributes

In SQC, a qualitative variable that can take only two values is called an attribute. The attribute most frequently used is that of conformance or non-conformance of units of output to the process specification. For example, in gauging, the GO gauge checks the maximum material limit of a component, whereas the NO GO gauge checks the minimum material limits. Thus, the inspection is restricted to identifying whether a component is within tolerance limits or not.

Control Chart for Attributes

The basic objective of SQC is to identify a parameter that is easy to measure and whose value is important for the quality of the process output, plot it in such a way that we can recognize non-random variations, and decide when to make adjustments to a process. These plots are popularly called *control charts*. A typical control chart has a central line, which may represent the mean value of measurements or acceptable percentage defective, and two control limits, *upper control limit* (UCL) and *lower control limit* (LCL). The two control limits are fixed based on statistically acceptable deviation from the mean value, due to random variations. If a process is in control, all observations should fall within the control limits. Conversely, observations that fall outside the control limits suggest that the process is out of control. This warrants further investigation to see whether some special causes can be found to explain why they fall outside the limits.

As already explained, a qualitative variable that can take only two values is called an *attribute*. The '*p* charts' are used for SQC of attributes. Let p = probability that tetra packs contain less than 0.495 l of milk.

Central line of p chart, CL $= p$ $\hspace{4cm}$ (11.1)

If there is a known or targeted value of p, that value should be used here. Otherwise, the p

value can be estimated by the overall sample fraction $p^{11} = \dfrac{\Sigma p_j}{k}$

where p_j is the sample fraction in the jth hourly sample and k is the total number of hourly samples.

In this case, $p = 0.03$

The standard deviation of sample proportion, $\sigma_p = \sqrt{pq/n}$ $\hspace{3cm}$ (11.2)
where $q = 1 - p$, and n is the sample size.

Control limits for p chart

$$\text{UCL} = p + 3\sqrt{\frac{pq}{n}} \qquad\qquad \text{LCL} = p - 3\sqrt{\frac{pq}{n}}$$

11.8.4 Statistical Quality Control by Variables

We have seen that SQC by attributes used the traditional GO and NO GO gauging principles. It judges what has already been done, but only by extension does it attempt to control what will be done. In contrast, SQC by variables determines the *natural capability* of the manufacturing process, be it machining, fabrication, assembly, or any other process involved in the production. It recognizes that all processes are imperfect and liable for small or large variations. These variations are called random variations. We need to establish the natural capability of a particular machine or process, because we need to provide realistic tolerances. An effective way to understand natural capability is by plotting the *normal distribution curve*.

Normal Distribution Curve

The most important continuous probability distribution is the normal distribution. The mathematician–astronomer Karl Gauss was instrumental in the development of normal distribution as early as the 18th century. In honour of his contribution to statistics, normal distribution is referred to as the *Gaussian distribution.*

The normal distribution has gained popularity in practical applications because it is close to observed frequency distributions of many phenomena such as physical processes (dimensions and yields) and human characteristics (weights, heights, etc.). Figure 11.5 illustrates the shape of a normal probability distribution.

The following features are unique to a normal probability distribution:
1. The curve has a single peak and is hence uni-modal. It has a typical bell shape.
2. The mean of a normally distributed population lies at the centre of its normal curve.
3. The curve is symmetrical about a vertical axis and, therefore, the median and mode of the distribution are also at the centre.
4. The two tails extend indefinitely and never touch the horizontal axis.

Standard deviation The normal distribution curve depends on process variability. One of the best tools to assess process variability is the *standard deviation*. This is expressed by the lower case Greek letter *sigma* (σ). Standard deviation is the square root of the mean average of squares of deviations of all the individual measurement values (X) from the mean or average value ($\overline{X}$).

Thus, $\sigma = \sqrt{\Sigma(X - \overline{X})^2 / n}$

Areas under normal curve Figure 11.6 illustrates the relationship between area under the normal curve and the distance from the mean measured in standard deviations. Accordingly,
1. approximately 68% of all values lie within ±1 standard deviation from the mean;
2. approximately 96.5% of all values lie within ±2 standard deviations from the mean; and
3. approximately 99.7% of all values lie within ±3 standard deviations from the mean.

Now, let us understand the significance of the aforementioned relationship. If we draw a random sample from a population wherein the values are normally distributed, it is easy to calculate the mean and standard deviation. If all observations in the sample fall within 3σ control limits, we would expect with 99.7% confidence that the process is *in control*. Conversely, observations that fall outside the 3σ control limits suggest that the process is *out of control* and warrants further investigation. $\overline{X}$ *chart*, the control chart for process means, and *R chart*, the control chart for process variability, are employed to determine whether a process is in control or not.

$\overline{X}$ Chart—Control Chart for Process Means

If a single sample of size n is drawn from the population, the control limits are easily determined. Let the values be $x_1, x_2, \ldots, x_n$, with the mean being $\overline{X}$ and the standard deviation σ.

CL $= \overline{X}$

UCL $= \overline{X} + 3\sigma$

LCL $= \overline{X} - 3\sigma$

However, in practice, a number of samples are drawn at different points of time from the same process, and sample means are calculated. The sample means have a sampling distribution with mean $\mu_{\overline{X}}$ and standard deviation $\sigma_{\overline{X}}$ given by $\sigma / \sqrt{n}$.

Therefore, the control chart can be drawn with the following centre line and control limits:

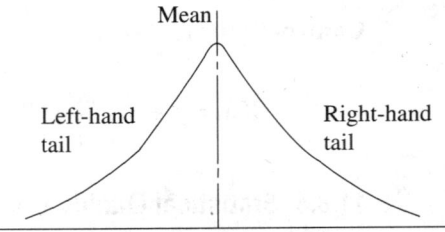

Fig. 11.5 Normal probability distribution

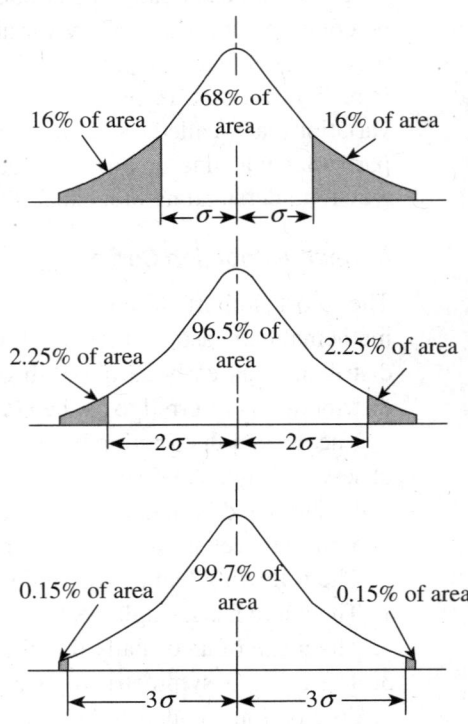

Fig. 11.6 Relationship between area under the normal curve and standard deviation

$$CL = \mu_{\bar{X}}$$

$$UCL = \mu_{\bar{X}} + 3\sigma_{\bar{X}}$$

$$LCL = \mu_{\bar{X}} - 3\sigma_{\bar{X}}$$

If a process is in control, all observations should fall within the control limits. On the other hand, any observation lying outside the control limits suggests that the process is out of control. It warrants further investigation to pin point the cause for this deviation. Even when all the observations are falling within control limits, the control chart can provide clues about an imminent possibility of the process going out of control. Figures 11.7–11.9 provide more insight into this aspect.

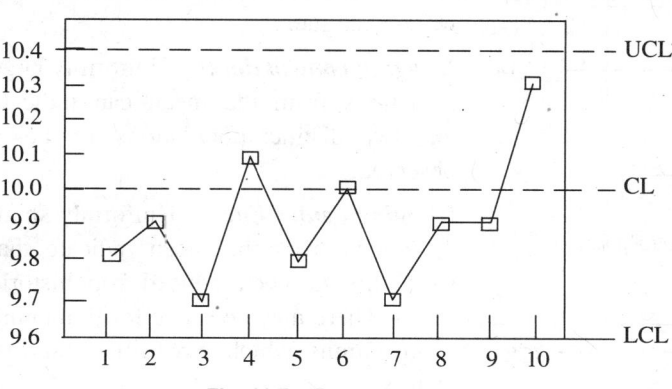

Fig. 11.7   Process in control

In Fig. 11.7, all observations fall within the control limits and there is no increasing or decreasing trend in values. The process is well under control. In Fig. 11.8, the fourth observation is an outlier, that is, it falls outside the UCL. The process is out of control. The production engineer should investigate the reason. Perhaps, the lathe setting may be wrong or the regular operator has been replaced by a temporary worker.

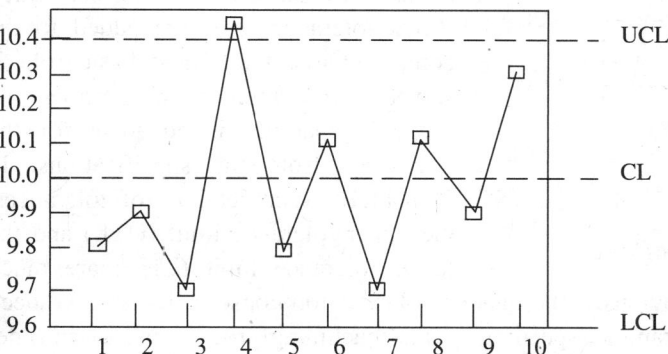

Fig. 11.8   Outliers beyond control limits: Process out of control

Figure 11.9 clearly indicates an increasing trend in the values. Even though all observations fall within control limits, they exhibit anything but random variation. We see a distinct pattern, which indicates imminent loss of process control. Movement of values towards one of the control limits is due to assignable causes, which needs to be investigated. The following are some of the typical trends/patterns that are observed, which indicate the imminent loss of process control.

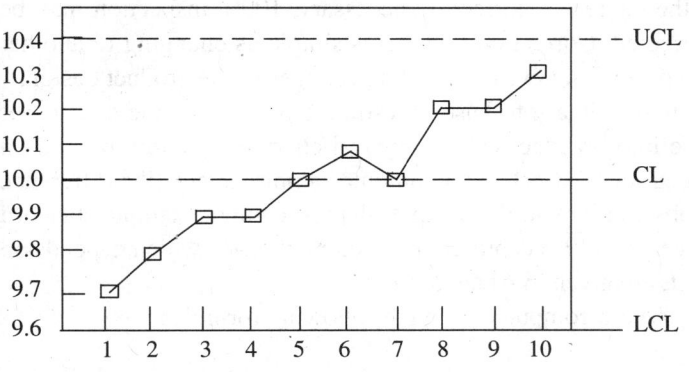

Fig. 11.9   Increasing trend: Process going out of control

Increasing or decreasing trends These indicate that the process mean may be drifting.

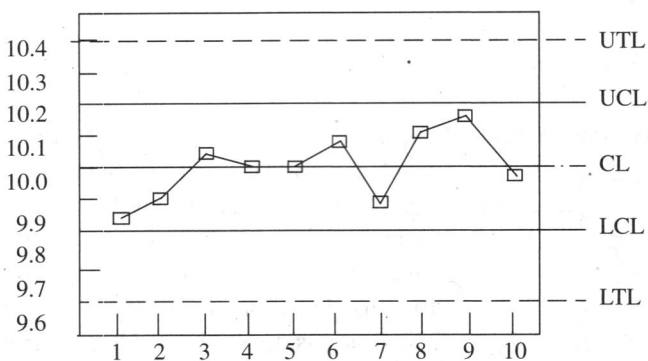

Fig. 11.10 Conservative quality control policy

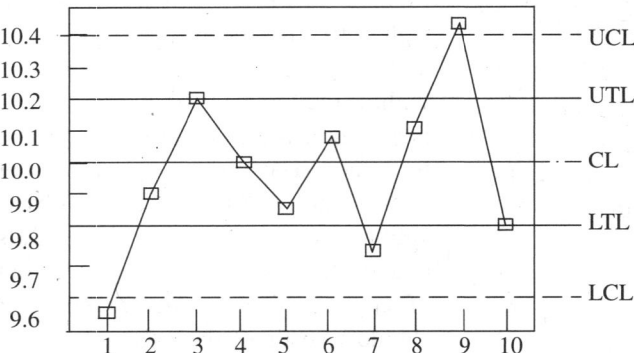

Fig. 11.11 Control limits outside tolerance limits

Jumps in levels around which the observations vary These indicate that the process mean may have shifted.

Cycles Regularly repeating cycles around the centre line indicate factors like worker fatigue.

Hugging control limits Uniformly large deviations from the mean can indicate that two distinct populations are being observed.

Hugging centre line Uniformly small deviations from the mean indicate that variability has been reduced from historic levels. There may be scope for tightening control limits, which may further improve quality.

It should be noted that UCL and LCL of the charts show the process capability. When tolerance lines are added to the control charts, their most basic role is served. They then show whether the process capability is adequate for the tolerances from the specifications. In Fig. 11.10, specifications of tolerances, the upper tolerance limit (UTL) and the lower tolerance limit (LTL), are much wider than the control limits. Obviously, the quality policy is too conservative and we need to consider improvements, for example, speeding up the machining process, use of an older machine that has lesser machine hour cost, or even a small stock size.

In Fig. 11.11, the control limits are outside the tolerance limits. However, two points exceed the control limits, but three points exceed the tolerance and two others are on the high limit of the tolerance. Depending on the degree of perfection necessary, 100% inspection may be required until the situation is corrected. Correction may be as simple as changing to another machine known to be more stable or may involve engineering changes in the product design.

In control charts, it has become customary to base an estimate of σ on $\overline{R}$, the average of sample ranges. This custom came into practice because control charts were often plotted on the factory floor, and it was a lot easier for workers to compute sample ranges (the difference between the highest and lowest observations in the sample) than to compute sample standard deviations. The relationship between σ on $\overline{R}$ is captured in a factor called d_2, which depends on n, the sample size. The values of d_2 are given in Annexure I.

The UCL and LCL for the X chart are computed with the following formulae:

$$UCL = \overline{X} + \frac{3\overline{R}}{d_2\sqrt{n}}$$

$$LCL = \overline{X} - \frac{3\overline{R}}{d_2\sqrt{n}}$$

R Charts—Control Charts for Process Variability

The $\overline{X}$ chart defines the UCL and LCL on the amount of variability we are willing to tolerate in the sample means. However, sample means have a moderating effect on variability and are therefore less variable than individual observations. However, quality concerns are addressed to individual observations like drive shaft diameters. In order to monitor the variability of individual observations, we use another control chart, known as the R chart. In an R chart, we plot the values of sample ranges for each of the samples. The central line is placed at $\overline{R}$ and the control limits are calculated from the sampling distribution.

Thus, standard deviation of the sampling distribution of R is, $\sigma_R = d_3\sigma$, where σ is the population standard deviation and d_3 is another factor depending on n. Values of d_3 are given in Table 1 of Annexure I. The control limits for R charts are computed as follows:

$$UCL = \overline{R} + \frac{3d_3\overline{R}}{d_2}$$

$$LCL = \overline{R} - \frac{3d_3\overline{R}}{d_2}$$

The range is only a convenient substitute for the variability of the process we are studying. It is simpler to calculate, plot, and understand in comparison to the standard deviation. However, the range considers only the highest and lowest values in a distribution and omits all other observations in the data set. Thus, it may ignore the nature of variation among all other observations and is mainly influenced by extreme values. In addition, since it measures only two values, the range can significantly change from one sample to the next in a given population.

Relative Precision Index

The control limits for both $\overline{X}$ and $\overline{R}$ charts do not consider the relationship between work tolerance and process variability. They do not provide any clue for the drifting of the process towards one side. In order to determine whether the process may be allowed to drift before corrective action is taken, the *relative precision index* (RPI) is used. This relates process variability to work tolerance.

$$RPI = \frac{\text{Work tolerance}}{\overline{R}}$$

Having calculated RPI, we need to refer to Table 11.4 to see whether, for a given sample size, the process is of low, medium, or high relative precision. If low, the process is unsatisfactory and likely to produce defective parts. A medium value indicates that the process is satisfactory but needs frequent review to ensure that it is under control. If the process is of high relative precision, a drift can be allowed to take place before the setting is adjusted.

Table 11.4 RPI based on sample size

Class	Low relative precision	Medium relative precision	High relative precision
Sample size n	RPI	RPI	RPI
2	Less than 6.0	6.0–7.0	Greater than 7.0
3	Less than 4.0	4.0–5.0	Greater than 5.0
4	Less than 3.0	3.0–4.0	Greater than 4.0
5 and 6	Less than 2.5	2.5–3.5	Greater than 3.5
State of production	Unsatisfactory, rejection of parts imminent	Satisfactory, if averages are within control limits	Satisfactory, if averages are within modified limits

11.9 TOTAL QUALITY MANAGEMENT

The ISO definition of *quality* is 'the degree to which a set of inherent characteristics (distinguishing features) fulfils requirements i.e., needs or expectations that are stated, generally implied or obligatory'. The term *quality control* denotes all those activities that are directed towards maintaining and improving quality. The following are the scope of QC:

1. Setting up quality targets
2. Appraisal of conformance
3. Initiating corrective actions
4. Planning for improvements in quality

While QC is primarily an operational issue, *quality management* has a broader scope. It is a more complex issue and involves leadership, customer-centric approach, organizational culture, and motivation of employees to internalize quality goals on strong footing. To manage all these issues, quality is seen as dealing not only with quality of products and services, but also with the performance of organization as a whole in its journey towards excellence. This concept is known as *total quality management*, popularly referred to as TQM.

The ISO definition of TQM is, 'total quality management is the management approach of an organization, centered on quality, based on the participation of all members and aiming at long-term success through customer satisfaction, and with benefits to all members of the organization and to society'. TQM has three basic tenets:

1. It is *total*, involving all departments/groups in the organization at all levels.
2. It relates to *quality* in the broader sense of organizational excellence and does not just refer to product quality.
3. It is a *management* function and not just confined to a technical discipline.

Perhaps no other management concept has contributed as much to the phenomenal industrial progress in modern times as the TQM. The concept of TQM, in fact, has emerged out of synergy between the Western and Eastern thinkers and management gurus. The TQM movement has been largely influenced by the *seven quality gurus*, from USA and Japan, and the patronage received by them in leading corporates like the Toyota Motor Company. The seven gurus of TQM are Edward Deming, Joseph M. Juran, Philip Crosby, and Tom Peters from USA, and Kaoru Ishikava, Shigeo Shingo, and Genichi Taguchi from Japan. Their contributions are summarized in Table 11.5.

Table 11.5 Quality gurus and their contributions

Quality guru	Main contributions
W. Edwards Deming	1. Became famous for his work on quality management in Japan, for its reconstruction after World War II
	2. Advocated a 14-point approach for ensuring excellence in quality
	3. Proposed the plan–do–check–act (PDCA) cycle for quality improvement activity
Joseph M. Juran	1. Published the famous *Quality Control Handbook* in 1951
	2. Developed the idea of the *quality trilogy*, with the elements quality planning, quality improvement, and QC
Philip Crosby	1. Promoted the concept of 'Do it Right First Time' and advocated *four absolutes* of quality management
	2. Introduced the concept of 'zero defects'
Tom Peters	1. Outlined the importance of *organizational leadership* for the success of TQM
	2. Prescribed *12 elements* that can usher in quality revolution
Kaoru Ishikawa	1. Introduced the famous '*cause and effect*' diagram for quality improvement
	2. Is the pioneer of *quality circles* movement in Japan
Shigeo Shingo	1. Popularized the Toyota Production System
	2. Developed the concept of '*poka-yoke*' or mistake proofing
	3. Is also famous for his innovations such as the *just in time system* and *single minute exchange of dies*
Genichi Taguchi	1. Applied statistical tools for solving manufacturing problems; Taguchi's method for robust design of experiments is a boon to researchers all over the world
	2. Has authored highly acclaimed books on quality engineering

The contributions of these management gurus have collectively come to be known as *total quality management*, popularly known to the manufacturing world as TQM. Table 11.6 lists the main elements in the concept of TQM.

Table 11.6 Main concepts of TQM

Concept	Main idea
Customer focus	Goal is to identify and meet customer needs.
Continuous improvement	A philosophy of never-ending improvement is involved.
Employee empowerment	Employees are expected to seek out, identify, and correct quality problems.
Use of quality tools	Ongoing employee training is required in the use of quality tools.
Product design	Products need to be designed to meet customer expectations.
Process management	Quality should be built into the process; sources of quality problems should be identified and corrected.
Managing supplier quality	Quality concepts must extend to a company's suppliers.

11.9.1 Customer Focus

The primary focus in TQM is to adopt a customer-centric approach. A perfectly designed and produced product is of little value unless it meets customer requirements. Assessing customer requirements is a rather complex issue and the organization should be in touch with customers and the market in order to understand their needs and preferences.

11.9.2 Continuous Improvement

Continuous improvement is a perpetual process that is carried day in and day out. While improvements of large magnitude are required once in a while, marginal improvements are the order of the day. Continuous improvement tools like kaizen would be of great help to organizations to develop and sustain the culture of continuous improvement.

Kaizen

Kaizen is the outcome of pioneering work by Edward Deming of USA in Japan. Deming was sent by the American government to Japan to help rebuild the country's productivity after World War II. He worked with a team of Japanese industrialists, and one of the tangible outcomes was the continuous improvement tool, called *kaizen*. The Japanese government instituted the prestigious *Deming award* in his honour for excellence in quality.

The Toyota Production System is known for kaizen, where all line personnel are expected to stop their moving production line in case of any abnormality and, along with their supervisor, suggest an improvement to resolve the abnormality that may initiate a kaizen. The cycle of kaizen activity can be defined as follows:

1. Standardize an operation and activities.
2. Measure the standardized operation (find cycle time and amount of in-process inventory).
3. Measure gauges against requirements.
4. Innovate to meet requirements and increase productivity.
5. Standardize the new, improved operations.
6. Continue cycle *ad infinitum*.

This is also known as the Shewhart cycle, Deming cycle, or PDCA. The kaizen technique received worldwide popularity and acceptance, thanks to the famous book by Masaaki Imai, *Kaizen: The Key to Japan's Competitive Success*.

11.9.3 Employee Empowerment

TQM has redefined the role of employees. It encourages employees to identify quality-related problems and propose their solutions. This enhances the sense of involvement of employees in organizational activities. The employees are either rewarded or their contributions acknowledged, which serves as a motivational tool. In order to perform this task, employees are often given training in quality measurement tools. Another interesting concept in TQM is the definition of *external* and *internal customers*. *External customers* are those who purchase the company's goods and services. *Internal customers* are employees of the organization who receive goods or services from others in the company. For example, the packaging department of an organization is an internal customer of the assembly department. Just as a defective item would not be passed on to an external customer, a defective item should not be passed on to an internal customer.

Quality circles is one of the innovative techniques developed by the Japanese to promote team building and employee involvement. It is an informal association of 8–10 employees working together in a particular section or department. They meet once in a fortnight or month after working hours, so as to not disturb the regular discharge of their duties. The meeting is coordinated by a senior among the group, and each employee is free to suggest improvements in the production method or QC. Good suggestions are accepted and implemented. After the successful implementation of the suggestion, the management of the company will acknowledge the contributions made by the employees.

11.9.4 Use of Quality Tools

The efficacy of QC and management is greatly enhanced, thanks to the seven tools of QC. The practitioner of TQM should be well versed with the use of these tools. These tools (illustrated in Fig. 11.12) are easy to understand and enable the investigation of quality issues in a systematic manner.

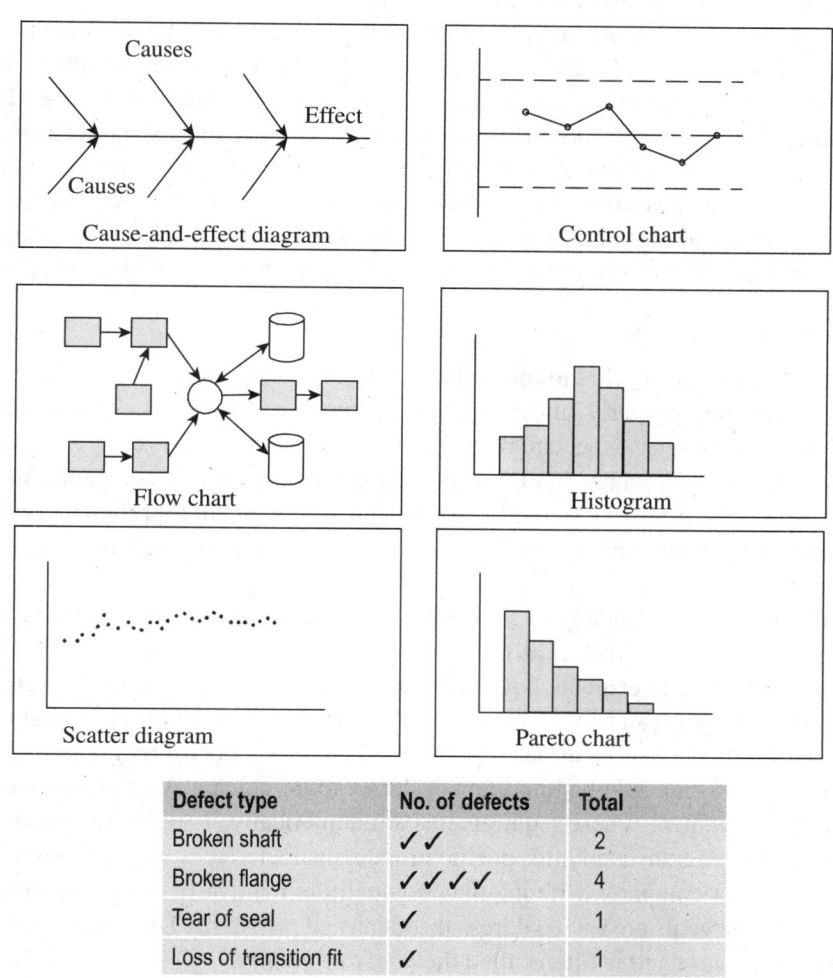

Defect type	No. of defects	Total
Broken shaft	✓ ✓	2
Broken flange	✓ ✓ ✓ ✓	4
Tear of seal	✓	1
Loss of transition fit	✓	1

Fig. 11.12 Seven tools of quality control

Cause-and-effect diagrams, also known as *fish-bone diagrams*, are useful in identifying potential causes for a particular quality-related problem. In the process of developing this diagram, the investigating team will be forced to think through all the possible causes associated with a quality problem. The 'head' of the fish is the quality problem, for example, a damaged key in a coupling or broken valves on a tire. The diagram is drawn so that the 'spine' of the fish connects the 'head' to the possible cause of the problem. These causes can be related to the machines, workers, measurement, suppliers, materials, etc. Each of these possible causes can then have smaller 'bones' that address specific issues related to each cause. For example, a problem with machines can be due to a need for adjustment, old equipment, or tooling problems. Similarly, a problem with workers can be related to lack of training, poor supervision, or fatigue.

While the flow chart is a schematic representation of the sequence of steps involved in an operation or process, the control chart enables the QC inspector to determine whether a manufacturing process is under control or not. A scatter diagram is useful in assessing the degree of correlation between two variables.

Pareto analysis is named after the famous Italian economist of the 19th century, Vilfredo Pareto. It aims to separate the most significant problems from the relatively less significant ones, so that more attention can be devoted to solving the major problems. This concept is often called the *80–20 rule*, because more attention is paid to those major causes, which usually account for 80% of defects. A histogram is a chart that shows the frequency distribution of observed values of a variable. For example, it graphically depicts the frequency of major quality issues over a period of time. A checklist is a list of common defects and the frequency of these defects. It is a simple way of compiling information and bringing out the most problematic area.

11.9.5 Product Design

The very essence of TQM is to maintain a customer-centric focus. The product design should be taken up after properly understanding customer requirements. This is easier said than done because customer expectations are varied and complex. However, a popular tool, called quality function deployment (QFD) is increasingly being used by companies to map customer requirements onto product specifications, the starting document for product design. QFD is also useful in enhancing communication between different functions such as marketing, operations, and engineering.

QFD enables the company to view the relationships among the variables involved in the design of a product like technical versus customer requirements. For example, the automobile designer would like to evaluate how changes in materials affect customer safety requirements. This type of analysis can be very beneficial in developing a product design that meets customer needs, yet does not create unnecessary technical requirements for production.

QFD begins by identifying important customer requirements, which typically come from the marketing department. These requirements are numerically scored based on their importance, and the scores are translated into specific product characteristics. Evaluations are then made of how the product compares with its main competitors relative to the identified characteristics. Finally, specific goals are set to address the identified problems. The resulting matrix looks like a picture of a house and is often called the *house of quality*.

11.9.6 Process Management

TQM attaches a lot of importance to ensuring quality of the manufacturing process also. The company should guarantee quality from the raw material stage to the finishing stage. Thanks to increased globalization, a company may source raw materials, components, and sub-assemblies from various parts of the world. Therefore, a fool-proof system should be in place to ensure that no defective items get into the production stream at any point in the value chain. One of the popular tools to ensure a defect-free process is *poka-yoke*, developed by Shigeo Shingo at Toyota Motor Company in Japan.

Poka-yoke

Poka-yoke, a Japanese word, means mistake proofing. It is a technique that prevents commitment of mistakes during production. The key is to eliminate defects in the production system by preventing errors or mistakes. *Poka-yoke* puts in place a device or mechanism that makes it impossible to commit a mistake. Some of the common errors made in manufacturing are as follows:

1. Use of out-of-calibration measuring instruments
2. Improper environmental conditions
3. Use of wrong or worn-out tools
4. Incorrect machine setting
5. Missed parts in assembly
6. Loose or improper parts in assembly

Most organizations are obviously aware of these problems, and a process control mechanism will be in place. However, they take the form of periodic inspection and calibration. There is a time lag between the occurrence of a problem and the problem being noticed. During this period, a number of defective parts would have been manufactured. *Poka-yoke* relies more on self-checks by operators before starting the operation to ensure proper conditions of equipment and tooling. *Poka-yoke* devices may be any one of the following forms:

1. Check lists to the machine operators, similar to the one used by airline pilots, to carry out pre-checks of the equipment and tooling before starting the process
2. Limits switches to prevent overrun of machine axes
3. Alarms on board the machines to alert the operator of signs of processes going out of control
4. Design of parts such that they cannot be assembled in the wrong way
5. Simple devices like assembly trays in which the worker places all sub-assemblies, components, and fasteners; this way, if any part is left out in the tray, it is a clear indication that there are missing parts in the assembly

In a large manufacturing company, it will not be feasible to implement *poka-yoke* for each and every operation. It is pragmatic to study the entire manufacturing process and identify critical processes based on Pareto analysis. Fool-proofing mechanisms are thought out and incorporated in the design of products and processes. Tools like quality circles may come in handy for *poka-yoke*, because it encourages innovative ideas from workers. Identification, incorporation, and success of *poka-yoke* should be documented through reports, as it can provide insights for further refinements and improvements.

11.9.7 Managing Supplier Quality

Traditionally, companies carried out inspection of the goods received from suppliers and, accordingly, accepted or rejected them. However, thanks to the philosophy of TQM and the quality measures enforced by ISO certification, companies are giving prime importance to ensuring that suppliers follow the best quality practices. In fact, if the suppliers follow sound quality practices, the inward inspection can be eliminated, thereby saving inspection cost and the resulting time delay. Today, many companies go a step further and put a resident engineer in the supplier's company to ensure that standard quality practices are followed. This is a win–win situation for both the buyer and the sub-contractor, since most of the suppliers are small-scale companies. These small-scale companies are short of quality manpower and the knowledge passed on by the large companies helps them adopt the latest production technology.

11.10 SIX SIGMA

Sigma (σ) is a Greek alphabet that is known to represent the standard deviation of a set of observations. From a QC point of view, it is a measure of degree of variation in the dimensions of a batch of manufactured components from the intended value. Lesser the variation, better the performance. Starting from the 1930s, many American companies started using variations up to three sigma levels as a good standard for their processes. On this basis, several techniques (e.g., control charts) were developed by Shewhart and others as a means of SPC. A three-sigma level of process control is associated with a 99.7% quality level, signifying three defects in every 1000 components, which is a fairly acceptable quality standard even today. However, Japanese companies, which have a fierce zeal to achieve zero defects, are known to not be content with the three-sigma level. We owe the evolution of the now famous *six sigma approach* to this never-ending zeal of the Japanese.

The origin of this concept can be traced to the Motorola Company in USA. Motorola was a popular company there, especially for the production of Quasar television sets. A Japanese company took over the Motorola television business in the early 1970s and introduced new methods for improving quality. These measures brought down the defects to an amazing 1/20th of the earlier defects rate. Motivated by this huge success, Bob Galvin, the CEO of Motorola, started his company's quality journey, which led to the development of the famous six sigma approach for TQM. Six sigma got worldwide attention when Motorola was awarded the prestigious *Baldrige award* for this innovative concept. This made six sigma the *new mantra* for success, and many leading companies started adopting it. Prominent among them were Jack Welsh's General Electric Company, Honda, Texas Instruments, and others.

In a normal distribution, there is a probability that 68.2% of the values will be within one-sigma limit from the mean, 95.5% within two-sigma limits, and 99.7% within three-sigma limits. We need to determine how much variation is acceptable based on process variability and, accordingly, fix the LTL and UTL. Now compare the standard deviation for each batch of production with the interval between the mean for

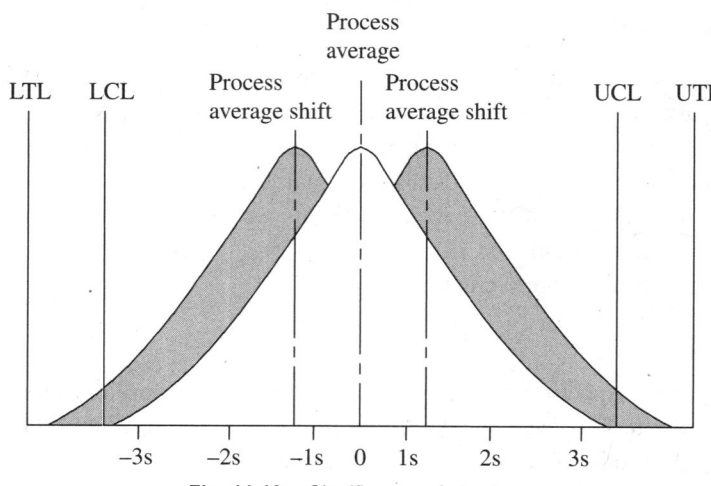

Fig. 11.13   Significance of six sigma

the batch and the LTL or UTL (refer to Fig. 11.13). If it is less, Congratulations! If not, we need to reduce the process variation, the sigma level.

The goal of six sigma is to reduce the standard deviation of the process to the point that six standard deviations (six sigma) can fit within the tolerance limits, which means improving the process capability, by employing engineering skills. In addition to LTL and UTL, another pair of limits should be plotted for any process, namely UCL and LCL. These values, as we know from the preceding sections of this chapter, mark the maximum and minimum inherent limits of the process, based on data collected from the process. If control limits are within the tolerance limits or align with them, the process is considered capable of meeting the specifications. If either or both the control limits are outside the tolerance limits, the process is considered incapable of meeting the dimensional specifications.

11.10.1 Six Sigma Approach

Six sigma is a project-based quality improvement methodology. It follows a five-step approach comprising *define*, *measure*, *analyse*, *improve*, and *control*, which has become popular with the acronym DMAIC. These steps are briefly described here:

Define The first step is to define the objective of the project based on customer requirements. Give priority to factors such as customer's expectations of quality, price, delivery, and having a low effort-to-impact ratio. In addition, try to meet organizational objectives in terms of increased market share, market leadership, and return on investment. The two other important considerations are the financial and time resources, and the probability of success.

Measure Measure the present state under the existing system and identify parameters to be measured to monitor the progress towards achievement of the objectives. It involves the following functions:
1. Select *critical-to-quality characteristics* for the product or process (the CTQC).
2. Define performance standards for CTQC.
3. Validate the measurement system for CTQC.
4. Establish the process capability of achieving CTQC.

Analyse Analyse the gap between the present status and the desired state. Examine the data and identify ways to reduce the gap between the current levels of performance with the desired level. Identify the sources of variation of CTQC and analyse potential causes.

Improve This calls for creativity and an innovative approach to find new methods to make products of a better quality at a lower cost. Generate possible solutions to the problems identified in the previous stage. Then test those possibilities using tools such as *design of experiments* and *failure mode and effects analysis* (FMEA). Finally, select the best solution and design an implementation plan.

Control Formalize the implementation of the new system through organizational policies, procedures, compensation, and incentive schemes. Continue to document and monitor the processes via well-defined metrics to assess and ensure their capability over time. The management should now be capable of making lasting and profitable changes and the organization should have a road map for staying on course.

Thus, the six sigma journey is a full-time trip and never ends, as long as the organization wishes to attain the pinnacle of quality and improve its brand equity as a high-quality, low-cost producer of goods and services.

11.10.2 Training for Six Sigma

Six sigma is a highly structured methodology and its success depends, to a large extent, on the people. It calls for initiative, commitment, and communication. Based on the experience gained in the implementation of six sigma at Motorola and General Electric, Michel Harry of Six Sigma Academy in the USA has developed a training module, which has become the standard training module all over the world. These training programmes, which are designed at two levels, are called *green belt* and *black belt*, terminologies borrowed from Judo and Karate. While the green belt programme is designed for project team members, the black belt is addressed to the needs of project leaders. The green belt certification is given after the candidate is found to be proficient in the concepts of six sigma, QFD, FMEA, lean management principles, process management, project management, and SPC. For black belt training, the candidate should be proficient in all the foregoing topics. In addition to the knowledge of business systems and processes, that of leadership, project planning, change management, etc., are mandatory.

In addition, the organization may have *master black belts* and *champions*. Master black belts train black belts and project team members, and provide guidance whenever needed. Champions are generally top managers. They serve as mentors and leaders; they support project teams, allocate resources, and remove barriers.

Six Sigma Project Implementation

Once the management has given its approval for the adoption of a six sigma strategy and spared the required human resource, the project leader should first identify a worthy project. Since six sigma is data driven, project selection should be based on availability of credible data. Once the project has been selected, a project plan should be developed. The project plan should address the following issues:
1. Problem to be addressed or the process to be improved
2. Budget and plan of resources required
3. Expected time frame and results
4. Methodology for implementation
5. Selection of team members

A steering committee should be constituted to monitor the implementation of the six sigma strategy. This committee should resolve all interdepartmental problems and may also appoint an auditor to assess the results. Finally, a comprehensive report should be prepared to document the entire process, encourage good performances, identify laggards, and suggest further improvements. It will also enthuse other functional departments to take up six sigma projects, thereby expanding the entire movement.

11.11 QUALITY STANDARDS

The last three decades, starting from the early 1980s, have seen a phenomenal growth in international trade and commerce. Many multinational companies such as Intel, General Motors, Honda, etc., have set up divisions in countries all over the world. More importantly, such companies source sub-assemblies and parts from all over the world. Today, a computer numerical control (CNC) machine tool manufacturer of Bangalore will procure the CNC controls from Japan, ball screws and linear guides from Holland, bearings from Germany, and most other parts from various companies all across India. In order to ensure top-of-the-line quality standards, the company should ensure that all the suppliers also conform to the best quality practices. However, inspecting the quality systems of all suppliers is an impossible task. This is where the ISO has stepped in and introduced a series of quality certifications, which is today popularly known as the *ISO certification*. ISO is an international organization whose purpose is to establish an agreement on international quality standards. It currently has members from 91 countries, including India. In 1987, the ISO published its first set of standards for quality management called ISO 9000. This consists of a set of standards and a certification process for companies. By receiving ISO 9000 certification, companies demonstrate that they have met the standards specified by the ISO. This provides assurance to the customers—both consumers and industrial clients—that the company has internationally accepted quality standards and systems in place.

Originally, these standards were designed for application in the manufacturing sector. On seeing the overwhelming response from industries all over the world, these standards were revised and made generic in nature so that these could be applied to any organization. In December 2000, the first major changes to ISO 9000 were made, introducing the following three new standards:

ISO 9000:2000   Quality management systems—fundamentals and standards   This standard provides the terminology and definitions used in the standards. It is the starting point for understanding the system of standards.

ISO 9001:2000   Quality management systems—requirements   This standard is used for the certification of a firm's quality management system and to demonstrate the conformity of quality management systems to meet customer requirements.

ISO 9004:2000   Quality management systems—guidelines for performance   This standard provides guidelines for establishing a quality management system. It focuses not only on meeting customer requirements but also on improving performance.

These three standards are the most widely used and apply to a majority of companies.

However, 10 more published standards and guidelines exist as part of the ISO 9000 family of standards, which are listed in Table 11.7.

11.11.1 Quality Management Principles of ISO 9000

At the core of ISO 9000 series of certification are the eight quality management principles, which provide the basis for quality management principles of ISO. These principles have been arrived at based on the pioneering work done by several management gurus highlighted in Section 11.8. These principles are described in the following paragraphs.

Customer focus   ISO strongly advocates a customer-centric approach. It stresses the need to understand customer requirements, communicate these requirements to those who matter in the

Table 11.7 List of published standards of ISO/TC 176

Standard	Scope
ISO 10002:2004	Quality management, customer satisfaction, guidelines for complaints handling in organization
ISO 10005:2005	Quality management systems, guidelines for quality plans
ISO 10006:2003	Quality management systems, guidelines for quality management in projects
ISO 10007:2003	Quality management systems, guidelines for configuration management
ISO 10012:2003	Measurement management systems, requirements for measurement processes and measuring equipment
ISO/TR 10013:2001	Guidelines for quality management system documentation
ISO 10014:2006	Quality management, guidelines for realizing financial and economic benefits
ISO 10015:1999	Quality management, guidelines for training
ISO/TR 10017:2003	Guidance on statistical techniques for ISO 9001:2000
ISO 10019:2005	Guidelines for selection of quality management system consultants and use of their services
ISO/TS 16949:2002	Quality management systems, particular requirements for the application of ISO 9001:2000 for automotive production, and relevant service part organization
ISO 19011:2002	Guidelines for quality and/or environmental management systems auditing

organization, and measure customer satisfaction in order to fine-tune the future strategies. It provides guidelines for strengthening customer relationship.

Leadership   The organization should have well-defined vision and mission statements, taking into account the needs and aspirations of all stakeholders. The top management should ensure that the organization's plans and activities strive to realize the vision. The leaders should set challenging goals and targets at every level and motivate the employees to deliver to the best of their abilities.

Process approach   The various activities are carried out systematically. More attention is paid towards utilizing resources such as men, materials, and money in the most optimal way. Duties and responsibilities are assigned at various levels, which make the personnel involved in the project accountable for their actions. Every operation, transaction, or process should be clearly defined and documented. This facilitates traceability and accountability.

Team approach   Team members are meticulously briefed about the objectives of ISO certification and the benefits that will accrue to the organization. Suggestions are sought and good ones are appreciated, which builds good team spirit. Proper training is given whenever required, which is a good enabler for improving individual performances.

Continuous improvement   Continuous improvement is an ongoing and perpetual part of ISO implementation. Formal systems should be put in place to ensure continuous improvement activities at all levels in the organization. Continuous improvements are sought in products, services, manufacturing, customer service, etc. Personnel should be provided training and adequate opportunities to come out with improvements.

Forging mutually beneficial relationships with suppliers   The organization and its suppliers are inter-dependant. A good implementation of ISO standards will be beneficial to both.

11.11.2 Implementation of ISO Standards

Implementation and maintenance of ISO standards require the commitment of management and staff at all levels in the organization. However, organizational leadership is the most deciding factor for its success. The top management should have the vision to realize the benefits of enforcing ISO standards and commit the necessary funds and manpower for its implementation. Since a customer-centric approach is built into these standards, the organization as a whole should be willing to value the customer as the king.

At an operational level, ISO standards demand strict adherence to documentation. Every operation, transaction, or process should be well defined and documented. Procedures and instructions should be documented if their non-observance can cause a risk of non-conformity in products and services. This is enforced to ensure consistent conformity with customer requirements, establishment of traceability, and providing evidence for evaluation of the quality management system.

The effectiveness of the standards need to be evaluated periodically to ensure that it is being properly implemented and complies with ISO 9001 requirements. In fact, ISO 9001 provides for the institutionalized audit of any quality management system. The organization will have internal auditors (purely on a voluntary basis), who will periodically audit various documents and processes. If any deviation is noticed, it is immediately brought to the attention of the concerned manager so that remedial measures can be taken up. The internal audit is followed by an audit by an accredited third party. Audits by both internal auditors and third party need to be carried out in pre-defined intervals as a mandatory part of ISO certification.

11.12 NUMERICAL EXAMPLES

Example 11.1 A company is engaged in packaging milk in half-litre tetra packs. Each half-litre tetra pack should ideally contain 0.5 l of milk. The customer demands that not more than 3% of tetra packs should contain less than 0.495 l. Filled tetra packs, in a sample size of 100, have been inspected twice a day for a week and the following data (Table 11.8) has been obtained. Draw the p chart and state your conclusion.

Table 11.8 Inspection data

Day	Hour	Fraction underfilled
Monday	8 a.m./4 p.m.	0.02/0.03
Tuesday	8 a.m./4 p.m.	0.01/0.04
Wednesday	8 a.m./4 p.m.	0.02/0.04
Thursday	8 a.m./4 p.m.	0.01/0.03
Friday	8 a.m./4 p.m.	0.03/0.04
Saturday	8 a.m./4 p.m.	0.02/0.04
Sunday	8 a.m./4 p.m.	0.03/0.05

Solution

Referring to this example, let $p = 0.03$, $n = 100$, then

$$\text{CL} = p = 0.03$$

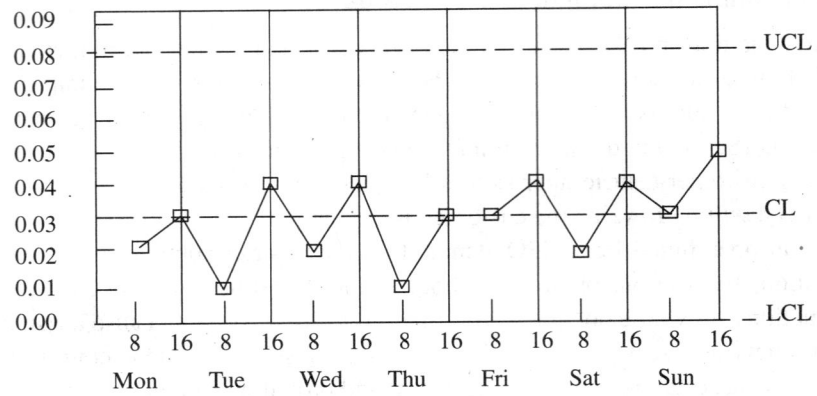

Fig. 11.14 p chart for Example 11.1

$$\text{UCL} = p + 3\sqrt{\frac{pq}{n}} = 0.03 + 3\sqrt{\frac{0.03(0.97)}{100}} = 0.081$$

$$\text{LCL} = p - 3\sqrt{\frac{pq}{n}} = 0.03 - 3\sqrt{\frac{0.03(0.97)}{100}} = -0.021$$

Since p can take values only between 0 and 1, the LCL is corrected to '0'.

Figure 11.14 illustrates the p chart for this problem. It is easily seen that all the data points are within control limits and therefore the process is under control.

Example 11.2 For the following cases, find CL, UCL, and LCL for a p chart based on the given information.

(a) $n = 144$, $p^{11} = 0.1$
(b) $n = 125$, 0.36 is the target value for p

Solution
$$\text{CL} = p^{11} = 0.1$$

$$\text{UCL} = p^{11} + 3\sqrt{\frac{pq}{n}} = 0.1 + 3\sqrt{\frac{0.1(0.9)}{144}} = 0.175$$

$$\text{LCL} = p^{11} - 3\sqrt{\frac{pq}{n}} = 0.1 - 3\sqrt{\frac{0.1(0.9)}{144}} = 0.025$$

$$\text{CL} = p = 0.36$$

$$\text{UCL} = p + 3\sqrt{\frac{pq}{n}} = 0.36 + 3\sqrt{\frac{0.36(0.64)}{125}} = 0.489$$

$$\text{LCL} = p - 3\sqrt{\frac{pq}{n}} = 0.36 - 3\sqrt{\frac{0.36(0.64)}{125}} = 0.231$$

Example 11.3 Components need to be turned to a diameter of 25 mm on a CNC turning centre. A sample of 15 components has been drawn and Table 11.9 shows the dimensions measured using a vernier calliper. Draw the $\bar{X}$ chart and give your recommendation on whether the lot can be accepted or not.

Table 11.9 Dimensions of components

Component	Dimension (mm)	Component	Dimension (mm)	Component	Dimension (mm)
1	25.02	6	25.07	11	25.05
2	24.89	7	25.11	12	24.98
3	24. 92	8	24.93	13	25.03
4	25.02	9	25.03	14	24.95
5	25.08	10	25.12	15	25.05

Solution

For $\bar{X}$ chart,

$$CL = \bar{X}$$
$$UCL = \bar{X} + 3\sigma$$
$$LCL = \bar{X} - 3\sigma$$

Where $\sigma = \sqrt{\Sigma(X_i - \bar{X})^2 / n}$

Now, $\bar{X} = \Sigma X_i / n = 25.017$.

It will be convenient to construct Table 11.10 in order to calculate $\bar{X}$ and σ.

Therefore, $\sigma = \sqrt{0.052 / 15} = 0.015$

Table 11.10 Calculating $\bar{X}$ and σ

Component	X_i	$X_i - \bar{X}$	$(X_i - \bar{X})^2$	Component	X_i	$X_i - \bar{X}$	$(X_i - \bar{X})^2$
1	25.02	−0.003	0	9	25.03	−0.013	0
2	24.89	0.127	0.002	10	25.12	−0.103	0.01
3	24.92	0.097	0.009	11	25.05	−0.033	0.001
4	25.02	−0.003	0	12	25.98	0.037	0.001
5	25.08	−0.063	0.004	13	24.03	−0.013	0
6	25.07	−0.053	0.003	14	24.95	0.067	0.004
7	25.11	−0.093	0.009	15	25.05	−0.033	0.001
8	24.93	0.087	0.008		$\Sigma(X_i - \bar{X})^2$		0.052

For the control chart,

$$CL = 25.017$$
$$UCL = \bar{X} + 3\sigma = 25.062$$
$$LCL = \bar{X} - 3\sigma = 24.972$$

The control chart is shown in Fig. 11.15. It is seen that four values each fall outside the UCL and LCL. The process is out of control, and further investigation needs to be carried out to determine the causes.

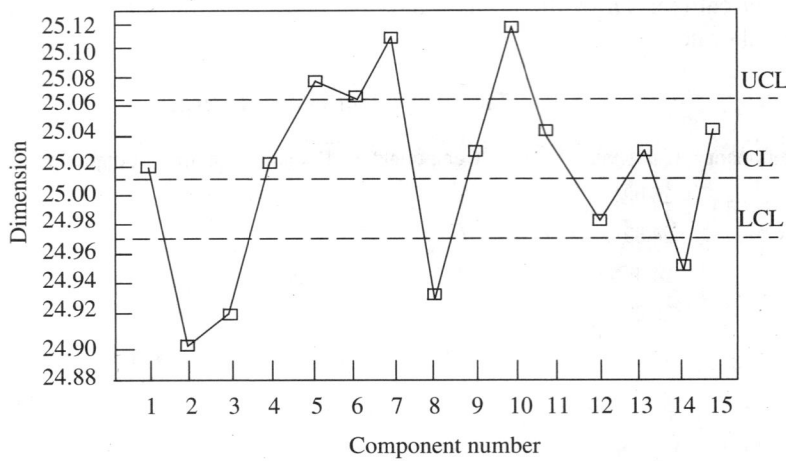

Fig. 11.15 $\overline{X}$ chart for Example 11.3

Example 11.4 Akhila is the QC engineer in a leading company that specializes in the manufacture of ball bearings for the automotive industry. She has been checking the output of 5 mm bearings used in the clutch assembly of motorcycles. For each of the last 5 hours, she has sampled five bearings, with the results shown in Table 11.11. Construct the $\overline{X}$ chart to help Akhila determine whether the production of 5 mm bearings is in control.

Table 11.11 Bearing observations

Hour	Bearing diameters (mm)				
1	5.03	5.06	4.86	4.9	4.95
2	4.97	4.94	5.09	4.78	4.88
3	5.02	4.98	4.94	4.95	4.80
4	4.92	4.93	4.90	4.92	4.96
5	5.01	4.99	4.93	5.06	5.01

Solution
No. of observations in each sample, $n = 5$
No. of samples taken, $k = 5$

We need to calculate the grand mean, $\overline{\overline{X}} = \dfrac{\Sigma X}{nXk} = \Sigma \overline{X}/k$

where $\overline{\overline{X}}$ is the grand mean, the sum of all observations, $\Sigma \overline{X}$ the sum of sample means, n the number of observations in each sample, and k the number of samples taken.
In this case, $n = 5$ and also $k = 5$.
Let us construct a simple table (Table 11.12) to find $\overline{X}$, $\Sigma \overline{X}$, and R (range).

$$\overline{\overline{X}} = \Sigma \overline{X}/k = 24.76/5 = 4.95$$

Table 11.12 Calculating $\Sigma\overline{X}$ and R

Hour	Bearing diameters (mm)					Mean $\overline{X}$	Range (R)
1	5.03	5.06	4.86	4.9	4.95	4.96	0.20
2	4.97	4.94	5.09	4.78	4.88	4.93	0.31
3	5.02	4.98	4.94	4.95	4.80	4.94	0.40
4	4.92	4.93	4.90	4.92	4.96	4.93	0.06
5	5.01	4.99	4.93	5.06	5.01	5.00	0.13
						$\Sigma\overline{X}$ = 24.76	ΣR = 1.1

This value is used as the centre line.

The UCL and LCL for the $\overline{X}$ chart are computed with the following formulae:

$$\text{UCL} = \overline{X} + \frac{3\overline{R}}{d_2\sqrt{n}}$$

$$\text{LCL} = \overline{X} - \frac{3\overline{R}}{d_2\sqrt{n}}$$

Now, $\overline{R} = \Sigma R/k = 1.1/5 = 0.22$. Referring to Table 1 in Annexure I, the value of d_2 for $n = 5$ is 2.326. Therefore,

$$\text{UCL} = 4.95 + \frac{3(0.22)}{2.326\sqrt{5}}$$

$$= 5.077$$

$$\text{LCL} = 4.95 - \frac{3(0.22)}{2.326\sqrt{5}}$$

$$= 4.823$$

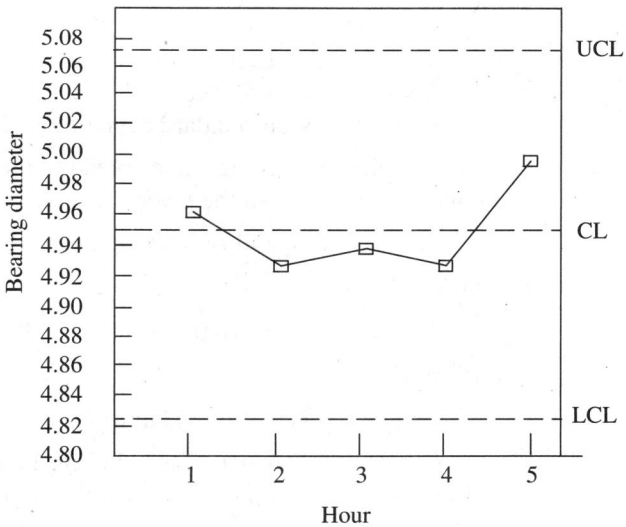

Fig. 11.16 $\overline{X}$ chart for Example 11.4

Now, we need to plot the $\overline{X}$ chart for hourly values of $\overline{X}$, as shown in Fig. 11.16. From the control chart, it is clear that the process is well under control.

Example 11.5 Components are being turned on a CNC lathe to a specification of 12.58 ± 0.05 mm. Five batches of five components each have been drawn for inspection at 1-hour intervals. The readings are tabulated in Table 11.13.

(a) Determine the process capability.

(b) Determine the three-sigma limits for the $\overline{X}$ chart.

(c) Draw the control chart and give your assessment.

 (Assume that normal distribution and d_2 for group size 5 is 2.326 from Table 1 in Annexure I.)

Solution

The values of $\overline{X}$ and R for the five batches are calculated first as shown in Table 11.14.

Table 11.13 Reading of components

Batch 1	Batch 2	Batch 3	Batch 4	Batch 5
12.62	12.63	12.62	12.61	12.59
12.60	12.56	12.56	12.66	12.58
12.62	12.60	12.57	12.62	12.57
12.61	12.59	12.58	12.61	12.59
12.65	12.60	12.63	12.60	12.56

Table 11.14 Values of $\overline{X}$ and R

	Batch 1	Batch 2	Batch 3	Batch 4	Batch 5
$\overline{X}$	12.62	12.60	12.59	12.62	12.50
R	0.05	0.07	0.07	0.06	0.03

$\overline{X} = 62.93/5 = 12.586 \, \text{mm}$

$\overline{R} = 0.28/5 = 0.056 \, \text{mm}$

$\sigma' = \text{population standard deviation} = \overline{R}/d_2 = 0.056/2.326 = 0.0241$

Process capability is the minimum spread of the measurement variation that includes 99.73% of the measurements from the given process, that is, $6\sigma'$.

$\therefore$ Process capability = $6 \times 0.0241 = 0.1446$ mm

3σ limits for $\overline{X}$ chart,

$$3\sigma_{\overline{X}} = 3\sigma'/\sqrt{n} = 3(0.0241)/\sqrt{5} = 0.0323$$

Therefore, UCL = $\overline{X} + 3\sigma_{\overline{X}} = 12.618 \, \text{mm}$

LCL = $\overline{X} - 3\sigma_{\overline{X}} = 12.554 \, \text{mm}$

The control chart is drawn as shown in Fig. 11.17.

It is clear that several points lie outside the control limits and the process is out of control.

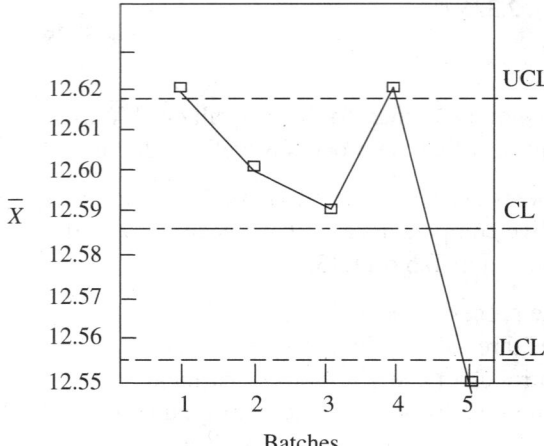

Fig. 11.17 $\overline{X}$ chart for Example 11.5

Example 11.6 Stanley Mendez, the QC manager, wishes to monitor the process variability of finish grinding operation of a 60-mm-diameter shaft and has instructed the QC inspector to do the investigation. Ramesh, the QC inspector, has collected five samples at one-hour intervals, which is shown in Table 11.15. Draw the R chart and give your conclusions.

Table 11.15 Diameter samples

Time	Diameter of component inspected							
	1	2	3	4	5	6	7	8
10 a.m.	61	62	56	59	60	63	61	57
11 a.m.	61	67	62	58	59	59	62	62
12 noon	57	58	60	62	58	57	60	63
1 p.m.	55	57	61	64	65	56	66	66
2 p.m.	54	59	62	57	66	68	64	65

Solution

First of all, we need to calculate $\overline{R}$ from the tabulated values as shown in Table 11.16.

Table 11.16 Calculation of $\overline{R}$

Time	Diameter of component inspected								Range ($\overline{R}$)
	1	2	3	4	5	6	7	8	
10 a.m.	61	62	56	59	60	63	61	57	7
11 a.m.	61	67	62	58	59	59	62	62	9
12 noon	57	58	60	62	58	57	60	63	6
1 pm	55	57	61	64	65	56	66	66	11
2 pm	54	59	62	57	66	68	64	65	14
								ΣR	47

The central line is given by $\overline{R} = \Sigma R/k = 9.4$, where k is the number of samples.
For a sample size of 8, the value of $d_2 = 2.847$ and $d_3 = 0.82$ (from Table 1 in Annexure I).

$$\text{UCL} = \overline{R} + \frac{3d_3\overline{R}}{d_2} = 9.4 + \frac{3(0.82)(9.4)}{2.847}$$

$$= 17.52$$

$$\text{LCL} = \overline{R} - \frac{3d_3\overline{R}}{d_2} = 9.4 - \frac{3(0.82)(9.4)}{2.847}$$

$$= 1.28$$

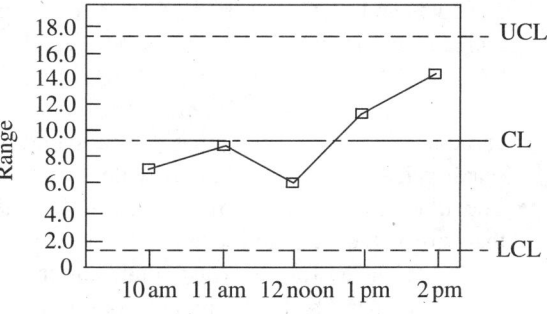

Fig. 11.18 R chart for Example 11.6

The control chart is shown in Fig. 11.18. It can be seen from the control chart that all values of the range fall within control limits and the process is under control.

A QUICK OVERVIEW

- Inspection is the scientific examination of work parts to ensure adherence to dimensional accuracy, surface texture, and other related attributes. It is an integral part of the quality assurance (QA) mechanism, which ensures strict adherence to the stated design intent.
- Most companies specify the limits of variability or tolerance based on several considerations apart from the purely engineering considerations proposed by ISO. Some of the other considerations are customer and market needs, quality of manpower, goals and policies of the management, and financial strength.
- The primary objective of *quality control* (QC) is to ensure conformance to design specifications in terms of dimensions, tolerances, surface texture, etc., at each stages of a manufacturing process. QC includes the activities from the suppliers, through production, and to the customers.
- There exists a clear distinction between QC and QA. While the former comprises a number of analytical measurements and tools used to assess the quality of products, the latter pertains to the overall management plan to guarantee the integrity of data. It is a comprehensive approach to build a scientific and reliable system to assure quality.
- Statistical techniques provide a reliable and economical method for ensuring QC, which is known as *statistical quality control* (SQC). There are two major categories in SQC: SQC by attributes and SQC by variables. An attribute is a characteristic that is of a yes or no type. The GO and NO GO inspection gauges are examples of attribute inspection. On the other hand, a variable is a characteristic that can be stated in scalar values along a continuous scale.
- The most economic sampling plan is the one that gives the required degree of protection to the producer for the least total amount of inspection per batch. There are four popular sampling methods: simple random sampling, constant interval sampling, stratified sampling, and cluster sampling.

- Control limits of p chart: $CL = p$,

$$UCL = p + 3\sqrt{\frac{pq}{n}}, LCL = p - 3\sqrt{\frac{pq}{n}}.$$

- Control limits for $\overline{X}$ *chart*: $CL = \overline{X}$, $UCL = \overline{X} + 3\sigma, LCL = \overline{X} - 3\sigma$.
- Control limits for $\overline{X}$ *chart* for a number of samples with sample mean $\mu_{\overline{X}}$ and standard deviation $\sigma_{\overline{X}}$: $CL = \mu_{\overline{X}}, UCL = \mu_{\overline{X}} + 3\sigma_{\overline{X}}, LCL = \mu_{\overline{X}} - 3\sigma_{\overline{X}}$.
- Control limits for R chart: $CL = \overline{R}$,

$$UCL = \overline{R} + \frac{3d_3\overline{R}}{d_2}, LCL = \overline{R} - \frac{3d_3\overline{R}}{d_2}$$

- Total quality management (TQM) is the management approach of an organization, centered on quality, based on the participation of all members and aiming at long-term success through customer satisfaction, and with benefits to all members of the organisation and to society. TQM has three basic tenets:
- It is *total*, involving all departments/groups in the organization at all levels.
- It relates to *quality* in the broader sense of organizational excellence and does not just refer to product quality.
- It is a *management* function and not just confined to a technical discipline.
- The goal of *six sigma* is to reduce the standard deviation of the process to the point that six standard deviations (six sigma) can fit within the tolerance limits, which means improving the process capability, by employing engineering skills.
- In 1987, the International Organisation for Standardization (ISO) published its first set of standards for quality management, called ISO 9000. This series consists of a set of standards and a certification process for companies. By receiving ISO 9000 certification, companies demonstrate that they have met the standards specified by the ISO. This provides an assurance to the customers—both consumers and industrial clients—that the company has internationally accepted quality standards and systems in place.

MULTIPLE-CHOICE QUESTIONS

1. Too close a fit or a narrow band of tolerance will demand
   (a) automated inspection methods
   (b) more precise manufacturing processes
   (c) less rejection
   (d) company policy

2. An example of an attribute gauge is a
   (a) plug gauge   (c) micrometer
   (b) slip gauge   (d) slide rule

3. An example of a variable-type gauge is
   (a) a ring gauge   (c) a dial indicator
   (b) a snap gauge   (d) an angle gauge

4. Suppose C_1 = cost of inspection and removal of non-conforming item, C_2 = cost of repair, and p = true fraction non-conforming, then 100% inspection is resorted to
   (a) if $p > C_1/C_2$ (c) if $p = C_1/C_2$
   (b) if $p > C_2/C_1$ (d) in none of these

5. Which of the following are the two chief categories of SQC?
   (a) SQC by attributes and SQC by inspection
   (b) SQC by variables and SQC by attributes
   (c) SQC by 100% inspection and SQC by measurement
   (d) SQC by variables and SQC by measurement

6. Which of the following is a valid statement about 'sampling'?
   (a) Inspection of samples before starting production
   (b) Division of large production into smaller lots
   (c) Inspection of 10% of manufactured parts
   (d) A representative portion of a group that is representative of the group

7. Which of the following is the most important information from the control charts?
   (a) Shows trends of process, that is, within control or going out of control.
   (b) Shows number and percentage of defects.
   (c) Correlates numbers and percentages.
   (d) Shows production record.

8. In a control chart, jumps in the level around which the observations vary indicate that
   (a) the process mean may be drifting

   (b) sampling scheme is not proper
   (c) the process mean may have shifted
   (d) all of these

9. In a control chart, uniformly small deviations from the mean indicate
   (a) mistake in calculation of mean
   (b) reduction in variability from historic levels
   (c) erroneous sampling procedure
   (d) none of these

10. In a control chart, if the UTL and LTL are much wider than the control limits, then
    (a) the quality is being compromised
    (b) the process capability appears to be low
    (c) the quality policy is optimal
    (d) the quality policy is too conservative

11. RPI is useful to determine
    (a) whether the process may be allowed to drift before corrective action is taken
    (b) precision of control limits
    (c) precision of sampling plan
    (d) whether the process capability is high or low

12. In the context of TQM, which of the following statements is wrong?
    (a) It is purely a technical discipline addressed at the operator level.
    (b) It has been largely influenced by the *seven quality gurus*.
    (c) It is a customer-centric approach.
    (d) It involves all departments/groups in the organization at all levels.

13. When all line personnel are expected to stop their moving production line in case of any abnormality and, along with their supervisor, suggest an improvement to resolve the abnormality, it may initiate a
    (a) *poka-yoke*   (c) kaizen
    (b) JIT   (d) Deming rule

14. The innovative techniques developed by the Japanese to promote team building and employee involvement is popularly known as
    (a) six sigma   (c) quality team
    (b) quality circle   (d) QFD

15. The terminology *house of quality* is associated with

(a) *poka-yoke*   (c) quality circles

(b) QFD   (d) none of these

16. *Poka-yoke*, a Japanese word, means
    (a) mistake proofing
    (b) team spirit
    (c) customer is the king
    (d) zero defects

17. The goal of six sigma is to reduce the standard deviation of the process to the point that six standard deviations (six sigma) can fit within
    (a) tolerance limits
    (b) control limits
    (c) process capability
    (d) range of observed values

18. Green belt programme in a six sigma system is designed for
    (a) environmental groups

(b) managers

(c) project team members

(d) customers

19. The standard used for the certification of a firm's quality management system is
    (a) ISO 9000:2000
    (b) ISO 9001:2000
    (c) ISO 10002:2004
    (d) ISO/TR 10013:2001

20. The tool used to separate out most significant problems from the relatively less significant ones is
    (a) FMEA
    (b) cause-and-effect diagram
    (c) scatter diagram
    (d) Pareto chart

REVIEW QUESTIONS

1. What is inspection? How do you analyse the economic justification for inspection?

2. What are the components of inspection cost?

3. Distinguish the following: size tolerance, geometric tolerance, and positional tolerance.

4. What are the functions of gauge control?

5. Distinguish between quality control and quality assurance.

6. With the help of a block diagram, discuss the role of quality control in a production cycle.

7. Distinguish between the following:
   (a) Random error and systematic error
   (b) SQC by attributes and SQC by variables

8. What is the significance of sampling in SQC?

Discuss four basic sampling methods.

9. What is the relevance of normal probability distribution for SQC?

10. What are the typical trends/patterns that are observed in control charts, which indicate the imminent loss of process control?

11. When are R charts preferred over $\overline{X}$ charts?

12. Highlight the contributions of the seven quality gurus to the development of TQM.

13. Write short notes on the following: poka-yoke, kaizen, quality circles, cause-and-effect diagram, and Pareto analysis.

14. Briefly discuss the QFD technique.

15. Explain the philosophy of 'six sigma'.

PROBLEMS

1. The 'Hot Meals' catering service is popular in the city of Bangalore for delivery of meals to offices on time. Sandeep Raj, the owner of the enterprise, always strives to deliver meals to customers within 30 minutes of leaving the kitchen. Each of his 10 delivery vehicles is responsible for delivering 15 meals daily. Over the past month, Sandeep has recorded the percentage of each day's 150 meals that were delivered on time (Table 11.17). Help Sandeep construct a p chart from the data.

   (a) How does your chart show that the attribute 'fraction of meals delivered on time' is out of control?

   (b) What action would you recommend to Sandeep?

Table 11.17

Day	% on time	Day	% on time	Day	% on time	Day	% on time
1	89.33	9	90.67	17	89.33	25	81.33
2	81.33	10	80.67	18	78.67	26	89.33
3	95.33	11	88.00	19	94.00	27	99.33
4	88.67	12	86.67	20	94.00	28	90.67
5	96.00	13	96.67	21	99.33	29	92.00
6	86.67	14	85.33	22	95.33	30	88.00
7	98.00	15	78.67	23	94.67		
8	84.00	16	89.33	24	92.67		

2. A firm manufacturing wiring harness for motor cycles has introduced p charts to control the quality of their product. One hundred harnesses out of their daily production were inspected every day for the month of June 2012, and the data is furnished in Table 11.18.

Table 11.18

Date	Number rejected	Date	Number rejected
1	14	16	23
2	22	17	14
3	25	18	6
4	15	19	7
5	20	20	33
6	14	21	Holiday
7	Holiday	22	17
8	12	23	34
9	24	24	11
10	10	25	16
11	17	26	25
12	35	27	Holiday
13	36	28	Holiday
14	Holiday	29	36
15	16	30	18

Determine the control limits and draw the control chart. State your conclusion.

3. The QC engineer of a company manufacturing pistons to the automotive industry has sampled eight pistons each, from the last 15 batches of 500 pistons. The measurement of interest is the diameter of the piston. The results have been recorded, with $\overline{X}$ and R being measured in centimetres, as shown in Table 11.19.

Table 11.19

Batch	$\overline{X}$	R	Batch	$\overline{X}$	R
1	15.85	0.15	9	15.83	0.19
2	15.95	0.17	10	15.83	0.21
3	15.86	0.18	11	15.72	0.28
4	15.84	0.16	12	15.96	0.12
5	15.91	0.14	13	15.88	0.19
6	15.81	0.21	14	15.84	0.22
7	15.86	0.13	15	15.89	0.24
8	15.84	0.22			

Draw the $\overline{X}$ chart and comment whether the process is in control or not. (For a sample size of 8, $d_2 = 2.847$.)

4. For the following cases, find CL, UCL, and LCL for an $\overline{X}$ chart based on the given information:
   (a) $n = 9$, $\overline{X} = 26.7$, $\overline{R} = 5.3$
   (b) $n = 17$, $\overline{X} = 138.6$, $\overline{R} = 15.1$

5. Sunil is the engineer in-charge of final testing of cars in a reputed automobile company before the cars are dispatched to the sales outlets. Due to the limited number of testing stations, he should ensure that the testing procedure will not take more time than necessary. For the last three weeks, he has randomly sampled the final test time for nine tests each day to get the results shown in Table 11.20, with $\overline{X}$ and R being measured in minutes:

Table 11.20

	Week 1		Week 2		Week 3	
	$\overline{X}$	R	$\overline{X}$	R	$\overline{X}$	R
Mon	11.6	14.1	9.5	12.6	11.4	12.1
Tue	17.4	19.1	12.7	17.0	16.0	21.1
Wed	14.8	22.9	17.7	12.0	11.0	13.5
Thurs	13.8	18.0	16.3	15.1	13.3	20.3
Fri	13.9	14.6	10.5	22.1	9.3	16.8
Sat	22.7	23.7	22.5	24.1	21.5	20.7
Sun	16.6	21.0	12.6	21.3	17.9	23.2

Construct an $\overline{X}$ chart to help Sunil determine whether the final vehicle test time process is in control or not.

6. For the following cases, find the CL, UCL, and LCL for an R chart based on the given data:
   (a) $n = 9$, $\overline{X} = 26.7$, $\overline{R} = 5.3$
   (b) $n = 17$, $\overline{X} = 138.6$, $\overline{R} = 15.1$

7. A cement-manufacturing company has installed an automated system to pack cement in bags of 50 kg. The QC inspector has tested the weight of five bags in each of the 12 batches of 1000 bags. Table 11.21 gives $\overline{X}$ and R measured in kg. Construct the R chart and give your conclusion as to whether the variability in weights is within control or not.

Table 11.21

Batch	1	2	3	4	5	6	7	8	9	10	11	12
R	50.5	49.7	50.0	50.7	50.7	50.6	49.8	51.1	50.2	50.4	50.6	50.7
	1.1	1.6	1.8	0.1	0.9	2.1	0.3	0.8	2.3	1.3	2.0	2.1

ANSWERS
Multiple-choice Questions

1. (b) 2. (a) 3. (c) 4. (a) 5. (b) 6. (d) 7. (a) 8. (c) 9. (b)
10. (d) 11. (a) 12. (a) 13. (c) 14. (b) 15. (b) 16. (a) 17. (a) 18. (c)
19. (b) 20. (d)

Problems

1. (a) $p^{11} = 0.898$, UCL $= 0.972$, LCL $= 0.824$
   (b) Five of the 30 days sampled have values of 'fraction on-time' below the LCL (being above the UCL is not a problem in this context)
   (c) Since percentage of meals delivered on time is out of control, Sandeep should investigate the reasons behind this. It might be a particular driver, or those days may have heavier traffic.
2. CL $= 0.2$, UCL $= 0.32$, LCL $= 0.08$. The process is out of control.
3. $\overline{X} = \Sigma X/k = 237.87/15 = 15.858$, $\overline{R} = 0.187$, UCL $= 16.029$, LCL $= 15.687$. Therefore, the process is under control.
4. (a) CL $= 26.7$, UCL $= 28.5$, LCL $= 24.9$
   (b) CL $= 138.6$, UCL $= 141.7$, LCL $= 135.5$
5. CL $= 14.9$, UCL $= 21.08$, LCL $= 8.72$. There are outliers on all the three Saturdays.
6. (a) CL $= 5.3$, UCL $= 9.62$, LCL $= 0.98$
   (b) CL $= 15.1$, UCL $= 24.49$, LCL $= 5.71$
7. $n = 5$, $d_4 = 2.114$, $d_3 = 0$; CL $= 1.367$, UCL $= 2.89$, LCL $= 0$

Annexure I—Control Chart Factors

Sample size, n	For $\bar{X}$ chart, d_2	For R chart, d_3	Sample size, n	For $\bar{X}$ chart, d_2	For R chart, d_3
2	1.128	0.853	14	3.407	0.763
3	1.693	0.888	15	3.472	0.756
4	2.059	0.880	16	3.532	0.750
5	2.326	0.864	17	3.588	0.744
6	2.534	0.848	18	3.640	0.739
7	2.704	0.833	19	3.689	0.734
8	2.847	0.820	20	3.735	0.729
9	2.970	0.808	21	3.778	0.724
10	3.078	0.797	22	3.819	0.720
11	3.173	0.787	23	3.858	0.716
12	3.258	0.779	24	3.895	0.712
13	3.336	0.770	25	3.931	0.708

PART II

Mechanical Measurements

- **Measurement Systems**
- **Transducers**
- **Measurement of Force, Torque, and Strain**
- **Measurement of Temperature**
- **Pressure Measurements**

Measurement Systems

After studying this chapter, the reader will be able to

- understand the effect of hysteresis in measuring instruments
- appreciate the importance of linearity
- explain some basic definitions in measurements
- appreciate the significance of system response
- elucidate the different functional elements of generalized measurement systems

12.1 INTRODUCTION

We know that measurement is defined as the quantification of a physical variable using a measuring instrument. During the process of measurement, a specific value is assigned to the unknown quantity after due comparison with a predefined standard. The measuring process is schematically represented in Fig. 12.1.

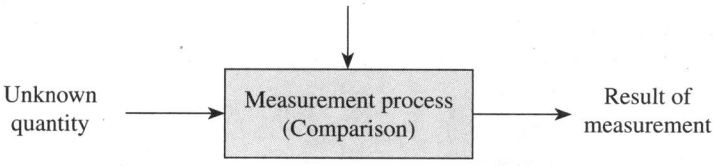

Fig. 12.1   Measuring process

It is pertinent to mention here that while making physical measurements, one needs to keep in mind that measurements are not completely accurate, as discussed in Chapter 1. It has to be remembered that the act of measurement will be complete when the uncertainty associated with measurements is also considered. In order to understand the uncertainty associated with measurement, we need to study a few more definitions, apart from those discussed in the first chapter. It is important to know various measurement attributes that affect the performance of a measuring instrument. The operating parameters within which an instrument performs the measurement process needs to be understood clearly.

12.2 SOME BASIC DEFINITIONS

In order to clearly understand the concepts of measurement, we will now look into a few more definitions.

12.2.1 Hysteresis in Measurement Systems

When the value of the measured quantity remains the same irrespective of whether the measurements have been obtained in an ascending or a descending order, a system is said to be free from hysteresis. Many instruments do not reproduce the same reading due to the presence of hysteresis. Slack motion in bearings and gears, storage of strain energy in the system, bearing friction, residual charge in electrical components, etc., are some of the reasons for the occurrence of hysteresis. Figure 12.2 shows a typical hysteresis loop for a pressure gauge. If the width of the hysteresis band formed is appreciably more, the average of the two measurements (obtained in both ascending and descending orders) is used. However, the presence of some hysteresis in measuring systems is normal, and the repeatability of the system is affected by it.

12.2.2 Linearity in Measurement Systems

It is desirable to design instruments having a linear relationship between the applied static input and the indicated output values, as shown in Fig. 12.3. A measuring instrument/system is said to be linear if it uniformly responds to incremental changes, that is, the output value is equal to the input value of the measured property over a specified range.

Linearity is defined as the maximum deviation of the output of the measuring system from a specified straight line applied to a plot of data points on a curve of measured (output) values versus the measurand (input) values. In order to obtain accurate measurement readings, a high degree of linearity should be maintained in the instrument or efforts have to be made to minimize linearity errors. A better degree of linearity renders the instrument to be readily calibrated. However, in practice, only an approximation of the linearity is achieved as there is always some small variance associated with the measuring system. Hence, the expected linearity of the input is usually specified as a percentage of the operating range.

A fuel gauge in an automobile is one such example of non-linearity. When the tank is completely filled with fuel, the gauge needle indicates a full tank. The needle continues to be in the nearly full position even after the consumption of a significant amount of fuel. However, after considerable travel, the needle appears to move rapidly towards the minimum fuel value.

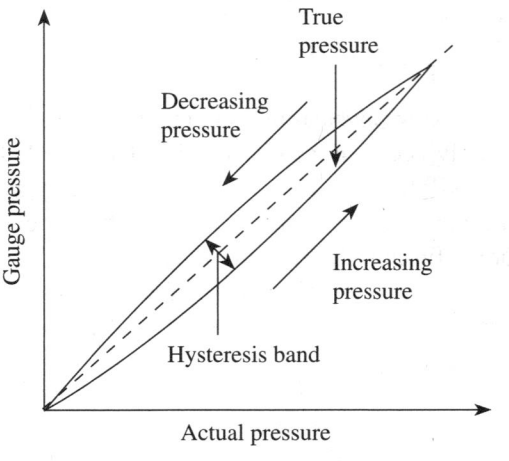

Fig. 12.2 Hysteresis in measurement systems

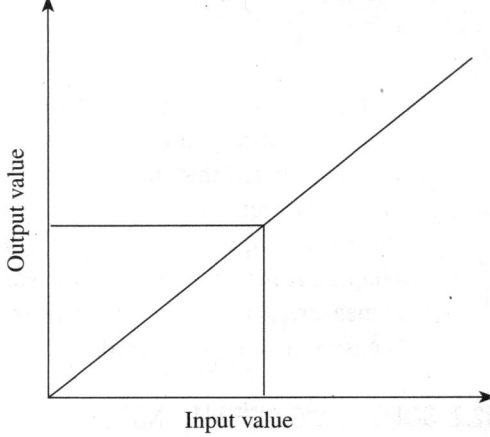

Fig. 12.3 Linearity in measurements

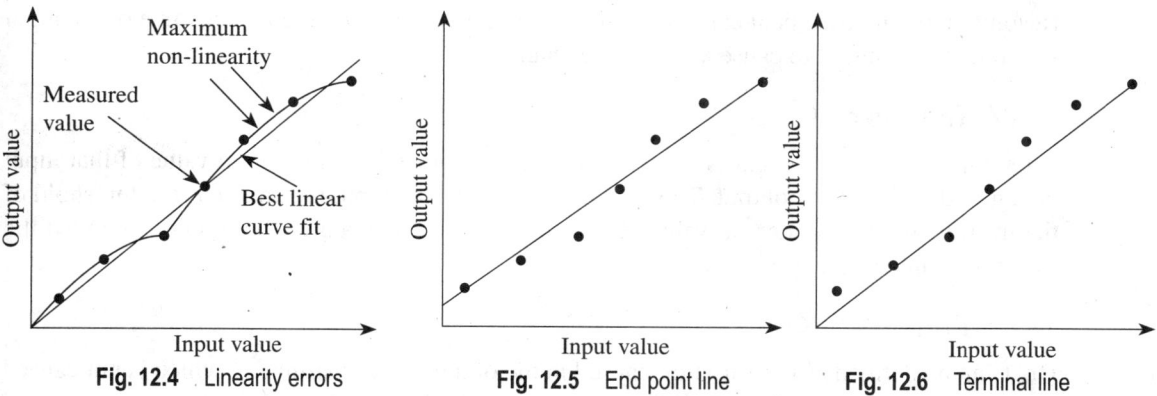

Fig. 12.4 Linearity errors   **Fig. 12.5** End point line   **Fig. 12.6** Terminal line

This may be due to the improper calibration of the fuel gauge at the maximum and minimum values of the operating range.

Before making any interpretation or comparison of the linearity specifications of the measuring instrument, it is necessary to define the exact nature of the reference straight line adopted, as several lines can be used as the reference of linearity. The most common lines are as follows:

Best-fit line   The plot of the output values versus the input values with the best line fit is shown in Fig. 12.4. The line of best fit is the most common way to show the correlation between two variables. This line, which is also known as the trend line, is drawn through the centre of a group of data points on a scatter plot. The best-fit line may pass through all the points, some of the points, or none of the points.

End point line   This is employed when the output is bipolar. It is the line drawn by joining the end points of the data plot without any consideration of the origin. This is represented in Fig. 12.5.

Terminal line   When the line is drawn from the origin to the data point at full scale output, it is known as terminal line. The terminal line is shown in Fig. 12.6.

Least square line   This is the most preferred and extensively used method in regression analysis. Carl Freidrich Gauss (1975) was the first to give the earliest description of the least square method. It is a statistical technique and a more precise way of determining the line of best fit for a given set of data points. The best-fit line is drawn through a number of data points by minimizing the sum of the squares of the deviations of the data points from the line of best fit, hence the name least squares. The line is specified by an equation relating the input value to the output value by considering the set of data points.

12.2.3 Resolution of Measuring Instruments

Resolution is the smallest change in a physical property that an instrument can sense. For example, a weighing machine in a gymnasium normally senses weight variations in kilograms, whereas a weighing machine in a jewellery shop can detect weight in milligrams. Naturally, the weighing machine in the jewellery shop has a superior resolution than the one at the gymnasium.

Resolution of an instrument can also be defined as the minimum incremental value of the input signal that is required to cause a detectable change in the output.

12.2.4 Threshold

If the input to the instrument is gradually increased from zero, a minimum value of that input is required to detect the output. This minimum value of the input is defined as the threshold of the instrument. The numerical value of the input to cause a change in the output is called the threshold value of the instrument.

12.2.5 Drift

Drift can be defined as the variation caused in the output of an instrument, which is not caused by any change in the input. Drift in a measuring instrument is mainly caused by internal temperature variations and lack of component stability. A change in the zero output of a measuring instrument caused by a change in the ambient temperature is known as thermal zero shift. Thermal sensitivity is defined as the change in the sensitivity of a measuring instrument because of temperature variations. These errors can be minimized by maintaining a constant ambient temperature during the course of a measurement and/or by frequently calibrating the measuring instrument as the ambient temperature changes.

12.2.6 Zero Stability

It is defined as the ability of an instrument to return to the zero reading after the input signal or measurand comes back to the zero value and other variations due to temperature, pressure, vibrations, magnetic effect, etc., have been eliminated.

12.2.7 Loading Effects

Any measuring instrument generally consists of different elements that are used for sensing, conditioning, or transmitting purposes. Ideally, when such elements are introduced into the measuring system, there should not be any distortion in the original signal. However, in practice, whenever any such element is introduced into the system, some amount of distortion occurs in the original signal, making an ideal measurement impossible. The distortion may result in wave form distortion, phase shift, and attenuation of the signal (reduction in magnitude); sometimes, all these undesirable features may combine to affect the output of the measurement.

Hence, loading effect is defined as the incapability of a measuring system to faithfully measure, record, or control the measurand in an undistorted form. It may occur in any of the three stages of measurement or sometimes it may be carried right down to the basic elements themselves.

12.2.8 System Response

One of the essential characteristics of a measuring instrument is to transmit and present faithfully all the relevant information included in the input signal and exclude the rest. We know that during measurements, the input rapidly changes with time, and hence, the output. The behaviour of the measuring system under the varying conditions of input with respect to time is known as the dynamic response.

There are two types of dynamic inputs: steady-state periodic quantity and transient magnitude. The magnitude of the steady-state periodic quantity has a definite repeating time cycle, whereas the time variation of the transient magnitude does not repeat. In some measurement applications, for the system to attain a steady state, enough time is available. In such situations, the transient characteristics of the system are not a matter of concern. In certain measurement systems, it is necessary to examine the transient behaviour of the physical variable under consideration. The design of the measurement system becomes more complicated when transient characteristics are considered.

When an input is given to a measuring instrument, a certain amount of time elapses before it indicates an output. This is because the measurement systems comprise one or more storage elements such as electrical inductance and capacitance, thermal and fluid capacitance, mass, and inertia. When an input is given to the measuring instrument, the system will not respond immediately because the energy storage elements do not allow a sudden flow of energy. The measuring instrument goes through a transient state before it finally attains a steady-state position. The following are the dynamic characteristics of a measurement system:

Speed of response One of most important characteristic of the measuring instrument, speed of response is defined as the speed with which the measuring instrument responds to the changes in the measured quantity. Some delay or lag is always associated with the measuring instrument, as it does not respond to the input instantaneously.

Measuring lag It is the time when an instrument begins to respond to a change in the measured quantity. This lag is normally due to the natural inertia of the measuring system. Measuring lag is of two types:

Retardation type In this case, the measurement system instantaneously begins to respond after the changes in the input have occurred.

Time delay type In this type, the measuring system begins to respond after a dead time to the applied input. Dead time is defined as the time required by the measuring system to begin its response to a change in the quantity to be measured. Dead time simply transfers the response of the system along the time scale, thereby causing a dynamic error. This type of measurement lag can be ignored as they are very small and are of the order of a fraction of a second. If the variation in the measured quantity occurs at a faster rate, the dead time will have an adverse effect on the performance of the system.

Fidelity It is defined as the degree to which a measurement system indicates the changes in the measured quantity without any dynamic error.

Dynamic error It is also known as a measurement error. It can be defined as the difference between the true value of a physical quantity under consideration that changes with time and the value indicated by the measuring system if no static error is assumed.

It is to be noted here that speed of response and fidelity are desirable characteristics, whereas measurement lag and dynamic error are undesirable.

12.3 FUNCTIONAL ELEMENTS OF MEASUREMENT SYSTEMS

We know that some physical quantities such as length and mass can be directly measured using measuring instruments. However, the direct measurement of physical quantities such

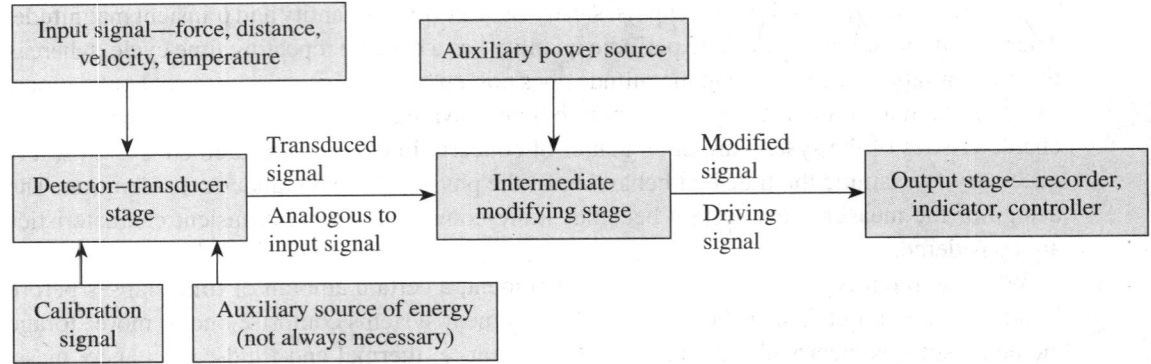

Fig. 12.7 Elements of a generalized measurement system

as temperature, force, and pressure is not possible. In such situations, measurements can be performed using a transducer, wherein one form of energy/signal that is not directly measurable is transformed into another easily measurable form. Calibration of the input and output values needs to be carried out to determine the output for all vales of input.

A measuring instrument essentially comprises three basic physical elements. Each of these elements is recognized by a functional element. Each physical element in a measuring instrument consists of a component or a group of components that perform certain functions in the measurement process. Hence, the measurement system is described in a more generalized method. A generalized measurement system essentially consists of three stages. Each of these stages performs certain steps so that the value of the physical variable to be measured is displayed as an output for our reference. Figure 12.7 schematically represents the generalized measurement systems. The three stages of a measurement system are as follows:

1. Primary detector–transducer stage
2. Intermediate modifying stage
3. Output or terminating stage

The primary detector–transducer stage senses the quantity to be measured and converts it into analogous signals. It is necessary to condition or modify the signals obtained from the primary detector–transducer stage so that it is suitable for instrumentation purposes. This signal is passed on to the intermediate modifying stage, wherein they are amplified so that they can be used in the terminating stage for display purposes. These three stages of a measurement system act as a bridge between the input given to the measuring system and its output.

12.4 PRIMARY DETECTOR–TRANSDUCER STAGE

The main function of the primary detector–transducer stage is to sense the input signal and transform it into its analogous signal, which can be easily measured. The input signal is a physical quantity such as pressure, temperature, velocity, heat, or intensity of light. The device used for detecting the input signal is known as a transducer or sensor. The transducer converts the sensed input signal into a detectable signal, which may be electrical, mechanical, optical, thermal, etc. The generated signal is further modified in the second stage. The transducer should have the ability to detect only the input quantity to be measured and exclude all other signals. The sensing process is schematically represented in Fig. 12.8.

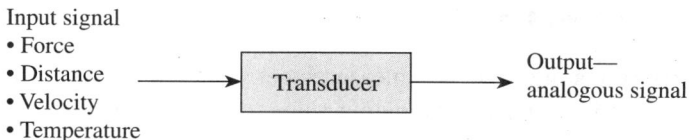

Input signal
- Force
- Distance
- Velocity
- Temperature

Transducer

Output—
analogous signal

Fig. 12.8 Sensing process

For example, if bellows are employed as transducers for pressure measurement, it should detect signals pertaining to only pressure and other irrelevant input signals or disturbances should not be sensed. However, in practice, the transducers used are seldom sensitive to only the signals of the quantity being measured.

12.5 INTERMEDIATE MODIFYING STAGE

In the intermediate modifying stage of a measurement system, the transduced signal is modified and amplified appropriately with the help of conditioning and processing devices before passing it on to the output stage for display. Signal conditioning (by noise reduction and filtering) is performed to enhance the condition of the signal obtained in the first stage, in order to increase the signal-to-noise ratio. If required, the obtained signal is further processed by means of integration, differentiation, addition, subtraction, digitization, modulation, etc. It is important to remember here that in order to obtain an output that is analogous to the input, the characteristics of the input signals should be transformed with true fidelity.

Table 12.1 Examples of the three stages of a generalized measurement system

Type	Detector–transducer stage	Intermediate modifying stage	Terminating stage
Mechanical	Spring mass Contact spindle Bourdan tube Bellows Proving ring	Gears Cams Linkages	*Indicators*: Moving pointer and scale Light beam and scale Electron beam scale CRO, liquid column
Hydraulic-pneumatic	Orifice meter Venturi meter Vane Propeller	Valves Dash pots	*Digital types*: Direct reading
Optical	Photo-electric cell Photographic film	Lens Mirrors Light levers Filters	*Recorders*: Digital printing Hot stylus and chart Inked pen chart, light beam and photographic film, computer recording Magnetic recording
Electrical	Resistance Capacitance Inductance Piezo-electric crystal Thermocouple	Amplifying systems Attenuating systems Filters Tele-metering systems	*Controlling devices*: All types

12.6 OUTPUT OR TERMINATING STAGE

The output or terminating stage of a measurement system presents the value of the output that is analogous to the input value. The output value is provided by either indicating or recording for subsequent evaluations by human beings or a controller, or a combination of both. The indication may be provided by a scale and pointer, digital display, or cathode ray oscilloscope. Recording may be in the form of an ink trace on a paper chart or a computer printout. Other methods of recording include punched paper tapes, magnetic tapes, or video tapes. Else, a camera could be used to photograph a cathode ray oscilloscope trace.

Table 12.1 gives some of the examples for the three stages of a generalized measurement system. Thus, measurement of physical quantities such as pressure, force, and temperature, which cannot be measured directly, can be performed by an indirect method of measurement. This can be achieved using a transduced signal to move the pointer on a scale or by obtaining a digital output.

A QUICK OVERVIEW

- When the value of the measured quantity remains the same irrespective of whether the measurements have been obtained in an ascending or a descending order, a system is said to be free from hysteresis. However, hysteresis in a system is normal, and repeatability of the system is affected by it.
- Resolution is the smallest change in a physical property that an instrument can sense. Resolution of an instrument can also be defined as the minimum incremental value of the input signal that is required to cause a detectable change in the output.
- Linearity is defined as the maximum deviation of the output of a measuring system from a specified straight line applied to a plot of data points on a curve of the measured (output) values versus the measurand (input) values. It is desirable to design instruments having a linear relationship between the applied static input and the indicated output values.
- The most common lines used as reference of linearity are best-fit line, end point line, terminal line, and least square line.
- The numerical value of the input to cause a change in the output is called the threshold value of the instrument.
- Drift can be defined as the variation caused in the output of an instrument, which is not caused by any change in the input. Drift in a measuring instrument is mainly caused by the internal temperature variations and lack of component stability.
- The behaviour of a measuring system under the varying conditions of input with respect to time is known as the dynamic response. There are two types of dynamic input: steady-state periodic quantity and transient magnitude.
- Zero stability is defined as the ability of an instrument to return to the zero reading after the input signal or measurand comes back to the zero value and other variations due to temperature, pressure, vibrations, magnetic effect, etc., have been eliminated.
- Loading effect is defined as the incapability of a measuring system to faithfully measure, record, or control the measurand in an undistorted form. It may occur in any of the three stages of measurement or sometimes may be carried right down to the basic elements themselves.

- The dynamic characteristics of a measurement system are speed of response, measuring lag, fidelity, and dynamic error.
- The generalized measurement system essen-tially comprises the following three stages: primary detector–transducer stage, intermediate modifying stage, and output or terminating stage.

MULTIPLE-CHOICE QUESTIONS

1. What is the correct definition of resolution?
   (a) Difference between accuracy and precision
   (b) Smallest graduation on the scale or dial
   (c) Visual separation between graduations
   (d) Difference between the instrument reading and actual dimension
2. Hysteresis of an instrument affects its
   (a) resolution   (c) repeatability
   (b) accuracy   (d) zero stability
3. Loading effect means
   (a) effect of the applied load on the instrument
   (b) impact of the applied load on the accuracy of the instrument
   (c) capability of the instrument to withstand the applied load
   (d) incapability of the instrument to measure the measurand in an undistorted form
4. In the intermediate modifying stage, the trans-duced signal is
   (a) amplified   (c) conditioned
   (b) filtered   (d) all of these
5. Which one of the following is not a dynamic characteristic?
   (a) Drift
   (b) Speed of response
   (c) Measuring lag
   (d) Fidelity
6. The behaviour of an instrument under varying conditions of input with time is known as
   (a) loading effect   (c) dynamic response
   (b) hysteresis   (d) linearity
7. Dead time is the time required by a measuring system to
   (a) respond to a change in the quantity to be measured
   (b) eliminate loading error

   (c) overcome zero drift
   (d) increase the threshold value
8. Threshold value corresponds to the minimum value of an input required to
   (a) reduce dynamic error
   (b) detect an output
   (c) increase drift
   (d) reduce dead time
9. Many instruments do not reproduce the same reading irrespective of whether the measure-ments have been obtained in an ascending or a descending order. This is due to
   (a) linearity error   (c) systematic error
   (b) drift error   (d) hysteresis
10. Fidelity is a/an
    (a) static character
    (b) dynamic character
    (c) neither a static nor a dynamic character
    (d) undesirable error
11. The ability of an instrument to return to zero immediately after the removal of input is known as
    (a) drift   (c) zero stability
    (b) hysteresis   (d) loading effect
12. The maximum deviation of the output of a measuring system from a specified straight line, applied to a plot of data points on a curve of the measured values versus the measured values is called
    (a) linearity
    (b) threshold
    (c) measurement lag
    (d) resolution
13. Dead time is a type of
    (a) drift error   (c) measurement lag
    (b) loading error   (d) threshold value

14. The degree to which a measurement system indicates the changes in the measured quantity without any dynamic error is called
    (a) loading effect
    (b) fidelity
    (c) drift error
    (d) measurement lag

15. Which of the following lines is known as the trend line?
    (a) End point line
    (b) Least square line
    (c) Terminal line
    (d) Best-fit line

REVIEW QUESTIONS

1. With a block diagram, explain the three stages of a generalized measurement system giving suitable examples.
2. Explain the different dynamic characteristics of a measuring system.
3. Define hysteresis and list the reasons for the presence of hysteresis in measuring systems.
4. Discuss the importance of linearity in measuring instruments.
5. Explain the following terms:
    (a) Drift
    (b) Threshold
    (c) Resolution
    (d) Zero stability
6. Discuss loading effect with respect to a measuring system.
7. With a neat sketch, explain the sensing process.

Answers to Multiple-choice Questions

1. (b)	2. (c)	3. (d)	4. (d)	5. (a)	6. (c)	7. (a)	8. (b)
9. (d)	10. (b)	11. (c)	12. (a)	13. (c)	14. (b)	15. (d)	

After studying this chapter, the reader will be able to

- understand the basic structure of several types of transducers
- elucidate the characteristics of several types of transducers
- appreciate the significance of system response
- explain the different functional elements of generalized measurement systems
- understand the inherent problems of mechanical systems
- appreciate the importance of electrical and electronic systems
- understand the working of a cathode ray oscilloscope and its importance

13.1 INTRODUCTION

We know that the generalized measuring system consists of three functional elements: primary detector-transducer stage, intermediate modifying stage, and output or terminating stage. Each stage performs certain functions so that the value of the physical variable to be measured (measurand) is displayed as output, as discussed in Chapter 12. In addition, a controlling function is required in many applications. For example, measurement systems employed in process control comprise a fourth stage called feedback control stage. The feedback control stage essentially consists of a controller that interprets the measured signal, depending on which decision is taken to control the process. Consequently, there is a change in the process parameter that affects the magnitude of the sensed variable. It is essential to note here that more the accuracy of measurement of the control variable, the better the accuracy of control. Hence, efforts have to be made towards accurate measurements before making any attempt to control the same. Figure 13.1 shows a simple schematic representation of a generalized measuring system.

13.2 TRANSFER EFFICIENCY

The detecting or sensing element of a measuring system first makes contact with the quantity to be measured, and the sensed information is immediately transduced into an analogous form. The transducer, which may be electrical, mechanical, optical, magnetic, piezoelectric, etc., converts the sensed information into a more convenient form. A transducer is a device that converts one form of energy into another form.

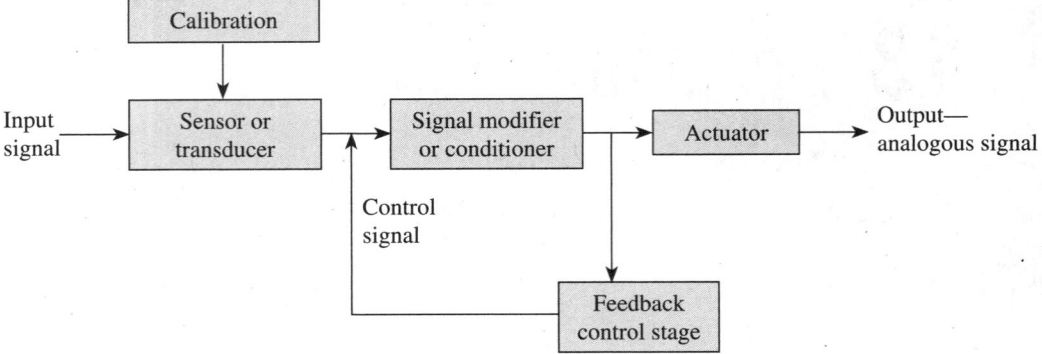

Fig. 13.1 Four stages of a generalized measuring system

The term transfer efficiency was first defined by Simpson (1955) for comparing the different devices used in the first stage. Depending on the information sensed and delivered by the sensor, transfer efficiency is defined as follows:

$$\text{Transfer efficiency} = \frac{I_{del}}{I_{sen}}$$

I_{del} is the information delivered by the pickup device and I_{sen} is the information sensed by the pickup device.

This ratio cannot be more than unity since the sensor device cannot generate any information on its own. Hence, it is desirable to have a sensor with very high transfer efficiency.

13.3 CLASSIFICATION OF TRANSDUCERS

Transducers are classified as follows:
1. Primary and secondary transducers
2. Based on the principle of transduction
3. Active and passive transducers
4. Analog and digital transducers
5. Direct and inverse transducers
6. Null and deflection transducers

13.3.1 Primary and Secondary Transducers

Transducers are generally defined as devices that transform values of physical quantities in the form of input signals into corresponding electrical output signals. The physical quantity may be heat, intensity of light, flow rate, etc. In terms of energy conversion, a transducer is defined as a device that transforms energy from one form to another. The energy may be in electrical, mechanical, or acoustical form. Transducers are devices that are used to transform the information sensed (signals) between two different physical domains. Transducers, which include the entire detector transducer or primary stage of the generalized measurement system, may comprise two important components:

Sensing or detecting element   The function of this element is to respond to a physical phenomenon or a change in the physical phenomenon. Hence it is termed a primary transducer.

Transduction element   The function of a transduction element is to transform the output obtained by the sensing element to an analogous electrical output. Hence it is termed a secondary transducer.

In addition to these, the primary stage also consists of auxiliary sources of energy, power amplifiers, and calibration sources.

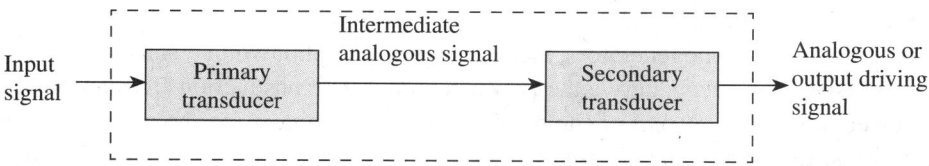

Fig.13.2   Primary detector transducer stage with primary and secondary transducers

Table 13.1   Typical mechanical primary detector transducers

Mechanical transducers	Typical operation
Contacting spindle, pin, or finger	Displacement to displacement
Elastic members Bourdon tube Bellows Diaphragm	Pressure to displacement
Proving ring Spring	Force to displacement
Mass Seismic mass	Forcing function to displacement
Pendulum scale	Force to displacement
Manometer	Pressure to displacement
Thermal Thermocouple	Temperature to electric current
Bimaterial	Temperature to displacement
Temperature stick	Temperature to phase
Hydropneumatic Static Float	Fluid level to displacement
Hydrometer	Specific gravity to displacement
Dynamic Orifice Venturi Pitot tube	Velocity to pressure
Vanes	Velocity to force
Turbines	Linear to angular velocity

An example of a primary detector transducer (see Table 13.1) stage comprising both these elements is the combination of the bourdon tube and the linear variable differential transformer (LVDT) (Figs 13.2 and 13.3). The bourdon tube, which acts as a detecting element, senses pressure and gives the output in the form of displacement. This displacement is further used to move the core of LVDT, and a voltage is obtained as output. Thus, the pressure is converted into displacement, which in turn is transduced into an analogous voltage signal. Thus, the bourdon tube acts as the primary sensing element and the LVDT as the secondary transducer.

13.3.2 Based on Principle of Transduction

This classification is based on how the input quantity is transduced into capacitance, resistance, and inductance values. They are known as capacitive, resistive, and inductive transducers. Transducers can also be piezoelectric, thermoelectric, magnetostrictive, electrokinetic, and optical.

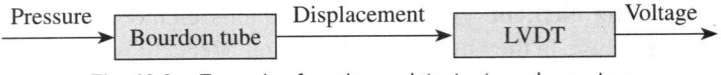

Fig. 13.3 Example of a primary detector transducer stage

13.3.3 Active and Passive Transducers

Active transducers are of a self-generating type, wherein they develop their own voltage or current output (Fig. 13.4). They do not need any auxiliary power source to produce the output. The energy required to produce the output is derived from the physical quantity being measured. Examples of active transducers are piezoelectric crystals (used for force or acceleration measurement), tachogenerators, thermocouples, and photovoltaic cells.

Passive transducers derive the power required for transduction from an auxiliary source of power (Fig. 13.5). A part of the power required for generating the output is derived from the physical quantity being measured. Since they receive power from an auxiliary source, they are termed externally powered transducers. Further, these transducers remain passive in the absence of an external power source; hence, they are called passive transducers. Resistive, capacitive, and inductive transducers are some of the examples of passive transducers.

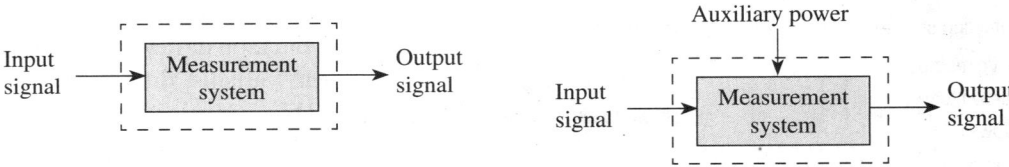

Fig. 13.4 Active transducer

Fig. 13.5 Passive transducer

13.3.4 Analog and Digital Transducers

Based on the output generated, that is, depending on whether the output generated is a continuous function of time or is in discrete form, transducers are classified as analog and digital transducers.

In case of analog transducers, the input quantity is converted into an analog output, which is a continuous function of time. LVDT, strain gauge, thermocouple, and thermistor are some examples of analog transducers.

If a transducer converts the input quantity into an electrical signal that is in the form of pulses, as output, it is called a digital transducer. These pulses are not continuous functions of time but are discrete in nature. In fact, it is easier to process and transmit data as a function of two numbers 0 and 1. Examples are shaft encoders, linear displacement transducers using conducting or non-conducting contacts, and opaque or translucent segments.

13.3.5 Direct and Inverse Transducers

When a measuring device measures and transforms a non-electrical variable into an electrical variable, it is called a direct transducer (Fig. 13.6). A thermocouple that is used to measure temperature, radiation, and heat flow is an example of a transducer.

If an electrical quantity is transformed into a non-electrical quantity, it is termed an inverse transducer (Fig. 13.7). A very common example of an inverse transducer is a piezoelectric crystal wherein a voltage is given as the input. This is because when a voltage is applied across its surfaces, its dimensions are changed causing a mechanical displacement.

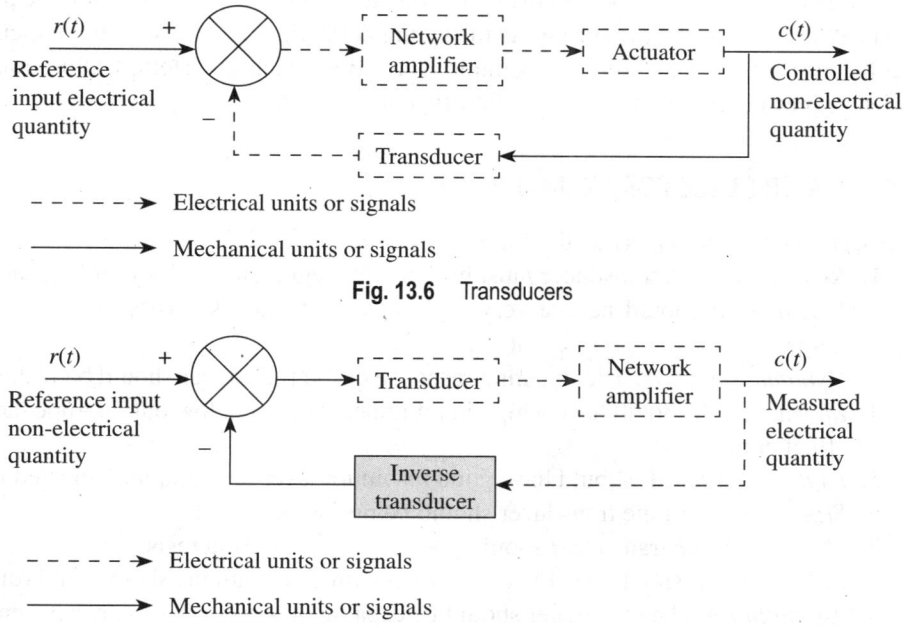

Fig. 13.6   Transducers

Fig. 13.7   Inverse transducers

13.3.6 Null- and Deflection-type Transducers

This classification is based on null or deflection principle. A null-type device works on the principle of maintaining a zero deflection by applying an appropriate known effect that opposes the one generated by the measured quantity. This type of transducer requires a detector capable of detecting an imbalance and a means to restore a balance either manually or automatically. (A null detector and a means of restoring the balance are necessary.) Since it is required to balance the deflection preferably at the zero level, in order to determine the numerical values, precise information of the magnitude of the opposing effect is essential. The major disadvantage of this method is that it cannot be used for dynamic measurements. Examples of null-type transducers include dead-weight pressure gauges and equal arm balances. In a dead-weight pressure gauge, which operates on the null method, standard weights are added to the platform of known weight and gravitational force can be used to balance the pressure force on the face of the piston. The balance of forces is indicated by the platform remaining stationary in between the two stops. The pressure required for the balance may be calculated as the weights added and the area of the piston are known.

A deflection-type transducer works on the principle that the measured quantity produces a physical effect that stimulates an analogous effect in the opposite direction in some parts of the measuring instrument. There is an increase in the opposing effect until a balance is restored, at which stage the measurement of deflection is carried out. A rudimentary pressure gauge is an example of a deflection transducer. It contains a piston that is driven by the fluid pressure to produce a resultant force. The piston acts as a primary transducer and a secondary transducer, which also acts as the variable conversion element. The force generated by the piston rod stimulates an opposing force in the spring due to an imbalance of forces on the piston rod. This causes a deflection of the spring. As the spring deflection increases, its force increases and thus a balance is achieved. Another example is a spring balance system. Calibration of the spring determines the accuracy in a deflection-type instrument.

13.4 QUALITY ATTRIBUTES FOR TRANSDUCERS

A transducer should possess the following qualities:
1. *Repeatability*: A transducer must have a high degree of accuracy and repeatability.
2. *Linearity*: It should have a very high degree of linearity within the specified operating range.
3. *Dynamic response*: The dynamic response of the transducer should be instantaneous.
4. *Impedance*: It should have a high input impedance and a low output impedance for loading effect elimination.
5. *High resolution*: It should have good resolution over the complete selected range.
6. *Size*: The size of the transducer should be preferably small.
7. *Hysteresis*: The transducer should possess low or no hysteresis.
8. *Robustness*: It should be able to endure pressure, vibrations, shocks, and rough handling.
9. *Adaptability*: The transducer should be capable of working in a corrosive environment.
10. *Sensitivity*: Cross-sensitivity of the transducer should be zero or minimum, and it should exhibit high sensitivity to the desired signal.
11. *Stability and reliability*: The degree of stability and reliability of the transducer should be high.
12. *Response*: It should have a requisite time domain specification of transient and frequency response.

13.5 INTERMEDIATE MODIFYING DEVICES

The primary detector transducer stage usually detects a mechanical quantity and transduces it into an electrical form. In the intermediate modifying stage of the measurement system, the transduced signal is further modified and amplified with the help of conditioning and processing devices, so that the signal is appropriate before passing it on to the output or terminating stage for display. The terminating-stage elements usually comprise indicators, recorders, display and data processing units, and control elements.

Signal conditioning is essential in order to enhance the condition of the signal obtained in the first stage by reducing the noise and filtering such that the signal-to-noise ratio is increased. If required, the obtained signal has to be processed further by means of attenuation, integration, differentiation, addition, subtraction, digitization, modulation, etc. It is important to remember here that, in order to obtain an output that is analogous to the input, the characteristics of the input signals should be transformed for display with true fidelity. Hence, the dynamic measurement of mechanical quantities imposes severe strain on the elements of the intermediate stage.

Table13.2   Typical electrical, electronic, and self-generating primary detector transducers

Electrical transducers	Typical operation
Resistance Contacting Variable area of conductor Variable length of conductor	Displacement to change in resistance
Variable dimensions of conductor	Strain to change in resistance
Variable resistivity of conductor	Air velocity to change in resistance
Inductance Variable coil dimensions Variable air gap Changing core material Changing coil positions Change core positions	Displacement to change in inductance
Moving coil Moving permanent magnet Moving core	Velocity to change in inductance
Capacitance Changing air gap Changing plate areas Changing dielectric	Displacement to change in capacitance
Self-generating Piezoelectric pickup Photovoltaic Moving coil generator Thermocouple	Displacement to voltage Intensity of light source to voltage Displacement to voltage Temperature or radiation to voltage
Electronic	Displacement to current
Photoelectric Photoconductive Photogenerative Photoemissive	Intensity of light source to voltage
Electrokinetic	Flow to voltage

One of the major requirements of intermediate modifying devices is large amplification coupled with good transient response. Mechanical, hydraulic, and pneumatic devices fulfil both these requirements inadequately and hence electrical and electronic devices are preferred (Table 13.2).

13.5.1 Inherent Problems in Mechanical Systems

We know that primary detector transducers transduce the input signal into mechanical displacement. This intermediate analogous signal is passed on to the secondary transducer and gets converted into an electrical form, which is more easily processed by the intermediate stage.

Design problems of considerable magnitude arise when the signals from the primary–secondary transducers are fed into mechanical intermediate elements. The mechanical intermediate elements, which include linkages, gears, cams, etc., are inadequate, especially in handling dynamic inputs.

13.5.2 Kinematic Linearity

When a linkage system is to be used as a mechanical amplifier, it should be designed in such a way that it provides the same amplification, which is often termed the gain, over its entire range of output. The gain should be linear, which would otherwise result in a poor amplitude response. Further, for recording the output, a straight line stylus movement should be provided; otherwise, the recorded trace will be distorted along the time axis. This would result in a poor phase response. Hence, it is essential to have a proper kinematic layout to ascertain linearity and straight line stylus movement. Additionally, due importance should be given to the control of tolerances on link dimensions and fixed point locations.

13.5.3 Mechanical Amplification

Mechanical amplification is defined, in terms of mechanical advantages, as follows:

Gain = Mechanical advantage

$$\text{Gain} = \frac{\text{Output displacement}}{\text{Input displacement}}$$

$$\text{Gain} = \frac{\text{Output velocity}}{\text{Input velocity}}$$

When mechanical amplification is used, frictional loading, inertial loading, elastic deformation, and backlash all contribute to errors. Errors resulting from inertial loading and elastic deformation can be grouped as systematic errors and those from frictional loading and backlash as random errors. Although we perceive this value to be more than unity, it need not be so.

13.5.3 Reflected Frictional Amplification

We know that any source of friction in the linkages, however small, will result in a force, which in turn gets magnified by an amount equal to the gain. This amplified force is reflected back to the input as a magnified load, which is numerically equal to the gain between the source of friction and the input. This effect is referred to as the *reflected frictional amplification*. If several such sources of friction are present in a system, all the sources would get amplified and reflected in the input. The total reflected frictional force is given by the following equation:

$$F_{tfr} = \Sigma A F_{fr}$$

F_{tfr} is the total reflected frictional force (in N) at the input of the system, A is the mechanical amplification or gain, and F_{fr} is the actual frictional force (in N) at its source.

An important limitation of mechanical intermediate devices is the losses due to friction. One has to understand that amplifications (attenuations) of only force and displacement signals take place, which is affected by an inverse relationship.

13.5.4 Reflected Inertial Amplification

Problems caused by inertial forces are similar to those caused by frictional forces. The effects of inertial forces are also amplified and reflected back to the input in proportion to the gain between their source and input. It is to be mentioned here that there will not be any loss of energy, because it is temporarily stored to be retrieved later. This effect is termed the *reflected inertial amplification*. The total reflected inertial force, which should be overcome at the input of the system, is as follows:

$$F_{tir} = \Sigma A \, \Delta F_{ir}$$

F_{tir} is the total reflected inertial force (in N) at the input of the system, A is the mechanical amplification or gain, and ΔF_{ir} is the increment of the inertial force (in N) at any point in the system.

It is interesting to note the differences between reflected frictional and inertial amplifications. The first difference is that frictional forces are concentrated, whereas inertial forces are distributed. The second difference is that frictional forces are static in nature, but inertial forces are dynamic. Furthermore, inertial forces vary with acceleration, which in turn directly varies with the square of velocity and hence becomes significant as speed increases and duration of signal interval decreases.

Therefore, it is necessary to determine the total reflected force, which is the sum of the reflected frictional and inertial forces:

$$F_r = F_{tfr} + F_{tir}$$

where F_r is the total reflected force (in N).

13.5.5 Amplification of Backlash and Elastic Deformation

Different components of mechanical systems are directly affected by transmitted and inertial loads that are being carried, causing elastic deformation in the parts. Backlash is the consequence of a temporary non-constraint in a linkage system. Clearances required for the components of the linkage system in order to attain the required mechanical fits where relative motion occurs cause backlash. It is the amount by which the width of a gear's tooth space exceeds the thickness of an engaging tooth measured at the pitch circle of the gears. In fact, backlash is the requisite clearance or play provided to accommodate manufacturing errors, enable space for lubrication, and allow for thermal expansion of components. In order to reduce gear backlash springs, anti-backlash gears are employed. Backlash can also be reduced with proper lubrication.

One of the consequences of backlash is lost motion. Lost motion occurs when an input given to any mechanism does not generate the analogous displacement at the output, which in turn results in a positional error and contributes to the uncertainty of a motion system.

At the output, both backlash and elastic deformation result in lost motion, which will be amplified by an amount equal to the gain between the source and the output. Hence, the lost motion is equal to the actual backlash or deformation multiplied by the gain between the source and the output. These two effects are termed *backlash amplification* and *elastic amplification*, respectively.

In order to assess the effect of lost motion due to elastic deformation or backlash on the system as a whole, it would be convenient to consider projected displacement losses ahead of the output instead of being reflected back to the input. The total projected displacement loss because of backlash is given by the following equation:

$$Y_{tbl} = \Sigma A Y_{bl}$$

Here, Y_{tbl} is the total projected displacement loss (in mm) due to backlash or clearances provided in mm, A is the mechanical amplification or gain, and Y_{bl} is the lost motion (in mm) due to backlash or any mechanical clearance.

Similarly, some displacement also occurs due to elastic deformation. Elastic deformation of the components is caused by the applied loads and forces carried by the linkage system. This deformation may be attributed to the applied writing load on the stylus, from frictional loads and especially from inertial loads, when the input is dynamic in nature. It is also important to remember here that since all the components of a mechanical system undergo elastic deformation due to applied load, it is distributed throughout the kinematic chain, while point sources cause backlash losses.

Assessing the total projected displacement loss at the output caused by elastic deformation, one gets the following equation:

$$Y_{tel} = \Sigma A Y_{el}$$

Here, Y_{tel} is the total projected displacement loss (in mm) due to backlash or clearances provided, A is the mechanical amplification or gain, and Y_{el} is the lost motion (in mm) due to backlash or any mechanical clearance.

The total projected displacement loss, Y_{pdl}, is given by the following equation:

$$Y_{pdl} = Y_{tbl} + Y_{tel} = \Sigma A Y_{bl} + \Sigma A Y_{el}$$

13.5.6 Tolerance Problems

One of the inherent problems of any mechanical system involving relative motion is the dimensional tolerance that needs to be provided in order to accommodate manufacturing errors. Further, these tolerances are inevitable because of the necessity of obtaining the required mechanical fits, providing space for lubrication, and allowing thermal expansion of components. These tolerances also cause lost motion. In order to minimize the effect of lost motion due to dimensional tolerance, the tolerance range has to be kept at a minimum level. However, it is to be emphasized here that lost motion due to tolerances cannot be totally eliminated.

13.5.7 Temperature Problems

One of the main attributes of any ideal measuring system is to only respond to the designed signal and ignore all other signals. Temperature variations adversely affect the operation of the measuring system and hence the concept of an ideal measurement has never been completely achieved. It is extremely difficult to maintain a constant-temperature environmental condition for a general-purpose measuring system. The only option is to accept the effects due to temperature variations and hence, methods to compensate temperature variations need to be devised.

Changes in dimensions and physical properties, both elastic and electrical, are dependent on temperature variations, which result in deviations known as *zero shift* and *scale error*.

Whenever a change occurs in the output at the no-input condition, it is referred to as zero shift. A zero shift is chiefly caused by temperature variations. It is a consequence of expansion and contraction due to changes in temperature, which results in linear dimensional changes. Zero indication is normally made on the output scale to correspond to the no-input condition, for most of the applications.

A very common example is setting the spring scales to zero at the no-input condition. Consider an empty pan of the weighing scale. If there is any temperature variation after the scale has been adjusted to zero, then the no-load reading will be altered. This change, which is due to the differential dimensional change between spring and scale, is termed a *zero shift*.

Temperature, especially when resilient load-carrying members are involved, affects scale calibration. Temperature variations alter the coil and wire diameters of the spring, and so does the modulus of elasticity of the spring material. The spring constant would change because of the temperature variations. This results in changed load–deflection calibration. This effect is referred to as *scale error*. Various methods can be employed in order to limit temperature errors:

1. Minimize temperature errors by proper and careful selection of materials and range of operating temperatures. The main reason for the occurrence of temperature errors is thermal expansion. When simple motion transmitting elements are considered, only thermal expansion causes temperature errors. Temperature errors are also caused when thermal expansion combines with modulus change when calibrated resilient transducer elements are considered. In case of electric resistance transducers, thermal expansion combines with resistivity change to cause temperature errors. In each of these cases, temperature errors can be minimized by appropriately choosing materials having low temperature coefficients. While selecting such materials, one needs to keep in mind that other requisite characteristics such as higher strength, low cost, and resistance to corrosion will not always be associated with minimum temperature coefficients. Hence, a compromise needs to be made.

2. Provide compensation by balancing the elements comprising inversely reacting elements or effects. This depends on the type of measurement system employed. In case of mechanical systems, a composite construction can be used to provide adequate compensation. A typical example is the composite construction of a balance wheel in a watch or clock. With the rise in temperature, the modulus of the spring material reduces and the moment of inertia of the wheel increases. The reason for this may be attributed to thermal expansion, which results in the slowing down of the watch. A bimetal element having appropriate features can be incorporated into the rim of the wheel to counter these effects such that the moment of inertia decreases with temperature so that it is enough to compensate for both expansion of the wheel spokes and change in modulus of the spring. When electrical systems are employed, compensation may be provided in the circuitry itself. Thermistors and resistance-type strain gauges are examples of this type.

3. Control temperature such that temperature problem is eliminated.

13.6 ADVANTAGES OF ELECTRICAL INTERMEDIATE MODIFYING DEVICES

From the preceding discussions, it is now clear that in mechanical systems, transient response characteristics, which are essential for dynamic measurement, suffer due to friction and inertia.

Further, mechanical systems become bulky and heavy when attempts are made to improve them or to provide them with required rigidity. Although hydraulic and pneumatic systems can be used to obtain higher signal power, these cannot be applied for quick-response control applications. In other words, hydraulic and pneumatic systems are suitable only for slow-response control applications. Hence, electrical systems are preferred as they possess the following merits:

1. Amplification or attenuation can be realized easily. Increased power output can be obtained by employing power amplifiers, which is not possible in mechanical systems because mechanical counterparts for power amplifiers are not available.
2. Effects of mass inertia and friction are minimized or are almost negligible.
3. An output power of almost any magnitude can be provided.
4. Remote indication of recording is possible. Remote telemetry and control are essential aspects of the aerospace R&D.
5. Transducers are commonly susceptible to miniaturization, especially in integrated circuits, which are being extensively employed in the field of instrumentation.

13.7 ELECTRICAL INTERMEDIATE MODIFYING DEVICES

One of the major functions of intermediate modifying devices is to transduce mechanical inputs into analogous electrical signals. In addition, these signals will be modified or conditioned such that they are capable of driving indicators and recorders in the terminating stage. In the intermediate modifying stage, amplification of either voltage or power, or both can be accomplished, depending on the requirement of the terminating stage. If the terminating device is an indicator, then amplification of voltage will suffice. Amplification of power is essential when a recorder is employed for driving a terminating device.

13.7.1 Input Circuitry

As discussed in Section 13.4.3, there are two types of electrical transducer devices: passive transducers, which require an auxiliary power source for energy, and active transducers, which are self-generating or powering and do not require any external power source. Simple bonded wire strain gauge and piezoelectric accelerometer are examples of passive and active transducers, respectively. Further, active transducers require minimum circuitry for the necessary transduction, whereas some special arrangements need to be provided when passive transducers are employed. The operating principle of the passive transducer used determines the types of arrangements to be provided. Broadly, the most common forms of input circuitry employed in transduction are as follows:

1. Simple current-sensitive circuits
2. Ballast circuits
3. Voltage-dividing circuits
4. Voltage-balancing potentiometer circuits

13.7.2 Simple Current-sensitive Circuits

A simple current-sensitive circuit shown in Figure 13.8 comprises a transducer, which may be in any one of the different forms of variable resistance elements. Let kR_t be the transducer resistance. R_t represents the maximum value of the transducer resistance and k represents a factor that depends on the magnitude of the input signal and varies between 0% and 100%. If a sliding contact resistor is used as a transducer element, the value of k varies through the complete range (0–100%) of the contact resistor. Let R_r represent the remaining resistance of the circuit and i_0 be the current flowing through the circuit.

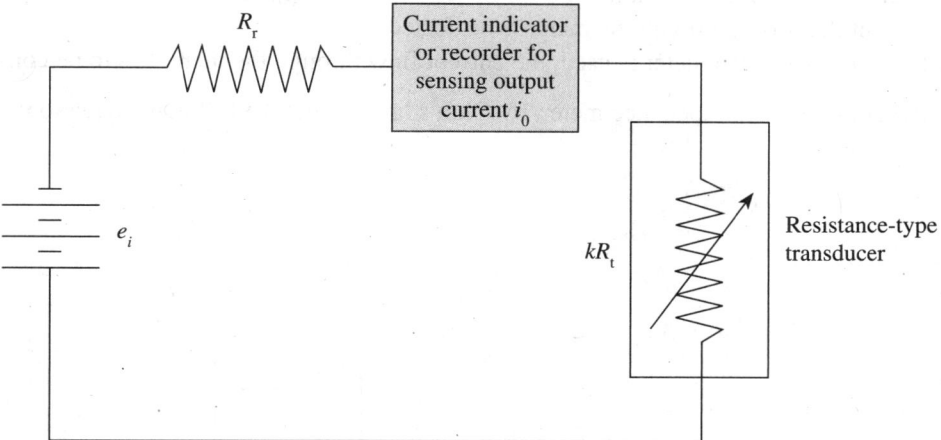

Fig. 13.8 Simple current-sensitive circuit

Using Ohm's law, the current indicated by the read-out circuit is given by the following equation:

$$i_0 = \frac{e_i}{kR_t + R_r}$$

This equation can be rewritten as follows:

$$\frac{i_0}{i_{max}} = \frac{i_0 R_r}{e_i}$$

$$= \frac{1}{1 + \left(\dfrac{R_t}{R_r}\right)k}$$

It can be observed here that the maximum current flows when $k = 0$, that is, when the current will be e_i/R_r. Further, the output variation or sensitivity is greater for higher value of transducer resistance R_t relative to R_r. It is to be mentioned here that since the output is a function of i_{max}, which in turn depends on e_i, the driving voltage has to be carefully controlled in order to maintain calibration. A drawback of this circuit is that the relationship between input and output is non-linear.

13.7.3 Ballast Circuit

A ballast circuit is a variation of the current-sensitive circuit (Fig. 13.9). A voltage-sensitive device is placed across the transducer in the circuit instead of a current-sensitive recorder or indicator. A ballast resistor R_b is introduced into the circuit. It is important to note that in the absence of R_b, the indicator will not indicate any change with variation in R_t and always show full source voltage. It is thus necessary to incorporate some value of resistance in the circuit to ensure its proper functioning. Depending on the impedance of the meter, two situations may arise:

1. In case some form of vacuum tube voltmeter, which has high impedance, was employed, the current that flows through the meter is neglected.
2. If a low impedance meter is used, the current flow through the meter has to be considered.

Assuming a high impedance meter, by Ohm's law, we get the following equation:

$$i = \frac{e_i}{R_b + kR_t}$$

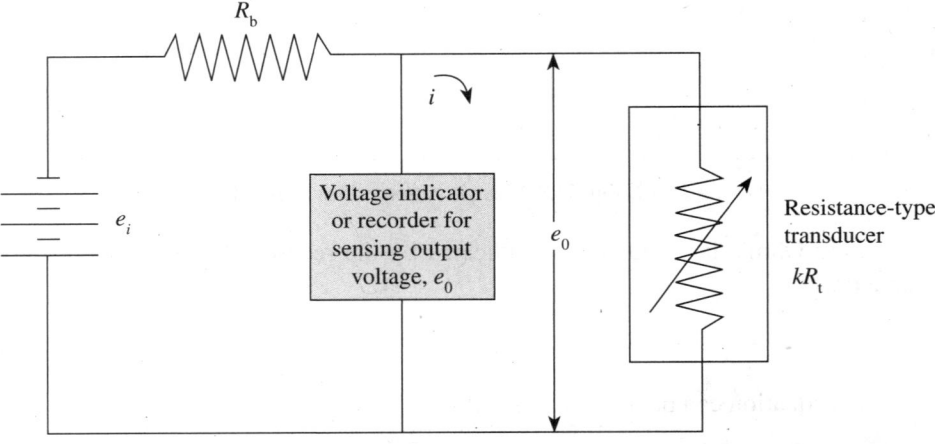

Fig. 13.9 Ballast circuit

Let e_0 be the voltage across kR_t (transducer), which is indicated or recorded by the output device. Then, the following equation holds true:

$$e_0 = i(kR_t) = \frac{e_i kR_t}{R_b + kR_t}$$

This equation can be rewritten as follows:

$$\frac{e_0}{e_i} = \frac{kR_t / R_b}{1 + (kR_t / R_b)}$$

For any given circuit, kR_t/R_b and e_0/e_i are the measures of input and output, respectively. Sensitivity or the ratio of change in output to change in input can be expressed as follows:

$$\eta = \frac{de_0}{dk} = \frac{e_i R_b R_t}{(R_b + kR_t)^2}$$

If the ballast resistance of various values is incorporated, we can change the sensitivity, which implies that there is an optimum value of R_b for the required sensitivity. We can obtain this value by differentiating with respect to R_b:

$$\frac{d\eta}{dR_b} = \frac{e_i R_t (kR_t - R_b)}{(R_b + kR_t)^3}$$

This derivative will be zero under the following two conditions:
1. Minimum sensitivity results when the ballast resistance $R_b = \infty$.
2. Maximum sensitivity results when the ballast resistance $R_b = kR_t$.

From these points it can be observed that full range can be utilized, and R_b, a constant value, cannot always be fixed at KR_t, a variable, and hence there must be a compromise. However, R_b may be chosen such that maximum sensitivity is obtained for a specific point in the range by equating its value to kR_t. A ballast circuit is sometimes used for dynamic measurement applications of resistance-type strain gauges, wherein the change in resistance is very small in comparison with the total gauge resistance. It is important to note here that this relationship also points out that an optimum value is obtained when the ballast resistance is equal to the gauge resistance. The major advantage of a voltage-sensitive circuit when compared with a current-sensitive circuit is that it is much simpler to carry out voltage measurement than current measurement. One of the drawbacks is that very careful control of voltage is essential, since a percentage variation in supply voltage e_i yields a greater change in the output than does a similar change in the value of k. The other disadvantage is that the relationship between output and input is non-linear.

13.7.4 Electronic Amplifiers

From the preceding discussions, it is clear that due to the inherent problems associated with mechanical intermediate devices, electrical or electronic systems are used to obtain the required amplification. Invariably, in the circuitry some form of amplification is provided during mechanical measurements. One of the factors that distinguishes electronic devices from electrical devices is assumed from the fact that, in some part of the electronic circuitry, electrons are made to flow through space in the absence of a physical conductor, which, in other words, implies the use of vacuum tubes. With rapid technological advancement, solid-state devices, diodes, transistors, etc., have come into existence, giving a wider perspective to the word electronics. The performance of amplifiers is assessed by good dynamic responses, minimum loading effects, no zero drift, and minimum noise.

The principle employed in the working of vacuum tubes is based on the fact that when a cathode is heated electrons are emitted. Emitted electrons get attracted to a positively charged plate, resulting in a flow of current through the plate circuit. A third element, called a grid, can be used to control the flow of current. The grid is introduced between the cathode and the plate such that it is duly charged negatively relative to the cathode. The negative voltage on the grid is termed bias.

The current flow in the plate circuit including the amplifier load can be regulated by the variations in the charge on the grid supplied by the input signal. Figure 13.10 represents a single-stage amplifier in its simplest form. It can be seen from the figure that A heats the filament, which in turn heats the cathode; B represents the plate supply, and the required bias voltage is supplied by C. Normally, in an amplifier, a common supply makes use of voltage dividers or dropping resistors to obtain various voltages.

If greater amplification is required, tubes with more elements can be incorporated; stages can be interconnected in such a way that the load for the input stage is a second stage, and so on.

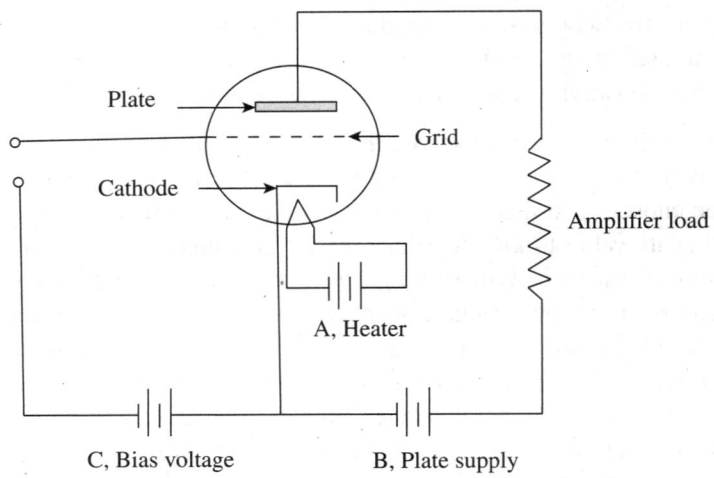

Fig. 13.10   Single-stage amplifier circuit

13.7.5 Telemetry

Telemetry is derived from the Greek words *tele* and *metron*, meaning remote and measure, respectively. Telemetry is measurement made at a distance, as suggested by its name. Distant measurement is usually performed using radio, hypersonic, or infrared systems and is hence referred to as wireless data communication. Data can also be communicated via telephone, computer network, optical link, etc. Telemetry finds wide application in allied fields such as meteorology, space science, water management, defence, space and resource exploration, rocketry, military intelligence, agriculture, medical science, and transportation, among others. It is a vital component of the intermediate measurement phase, especially in flight testing, and missile and space telemetry systems.

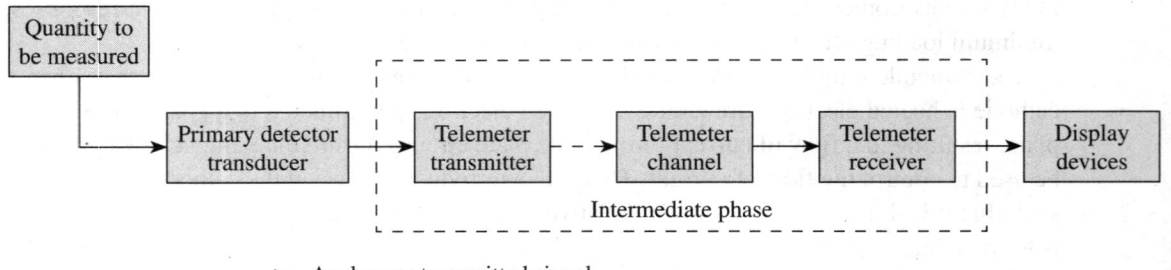

Fig. 13.11   General telemetering system

Figure 13.11 depicts a block diagram of a general telemetering system. The intermediate phase comprises three important constituents: telemeter transmitter, telemeter channel, and telemeter receiver. The telemeter transmitter converts the output of a primary detector into an analogous signal, which is transmitted over the telemeter channel. The telemeter receiver, which is placed at a remote location, converts the transmitted signal into an appropriate quantity. The functions of the primary detector transducer and terminating devices remain the same as those in a generalized measurement system.

Figure 13.12 represents a basic telemetering system. A telemetering system normally uses sub-carrier oscillators (SCOs). Frequencies of an SCO are regulated by the input signals through suitable transducer elements. The frequency of each SCO is modulated by the magnitude of the input signal, and thus a range of audio frequencies can be used. Since it is difficult and expensive to maintain a separate path for each source of information acquired, a multiplexer or commutator is used. Outputs obtained from various SCOs are then coded, merged, and transferred to a phase-modulated transmitter, which in turn relays the mixed information to the remote receiving station. Each SCO can handle multiple input data acquired by the different transducers.

Pulse code modulation (PCM) is the preferred telemetry format as it has many advantages, such as multiple data handling, integration of digital data, improved accuracy, and noise reduction, when compared to pulse amplitude modulation. After each signal is coded in PCM format, a set of these signals are time-division multiplexed into a single SCO. Care must be taken such that the time allotted for one cycle during time-division multiplexing is much smaller when compared to the rate of change of each input signal. This prevents loss of information. A unique set of synchronization pulses is added to identify and decommutate the original inputs and their values.

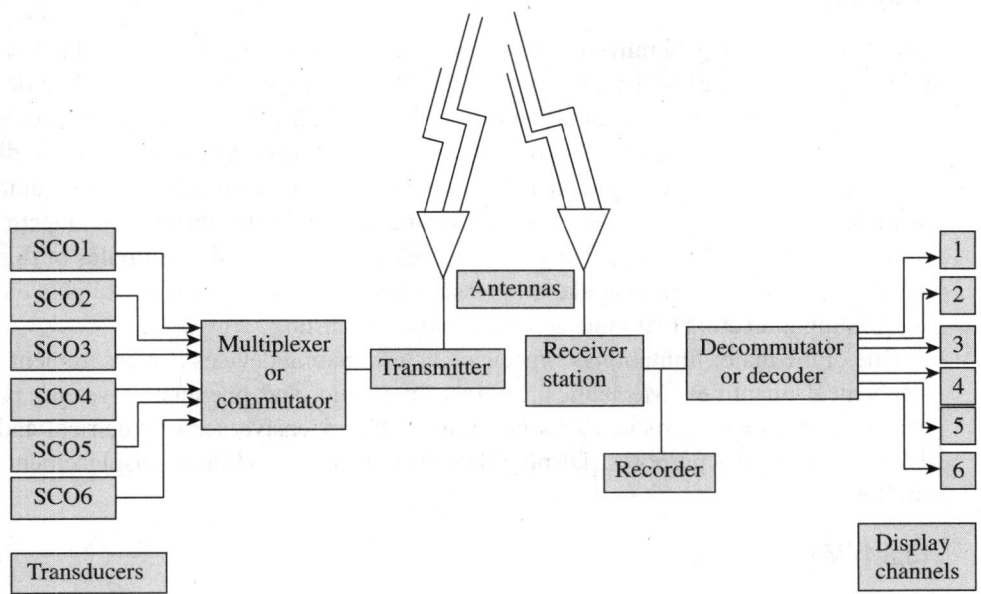

Fig. 13.12   Basic telemetering system

The information transmitted through the various SCOs is processed at the receiving station. The various sub-carrier frequencies are segregated and decoded by employing a decommutator, demultiplexer, or a discriminating circuit. The separated signals are distributed to appropriate output channels. It is important to note that the information or test data acquired is first recorded so that they are not lost. Further, it is very much essential to synchronize the information that is coming out of the multiplexer exactly with the decommutator, since time-division multiplexing is employed while collecting the data, which is based on precise timing. This ensures proper information flow, and there is no mix up of information; for example, information on temperature is not interpreted as pressure information.

There are certain advantages and disadvantages associated with telemetering systems. However, the advantages outweigh the disadvantages. The advantages are as follows:

1. Many channels can be monitored continuously without human intervention or attention.
2. In case of any destruction of the telemetry system, the entire telemetered data can be recovered, since the complete data, up to the final moment is recorded. This is very important, especially in applications such as missile testing where the test piece may not be recoverable.
3. There is no time limit for practical recording of data.
4. Miniaturization of the components is possible and hence low weight.
5. Recognition of surpassing safety limits is instantaneous and corrective action can be taken.

Some of the disadvantages are as follows:

1. The telemetering system is complex in nature and costly.
2. There is a very high chance of picking up some unwanted signals and hence more noise.
3. Extra processing of data, which is essential, may lead to a greater possibility of error.

13.8 TERMINATING DEVICES

The final stage of a generalized measurement system is the presentation of the processed data. It is an important part of the system wherein the human observer or controlling device should be able to comprehend the output and interpret it suitably. The data can be presented by the terminating devices, usually in two ways, either by relative displacement or in digital form. Displacement is indicated by scale and pointer, light beam and scale, liquid column and scale, pen tracing on a chart, etc. An odometer in an automobile speedometer, an electronic decade counter, a rotating drum mechanical counter, etc., are some of the examples of digital display devices. In some applications like oil pressure lights in automobiles, pilot lights on equipment are limiting indicators that indicate either on/yes or off/no.

One of the major limitations of mechanical terminating devices is measurement of dynamic mechanical quantities. Mechanical, optical, hydraulic, and pneumatic systems possess very poor response characteristics. This necessitates the extensive use of electrical and electronic devices for display purposes. Display devices that indicate relative displacement are widely employed.

13.8.1 Meter Indicators

Meter indicators are generally preferred for measurements involving static and steady-state dynamic indications. These are not generally applicable for measurements that are transient in nature due to the high inertia of the meter movement. Meter indicators can be categorized as follows:

1. Simple D'Arsonval type to measure current or voltage
2. Volt–Ohm milliammeters or multimeters
3. Vacuum tube voltmeters

Simple D'Arsonval Type to Measure Current or Voltage

Professor Jacques-Arsène D'Arsonval, a French Physicist, developed the D'Arsonval movement in 1882, which is most widely employed in electric meters for the measurement of current or voltage. In most instruments, the permanent-magnet moving-coil meter movement, which is known as D'Arsonval movement, is the basic movement. It is a current-sensitive device and hence current will flow, regardless of whether it is used as a voltmeter or a current meter. If the current flow is smaller, the loading on the circuit used for measurement will be lower, for most of the applications.

It essentially consists of the following components: permanent magnet, coil, hair springs, pointer, and scale. The coil assembly is mounted on a shaft, which in turn is hinged on a hair spring, as shown in Fig. 13.13. Rotation of the shaft is regulated by two hair springs fixed on either end of the shaft. The rotation is due to the resistance offered by the springs to the movement of the coil when current passes through it. Apart from making electrical connections to the coil, the hair springs are also used to make the needle return to its original position, if there is no current in the circuit. The entire coil assembly is placed in a magnetic field.

To concentrate the magnetic fields, initially the iron core is placed inside the coil. The curved pole pieces are then attached to the magnet to ascertain if the torque increases as the current increases. This means that electromagnetic energy is used to move the meter. In other words, as the electric current is made to flow through the coil, the two magnetic fields interact, thus producing a torque on the pivoted assembly. This results in the displacement of the pointer on the scale, which is calibrated in terms of electric current.

In a DC voltmeter circuit, a series resister, called a multiplier, is employed to limit the flow of current through the meter. The total voltmeter resistance should be much greater than the resistance of the circuit under consideration, in order to minimize the circuit loading.

In case of a DC current meter, a shunt resistor is used to keep the resistance of the meter at a minimum level. In order to achieve higher sensitivity, it is essential to obtain meter movements that provide larger deflections for the given current flow through the meter for both these cases.

Volt–Ohm Milliammeters or Multimeters

The typical volt–ohm milliammeter employs switching provisions for connecting multipliers, shunt resistors, and rectifier circuits. Further, using an internal battery energy source, resistance can be measured.

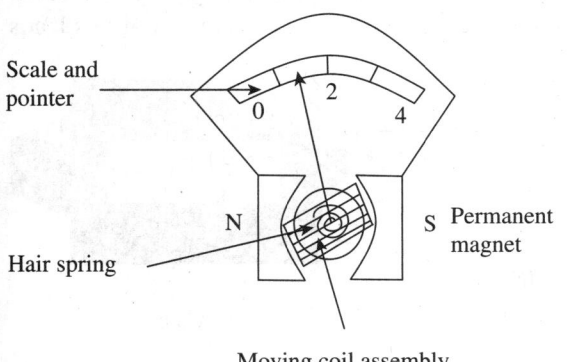

Fig. 13.13 D'Arsonval meter

The current flowing through the resistor can easily be determined by switching over to the ohm-meter function, wherein the leads can be connected to the unknown resistance that causes the meter movement. The advantage is that direct measurement is possible since the current flow indication is calibrated in terms of resistance.

Vacuum Tube Voltmeters

The vacuum tube voltmeter, popularly known as VTVM, is basically used to measure voltage. It comprises one or more electronic tubes in its circuitry, which is useful in the amplification of the input and for rectification. To control the grid of the tube, proper multiplier resistors are employed through which the AC or DC input is given to the meter. This results in a change in the plate current, which is indicated by a proportional meter deflection. In case the input given to the VTVM is from an AC source, a rectifier leads the meter.

The input resistance of a VTVM is usually very high, when compared to that of a simple meter, due to high grid circuit resistance. This minimizes the loading of the signal source. However, this advantage is only to a limited extent because it gives the instrument the capability of sensing very low energy voltages, which makes it vulnerable to various electrostatic voltages or noises. Although shielding can prevent this unwarranted picking up of signals, it is not preferred. Another advantage of a VTVM, apart from high impedance, is that amplification can be obtained without affecting sensitivity. This clearly permits the use of relatively rugged meter movements.

13.8.2 Mechanical Counters

Mechanical counters are employed in traditional automobile odometers. These decade drum-type counters can be directly coupled to a rotating input shaft. These may also be actuated by using linkage and ratchet and pawl mechanisms, which provide oscillating or reciprocating motions. The use of solenoids for actuating mechanical counters by generating electrical pulses is also prevalent. The energy for counters that are electrically actuated can be derived from a simple switch, a relay, photocells, or any source of pulse. These can supply a power of about several watts. A drum-type mechanical counter is shown in Fig. 13.14.

For high-speed counting applications, mechanical counters are inadequate. Electronic counters basically comprise decade counters, which count from 1 to 10. After every 10 counts, a pulse is generated, prompting it to change by a unit. In this way, one can count units, tens, hundreds, and so on, depending on the requirement. The advancement of this leads to multipurpose electronic counters, which are capable of measuring time periods, frequency, etc.

13.8.3 Cathode Ray Oscilloscope

A cathode ray oscilloscope (CRO) is probably the most popular terminating device, which finds various applications in mechanical measurements. A CRO is a voltage-sensitive device that is used to view,

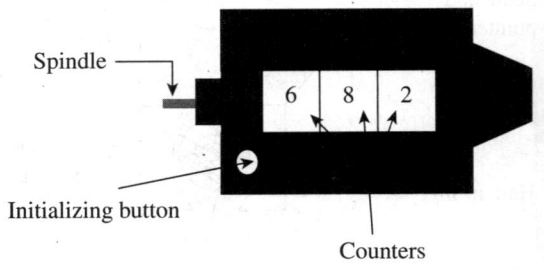

Fig. 13.14 Mechanical counter

measure, and analyse waveforms. Its advantage is that a beam of electrons with low inertia strikes the fluorescent screen, generating an image that can rapidly change with varying voltage inputs to the system. The cathode ray tube (CRT) is the basic functional unit of CRO. The working of CRT is explained in this section.

The electron gun assembly comprises a heater, a cathode, a control grid, and accelerating anodes. When a cathode is heated, electrons are generated. The grid provides the necessary control of flow of electrons that determines the spot intensity. The accelerating anodes, which are positively charged, provide the necessary striking velocity to the emitted electron stream.

The electron beam, after gaining necessary acceleration, passes through horizontal and vertical deflection plates, which provide the basic movements in the X and Y directions, respectively, depending on the voltages applied over these plates. In order to facilitate free movement of emitted electrons, vacuum is created within the tube. The different components of a CRT are shown in Fig. 13.15.

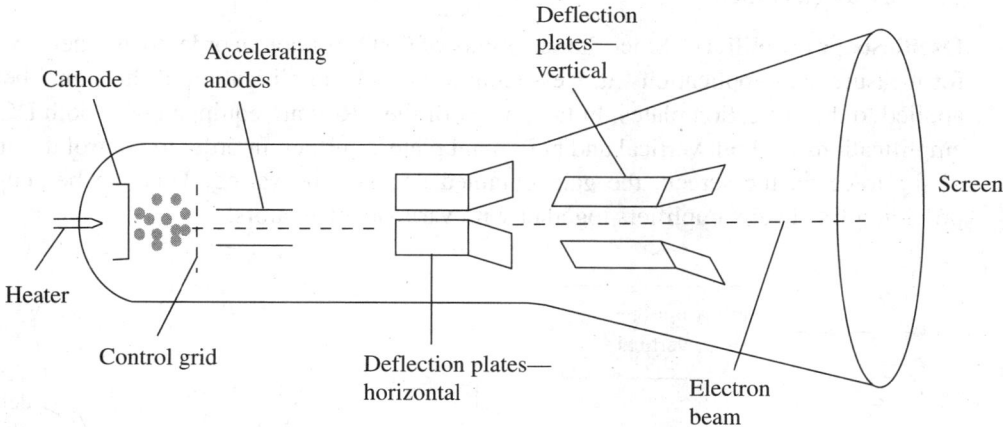

Fig. 13.15 Different components of a CRT

An internally generated sawtooth voltage shown in Fig. 13.16 is basically a time-based function, which is normally used in CRO as a horizontal input voltage. A time base on the X axis is generated from this ramp voltage where the electrons move from the extreme left to the extreme right in the middle of the screen. Hence, when the CRO is switched on, we always see a single horizontal line. The waveforms that are generally displayed by the CROs vary with time. This integral sawtooth is also known as sweep oscillator or sweep voltage. The voltage that is applied to the vertical deflection plates is the input that has to be analysed. This input voltage moves the electrons vertically with respect to the time base that is provided by the sawtooth wave form, thus forming an image. In order to obtain a stable or stationary image that is continuous, the frequency of the waveform has to be maintained at an optimally high level. Thus, CROs provide visual representation of the dynamic quantities. Further, by employing appropriate transducers, we can transduce other physical quantities such as current, strain, acceleration, and pressure into voltages to produce images.

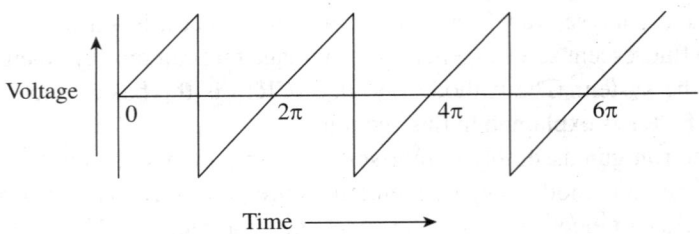

Fig. 13.16 Sawtooth waveform

A typical general-purpose CRO is illustrated in Fig. 13.17. Depending on the applications, there may be many variations in CROs.

Unique Features of CRO

The following are the unique features of a CRO, apart from the sawtooth oscillator, which has been discussed earlier.

Oscilloscope amplifiers Since the sensitivity of CROs is poor, in order to use them extensively for measurement applications, it is essential to provide amplification of the signal before it is applied to the deflection plates. In fact, many of the CROs are equipped with both DC and AC amplifications for both vertical and horizontal plates. Further, in order to control the amplitude of the trace on the screen, the gain obtained needs to be varied. This can be achieved by providing fixed gain amplifiers together with variable attenuators.

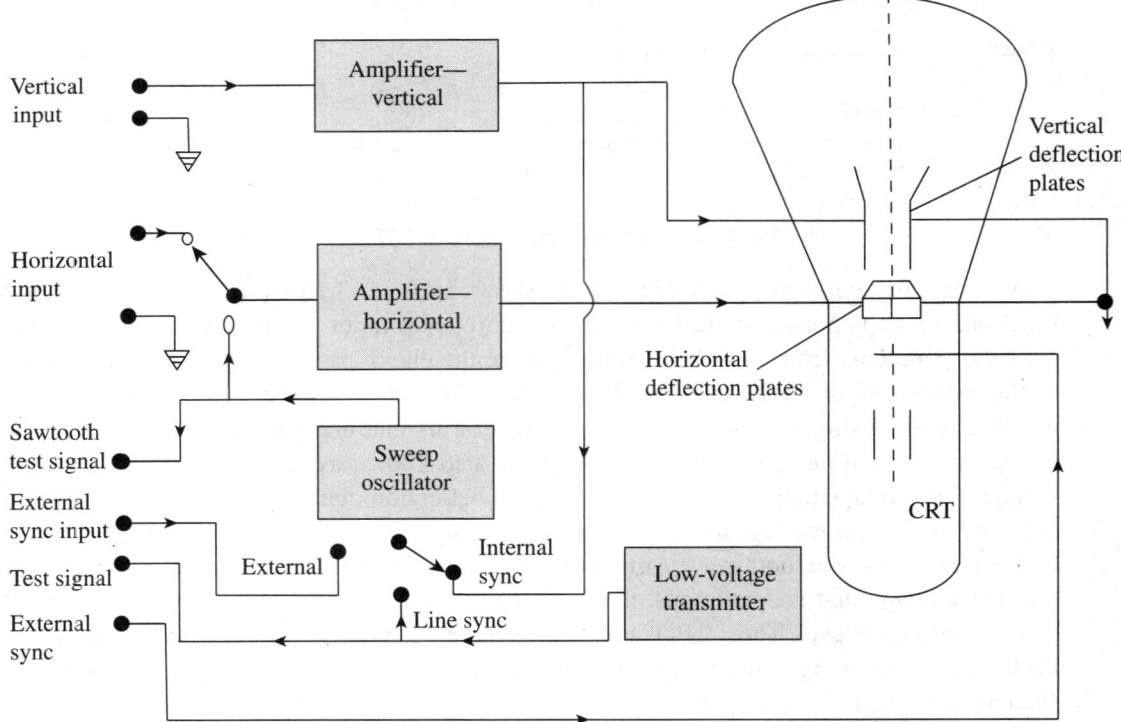

Fig. 13.17 Block diagram of a general-purpose CRO

Synchronization   We know that to obtain a stable or stationary image, which is continuous, it is essential to maintain the frequency of the waveform at an optimally high level. In other words, if the voltage to be applied to the vertical deflection plates (i.e., Y axis) is required to be fixed at, say, 60 Hz, in order to obtain a continuous image the sawtooth frequency applied to the X axis has to be fixed at 60 Hz; otherwise it would result in a distorted image.

However, irrespective of the type of sweep used, synchronization is a must in order to ensure a stationary image, since the input voltage is a time-dependent function. Synchronization is essential to avoid drifting of the image across the screen.

Intensity or Z-modulation   We know that the flow of electrons can be controlled by controlling the grid voltage. In order to obtain bright and dark shades in the fluorescent image generated by electron beams, positive and negative voltage can be correspondingly applied to the grid of the CRT. In addition, to modulate the brightness further, voltage from an external source can be used. This is achieved by providing a terminal in the front panel or a connection at the back of the instrument. This is termed intensity or Z-modulation.

External horizontal input   One need not essentially use a sweep oscillator for giving horizontal input. In such a case, provisions for input terminals in order to connect extra sources of voltage are made. This aids in comparing voltages, frequencies, and phase relationships.

13.8.4 Oscillographs

The D'Arsonval meter movement, which has been discussed in Section 13.8.1, is used as the working principle of oscillographs. The basic difference between a CRO and an oscillograph is that in case of CRO the output is visual in nature, whereas in case of oscillographs it is traced on a paper directly. In either case, the electrical signal is transformed into mechanical displacement. Another important difference between the two is with respect to the impedance. While the former has an input of high impedance, the latter is a low-impedance device. Two types of oscillographs are available.

Direct Writing-type Oscillographs

This essentially consists of a stylus that traces an image on the moving paper by establishing direct contact. The stylus used for the purpose of tracing or writing is current sensitive. Further, to enable smooth movement of the paper, a paper-transporting mechanism has to be provided. The recording may be realized through the use of ink, by moving the heated stylus on a treated paper, or by using a stylus and pressure-sensitive paper. The recording, which can be obtained by moving the paper under the stylus, and which is deflected by the input signal, is a time-based function of the input signal. Since recording is accomplished by the movement of stylus on paper, a frictional drag is created, which necessitates more torque.

Light Beam Oscillographs

This essentially comprises a paper drive mechanism, a galvanometer, and an optical system for transmitting current-sensitive galvanometer rotation to a displacement that can be recorded on a photographic film or paper. The galvanometer of the oscillograph works on the D'Arsonval principle. Depending on the input, which again depends on various parameters, mass moment of inertia of the coil is most important. Further, sensitivity of the oscillograph is a function of the number of turns in the coil. It is essential to keep the mass moment of inertia at a minimum level; thus, the actual design is a compromise.

13.8.5 *XY* Plotters

The *XY* plotters generate Cartesian graphs, which are obtained by two DC inputs, one in *X* and the other in *Y* axis. It basically consists of two self-balancing potentiometers, one to control the position of the paper and the other to control the position of the pen. In some *XY* plotters, the paper does not move while two self-balancing potentiometers are used to obtain the movement of the pen in the *X* and *Y* directions.

In *XY* plotters, the output is generated by plotting one emf as a function of the other emf. Further, the emf, which is the output signal of transducers, may be a value of displacement, force, pressure, strain, or any other parameter. Thus, by transducing appropriate parameters, plots for any combination of parameters may be obtained. A pair of servo systems enables the movement of the recording pen in both the axes. The paper chart remains stationary.

A QUICK OVERVIEW

- Depending on the information sensed and delivered by the sensor, transfer efficiency is defined as follows:

$$\text{Transfer efficiency} = \frac{I_{del}}{I_{sen}}$$

Here, I_{del} is the information delivered by the pickup device and I_{sen} is information sensed by the pickup device.

- Active transducers are of a self-generating type wherein they develop their own voltage or current output. They do not need any auxiliary power source to produce output. Passive transducers derive power required for transduction from an auxiliary source of power. A part of the power required for generating the output is derived from the physical quantity being measured.
- When a measuring device measures and transforms a non-electrical variable into an electrical variable, it is called a transducer. A thermocouple, which is used to measure temperature, radiation, and heat flow, is an example of a transducer. If an electrical quantity is transformed into a non-electrical quantity, it is termed an inverse transducer.

- A null-type device works on the principle of maintaining a zero deflection by applying an appropriate known effect that opposes the one generated by the measured quantity. This type of transducer requires a detector capable of detecting an imbalance and a means to restore a balance either manually or automatically. A deflection-type transducer works based on the principle that the measured quantity produces a physical effect that stimulates an analogous effect in the opposite direction in some parts of the measuring instrument. There is an increase in the opposing effect until a balance is restored, at which stage the measurement of deflection is carried out.
- Reflected frictional amplification, reflected inertial amplification, amplification of backlash and elastic deformation, tolerance, and temperature problems are the major problems associated with the mechanical systems.
- Telemetry is measurement made at a distance, as suggested by its name. Distant measurement is usually performed using radio, hypersonic, or infrared systems and hence referred to as wireless data communications. Data can also be

communicated via telephone, computer network, optical link, etc.

- In most of the instruments, the permanent-magnet moving-coil meter movement, which is known as D'Arsonval movement, is the basic movement. It is a current-sensitive device and hence current will flow, regardless of whether it is used as a voltmeter or a current meter.

- A cathode ray oscilloscope (CRO) is probably the most popular terminating device, which finds various applications in mechanical measurements. A CRO is a voltage-sensitive device that is used to view, measure, and analyse waveforms.

- Due to poor sensitivity of the CROs, it is essential to provide amplification of the signal (using an oscillation amplifier) before it is applied to the deflection plates in order to use them extensively for measurement applications.

- Synchronization is a must in order to ensure a stationary image, since the input voltage is a time-dependent function irrespective of the type of sweep employed. Synchronization is essential to avoid drifting of the image across the screen.

- In order to obtain bright and dark shades in the fluorescent image generated by electron beams, positive and negative voltage can be applied correspondingly to the grid of the cathode ray tube. In order to modulate the brightness further, voltage from an external source can be used. This is achieved by providing a terminal in the front panel or a connection at the back of the instrument. This is termed intensity or Z-modulation.

MULTIPLE-CHOICE QUESTIONS

1. Which of the following transducers is a passive transducer?
   (a) Photovoltaic cell
   (b) Piezoelectric
   (c) Thermocouple
   (d) Thermistor

2. Which of the following is an inverse transducer?
   (a) Piezoelectric crystal
   (b) Thermocouple
   (c) Capacitive transducer
   (d) LVDT

3. Thermocouple is
   (a) a passive transducer
   (b) an inverse transducer
   (c) an active transducer
   (d) both active and passive transducers

4. A consequence of backlash is
   (a) gain in amplification
   (b) lost motion
   (c) conditioned signal
   (d) attenuation of signal

5. Zero shift is the result of
   (a) temperature variations
   (b) backlash
   (c) hysteresis

   (d) tolerance problems

6. Temperature problems can be limited by
   (a) selection of proper materials
   (b) providing temperature compensation
   (c) proper control of temperature
   (d) all of these

7. The function of a multiplexer is
   (a) mixing and transferring of information
   (b) performing multiple functions
   (c) calibrating telemetry system
   (d) attenuation of signals

8. PCM stands for
   (a) pulse code multiplexer
   (b) pulse code modulation
   (c) pulse character multiplexer
   (d) pulse course modulation

9. VTVM has
   (a) high impedance   (c) zero impedance
   (b) low impedance   (d) negative impedance

10. Synchronization is required
    (a) for obtaining a moving image
    (b) for amplifying the image
    (c) for not obtaining any image
    (d) for obtaining a stationary image

REVIEW QUESTIONS

1. Define transfer efficiency.
2. Classify transducers.
3. Distinguish between the following:
   (a) Active and passive transducers
   (b) Transducers and inverse transducers
4. With an example, explain a primary detector transducer stage.
5. Explain null- and deflection-type transducers.
6. List the quality attributes of transducers.
7. Explain the inherent problems associated with mechanical systems.
8. With a chart explain the typical operation of transduction in electrical, electronic, and self-generating transducers.
9. List the advantages of electrical transducers.
10. Explain a simple current-sensitive circuit.
11. With a neat circuit diagram, explain the working of a ballast circuit. In addition, derive an expression for the same.
12. Explain a single-stage electronic amplifier with a neat circuit.
13. What is telemetry? With the help of a neat block diagram, explain the working of a telemetry system.
14. With the help of a sketch, explain a simple D'Arsonval type to measure current or voltage.
15. Explain the working of a VTVM.
16. With a neat diagram, explain the working of mechanical counters.
17. Explain the working of a CRT with a neat figure.
18. With a block diagram, explain the working of a general-purpose CRO.
19. Explain the unique features of a CRO.
20. Write short notes on the following:
    (a) Direct writing-type oscillographs
    (b) Light beam oscillographs
    (c) *XY* plotters

Answers to Multiple-choice Questions

1. (d)	2. (a)	3. (c)	4. (b)	5. (a)	6. (d)	7. (a)	8. (b)
9. (a)	10. (d)						

Measurement of Force, Torque, and Strain

After studying this chapter, the reader will be able to

- understand the basics and describe methods of force, torque, and strain measurements
- explain the direct method of force measurement
- differentiate between the different types of elastic members for force measurements
- discuss the different electrical pressure transducers
- elucidate the various types of dynamometers used for torque measurements
- appreciate the different aspects of strain measurement systems
- explain gauge factor measurement
- describe temperature compensation

14.1 INTRODUCTION

We all know that force is defined as the product of mass and acceleration, as per Newton's second law of motion. Force is a vector quantity whose unit of measurement is Newton (N); 1 N is equivalent to the force required to accelerate a mass of 1 kg at a rate of 1 m/s^2, that is, 1 N = 1 kg × 1 m/s^2. The standards of mass and acceleration determine the standard of force. It is a well-known fact that the fundamental unit of mass is kilogram, which is traceable to the international prototype of mass. The international standard for mass is international kilogram, which is equivalent to the mass of a cylinder of platinum–iridium maintained at Sevres, France. Although acceleration is not a fundamental unit, it is derived from two other well-defined and internationally accepted fundamental quantities: length and time. The acceleration due to gravity g can be adopted as a suitable standard and is approximately equal to 9.81 m/s^2.

Torque is defined as the force, which, besides exerting effects along its line of action, may also exert a rotating force relative to any axis excepting those intersecting the line of action. It can also be termed as a couple or twisting force. It is determined by measuring the force at a known radius r and is expressed as $T = F \times r$. The metric unit of measurement of torque is Newton metre (N m), which is a force multiplied by the distance of this force from an axis of rotation. Torque is closely related to the measurement of force, and since force and length that define torque conform to international standards, separate standards are not required for torque.

14.2 MEASUREMENT OF FORCE

The methods for measuring force can be classified into two basic categories: direct and indirect. In case of direct methods, a direct comparison is made between an unknown force and the known gravitational force on a standard mass. For this purpose, a beam balance may be employed wherein masses are compared. In this case, the beam neither attenuates nor amplifies. Indirect comparison is made by a calibrated transducer that senses gravitational attraction or weight. Sometimes, the deformation due to a force applied on an elastic member is measured.

14.2.1 Direct Methods

Direct methods involve the comparison of an unknown force with a known gravitational force on the standard mass. A force is exerted on a body of mass m due to the earth's gravitational field, which can be represented by the following equation:

$$W = mg$$

Here m is the standard mass, g is the acceleration due to gravity, and W is the weight of the body.

It is imperative to know the values of mass and acceleration due to gravity accurately in order to determine the force acting on the body. With the help of an analytical balance, a direct comparison can be drawn between an unknown force and the gravitational force. The following section describes the working of an analytical balance.

Analytical Balance

An analytical balance, also known as an equal arm balance, is probably the simplest force-measuring system. As discussed in Section 14.2, an unknown force is directly compared with a known gravitational force. Comparison of masses is carried out by attaining some kind of beam balance by employing a null balance method. It is sufficient to only find out the magnitude since the unknown force and the gravitational force act in directions parallel to each other. The working principle is illustrated schematically in Fig. 14.1.

The rotation of the balance arm is about the knife edge point or fulcrum O. Figure 14.1 shows the balance in an unbalanced condition. The distance between the fulcrum and the centre of gravity point CG is d_G. Let W_B denote the weight of the balance arms and pointer; W_1 and W_2 are two weights acting on either side of the balance. When the two weights W_1 and W_2 are equal, angle θ will be zero, and hence the weight of the balance arms and pointer will

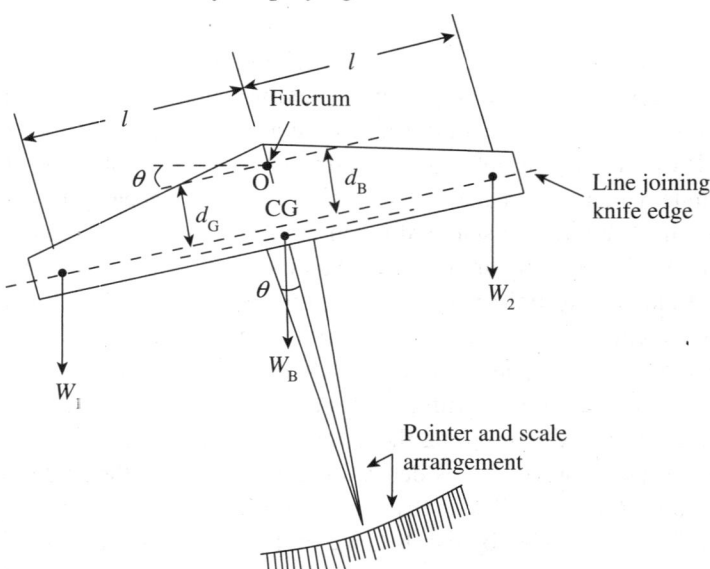

Fig. 14.1   Analytical balance

not influence the measurements.

Deflection per unit unbalance gives a measure of the sensitivity of the balance. The difference between the two weights, that is, $W_1 - W_2$, gives the unbalance. Let this difference be ΔW.

Hence, sensitivity $S = \dfrac{\theta}{(W_1 - W_2)} = \dfrac{\theta}{\Delta W}$ (14.1)

When the balance is at equilibrium, the following equation is realized:

$$W_1(l \cos \theta - d_B \sin \theta) = W_2 (l \cos \theta + d_B \sin \theta) + W_B\, d_G \sin \theta \qquad (14.2)$$

When the angle of deflection is small, $\sin \theta = \theta$ and $\cos \theta = 1$. For such small angles of deflection, Eq. (14.1) can be modified as follows:

$$W_1 (l - d_B \theta) = W_2 (l + d_B \theta) + W_B\, d_G\, \theta$$

Therefore, $\theta = \dfrac{l\,(W_1 - W_2)}{(W_1 + W_2)\, d_B + W_B\, d_G}$ (14.3)

Hence, sensitivity $S = \dfrac{\theta}{(W_1 - W_2)} = \dfrac{l}{(W_1 + W_2)\, d_B + W_B\, d_G}$ (14.4)

When in equilibrium, $W_1 = W_2 = W$.

$$S = \dfrac{l}{2W d_B + W_B\, d_G} \qquad (14.5)$$

If balances are constructed such that d_B, the distance between the fulcrum and the line joining the knife edges, is zero, then equation (14.5) becomes as follows:

$$S = \dfrac{l}{W_B\, d_G} \qquad (14.6)$$

Hence, it can clearly be seen from Eq. (14.6) that the sensitivity is independent of the applied weights W_1 and W_2. There may be errors in measurement due to the buoyancy effects of air exerted on the weights, which can be eliminated through proper care. In addition, to enhance sensitivity, it is necessary to increase l and decrease W_B and d_G, which again affect the stability. Hence, a trade-off needs to be made between the stability and sensitivity of the balance. It is pertinent to mention here that for optimal performance, the design and operation of the balance is crucial. The disadvantage associated with an analytical balance is that it necessitates the use of a set of weights that are at least equal to the maximum weight to be measured. However, it can be mentioned here that an equal-arm balance system is not appropriate for the measurement of large weights.

Platform Balance

For applications involving measurement of large weights, a platform balance or multiple-lever system is preferred. A platform balance is schematically represented in Fig. 14.2.

In a platform balance, two smaller weights, poise weight W_x and pan weight W_y, are used for the measurement of a large weight W. The initial zero balance can be set by an adjustable

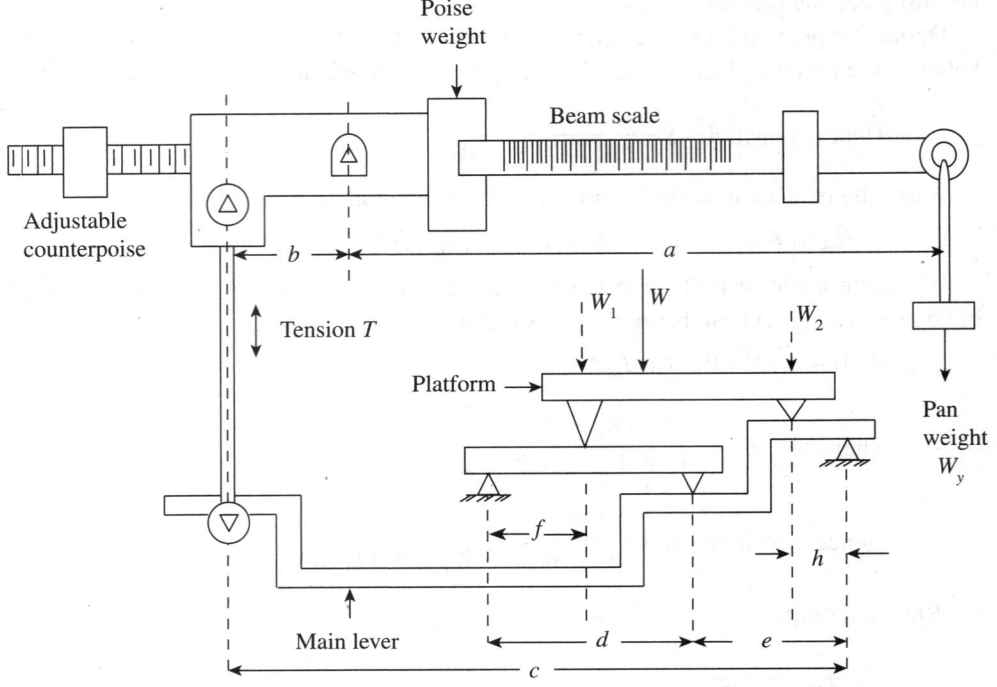

Fig. 14.2 Platform balance

counterpoise. The poise weight W_x is set to zero of the beam scale, and then the counterpoise is adjusted for obtaining the initial balance before applying the unknown weight W on the platform. For ease of analysis, it can be assumed here that the two weights W_1 and W_2 are substituted by W. Since W_x has already been adjusted to zero, the entire unknown weight W is balanced by the pan weight W_y.

It can be seen that $T \times b = W_y \times a$ (14.7)

and $T \times c = W_1(f/d)e + W_2 \times h$ (14.8)

The linkage proportion can be arranged such that $h/e = f/d$.

Hence, we have $T \times c = (W_1 + W_2)h = Wh$ (14.9)

It is clear from these discussions that weight W can be placed anywhere on the platform and its position with respect to the knife edges of the platform is not significant.

From Eqs (14.7) and (14.9), we obtain the following equation:

$$\frac{W_y a}{b} = \frac{Wh}{c}$$ (14.10)

This gives a new equation.

$$W = \frac{a}{b}\frac{c}{h}W_y = S\,W_y$$ (14.11)

where S is known as the scale multiplication ratio, which is given by another equation:

$$S = \frac{a}{b}\frac{c}{h} \tag{14.12}$$

The multiplication ratio gives an indication of the weight that should be applied to the pan in order to balance the weight on the platform. This means that if the multiplication factor is 100, then a weight of 1 kg applied on the pan can balance a weight of 100 kg placed on the platform. Let us assume that the beam is divided with a scale of m kg per scale division, then a poise movement on the y scale division has to produce the same result as a weight W_x placed on the pan at the end of the beam; hence, $W_x y = mya$. Therefore, we obtain the following equation:

$$m = W_x/a \tag{14.13}$$

The required scale divisions on the beam for any poise weight W_x is determined by this relationship. The length of the beam scale a is expressed in terms of scale divisions. The beam is balanced by appropriate combinations of pan weights and an adjustment of the poise weight along the calibrated beam scale.

14.3 ELASTIC MEMBERS

For determining the applied force, many force-measuring transducers employ various mechanical elastic members or their combinations. Application of a load or force on these members causes an analogous deflection. This deflection, which is usually linear, is measured either directly or indirectly by employing secondary transducers. This displacement is then converted into an electrical signal by the secondary transducers. The strain gauge is the most popular secondary transducer employed for the measurement of force. Figure 14.3 depicts the use of a strain gauge for the determination of applied force.

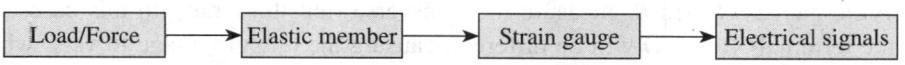

Fig. 14.3 Use of a strain gauge for force determination

The strain gauge operates on a simple principle. When force is applied on an elastic member, a steel cylindrical rod in this case, the dimensions undergo a change. If the strain gauge is bonded to the cylinder, the gauge is stretched or compressed, resulting in a change in its length and diameter. The dimension of the strain gauge changes due to a change in resistance. This change in resistance or output voltage of the strain gauge gives a measure of the applied force. Load cells and proving rings that employ strain gauges are the two most common instruments used for force measurement.

14.3.1 Load Cells

Elastic members are generally used for the measurement of force through displacement measurement. When an elastic member is combined with a strain gauge and used for the measurement of force, it is termed as a load cell. In load cells, elastic members act as primary transducers and strain gauges as secondary transducers. A load cell is used in an indirect method of force measurement where force or weight is converted into an electrical signal. Load cells are used extensively for the measurement of force.

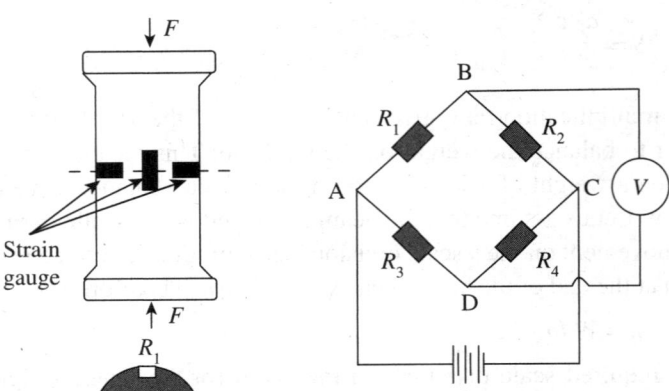

Fig. 14.4 Strain gauge arrangement for a load cell

A load cell comprises four strain gauges; two of these are used for measuring the longitudinal strain while the other two for measuring the transverse strain. The four strain gauges are mounted at 90° to each other, as shown in Fig. 14.4.

Two gauges experience tensile stresses while the other two are subjected to compressive stresses. At the no-load condition, resistance in all the four gauges will be same. The potential across the two terminals B and D are same. The Wheatstone bridge is now balanced and hence output voltage is zero When the specimen is stressed due to the applied force, the strain induced is measured by the gauges. Gauges R_1 and R_4 measure the longitudinal (compressive) strain, while gauges R_2 and R_3 measure the transverse (tensile) strain. In this case, voltages across the terminals B and D will be different, causing the output voltage to vary, which becomes a measure of the applied force upon calibration.

The longitudinal strain developed in the load cell is compressive in nature and given by the following relation:

$$\varepsilon_1 = -\frac{F}{AE}$$

Here, F is the force applied, A is the cross-sectional area, and E is the Young's modulus of elasticity.

Gauges 1 and 4 experience this strain. Strain gauges 2 and 3 would experience a strain given by the following relation:

$$\varepsilon_2 = -\frac{\gamma F}{AE}$$

Here, γ is the Poisson's ratio.

This arrangement of mounting of gauges compensates the influence of bending and temperature. In fact, a complete compensation is obtained if the gauges are mounted symmetrically.

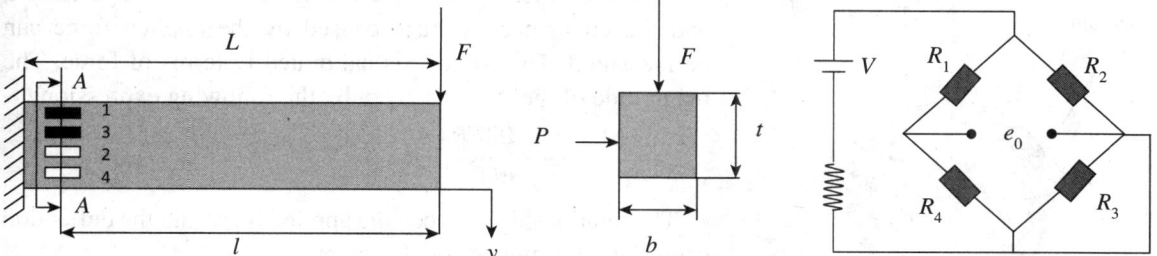

Fig. 14.5 Cantilever beam for force measurement

Fig. 14.6 Strain gauge arrangement for force measurement

14.3.2 Cantilever Beams

A cantilever beam is one of the simple devices used for the measurement of force. One end of the cantilever is fixed and a force F is applied on the free end, as shown in Fig. 14.5. Four strain gauges are bonded near the fixed end to measure the strain induced by the application of force. Two strain gauges are fixed on the top for measuring tensile strain $+\varepsilon$ and two at the bottom side for measuring compressive strain $-\varepsilon$. The strain developed at the fixed end is determined using the following equation:

$$\varepsilon = \frac{6Fl}{Ebt^2}$$

Here, l is the length of the beam, E is the Young's modulus of the material, b is the width of the beam, and t is the thickness of the cantilever.

In Fig. 14.6, strain gauges R_1 and R_3 measure tensile strain, whereas compressive strain is measured by R_2 and R_4. Strains measured by all the four gauges are of equal magnitude. Apart from the measurement of force, a load cell finds application in vehicle weighbridges and cutting tool force dynamometers.

14.3.3 Proving Rings

One of the most popular devices used for force measurement is the proving ring. In order to measure the displacement caused by the applied pressure, a displacement transducer is connected between the top and bottom of the ring. Measurement of the relative displacement gives a measure of the applied force. A proving ring can be employed for measuring the applied load/force, with deflection being measured using a precise micrometer, a linear variable differential transformer (LVDT), or a strain gauge. When compared to other devices, a proving ring develops more strain owing to its construction. A proving ring, which is made up of steel, can be used for measuring static loads, and is hence employed in the calibration of tensile testing machines. It can be employed over a wide range of loads (1.5 kN to 2 MN).

A proving ring comprises a circular ring having a rectangular cross-section, as shown in Fig. 14.7. It has a radius R, thickness t, and an axial width b. The proving ring may be subjected to either tensile or compressive forces across its diameters. The two ends between which force is measured are attached with structures. Four strain gauges are mounted on the walls of the proving ring, two on the inner walls, and two other on the outer walls. The applied force induces a strain (compressive) $-\varepsilon$ in gauges 2 and 4, while gauges 1 and 3 undergo tension

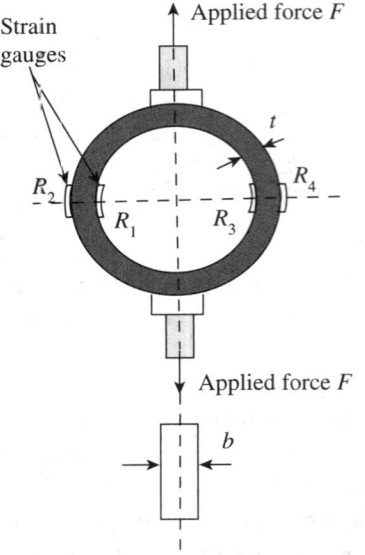

Strain gauges

Applied force F

R_2 R_1 R_3 R_4

t

Applied force F

b

Fig. 14.7 Proving ring

+ε. The four strain gauges are connected to the bridge circuit, and the unbalanced voltage caused by the applied force can be measured. This voltage is calibrated in terms of force. The magnitude of the strain is given by the following expression:

$$\varepsilon = \frac{1.08FR}{Ebt^2}$$

The relationship between the applied force and the deflection caused by the applied force is given by

$$\delta_y = \left(\frac{\pi}{2} - \frac{4}{\pi}\right)\frac{Fd^3}{16EI}$$

Here, E is the Young's modulus, I is the moment of inertia, F is the force, d is the outside diameter of the ring, and δ_y is the deflection.

14.3.4 Differential Transformers

An LVDT can be employed in a load cell to measure the applied force. The working of an LVDT has been discussed in detail in Chapter 16 (refer to Section 16.7.3). An LVDT can be used in conjunction with a primary transducer to measure force, pressure, weight, etc. It can generate an AC output voltage as a function of the displacement of the magnetic core caused by the applied force. The frictionless movement and low mass of the core provide advantages such as high resolution and low hysteresis, making the LVDT an obvious choice for dynamic measurement. LVDTs can be employed as alternatives to strain gauges or micrometers.

14.4 MEASUREMENT OF TORQUE

Measurement of torque is important for the following reasons:
1. It is necessary to obtain load information for the analysis of stress or deflection. In this case, torque T is determined by the measurement force F at a known radius r using the following relationship:
   $T = Fr$ (in N m)
2. Measurement of torque is essential in the determination of mechanical power. Mechanical power is nothing but the power required to operate a machine or the power developed by the machine, and is computed using the following relation:
   $P = 2\pi NT$
   Here, N is the angular speed in resolution per second.
   Torque-measuring devices employed for this purpose are popularly known as dynamometers. Dynamometers are used to measure torque in internal combustion machines, small steam turbines, pumps, compressors, etc.
3. Measurement of torque is important for evaluating the performance characteristics of machines.
   The nature of the machine to be tested determines the kind of dynamometer to be used for torque measurement. If the machine generates power, the dynamometer employed must

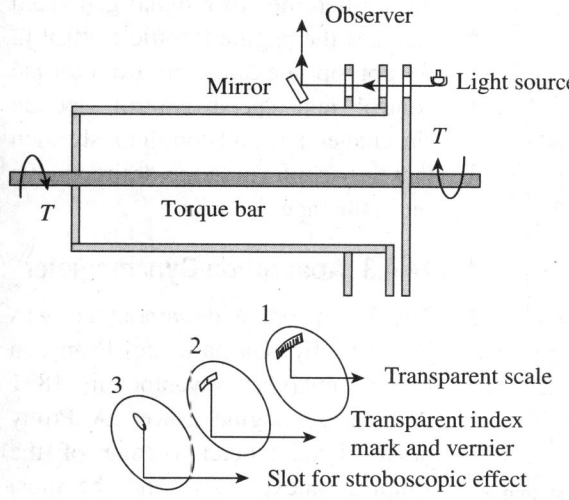

Fig. 14.8 Torsion-bar dynamometer

have the capability of absorbing that power. These dynamometers are called absorption dynamometers, which are particularly useful for measuring the power or torque developed by power sources such as engines or electric motors.

In case the machine is a power absorber, the dynamometer must have the capacity to drive it; such a dynamometer is known as a driving dynamometer. Driving dynamometers are therefore useful in determining performance characteristics of devices such as pumps and compressors. The third type is the transmission dynamometer. These are passive devices placed at an appropriate location within the machine or between machines. They are used to sense torque at that particular location. These are sometimes known as torque meters.

14.4.1 Torsion-bar Dynamometer

Elastic deflection of the transmitting element may be used for the measurement of torque, which can be achieved by measuring either a gross motion or a unit strain. The main problem associated with either case is the difficulty in reading the deflection of the rotating shaft. A torsion-bar dynamometer is also known as a torsion-bar torque meter, which employs optical methods for deflection measurement, as shown in Fig. 14.8.

Calibrated scales are used to read the relative angular displacement of the two sections of the torsion bar. This is possible because of the stroboscopic effect of intermittent viewing and persistence of vision. Transmission dynamometers, which employ this principle, are available in ranges up to 60,000 m kgf and 50,000 r/min, having an error of ±0.25%.

Replacing the scales on disks 1 and 2 with sectored disks, which are alternately transparent and opaque sectors, and the human eye with an electro-optical transducer, a version having an electrical output is obtained. When there is no torque, the sectored disks are positioned to give a 50% light transmission area. The area of proportionality increases with positive torque and decreases with negative torque, thus giving a linear and direction-sensitive electric output.

14.4.2 Servo-controlled Dynamometer

Torque and speed are measured under actual driving conditions of an automobile engine; tape recordings of such an exercise of an engine are obtained and then simulated under laboratory conditions. Engine speed and torque are controlled by two feedback systems. A servo-controlled dynamometer is schematically represented in Fig. 14.9.

The actual speed signal generated by the tachometer generator, from the dynamometer, is compared with the preferred speed that is set in the tape recorder. If the actual and the preferred speeds are not the same, the dynamometer control is automatically adjusted until they are equal. The load cell on the dynamometer measures the actual torque from the engine, and is compared with the preferred torque that is set in the tape recorder. If these two values

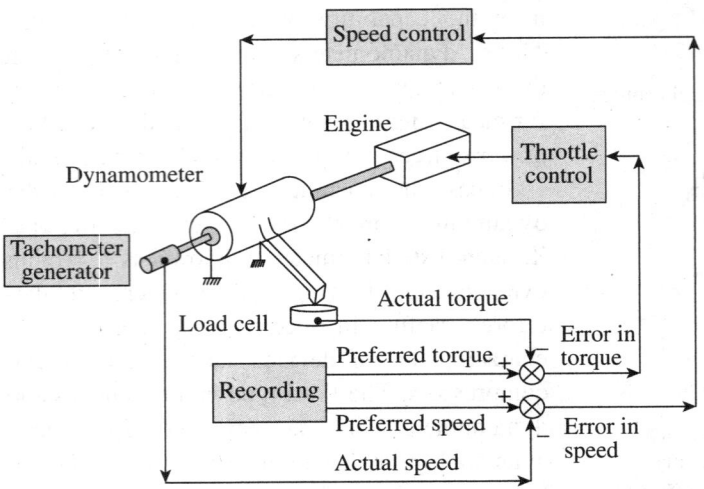

Fig. 14.9 Schematic representation of a servo-controlled dynamometer

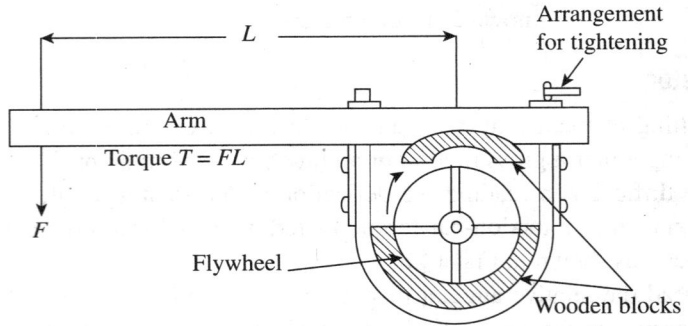

Fig. 14.10 Prony brake dynamometer

differ, then the error signal generated actuates the engine throttle control in the appropriate direction. Both torque control and speed control operate simultaneously and continuously such that they conform to the desired value set in the tape recorder.

14.4.3 Absorption Dynamometer

The Prony brake dynamometer was invented by Gaspard de Prony, a French engineer inventor, in 1821 to measure engine power. A Prony brake dynamometer is one of the simplest, inexpensive, and the most popular absorption dynamometers. It is a mechanical type of device that depends on dry friction wherein mechanical energy is converted into heat.

As illustrated in Fig. 14.10, a Prony brake dynamometer comprises two wooden blocks that are mounted on either side of the fly wheel in diagrammatically opposite directions. The fly wheel is attached to the shaft whose power needs to be determined. A lever arm is fixed to one block and the other arm is connected to an arrangement provided to tighten the rope. Tightening of the rope is performed in order to enhance the frictional resistance between the blocks and the flywheel.

The torque exerted by the Prony brake is given by the following equation:

$$T = FL$$

Here, force F is measured by conventional force-measuring instruments such as load cells or balances.

The power dissipated in the brake is then calculated by the following equation:

$$P = \frac{2\pi NT}{60}$$

or $$P = \frac{2\pi NFL}{60} \, W$$

Here, P is the dissipated power in watts, L is the length of the lever arm in metres, N is the angular speed in revolution per minute, and F is the force in Newton.

Although a Prony brake is inexpensive, it is inherently unstable. It is difficult to adjust or maintain a specific load. The following are certain limitations associated with a Prony brake dynamometer:

1. Due to wear of the wooden blocks, there will be variations in the coefficients of friction between the blocks and the flywheel. This necessitates tightening of the clamp. This makes the system unstable, and large powers cannot be measured particularly when used for longer periods.
2. The coefficients of friction decrease due to excessive rise in temperature, which may result in brake failure. Therefore, cooling is required in order to limit the temperature rise. Water is supplied into the hollow channel of the flywheel to provide cooling.
3. Due to variations in coefficients of friction, there may be some difficulty in taking readings of force F. The measuring arrangement may be subjected to oscillations, especially when the machine torque is not constant.

14.5 MEASUREMENT OF STRAIN

The problem of determining stresses acting on a body has been an important and interesting aspect of materials engineering. Prior to the invention of electrical resistance, strain gauge extensometers were extensively employed for the determination of strain. These were associated with a number of problems. One of the most important problems was their bulk, which was a major constraint in its use. Due to lack of precise knowledge of stress conditions of the material or structure, many engineering problems were solved on a theoretical basis, by employing trial and error methods and by assuming a large factor of safety.

In 1856, Lord Kelvin established the principle of electrical resistance strain gauges when he demonstrated that when stress is applied to a metal wire, apart from undergoing changes in both length and diameter, there will be some changes in the electrical resistance of the wire.

When external forces act on a body, the body is in a condition of stress and strain. Since direct measurement of stress is not possible, its effects, such as change of shape of the body or change in length, are measureable, thus providing a known relationship between stress and strain. If adequate information is available, stresses acting on a body can be computed.

When a force or load is applied to a body, it undergoes some deformation. This deformation per unit length is known as unit strain or simply strain, denoted by ε, and is given by the following equation:

$$\varepsilon = \frac{\delta l}{l}$$

Here, δl is the change in the length of the body and l is the original length of the body.

Strain gauges are essentially employed for two different purposes:

1. To determine the state of strain existing at a point on a loaded member for carrying out strain analysis
2. To act as a strain-sensitive transducer element in the measurement of quantities such as force, pressure, displacement, and acceleration

Normally, measurements are made over the shortest gauge lengths. The change in length measured over a finite length does not give the value of strain at a fixed point but rather gives the average strain over the entire length. A magnification system is essential since the change in

length is over a small gauge length. Two types of strain gauges are employed:
1. Mechanical strain gauges
2. Electrical strain gauges

14.5.1 Mechanical Strain Gauges

The Berry strain gauge extensometer, shown in Fig. 14.11, is used in civil engineering for structural applications. This gauge was designed by Professor Herman C. Berry of the University of Pennsylvania in 1910. It is employed for measuring small deformations under linear strain conditions over gauge lengths up to 200 mm.

The strain obtained is magnified by employing mechanical means. Lever systems are employed for amplifying the displacements measured over the gauge lengths. Mechanical strain gauges comprise two gauge points: one is a fixed point and the other is connected to the magnifying lever. Both points are placed on the specimen. The displacement caused is magnified by the lever and is indicated in the dial indicator. Strain is determined by dividing the measured displacements over the gauge length.

Earlier extensometers used a single mechanical lever for obtaining a magnification of 10 to 1. This was sufficient to work on a long gauge length. Recent extensometers, which are used for short gauge length, employ compound levers (dial gauges) having a higher magnification of 2000 to 1.

One of the main advantages of a mechanical strain gauge is that it contains a self-contained magnification system. There is no need to use auxiliary equipment in a mechanical strain gauge. It is best suited for conducting static tests. The following are some obvious disadvantages associated with a mechanical strain gauge:
1. Its response is slow due to high inertia, and friction is more.
2. Automatic recording of the readings is not possible.
3. It cannot be used for dynamic strain measurements and varying strains.

14.5.2 Electrical Strain Gauges

In electrical strain gauges, a change in strain produces a change in some electrical characteristic

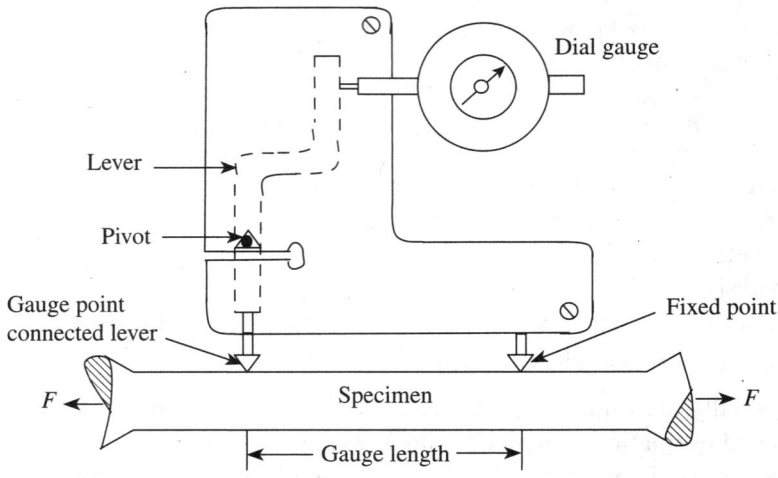

Fig. 14.11 Berry strain gauge extensometer

such as capacitance, inductance, or resistance. Magnification of the measured strain can be attained by employing an auxiliary equipment. The main advantages of measurement of strain using electrical methods are their high sensitivity and ability to respond to dynamic strains. The following are the advantages of these strain gauges:

1. They are simple in construction.
2. Strain gauges can be calibrated in terms of quantities such as force, displacement, pressure, and acceleration.
3. They are very sensitive.
4. Linear measurement is accomplished.
5. They provide good output for indicating and recording purposes.
6. Strain gauges are inexpensive and reliable.

There are two types of resistance strain gauges: unbonded and bonded.

Unbonded Resistance Strain Gauges

In case of unbonded resistance strain gauges, the electrical resistance element or the grid is not supported. A fine wire is stretched out between two or three points, which may become part of a rigid base that itself is strained, as shown in Fig. 14.12.

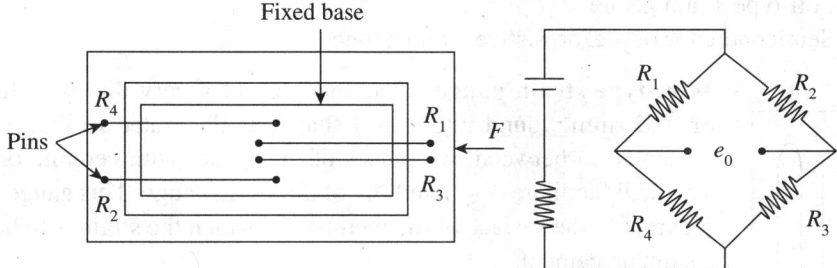

Fig. 14.12 Unbonded resistance strain gauge

The resistance strain gauge shown in Fig. 14.12 can be used to measure compressive strain as it is preloaded. The four resistance wires are connected in such a manner that they act as a full bridge. Unbonded strain gauges are used more often as elements of force and pressure transducers, and also in accelerometers. They are seldom used for strain measurement.

Bonded Resistance Strain Gauges

When a strain gauge, which is an appropriately shaped piece of resistance material, is bonded to the surface of the material to be tested, it is known as a bonded strain gauge. As shown in Fig. 14.13, a thin wire in the form of a grid pattern is cemented in between thin sheets of

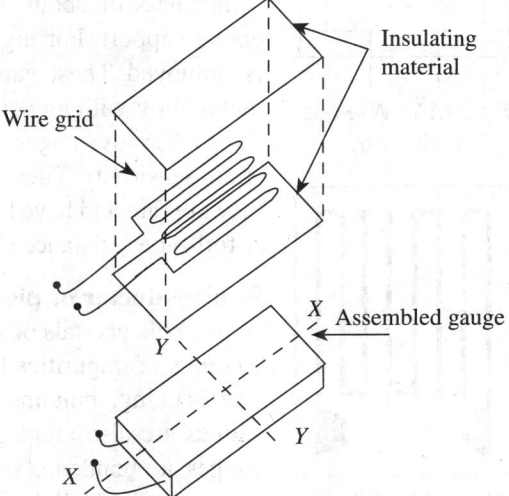

Fig. 14.13 Exploded view of a bonded resistance strain gauge

insulating materials such as paper or plastic. The strain gauge may also be a thin metal foil. The gauges are bonded to the surface under study using a thin layer of adhesive, and waterproofing is provided by applying a layer of wax or lacquer on the gauge. The strain experienced by the grid and the surface under study to which the grid is bonded will be the same. The gauge is highly sensitive in the axial direction, and in the transverse direction strain occurs due to the Poisson's ratio effect; hence, some small error may be present in the measurement of strain. However, cross-sensitivity can be completely eliminated by making the ends of the foil gauges thick. A cluster of three or more strain gauges, known as rosette gauges, can be employed if the direction of the principal strain is unknown.

Rosette gauges are normally used in the measurement of strain in complex parts. These gauges comprise three or four separate grids having different orientations and can be cemented to the surface of the part. The resultant strain can then be gathered and recorded. From the data gathered, the true magnitudes and directions of the significant surface strains can be computed.

The resistance type of electric strain gauges is the most popular device used for measuring strain. In this case, electrical resistance of the gauge proportionately changes with the strain it is subjected to. These gauges exist in the following forms:

1. Wire-type strain gauge
2. Foil-type strain gauge
3. Semiconductor or piezoresistive strain gauge

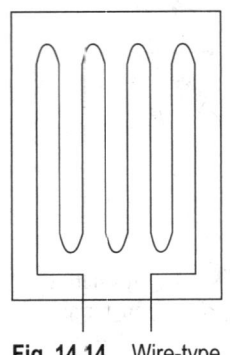

Fig. 14.14 Wire-type strain gauge

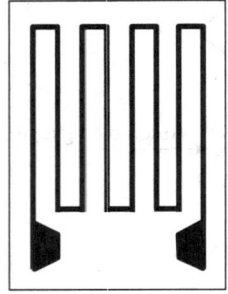

Fig. 14.15 Foil-type resistance strain gauge

Wire-type strain gauge It is made up of a very thin wire having a diameter of 0.025 mm wound into a grid shape, as illustrated in Fig. 14.14. The grid is cemented in between two pieces of thin paper with ceramic or plastic backing. This will facilitate easy handling of the strain gauge. This gauge is further bonded securely to the surface of the member on which the strain is to be measured using a suitable cement.

Foil-type resistance strain gauge A foil-type resistance strain gauge has a thickness of about 0.005 mm. It comprises a metal foil grid element on an epoxy support. For high-temperature applications, epoxy filled with fibre glass is employed. These gauges are manufactured by printing them on a thin sheet of metal alloy with an acid-resistant ink and the unprinted portion is etched away. They offer advantages such as improved hysteresis, better fatigue life, and lateral strain sensitivity. They can be fixed on fillets and sharply curved surfaces because they are thin and have higher flexibility. These are also known as metallic gauges. A foil-type resistance strain gauge is shown in Fig. 14.15.

Semiconductor or piezoresistive strain gauge Semiconductor gauges are cut from single crystals of silicon or germanium. This type of gauge contains precise amounts of impurities like boron, which impart certain desirable characteristics. The backing, bonding materials, and mounting techniques used for metallic gauges are also employed here. When the material onto which semiconductor gauges are bonded is strained, the current in the semiconductor material changes correspondingly. When the electrical resistance of the gauge increases due to the tensile strain, it is known as a positive or p-type semiconductor gauge. When

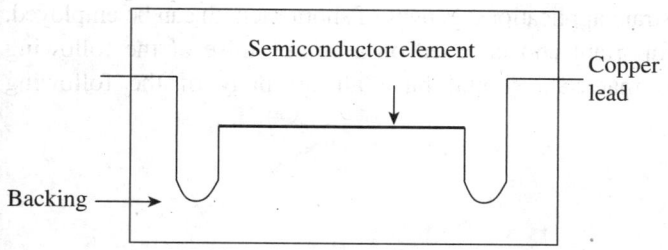

Fig. 14.16 Semiconductor gauge

the electrical resistance of the gauge decreases in response to tensile strain, it is called a negative or n-type semiconductor gauge. A semiconductor gauge is shown in Fig. 14.16.

A semiconductor gauge comprises a rectangular filament having a dimension of $10 \times 0.25 \times 0.05$ mm. It is essential to make them as thin as possible because breaking stresses of the material increase as its cross-sectional area decreases. Without fracturing the gauge, it can be bent to a much smaller radius of curvature. These have certain advantages: they show high strain sensitivity and can measure small strain variations accurately. Semiconductor gauges also possess certain disadvantages:

1. The output obtained in a semiconductor gauge is non-linear in nature.
2. The strain sensitivity of the gauge is temperature dependent.
3. It is more fragile when compared to a wire or foil element.
4. It is more expensive than an ordinary metallic gauge.

14.6 STRAIN GAUGE MATERIALS

Strain gauge sensing materials can be divided into two categories: metallic and semiconductor materials. The following are the five most popular metals used in the manufacture of strain gauges:

Constantan It is an alloy of copper and nickel having a low and controllable temperature coefficient. It finds applications in static and dynamic measurements.

Nichrome It is a nickel–chrome alloy that finds applications in static and dynamic measurements.

Dynaloy It is made up of nickel and iron, which has a high gauge factor, and exhibits high resistance to fatigue. It can be used for dynamic strain measurements.

Stabiloy It is a modified nickel–chrome alloy. It has a wide compensation range, from cryogenic to elevated temperatures, and good fatigue life.

Platinum alloy It is an alloy of platinum and tungsten and displays unusual stability and fatigue life at elevated temperatures. It is used for dynamic and static tests up to a temperature of 800 and 650 °C, respectively.

A specially processed silicon crystal, which is cut into filaments, is used for manufacturing semiconductor gauges. The filaments exhibit either p (positive) or n (negative) characteristics. When compared to metallic sensing elements, the gauge factor is much higher in semiconductor materials. Semiconductor materials are highly sensitive and provide a high output.

Resistances of commercially available strain gauges vary from 60 to 5000 Ω. However, the strain gauge of 120 Ω is considered to be a standard because most of the commercially available strain-measuring equipment are designed for 120 Ω. The safe current carrying range is between 25 and 50 mA. If stress distribution of the specimen is reasonably uniform, then a narrow gauge of 10–

30 mm length can be used. For high strain applications, gauges of shorter length can be employed.

Selection of grid materials is important and is based on a compromise of the following desirable characteristics. The grid materials should have all or many of the following characteristics:

1. High gauge factor
2. High resistivity
3. Low temperature sensitivity
4. High electrical stability
5. High yield point
6. High endurance limit
7. Good workability
8. Good solderability or weldability
9. Low hysteresis
10. Low thermal emf when joined with other materials
11. Good corrosion resistances

14.7 BACKING OR CARRIER MATERIALS

Backing materials provide support, dimensional stability, and some mechanical protection for fragile and delicate strain-sensing elements. Basically, two types of carrier materials are used.

Permanent carriers   In addition to supporting the gauge, permanent carriers provide electrical insulation between the grid and the surface of the part under study. These carriers are manufactured from organic materials. They are thin and flexible enough to sustain mechanical contact with the specimen irrespective of the surface contour, temperature gradient, or transient conditions.

Temporary carriers   Organic carriers may not be appropriate for certain environmental conditions. In such cases, temporary carriers are employed. Gauges that are used for high-temperature applications employ temporary carriers, which provide temporary backing. The backing is removed once the grid is mounted. The advantage of using temporary carriers is that even though there is an increase in specimen stiffness, there is an improvement in mechanical and thermal contacts. The following are the desirable characteristics of a backing material:

1. The backing material should be very thin.
2. It should have high mechanical strength so that it is strong enough to transmit force from the structure under study to the sensing resistance wire.
3. It should have high dielectric strength, that is, it should be an insulator of electricity.
4. The backing material should not absorb humidity, that is, it should be non-hygroscopic.
5. It should be compatible with the adhesive material used to bond it to the structure under study.
6. It should have minimum temperature restriction, that is, temperature changes should not affect it.

The most commonly used backing materials are thin papers, phenolic-impregnated papers, epoxy-type plastic films, and epoxy-impregnated fibre glass. Most foil gauges intended for a moderate range of temperature (−100 to 200 °F) employ an epoxy film backing.

14.8 ADHESIVES

Adhesives or bonding materials are the most commonly employed agents to bond the strain gauge to the specimen or the test area in order to transmit strain to the gauge carrier and later on to the strain-sensing element with true fidelity. They provide some additional insulation. The strain gauge is fixed on the structure under consideration using an adhesive or a paste. The adhesives are also called cements. Installation of a strain gauge is probably the most critical aspect of strain measurement, which is least emphasized. If not properly installed, it may give erroneous readings, thus degrading the validity of the test. The following are the desirable characteristics of a strain gauge adhesive:

1. It should have high mechanical strength so that it would have good shear strength to transmit force from the structure under study to the sensing resistive wire.
2. It should possess high creep resistance.
3. It should have high dielectric strength.
4. It should possess minimum temperature restrictions, that is, temperature changes should not affect it.
5. It should spread easily and provide good bonding adhesion. It should be easy to apply.
6. It should be non-hygroscopic, which means that the bonding material should not absorb moisture. If moisture is absorbed, the cement swells and causes large zero shifts.
7. It should go along with the backing material so that the latter is bonded rigidly to the structure under study.
8. It should have the capacity to be set up fast.
9. It should be economically viable and easily available.

Epoxy and polyamide adhesives are commonly used. Table 14.1 gives the recommended adhesives with temperature ranges.

14.9 PROTECTIVE COATINGS

In order to prevent damage that may be caused to the relatively delicate gauge installation, it is essential to use protective coatings. There are chances of mechanical damage, chemical damage, or damage by other environmental conditions. The following factors may influence the selection of coatings:

1. Test environment
2. Test duration
3. Accuracy requirements

Table 14.1 Recommended adhesives with temperature ranges

Adhesive	Temperature range (°F)
Nitro cellulose	−300 to 180
Cyanoacrylate	−100 to 200
Phenolic	−400 to 400
Ceramic	−400 to 1000

The different materials employed for coating strain gauges are as follows:

1. Micro-crystalline wax from −75 to 70 °C
2. Silicones up to 420 °C
3. Synthetic rubber coatings, −75 to 220 °C
4. Plastic coatings, which are of two types: vinyl coatings (used at the cryogenic temperature range) and epoxy coatings (used at a higher operating range)
5. Metal foil type

14.10 BONDING OF GAUGES

The following are the steps to be followed while bonding the gauge to the specimen:
1. Wash hands with soap and water. The working desk area should be kept clean and all related tools should be washed with a solvent or degreasing agent.
2. The folder containing the gauge should be opened carefully. Bare hands should not be used to grasp the gauge. Tweezers can be used to hold the gauge. Do not touch the grid.
3. Place the gauge on a clean working area with the bonding side down.
4. Use a proper length, say, about 15 cm, of cellophane tape to pick up the strain gauge and transfer it to the gauging area of the specimen. Align the gauge with the layout lines. Press one end of the tape to the specimen, and then smoothly and gently apply the whole tape and gauge into position.
5. Lift one end of the tape such that the gauge does not contact the gauging area and the bonding site is exposed. Apply the catalyst evenly and gently on the gauge.
6. Apply enough adhesive to provide sufficient coverage under the gauge for proper adhesion. Some iteration may be required in determining 'sufficient' coverage. Place the tape and gauge back onto the specimen smoothly and gently. Immediately place the thumb over the gauge and apply firm and steady pressure on it for at least one minute.

14.11 GAUGE FACTOR

The gauge factor is the most important parameter of strain gauges. It is a measure of the amount of resistance change for a given strain and therefore serves as an index of the strain sensitivity of the gauge. Mathematically, it is expressed as follows:

$$F = \frac{\Delta R / R}{\Delta L / L}$$

Here, F is the gauge factor, ΔR is the change in resistance, and ΔL is the change in length. Both ΔR and ΔL are due to straining of the gauge along the surface to which it is bonded by the application of force. R and L are the initial resistance and length of the strain gauge, respectively.

A higher gauge factor makes the gauge more sensitive, and the electrical output obtained for indication and recording purposes will be greater. The gauge factor is normally supplied by the manufacturer and may range from 1.7 to 4, depending on the length of the gauge. Thus, a means is available for measuring the change of resistance, and hence strain can be determined. Although a higher gauge factor increases sensitivity, in case of metallic gauges it is limited owing to the relatively low resistivity of the metals. This limitation can be overcome by using semiconductor strain gauges, which have gauge factors of the order of ± 100 or more.

14.12 THEORY OF STRAIN GAUGES

The change in the resistance value obtained by straining the gauge is attributed to the elastic behaviour of the material. If an elastic material is strained by applying a tensile load in the axial direction, its longitudinal dimension increases and the lateral dimension reduces as a function of the Poisson's ratio. The resistance of the elastic material is given by the following equation:

$$R = \frac{\rho L}{A} \tag{14.14}$$

Here, L is the length of the wire, A is the cross-sectional area, and ρ is the resistivity of the material.

Assume that the initial length is L having a cross-sectional area of CD^2, where C is the proportionality constant and D is a sectional dimension; C = 1 for a square cross-section and C = $\pi/4$ for a circular cross-section. In order to determine the effect of change in resistance ΔR, which depends upon the physical quantities of the material, Eq. (14.14) must be differentiated.

$$dR = \frac{CD^2(Ld\rho + \rho dL) - 2C\rho DLdD}{(CD^2)^2}$$

$$dR = \frac{1}{CD^2}\left[\left(Ld\rho + \rho dL\right) - 2\rho L\frac{dD}{D}\right] \tag{14.15}$$

Dividing Eq. (14.15) by Eq. (14.14), we get the following equation:

$$\frac{dR}{R} = \frac{dL}{L} - \frac{2dD}{D} + \frac{d\rho}{\rho} \tag{14.16}$$

Dividing throughout by dL/L, equation (14.16) becomes as follows:

$$\frac{dR/R}{dL/L} = 1 - \frac{2dD/D}{dL/L} + \frac{d\rho/\rho}{dL/L} \tag{14.17}$$

We now conclude the following:

Axial strain = $dL/L = \varepsilon_a$

Lateral strain = $dD/D = \varepsilon_L$

Poisson's ratio $v = -\dfrac{dD/D}{dL/L}$

Substituting these values, we get the basic relation for the gauge factor F:

$$F = \frac{dR/R}{dL/L} = \frac{dR/R}{\varepsilon_a} = 1 + 2v + \frac{d\rho/\rho}{dL/L}$$

This is the basis for a resistance strain gauge.

Here, 1 indicates the change in resistance due to change in length, $2v$ indicates the change in resistance due to change in area, and $\left(\dfrac{d\rho/\rho}{dL/L}\right)$ specifies the change in resistance due to the piezoresistance effect. The piezoresistance effect signifies the dependency of resistivity ρ on the mechanical strain.

14.13 METHODS OF STRAIN MEASUREMENT

An electrical resistance-type strain gauge predominantly employs a highly sensitive resistance bridge arrangement for strain measurement. As can be seen from Fig. 14.17, it comprises four resistance arms, an energy source, and a detector. The two methods used for strain measurement are null method and deflection method.

It can be assumed that in order to balance the bridge, the resistances have been balanced, that is, $e_{BD} = 0$. To attain this, it is necessary that the ratio of the resistances of any two adjacent arms and the ratio of the resistance of the remaining arms taken in the same sense are equal;

$$\frac{R_1}{R_4} = \frac{R_2}{R_3}$$

In case any one of the resistances change, the meter across BD will indicate a change in the voltage, which unbalances the bridge. This change in the meter reading can be made use of to compute the change in resistance. Since the change in resistance is indicated by the deflection of the meter, this method is known as the deflection method. In this method, readings of the change in strain can be noted down continuously without any rebalancing. This method is not accurate because of the tendency of the bridge output to become non-linear with the change in strain.

In the null method, after each successive change in resistance, the bridge is balanced so that the meter is brought back to zero, thus nullifying the effect of any change in resistance. The null method is superior to the deflection method because the strain indicators calibrated under null balance conditions exhibit linearity and high accuracy.

14.14 STRAIN GAUGE BRIDGE ARRANGEMENT

The arrangement of the resistance bridge shown in Fig. 14.18 is suitable for use with strain gauges, as the adjustment to null position for zero strain is very easy. In addition, it provides a means for effectively minimizing or eliminating the temperature effects. Arm 1, shown in Fig. 14.18, includes a strain-sensitive gauge mounted on the test item. Arm 2 contains an identical gauge mounted on a piece of unstrained material, which is almost a replica of the test material, placed near the test location in order to maintain the same temperature. The arrangement also consists of arms 3 and 4, which act as fixed resistors for providing good stability. For balancing the resistance of the bridge, the arrangement has portions of slide wire resistance D.

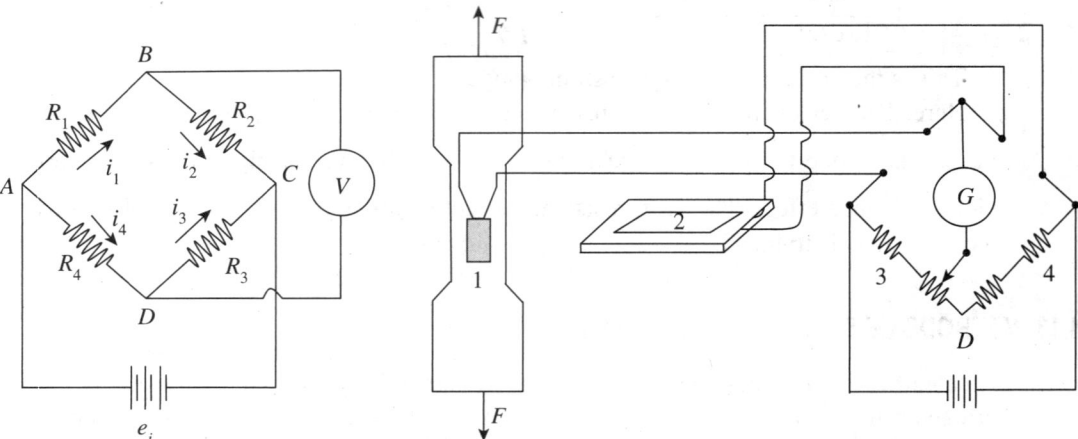

Fig. 14.17 Basic Wheatstone bridge   **Fig. 14.18** Arrangement of a resistance bridge for strain measurement

14.15 TEMPERATURE COMPENSATION IN STRAIN GAUGES

One of the most important and critical factors in the use of resistance strain gauges is temperature sensitivity. Although compensation is provided in the electrical circuitry, for a majority of the applications, the problem is not eliminated completely. The following three factors are involved:

1. The gauge is unable to differentiate the strain resulting from the differential expansion existing between the grid support and the proper grid from the load strain.
2. The resistivity r changes with the change in temperature.
3. The strong magnetic field has an influence on the gauge performance.

Temperature changes affect the sensitivity of resistance-type strain gauges. These strain gauges are influenced by variations in temperature in the following ways:

1. Temperature variation due to creep affects the gauge factor of the strain gauge.
2. Due to temperature variation, the resistance of the strain gauge element varies.
3. Owing to the differential expansion between the test member and the strain gauge bonding material, strains may be induced in the gauge.

Compensation may be provided for eliminating the effects of temperature in the following ways:

1. By the use of an adjacent-arm balancing or compensating gauge
2. By means of self-temperature compensation

14.15.1 Adjacent-arm Compensating Gauge

In Fig. 14.17, arms 1 and 2 are in similar conditions. The initial arm balance is given by the following relation:

$$\frac{R_1}{R_2} = \frac{R_3}{R_4}$$

If the gauges contained in arms 1 and 2 are similar and mounted on identical materials, and if their shift in resistance ΔR_t due to temperature effect is the same, then the following equation is obtained:

$$\frac{R_1 + \Delta R_t}{R_2 + \Delta R_t} = \frac{R_3}{R_4}$$

It can be observed here that the change in temperature has not affected the output, and the bridge remains in balance. The compensating gauge that is used to only complete the bridge circuit and to balance the output temperature component is called a dummy gauge.

14.15.2 Self-temperature Compensation

In some situations, the adjacent-arm compensating or dummy gauge may not provide complete compensation. Sometimes, due to the presence of large temperature gradients in the test specimens, it may not be possible to maintain identical conditions between two gauges. In such instances, self-temperature compensation may be employed. A completely self-temperature-compensated gauge displays zero change of resistance with changes in temperature when bonded to a specified material. Gauges made from selected melt alloys are the most widely used self-temperature-compensated gauges. During manufacture, it is essential that the coefficient of linear expansion of both the gauge and the test material is matched by a suitable heat treatment process. The matching can only be carried out over a limited range of temperatures. In order to overcome this limitation, it is essential to calibrate some selected gauges from each batch produced. Another method is that, during the formation of the grid, two different wires, having a positive and a negative gauge factor, are connected in series. The advantage is that due to temperature variation, an increase in resistance in one wire is almost compensated by a decrease in resistance in the other wire. However, the problem of matching still persists, especially for use in the case of some specified materials.

A QUICK OVERVIEW

- Force is defined as the product of mass and acceleration, as per Newton's second law of motion. Force is a vector quantity whose unit of measurement is Newton (N), which is equivalent to the force required to accelerate a mass of 1 kg at a rate of $1\,\text{m/s}^2$, that is, $1\,\text{N} = 1\,\text{kg} \times 1\,\text{m/s}^2$.
- The acceleration due to gravity g can be adopted as a suitable standard and is approximately equal to $9.81\,\text{m/s}^2$.
- Torque is defined as the force, which, besides exerting effects along its line of action, may also exert a rotating force relative to any axis excepting those intersecting the line of action. Torque can also be termed as a couple or twisting force. It is determined by measuring the force at a known radius r and is expressed as $T = F \times r$.
- Direct methods involve the comparison of an unknown force with a known gravitational force on the standard mass. A force is exerted on a body of mass m due to the earth's gravitational field, which can be represented by the following equation:

  $W = mg$

  Here, m is the standard mass, g is the acceleration due to gravity, and W is the weight of the body.
- Elastic members are generally used for the measurement of force through displacement measurement. When an elastic member is combined with a strain gauge and used for the measurement of force, it is termed a load cell. In load cells, elastic members act as primary transducers and strain gauges as secondary transducers.

- Torque and speed are measured under actual driving conditions of an automobile engine; tape recordings of such an exercise of an engine are obtained and then simulated under laboratory conditions. Engine speed and torque are controlled by two feedback systems.
- A Prony brake dynamometer is one of the simplest, inexpensive, and the most popular absorption dynamometer. It is a mechanical type of device that depends on dry friction wherein mechanical energy is converted into heat.
- In 1856, Lord Kelvin established the principle of electrical resistance strain gauges when he demonstrated that when stress is applied to a metal wire, apart from undergoing changes in both length and diameter, there will be some changes in the electrical resistance of the wire.
- In electrical strain gauges, a change in strain produces a change in some electrical characteristic such as capacitance, inductance, or resistance. Magnification of the measured strain can be attained by employing auxiliary equipment. The main advantages of measurement of strain using electrical methods are their high sensitivity and ability to respond to dynamic strains.
- When a strain gauge, which is an appropriately shaped piece of resistance material, is bonded to the surface of the material to be tested, it is known as a bonded strain gauge.
- A cluster of three or more strain gauges, known as rosette gauges, can be employed if the direction of principal strain is unknown. Rosette gauges are normally used in the measurement of

strain in complex parts.
- A gauge factor is a measure of the amount of resistance change for a given strain and therefore serves as an index of the strain sensitivity of the gauge. Mathematically, it is expressed as follows:

$$F = \frac{\Delta R / R}{\Delta L / L}$$

Here F is the gauge factor, ΔR is the change in the resistance, and ΔL is the change in length. Both ΔR and ΔL are due to straining of the gauge along the surface to which it is bonded by the application of force. R and L are the initial resistance and length of the strain gauge, respectively.

MULTIPLE-CHOICE QUESTIONS

1. Strain gauge rosettes are used when the direction of the
   (a) principal stress is unknown
   (b) hoop stress is known
   (c) principal stress is known
   (d) hoop stress is unknown
2. Dummy strain gauge are used for
   (a) bridge sensitivity enhancement
   (b) temperature compensation
   (c) determining gauge factor
   (d) calibration of strain gauges
3. Strain gauges cannot be used at high temperatures because of
   (a) inferior strain gauge materials
   (b) gauge not being properly welded
   (c) a problem with gauge factor
   (d) decomposition of cement and carrier materials
4. Which of the following equations is correct?
   (a) $\dfrac{dR/R}{dL/L} = \dfrac{dR/R}{\varepsilon_a} = 1 - 2\upsilon + \dfrac{d\rho/\rho}{dL/L}$

   (b) $\dfrac{dL/L}{dR/R} = \dfrac{dR/R}{\varepsilon_a} = 1 - 2\upsilon + \dfrac{d\rho/\rho}{dL/L}$

   (c) $\dfrac{dL/L}{dR/R} = \dfrac{dR/R}{\varepsilon_a} = 1 + 2\upsilon + \dfrac{d\rho/\rho}{dL/L}$

   (d) $\dfrac{dR/R}{dL/L} = \dfrac{dR/R}{\varepsilon_a} = 1 + 2\upsilon + \dfrac{d\rho/\rho}{dL/L}$

5. In a platform balance, the unknown weight is placed at
   (a) a specified position only
   (b) any position with certain conditions
   (c) any position without any conditions
   (d) a corner only
6. Buoyancy effects are found in
   (a) platform balance
   (b) analytical balance
   (c) electromagnetic-type balance
   (d) pendulum-type balance
7. Which of the following balances is not appropriate for large weights?
   (a) Analytical balance
   (b) Platform balance
   (c) Electromagnetic-type balance
   (d) Pendulum-type balance
8. When an elastic member is used in conjunction with a strain gauge, it is called a
   (a) proving ring
   (b) absorption dynamometer
   (c) Prony brake
   (d) load cell
9. Tape recordings are used in
   (a) servo-controlled dynamometers
   (b) torsion-bar dynamometer
   (c) Prony brake dynamometer
   (d) piezoelectric transducers
10. In servo-controlled dynamometers,
    (a) speed is controlled
    (b) torque is controlled
    (c) both speed and torque are controlled
    (d) strain is controlled
11. In a Prony brake dynamometer,
    (a) mechanical energy is converted into heat
    (b) mechanical energy is converted into electrical energy
    (c) electrical energy is converted into torque
    (d) there is no need for energy conversion

12. In a p-type semiconductor gauge, as tensile strain increases
    - (a) gauge resistance decreases
    - (b) gauge resistance increases
    - (c) gauge resistance increases and then decreases
    - (d) there is no effect on gauge resistance
13. In torsion-bar dynamometers, deflection measurement is carried out by
    - (a) mechanical methods
    - (b) electrical methods
    - (c) electronic methods
    - (d) optical methods
14. In torsion-bar dynamometers, to read the deflection of the rotating shaft
    - (a) hygroscopic effect is used
    - (b) telescopic effect is used
    - (c) stroboscopic effect is used
    - (d) microscopic effect is used

REVIEW QUESTIONS

1. With the help of a neat sketch, explain the working of an analytical balance.
2. Describe a platform balance with a neat diagram.
3. Explain, with a neat sketch, determination of force using a load cell.
4. Explain the working of a proving ring with a neat sketch.
5. Explain the working of a torsion-bar dynamometer.
6. Discuss a servo-controlled dynamometer with a neat schematic diagram.
7. Explain the working of the Berry strain gauge extensometer.
8. Explain the following:
   - (a) Unbonded strain gauge
   - (b) Bonded strain gauge
9. Explain the three types of bonded strain gauges with a neat sketch.
10. What are rosette gauges? Explain briefly.
11. Write a note on strain gauge materials.
12. Enlist the desirable characteristics of grid materials.
13. Explain the different types of carrier materials.
14. Mention the desirable characteristics of backing materials.
15. Why are adhesives required? What are the desirable characteristics of adhesives?
16. Write short notes on the following:
    - (a) Protective coatings on strain gauges
    - (b) Bonding of strain gauges
17. What is a gauge factor? Explain its importance.
18. Deduce an expression for the gauge factor.
19. Explain the different methods of strain measurement.
20. Explain how the strain gauge is influenced by temperature variations.
21. Explain the different methods of strain gauge compensation.

Answers to Multiple-choice Questions

1. (a)   2. (b)   3. (d)   4. (d)   5. (c)   6. (b)   7. (a)   8. (d)
9. (a)   10. (c)   11. (a)   12. (b)   13. (d)   14. (c)

Measurement of Temperature

After studying this chapter, the reader will be able to

• understand the basics of temperature measurement
• describe the methods of temperature measurements
• explain thermocouples and the different laws of thermocouples
• elucidate the different resistance temperature detectors
• comprehend calibration of liquids in glass thermometers
• discuss bimetallic strip thermometers
• throw light on pyrometers

15.1 INTRODUCTION

We know that temperature is a physical property of a material that gives a measure of the average kinetic energy of the molecular movement in an object or a system. Temperature can be defined as a condition of a body by virtue of which heat is transferred from one system to another. It is pertinent to mention here that both temperature and heat are different. Temperature is a measure of the internal energy of a system, whereas heat is a measure of the transfer of energy from one system to another. Heat transfer takes place from a body at a higher temperature to one at a lower temperature. The two bodies are said to be in thermal equilibrium when both of them are at the same temperature and no heat transfer takes place between them. The rise in temperature of a body is due to greater absorption of heat, which increases the movement of the molecules within the body.

The first thermometer was developed by Galileo Galilei in the 17th century, which has undergone significant improvement with the advancement of science and technology; present-day thermometers are capable of measuring temperatures more accurately and precisely. In 1724, D.G. Fahrenheit, a German physicist, contributed significantly to the development of thermometry. He proposed his own scale, in which 32° and 212° were considered the freezing point and boiling point of water, respectively. The Swedish physicist Anders Celsius, in 1742, developed the mercury-in-glass thermometer. He identified two points, namely the melting point of ice and the boiling point of water, and assigned 0° and 100°, respectively, to them. He made 100 divisions between these two points. In 1859, William John Macquorn Rankine, a Scottish physicist, proposed an absolute or thermodynamic scale, known as Rankine scale when, after investigating the changes in thermal energy with changes in temperature, he came to a conclusion that the theoretical temperature of each of the substances was the same at zero thermal energy level. According to him, this temperature was approximately equal to −460 °F.

William Thomson, first Baron Kelvin, popularly known as Lord Kelvin, a British physicist, introduced a new concept, known as the Kelvin scale, in the mid-1800s. He suggested 0 K as the absolute temperature of gas and 273 K as the freezing point of water. A comparison between Kelvin, Celsius, and Fahrenheit scales with respect to absolute zero, and boiling and freezing points of water is shown in Table 15.1. Although human beings generally perceive temperature as hot, warm (neutral), or cold, from an engineering perspective, a precise and accurate measurement of temperature is essential.

Table 15.1 Comparison of temperature scales

Scales	Water boils	Water freezes	Absolute zero
Kelvin	373.16 K	273.16 K	0 K
Celsius	100 °C	0 °C	−273.16°C
Fahrenheit	212 °F	32 °F	−459.7 °F

The scales used to measure temperature can be divided into relative scales [Fahrenheit (°F) and Celsius (°C)] and absolute scales [Rankine (°R) and Kelvin (K)]. The various temperature scales are related as follows:

$$F = 1.8C + 32$$
$$C = (F - 32)/1.8$$
$$R = F + 460$$
$$K = C + 273$$

15.2 METHODS OF MEASURING TEMPERATURE

Measurement of temperature cannot be accomplished by direct comparison with basic standards such as length and mass. A standardized calibrated device or system is necessary to determine temperature. In order to measure temperature, various primary effects that cause changes in temperature can be used. The temperature may change due to changes in physical or chemical states, electrical property, radiation ability, or physical dimensions. The response of the temperature-sensing device is influenced by any of the following factors:
1. Thermal conductivity and heat capacity of an element
2. Surface area per unit mass of the element
3. Film coefficient of heat transfer
4. Mass velocity of a fluid surrounding the element
5. Thermal conductivity and heat capacity of the fluid surrounding the element

Temperature can be sensed using many devices, which can broadly be classified into two categories: contact- and non-contact-type sensors. In case of contact-type sensors, the object whose temperature is to be measured remains in contact with the sensor. Inference is then drawn on the assessment of temperature either by knowing or by assuming that the object and the sensor are in thermal equilibrium. Contact-type sensors are classified as follows:

1. Thermocouples
2. Resistance temperature detectors (RTDs)
3. Thermistors
4. Liquid-in-glass thermometers
5. Pressure thermometers
6. Bimetallic strip thermometers

In case of non-contact-type sensors, the radiant power of the infrared or optical radiation received by the object or system is measured. Temperature is determined using instruments such as radiation or optical pyrometers. Non-contact-type sensors are categorized as follows:

1. Radiation pyrometers
2. Optical pyrometers
3. Fibre-optic thermometers

15.3 THERMOCOUPLES

Thermocouples are active sensors employed for the measurement of temperature. The thermoelectric effect is the direct conversion of temperature differences to an electric voltage. In 1821, Thomas Johan Seebeck discovered that when two dissimilar metals are joined together to form two junctions such that one junction (known as the hot junction or the measured junction) is at a higher temperature than the other junction (known as the cold junction or the reference junction), a net emf is generated. This emf, which also establishes the flow of current, can be measured using an instrument connected as shown in Fig. 15.1. The magnitude of emf generated is a function of the junction temperature. It is also dependent on the materials used to form the two junctions. The thermoelectric emf is a result of the combination of two different effects—the Peltier effect and the Thomson effect.

The French physicist Jean Charles Athanase Peltier discovered that if two dissimilar metals are connected to an external circuit in a way such that a current is drawn, the emf may be slightly altered owing to a phenomenon called Peltier effect. A potential difference always exists between two dissimilar metals in contact with each other. This is known as the Peltier effect.

Thomson found out that the emf at a junction undergoes an additional change due to the existence of a temperature gradient along either or both the metals. The Thomson effect states that even in a single metal a potential gradient exists, provided there is a temperature gradient.

Both these effects form the basis of a thermocouple, which finds application in temperature measurement. The flow of current through the circuit is spontaneous when two dissimilar metals are joined together to form a closed circuit, that is, a thermocouple, provided one junction is maintained at a temperature different from the other. This effect is termed the Seebeck effect.

In Fig. 15.1, if temperatures at the hot junction (T_1) and the cold junction (T_2) are equal and at the same time opposite, then there will not be any flow of current. However, if they are unequal, then the emfs will not balance and hence current will flow. It is to be mentioned here that the voltage signal is a function of the junction temperature at the measured end and the voltage increases as the temperature rises. Variations in emf are calibrated in terms of temperatures; the devices employed to record these observations are termed thermocouple pyrometers.

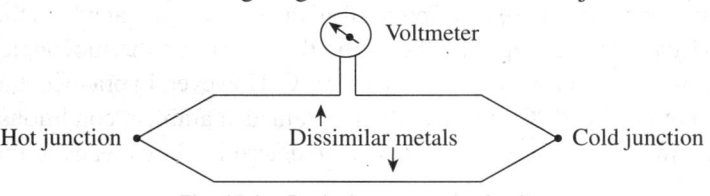

Fig. 15.1 Basic thermocouple circuit

15.3.1 Laws of Thermocouples

Apart from the Peltier and Thomson effects, which form the basis of thermoelectric emf generation, three laws of thermocouples that govern this phenomenon are required to be studied in order to understand their theory and applicability. They also provide some useful information on the measurement of temperature.

Law of Homogeneous Circuit

This law states that a thermoelectric current cannot be sustained in a circuit of a single homogenous material, regardless of the variation in its cross section and by the application of heat alone. This law suggests that two dissimilar materials are required for the formation of any thermocouple circuit.

Law of Intermediate Metals

If an intermediate metal is inserted into a thermocouple circuit at any point, the net emf will not be affected provided the two junctions introduced by the third metal are at identical temperatures. This law allows the measurement of the thermoelectric emf by introducing a device into the circuit at any point without affecting the net emf, provided that additional junctions introduced are all at the same temperature.

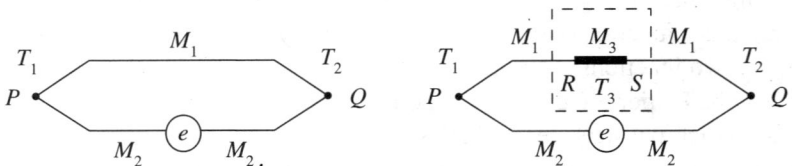

Fig. 15.2 Law of intermediate metals

It is clear from Fig. 15.2 that when a third metal, M_3, is introduced into the system, two more junctions, R and S, are formed. If these two additional junctions are maintained at the same temperature, say T_3, the net emf of the thermocouple circuit remains unaltered.

Law of Intermediate Temperatures

If a thermocouple circuit generates an emf e_1 when its two junctions are at temperatures T_1 and T_2, and e_2 when the two junctions are at temperatures T_2 and T_3, then the thermocouple will generate an emf of $e_1 + e_2$ when its junction temperatures are maintained at T_1 and T_3 (Fig. 15.3).

This law pertains to the calibration of the thermocouple and is important for providing reference junction compensation. This law allows us to make corrections to the thermocouple readings when the reference junction temperature is different from the temperature at which the thermocouple was calibrated. Usually while preparing the calibration chart of a thermocouple, the reference or cold junction temperature is taken to be equal to 0 °C. However, in practice, the reference junction is seldom maintained at 0 °C; it is usually maintained at ambient conditions. Thus, with the help of the third law, the actual temperature can be determined by means of the calibration chart.

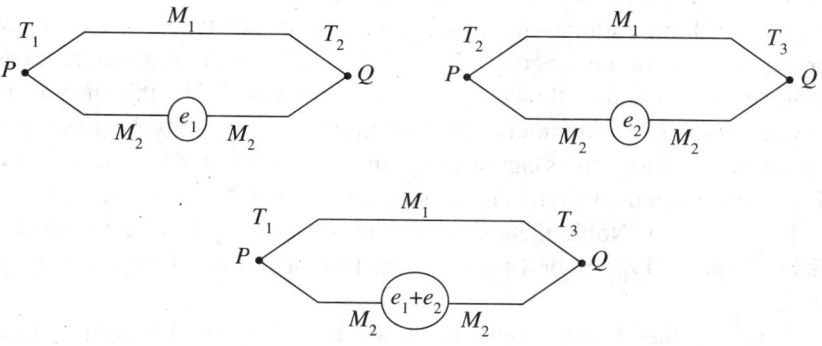

Fig. 15.3 Law of intermediate temperatures

15.3.2 Thermocouple Materials

Theoretically, any two different materials can be used to form a thermocouple. However, only a few are suitable for temperature measurement applications. Combinations of different thermocouple materials and their temperature range are given in Table 15.2. Base metal-type thermocouples like copper–constantan pose high resistance to condensed moisture corrosion. The iron–constantan type is essentially an inexpensive thermocouple capable of enduring oxidizing and reducing atmospheres. The chromel–alumel thermocouple can resist an oxidizing atmosphere.

Table 15.2 Temperature range of various thermocouple materials

Type	Thermocouple materials	Temperature range (°C)
Base metal type		
T	Copper (40%)–constantan (60%)	−200 to 350
J	Iron–constantan	−150 to 750
E	Chromel–constantan (57% Cu, 43% Ni)	−200 to 1000
K	Chromel (90% Ni, 10% Cr)–Alumel (94% Ni, 2% Al, 3% Mn, 1% Si)	−200 to 1300
Rare metal type		
S	Platinum (90%)–rhodium–platinum (10%)	0–1500
R	Platinum–rhodium (87% Pt, 13% Rh)–platinum	0–1500

Thermocouple materials are divided into base metal type and rare, noble, or precious metal type. Platinum (platinum–rhodium) thermocouples are called noble thermocouples, and all other thermocouples belong to the base metal type. In a reducing atmosphere, these thermocouples are used with protection. Material combinations such as tungsten–tungsten–rhenium, iridium–tungsten, and iridium–iridium–rhodium are called special types of thermocouples and are used for a high temperature range of 1500–2300 °C. For high-temperature measurements, the thermocouple wire should be thicker. However, an increase in the thickness of the wire lowers the time of response of the thermocouple to temperature variations. Depending on the range of temperature that thermocouples can measure, they are designated by a single letter and grouped accordingly. Base metals, which can measure up to 1000 °C, are designated as Type K, Type E, Type T, and Type J. Noble metals, which can measure up to a temperature of 2000 °C, is classified as Type R, Type S, or Type B. Refractory metals are designated as Type C, Type D, or Type G.

The choice of the thermocouple materials is influenced by several factors. Different combinations of thermocouple materials should possess the following characteristics in order to be used for temperature measurement:

1. Capable of producing a reasonable linear temperature–emf relationship
2. Able to generate sufficient thermo-emf per degree temperature change to facilitate detection and measurement
3. Capable of withstanding persistent high temperatures, rapid temperature variations, and the effects of corrosive environments
4. Good sensitivity to record even small temperature variations
5. Very good reproducibility, which enables easy replacement of the thermocouple by a similar one without any need for recalibration
6. Good calibration stability
7. Economical

15.3.3 Advantages and Disadvantages of Thermocouple Materials

The following are some distinct advantages that merit the use of thermocouples:

1. Temperature can be measured over a wide range.
2. Thermocouples are self-powered and do not require any auxiliary power source.
3. A quick and good response can be obtained.
4. The readings obtained are consistent and hence are consistently repeatable.
5. Thermocouples are rugged, and can be employed in harsh and corrosive conditions
6. They are inexpensive.
7. They can be installed easily.

However, thermocouples also have certain disadvantages, which are listed as follows:

1. They have low sensitivity when compared to other temperature-measuring devices such as thermistors and RTDs.
2. Calibration is required because of the presence of some non-linearity.
3. Temperature measurement may be inaccurate due to changes in the reference junction temperature; hence thermocouples cannot be employed for precise measurements.
4. For enhancing the life of thermocouples, they should be protected against contamination and have to be chemically inert.

15.3.4 Thermopiles

An extension of thermocouples is known as a thermopile. A thermopile comprises a number of thermocouples connected in series, wherein the hot junctions are arranged side by side or in a star formation. In such cases, the total output is given by the sum of individual emfs. The advantage of combining thermocouples to form a thermopile is that a much more sensitive element is obtained. For example, a sensitivity of 0.002 °C at 1 mV/°C can be achieved with a chromel–constantan thermopile consisting of 14 thermocouples. If n identical thermocouples are combined to form a thermopile, then the total emf will be n times the output of the single thermocouple.

For special-purpose applications such as measurement of temperature of sheet glass, thermopiles are constructed using a series of semiconductors. For average temperature measurement, thermocouples can be connected in parallel. During the formation of a thermopile, one has to ensure that the hot junctions of the individual thermocouples are properly insulated from one another.

Figures 15.4(a) and 15.4(b) illustrate, respectively, a thermopile having a series connection and one having a star connection.

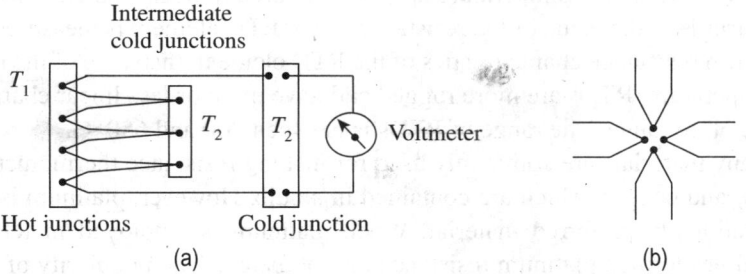

Fig. 15.4 Thermopiles (a) Series connection (b) Star connection

15.4 RESISTANCE TEMPERATURE DETECTORS

Thomas Johan Seebeck, in 1821, discovered thermoelectric emf. In the same year, Sir Humphrey Davy showed that the resistivity of metals is highly dependent on temperature. In 1871, Sir William Siemens proposed the use of platinum as the primary element in resistance thermometers. Platinum is extensively used in high-accuracy resistance thermometers because it is capable of withstanding high temperatures and, at the same time, can sustain excellent stability and exhibit good linearity. The first classical RTD was constructed by C.H. Meyers in 1932, using platinum. A helical coil of platinum was wound on a crossed mica web, and the entire assembly was mounted inside a glass tube. The advantage of this type of construction is that the strain on the wire can be minimized and its resistance can be maximized. Slow thermal response time and fragility of the structure limited its application due to poor thermal contact between platinum and the measured point. Technological advancement led to the development of more rugged RTDs later. The International Practical Temperature Scale was developed in 1968, and pure platinum RTDs have been used as the standard instruments for interpolating between fixed points of the scale. The triple point of water (0.01 °C), boiling point of water (100 °C), triple point of hydrogen (13.81 K), and freezing point of zinc (419.505 °C) are some of the fixed points.

RTDs are also known as resistance thermometers. The American Society for Testing and

Materials has defined the term resistance thermometer as follows: RTD is 'a temperature-measuring device composed of a resistance thermometer element, internal connecting wires, a protective shell with or without means for mounting a connection head, or connecting wire or other fittings, or both'.

We know that the electrical conductivity of a metal is dependent on the movement of electrons through its crystal lattice. An RTD is a temperature sensor that works on the principle that the resistance of electrically conductive materials is proportional to the temperature to which they are exposed. Resistance of a metal increases with an increase in temperature. Hence, metals can be classified as per their positive temperature coefficient (PTC).

When temperature measurement is performed by a resistance thermometer using metallic conductors, it is called a resistance temperature detector (RTD); on the other hand, semiconductors used for temperature measurement are called thermistors.

We know that an RTD measures temperature using the principle that the resistance of a metal changes with temperature. In practice, the RTD element or resistor that is located in proximity to the area where the temperature is to be measured transmits an electrical current. Then, using an instrument, the value of the resistance of the RTD element is measured. Further, on the basis of known resistance characteristics of the RTD element, the value of the resistance is correlated to temperature. RTDs are more rugged and have more or less linear characteristics over a wide temperature range. The range of RTDs is between 200 and 650 °C.

Many materials are commonly used for making resistance thermometers, such as platinum, nickel, and copper, which are contained in a bulb. However, platinum is the most popular and internationally preferred material. When platinum is employed in RTD elements, they are sometimes termed platinum resistance thermometers. The popularity of platinum is due to the following factors:

1. Chemical inertness
2. Almost linear relationship between temperature and resistance
3. Large temperature coefficient of resistance, resulting in readily measurable values of resistance changes due to variations in temperature
4. Greater stability because the temperature resistance remains constant over a long period of time

Selection of a suitable material for RTD elements depends on the following criteria:

1. The material should be ductile so that it can be formed into small wires.
2. It should have a linear temperature-versus-resistance graph.
3. It must resist corrosion.
4. It should be inexpensive.
5. It should possess greater stability and sensitivity.
6. It must have good reproducibility.

RTDs essentially have the following three configurations:

1. A partially supported wound element: A small coil of wire inserted into a hole in a ceramic insulator and attached along one side of that hole
2. Wire-wound RTD: Prepared by winding a platinum or metal wire on a glass or ceramic bobbin and sealed with a coating on molten glass known as wire-wound RTD elements (Fig. 15.5)
3. Thin film RTD: Prepared by depositing or screening a platinum or metal glass slurry film onto a small flat ceramic substrate called thin film RTD elements

The general construction of a resistance thermometer is shown in Fig. 15.6. It comprises a number of turns of resistance wire wrapped around a solid silver core. Transmission of heat takes place quickly from the end flange through the core to the winding.

The thin film element used for temperature sensing is manufactured by depositing a very thin (around 10–100 Å) layer of platinum on a ceramic substrate. The platinum layer is coated with epoxy or glass, which protects the deposited platinum film and also acts as a strain reliever for external lead wires. During the early stages of development, thin film sensors were unreliable due to their instability and susceptibility to mechanical failure resulting in the breakage of lead wires. Thin film RTD is the most rugged of the three RTD elements and is preferred for its increased accuracy over time and improved reliability. A thin film responds faster due to its low thermal mass and ease of assembly into smaller packages. Figure 15.7 shows a thin film RTD.

Compared to other types of temperature sensors, RTDs have the following advantages:

1. The resistance versus temperature linearity characteristics of RTDs are higher.
2. They possess greater accuracy (as high as ±0.1 °C). Standard platinum resistance thermometers have ultra-high accuracy of around ±0.0001 °C.
3. They have excellent stability over time.
4. Resistance elements can be used for the measurement of differential temperature.
5. Temperature-sensitive resistance elements can be replaced easily.
6. RTDs show high flexibility with respect to the choice of measuring equipment, interchangeability of elements, and assembly of components.
7. Multiple resistance elements can be used in an instrument.
8. RTDs have a wide working range without any loss of accuracy and, at the same time, can be employed for small ranges also.
9. They are best suited for remote indication applications.
10. It is possible to operate indicators, recorders, or controllers.

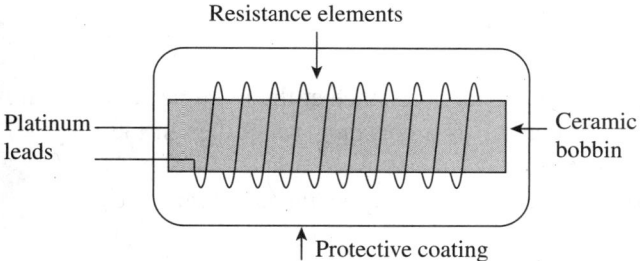

Fig. 15.5   Wire-wound RTD

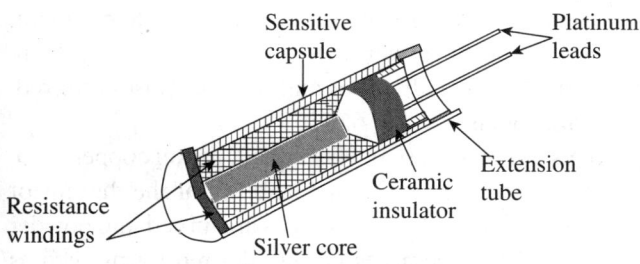

Fig. 15.6   General construction of an RTD

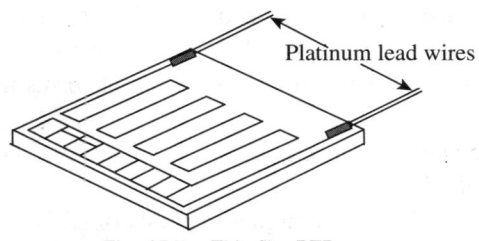

Fig. 15.7   Thin film RTD

RTDs are also associated with some disadvantages. They are as follows:

1. The use of platinum in RTDs makes them more expensive than other temperature sensors.
2. The nominal resistance is low for a given size, and the change in resistance is much smaller than other temperature sensors.

3. Although its temperature sensitivity is high, it is less than that of thermistors.

15.5 THERMISTORS

Semiconductors that are used to measure temperature are called thermistors. When a thermistor is employed for temperature measurement, its resistance decreases with increase in temperature. The valence electrons, which are mutually shared by the metal atoms, move continuously and freely through the metal during their movement from atom to atom. The vibration in the crystal lattice of atoms increases with the increase in temperature. The free movement of electrons becomes restricted due to an increase in the volume of space occupied by the atoms. In case of thermistors, the valence electrons are attached more firmly to the atoms; some of the electrons are detached and flow due to the increase in temperature, which decreases electrical resistance facilitating the easy flow of electrons. Materials used in thermistors for temperature measurements have very high temperature coefficients (8–10 times higher than platinum and copper) and high resistivity (higher than any pure metal). Thus, they are very sensitive to small variations in temperature and respond very quickly. The relationship between temperature and resistance is given by the following equation:

$$R = R_R \, e^{\beta\left(\frac{1}{T} - \frac{1}{T_R}\right)}$$

Here, R is the resistance at temperature T, R_R is the resistance at the reference temperature T_R, e is the base of the Napierian logarithm, and β is a constant, which lies in the range of 3000–4600 K depending on the composition.

The temperature coefficient of resistance is given by the following equation:

$$\frac{dR/dT}{R} = \frac{\beta}{T^2}$$

The temperature coefficient of platinum at 25 °C is +0.0036/K and, for thermistors, it is generally around −0.045/K, which is more than 10 times sensitive when compared to platinum. A variety of ceramic semiconductor materials qualify as thermistor materials. Among them, germanium containing precise proportions of arsenic, gallium, or antimony is most preferred. The temperature measurement range of thermistors is −250 to 650 °C.

Thermistors are also produced using oxides of manganese, nickel cobalt, nickel copper, iron, zinc, titanium, and tin. In order to attain better reproducibility and stability of the thermistor characteristics, some chemically stabilizing oxides are added. The oxides are milled into powder form and mixed with a plastic binder, which are then compressed into desired forms such as disks or wafers. Disks are formed by compressing the mixtures using pelleting machines, and the wafers are compression moulded. They are then sintered at high temperatures to produce thermistor bodies. Depending on their intended application, leads are then added to these thermistors and coated if necessary. To achieve the required stability, the thermistors so formed are subjected to a special ageing process. Figure 15.8 illustrates the different forms in which thermistors can be made.

The variation of temperature with voltage and resistance is shown in Fig. 15.9. The use of thermistors as temperature sensors has several advantages:

1. Thermistors possess very high sensitivity, which is much higher than that of RTDs and thermocouples, and hence have the capability to detect very small changes in temperature.

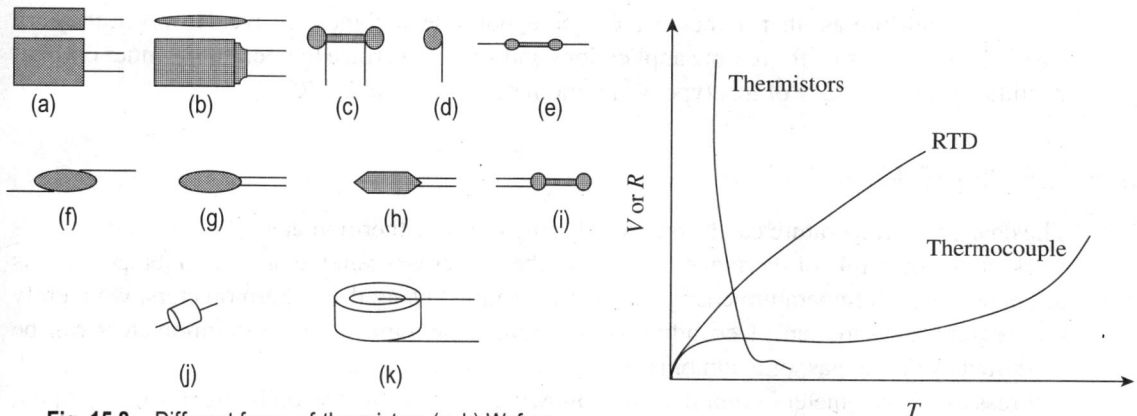

Fig. 15.8 Different forms of thermistors (a, b) Wafers (c–e) Rods (f–i) Beads (j, k) Disks

Fig. 15.9 Comparison of temperature characteristics

2. Their response is very fast, and hence, they are employed for precise control of temperature.
3. They are inexpensive.

Thermistors also have certain disadvantages:
1. They have highly non-linear resistance temperature characteristics.
2. The temperature range is narrow.
3. Low fragility is often a problem.
4. High-temperature performance of thermistors is not good and they exhibit instability with time.
5. They are prone to self-heating errors.

15.6 LIQUID-IN-GLASS THERMOMETERS

The liquid-in-glass thermometer is the most popular and is widely used for temperature measurement. It comprises a bulb that contains a temperature-sensing liquid, preferably mercury. Alcohol and pentane, which have lower freezing points than mercury and do not contaminate if the bulb is broken, are also used. Since alcohol has a better expansion coefficient than mercury, it is also used. A graduated capillary tube is connected to the bulb. At the top of the capillary, a safety or expansion bulb is provided. Figure 15.10 shows a liquid-in-glass thermometer. A range cavity is provided just above the bulb to accommodate the range variation. The walls of the bulb should be thin in order to facilitate quick transfer of heat. Further, for the response to be quick, the volume of liquid should be small. However, the larger the volume of the liquid, the higher the sensitivity. Since speed of response depends on the volume of the liquid, a compromise needs to be made between sensitivity and response.

The entire assembly is enclosed in a casing to provide protection from breakage. An extra-long stem may be provided to facilitate easy dipping into hot liquids. Calibration of thermometers has to be carried out for better

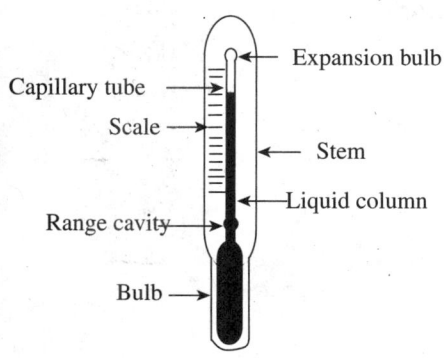

Fig. 15.10 Liquid-in-glass thermometer

results. Liquid-in-glass thermometers are simple, portable, and inexpensive. However, they are fragile and not suitable for remote applications and sensing surface temperature. Under optimal conditions, the accuracy of this type of thermometers is around 0.1 °C.

15.7 PRESSURE THERMOMETERS

The change in temperature can be measured using pressure thermometers. These thermometers work on the principle of thermal expansion of the matter wherein the change in temperature is to be measured. Temperature change can be determined using these thermometers, which rely on pressure measurement. Depending on the filling medium, pressure thermometers can be classified as liquid, gas, or a combination of liquid and its vapour.

Pressure thermometers comprise the following components: a bulb filled with a liquid, vapour, or gas; a flexible capillary tube; and a bourdon tube. Due to variation in temperature, the pressure and volume of the system change and the fluid either expands or contracts. This causes the bourdon tube to move or uncoil, which actuates the needle on the scale, thus providing a measure of the temperature. A typical pressure thermometer is shown in Fig. 15.11.

Pressure thermometers are extensively preferred because of the following advantages:
1. Direct reading or recording is possible since pressure thermometers give adequate force output.
2. Pressure thermometers are less expensive than other systems.

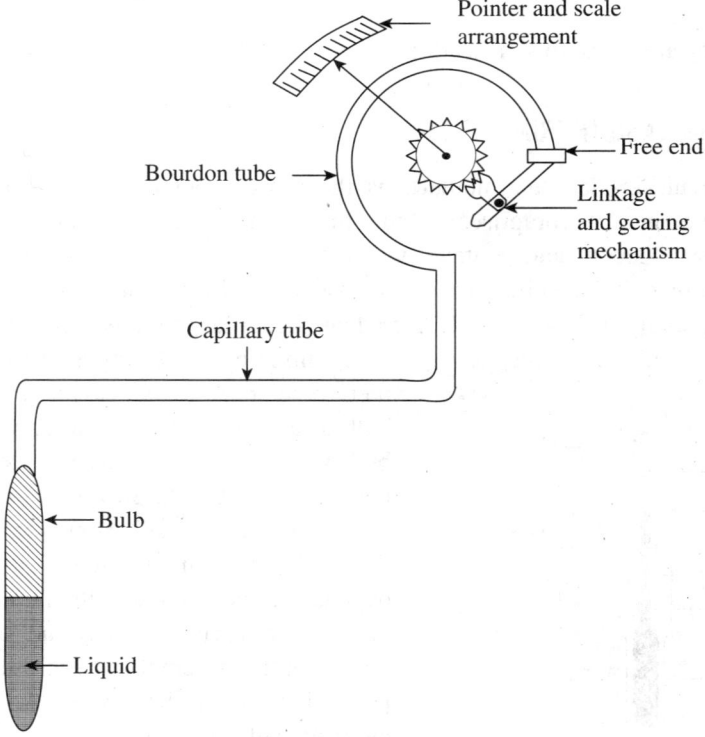

Fig. 15.11 Pressure thermometer

3. They show a superior dynamic response when compared to liquid-in-metal thermometers.
4. They deliver a high-speed response.
5. Their maintenance is easy.

However, they suffer from limited linearity and are prone to output errors. In case of liquid- or gas-filled thermometers, the tube and the filling are temperature sensitive, and if calibration conditions differ along the tube, an error is induced. This error can be minimized by increasing the ratio of volumes of the bulb and the tube. Increasing the bulb size reduces the time response of the system but does not degrade the response. The presence of a pressure gradient resulting from the elevation difference between the bulb and the bourdon tube also contributes to errors.

15.8 BIMETALLIC STRIP THERMOMETERS

John Harrison, who invented bimetallic thermometers, made the first bimetallic strip in 1759 to compensate for changes in the balance spring induced by temperature. A bimetallic strip thermometer works on the well-known principle that different metals expand and contract to different degrees, depending on the coefficient of expansion of the individual metals. For example, if two strips of two different metals (steel and copper) are firmly welded, riveted, or brazed together and subjected to temperature changes, either cooling or heating, the degree of contraction or expansion of the metals differ depending on their coefficient of expansion. The metal strips tend to bend owing to their different coefficients of expansion; the contraction or expansion of one strip will be greater than that of the other. The difference in the expansion of two metals, which makes the strip bend, is a measure of temperature, and since two different metal strips are employed it is called a bimetallic strip thermometer. Figure 15.12 shows the principle of a bimetallic strip.

Bimetallic strips are manufactured in different shapes: cantilever type, flat form, U form, and helical and spiral shapes. In bimetallic strips, the lateral displacement in both the metals is much larger than the small longitudinal expansion. This effect is made use of in mechanical and electrical devices. For industrial use, the strips are wrapped around a spindle into a helical coil. Due to its coil form, the length of the bimetallic strip increases, which in turn increases its sensitivity. Bimetallic strip thermometers are preferred for their ruggedness and availability in

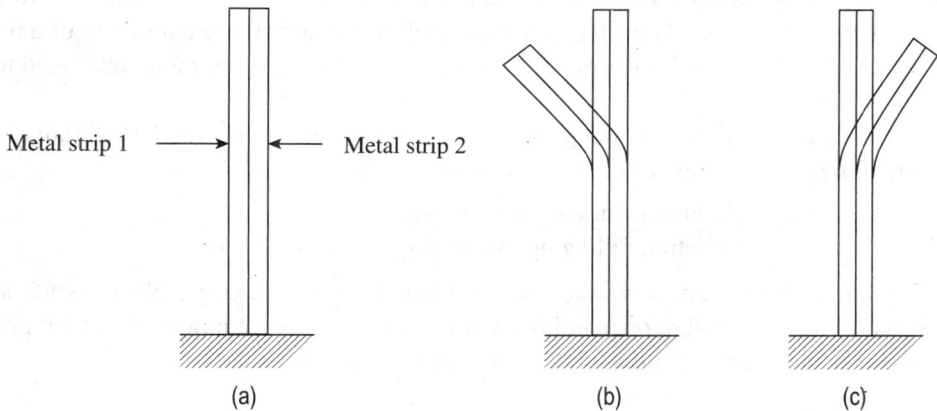

Metal strip 1 → ← Metal strip 2

(a)   (b)   (c)

Fig. 15.12 Deflection of a bimetallic strip (a) Normal condition (b) Cold condition (c) Hot condition

suitable forms. These thermometers are used for sensing temperature of hot water pipes, steam chambers, etc. They are also used in temperature compensation clocks and circuit breakers.

15.9 PYROMETRY

In the preceding sections, contact-type temperature measurement devices have been discussed. If the temperature of a very hot body has to be measured, contact-type temperature-measuring devices will not be suitable, because they are liable to be damaged when they come in contact with the hot body. Hence, the use of non-contact-type temperature-measuring devices becomes imperative. When such devices are employed for high-temperature measurement, the distance between the source of the temperature and the instrument has no effect on the measurement. These non-contact-type devices are called pyrometers. The term pyrometer is of Greek origin, wherein pyro stands for 'fire' and metron means 'to measure'. Pyrometer was first invented by the great English potter Josiah Wedgwood in the early 1780s, for which he received the Fellowship of the Royal Society in 1783. Measurements of temperature are carried out either by measuring energy radiated by a hot body or by colour comparison. Pyrometers are classified into two distinct categories: total radiation pyrometers and optical pyrometers.

It is very well known that conduction, convection, and radiation are the three modes of heat transfer. It is also a fact that all bodies above absolute zero radiate energy, and radiation does not require any medium. When heat is radiated by a body at a given intensity, radiation is emitted across a spectrum of wavelengths. Radiation intensity is directly proportional to temperature. When radiation intensity falls, there is a corresponding increase in the wavelength of radiation. On the other hand, the wavelength decreases when there is an increase in the intensity of radiation.

A body that is capable of absorbing all the radiations that fall on it and does not reflect or transmit any radiation is called a black body. A black body is an ideal thermal radiator. The capacity of radiation absorption of a body depends on its surface condition. Under similar conditions, a black body absorbs and loses heat more quickly than a non-black body. The radiation absorption of a body is much superior when its colour is black and when it has a matt or rough surface finish. The heat radiated by a black body is a function of temperature.

When a body radiates less heat than a black body, the ratio of this radiation to the black body radiation is known as the total emissivity of the surface. If the emissivity of a black body radiation is 1, then other bodies that transmit some of the incoming radiation is said to have an emissivity of less than 1.

If E is the radiation impinging on a body (W/m^2), α is the fraction of radiation power absorbed by the body, and e is the total emissivity, then

$$e = \frac{\alpha E}{E} = \frac{\text{Radiation emitted from a body}}{\text{Radiation falling upon the body}}$$

The Stefan–Boltzmann law states that the total energy radiated by a black body is a function of its absolute temperature only and is proportional to the fourth power of that temperature.

If E is the total energy radiated per unit area per unit time, then

$$E \propto T_a^{\,4} \text{ or } E = \sigma T_a^{\,4}$$

Here, σ is the Stefan's constant, which is equal to 50×10^{-8} W K^4/m^2 and T_a is the absolute temperature in K.

According to the Stefan–Boltzmann law, if two ideal radiators are considered, the pyrometer not only receives heat from the source, but also radiates heat to the source. Therefore, the net rate of exchange of energy between them is given by

$$E = \sigma \, (T_p^{\,4} - T_s^{\,4})$$

where T_p and T_s are the temperatures of the pyrometer and the radiating source, respectively.

Further, $T_p^{\,4}$ is small when compared to T_s and can be neglected.

If e is the emissivity of the body at a given temperature, then

$$E = e\sigma T_a^{\,4}$$

If the emissivity of the surface is known, then the temperature of the body can be determined provided it is kept in the open. If T_t is the true temperature and T is the apparent temperature taken with a radiation pyrometer positioned in such a way that the body is in full view, then

$$E = \sigma T^4$$

The pyrometer receives energy that is equivalent to the energy emitted by a perfect black body radiating at a temperature T. This temperature is lower than the temperature of the body under consideration. We know that the energy emitted by a non-black body is given by the equation $E = e\sigma T_a^{\,4}$.

Hence, $eT_a^{\,4} = T^4$

With the help of this equation, the value of T_a can be determined.

15.9.1 TOTAL RADIATION PYROMETER

A total radiation pyrometer gives a measure of temperature by evaluating the heat radiation emitted by a body. All the radiations emitted by a hot body or furnace are measured and calibrated for black-body conditions. A total radiation pyrometer (shown in Fig. 15.13) basically comprises an optical system that includes a lens, a mirror, and an adjustable eyepiece. The heat energy emitted from the hot body is focused by an optical system onto the detector. The heat energy sensed by the detector, which may be a thermocouple or a thermopile, is converted to its analogous electrical signal and can be read on a temperature display device.

The pyrometer has to be aligned properly such that it is in line with the furnace or hot body and is placed as close to it as possible. This is essential to minimize the absorption of radiation by the atmosphere. Radiation pyrometers find applications in the measurement of temperature in corrosive environments and in situations where physical contact is impossible. In addition, radiations of moving targets and invisible rays can also be measured. It is also used for temperature measurement when sources under consideration have near-black body conditions.

The following are the advantages of radiation pyrometers:
1. It is a non-contact-type device.
2. It gives a very quick response.
3. High-temperature measurement can be accomplished.

Radiation pyrometers also have certain disadvantages. They are as follows:

1. Errors in temperature measurement are possible due to emission of radiations to the atmosphere.
2. Emissivity errors affect measurements.

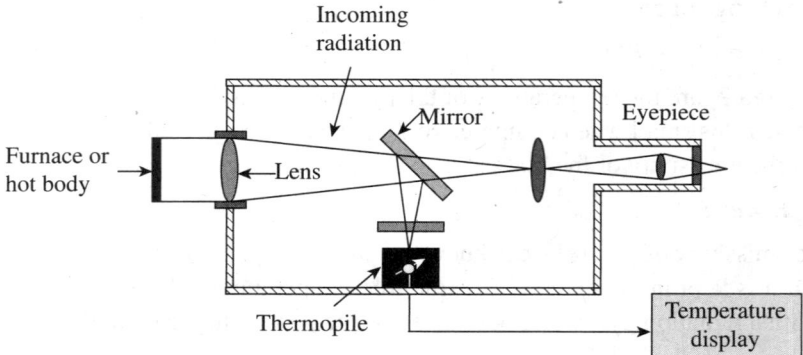

Fig. 15.13   Total radiation pyrometer

15.9.2 Optical Pyrometer

Optical pyrometers work on the disappearing filament principle. In order to measure temperature, the brightness generated by the radiation of the unknown source or hot body whose temperature is to be determined is compared with that of the reference lamp. The brightness of the reference lamp can be adjusted so that its intensity is equal to the brightness of the hot body under consideration. The light intensity of the object depends on its temperature, irrespective of its wavelength. A battery supplies the current required for heating the filament. The current flowing through the filament is adjusted by means of a rheostat and an ammeter is used to measure it. The current passing through the circuit is proportional to the temperature of the unknown source. An optical pyrometer is schematically represented in Fig. 15.14.

An optical pyrometer essentially consists of an eyepiece, by means of which the filament and the source are focused so that they appear superimposed, enabling a clear view for the observer. Between the eyepiece and the reference lamp, a red filter is positioned, which helps narrow

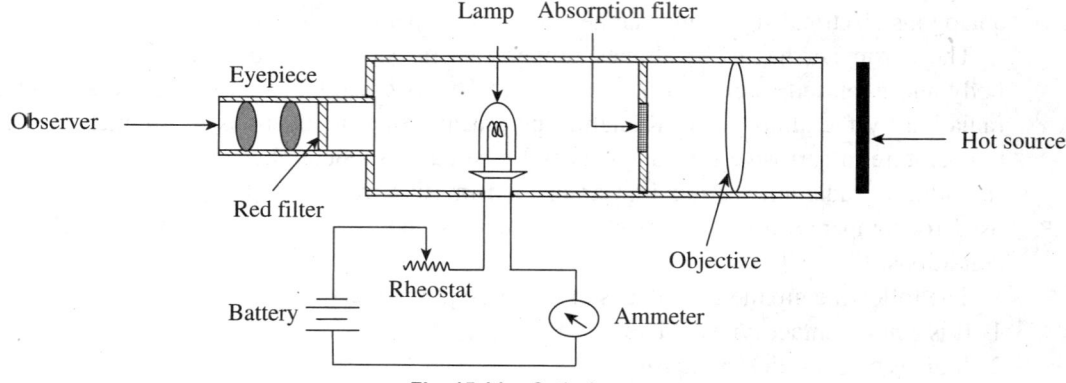

Fig. 15.14   Optical pyrometer

down the wavelength band and attain monochromatic conditions. An absorption filter helps operate the lamp at reduced intensity, thereby enhancing the life of the lamp. The current in the reference lamp can be varied by operating the rheostat and can be measured using the ammeter. Thus, the intensity of the lamp can be altered. The following three situations (represented in Fig. 15.15) arise depending on the current passing through the filament or lamp:

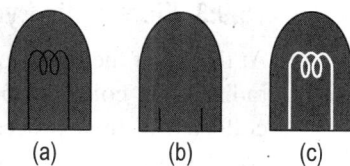

(a)   (b)   (c)

Fig. 15.15 Disappearing filament principle (a) Current passing though the filament is low (b) Current passing though the filament is exact (c) Current passing though the filament is high

1. When the current passing through the filament is very low, the radiation emitted by the filament is of a lesser intensity than that of the source, and the filament appears dark against a bright backdrop, as shown in Fig. 15.15(a).
2. When the current passing through the filament is exact, the intensity of the radiation emitted by the filament is equal to that of the source and hence the filament disappears into the background, as shown in Fig. 15.15(b).
3. When the current passing through the filament is very high, the radiation emitted by the filament is of higher intensity than that of the source and the filament appears brighter than the background, as shown in Fig. 15.15(c).

When the filament disappears, the current that flows in the circuit is measured. The value at which the filament disappears is a measure of the temperature of the radiated light in the temperature source, when calibrated.

In another form of optical pyrometer, the red wavelength from the hot body is compared with the radiation of a similar wavelength emitted by an electric lamp after calibration. The advantages associated with optical pyrometers are as follows:

1. They are simple in construction and portable.
2. Optical pyrometers are flexible and easy to operate.
3. They provide very high accuracy of up to ±5 °C.
4. Since they are non-contact-type sensors, they are used for a variety of applications.
5. They can be used for remote-sensing applications, since the distance between the source and the pyrometer does not affect the temperature measurement.
6. Optical pyrometers can be employed for both temperature measurement and for viewing and measuring wavelengths that are less than 0.65 μm.

The following are the disadvantages of optical pyrometers:

1. Optical pyrometers can be employed for measurement only if the minimum temperature is around 700 °C, since it is based on intensity of light.
2. Temperature measurement at short intervals is not possible.
3. Emissivity errors may affect measurement.
4. Optical pyrometers are used for the measurement of clean gases only.

Optical pyrometers can be employed to measure temperatures of liquid metals and materials that are heated to a high temperature. This method is useful in situations where physical contact is impossible, for example, for determining the temperature of molten metals and materials that are heated to a high temperature. Temperature of furnaces can be measured easily.

15.9.3 Fibre-optic Pyrometers

At the tip of the fibre optics' free end, a temperature-sensing component is placed. The desired radiation is collected by connecting a measuring system to the other end. The information collected is then processed into a temperature value. The fibre-optic pyrometer essentially comprises a fibre-optic cable and an array of components such as probes, sensors or receivers, terminals, lenses, couplers, and connectors. A fibre-optic cable is used to transmit radiation from the black box cavity to a spectrometric device that computes the temperature. The operating temperature of the fibre optics can be raised as they do not possess any electronic components and hence do not require any cooling effort. Fibre optics can be used up to a temperature of 300 °C and are available for wavelengths of 1 and 1.6 µm. With the advent of technology, replacement of fibre-optic cables and optic systems can be carried out without any recalibration. These find applications in procedures such as induction heating and welding. Fibre-optic pyrometers are basically employed in applications where strong electrical and magnetic interference fields act.

The following are the advantages of fibre optics:
1. These can be employed when the path of sight to the target is unclear.
2. When accuracy is a critical parameter, these can be used for measurement.
3. These can be used when the target whose temperature is to be determined is subjected to a physical or chemical change.
4. Temperature as low as 100 °C can be measured.
5. Fibre-optic pyrometers do not carry any electrical current and can hence be safely employed in explosive and hazardous locations.

15.9.4 Infrared Thermometers

It is a well-known fact that every material or matter whose temperature is above absolute zero emits infrared radiations depending on the temperature. Infrared radiations are invisible to the human eye and can be sensed as heat. An infrared thermometer is a non-contact-type sensor that can detect infrared radiation from a heated body. We know that the radiation emitted by an object or a heated body has different wavelengths. Radiations that have longer wavelengths than visible light are known as infrared radiations; these radiations possess less energy and are less harmful. A part of the infrared energy radiated by an object is detected by an infrared sensor. It essentially measures the amount of radiation emitted by an object. Infrared thermometers are ideally suited for high-temperature measurement. The surface of the object begins to radiate when it attains a temperature of around 500–600 °C. The infrared energy increases with increase in temperature. The practical wavelength range of infrared radiations is between 0.7 and 20 µm, but normally radiations in the wavelength range 0.7–14 µm are employed for measurement.

An infrared thermometer comprises a lens through which the infrared wave is focused on the detector.

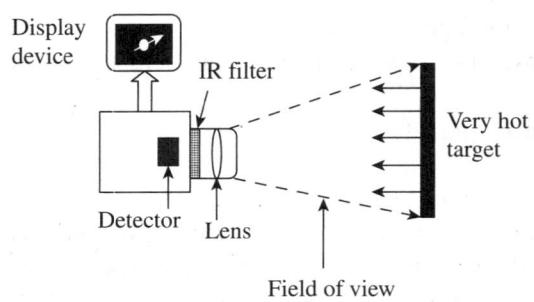

Fig. 15.16 Principle of infrared measurement

The infrared energy is absorbed and converted into an electrical signal by the detector. The amount of radiation striking the detector determines the electrical output signal. The amplified electrical output signal is then displayed on the screen in terms of temperature units. Variation due to ambient temperature is adequately compensated before display. The principle of infrared measurement is schematically represented in Fig. 15.16.

In order to assess the temperature of the target accurately, care needs to be taken to place the heated body or target in such a way that it is in the complete field of view of the instrument. The field of view can be defined as the angle of vision at which the instrument functions. The infrared instrument determines the average temperature of all the surfaces exposed within the field of vision. Temperature measurement may not be accurate if the temperature of the background surfaces is different from that of the surfaces of the target under consideration. The infrared thermometer provides better accuracy if emissivity of the material is known. Higher the emissivity of the material, higher the accuracy of measurement. Reflection from other surfaces also contributes to measurement errors.

An infrared thermometer facilitates the measurement of temperatures of moving objects. In situations where objects are placed in vacuum or in a controlled atmosphere, or the distance between the source of the temperature and the instrument of measurement is large, an infrared device is most useful. It is particularly helpful when the objects are in inaccessible areas/ hazardous conditions, wherein non-contact measurement is the only option available for determining temperature. Ambient conditions such as temperature variations and percentage of CO_2 (since it absorbs infrared radiations) in the atmosphere also affect accuracy. However, infrared thermometers have limited accuracy.

A QUICK OVERVIEW

- Temperature can be defined as a condition of a body by virtue of which heat is transferred from one system to another. Temperature is a measure of the internal energy of a system, whereas heat is a measure of the transfer of energy from one system to another.
- Temperature can be sensed using many devices, which can broadly be classified into two categories: contact- and non-contact-type sensors.
- Thomas Johan Seebeck discovered that when two dissimilar metals are joined together to form two junctions such that one junction (called the hot junction or the measured junction) is at a higher temperature than the other junction (known as the cold junction or the reference junction), a net emf is generated.
- Jean Charles Athanase Peltier discovered that if two dissimilar metals are connected to an external circuit in a way such that a current is drawn, the emf may be slightly altered owing to a phenomenon called Peltier effect. A potential difference always exists between two dissimilar metals in contact with each other. This is known as the Peltier effect.
- Thomson found out that the emf at a junction undergoes an additional change due to the existence of temperature gradient along either or both the metals. The Thomson effect states that even in a single metal, a potential gradient exists, provided there is a temperature gradient.
- Law of homogeneous circuit states that a thermoelectric current cannot be sustained in a circuit of a single homogenous material, regardless of the variation in its cross-section and by the application of heat alone. This law suggests that two dissimilar materials are

- required for the formation of any thermocouple circuit.
- Law of intermediate metals states that if an intermediate metal is inserted into a thermocouple circuit at any point, there will not be any effect on the net emf, provided the new two junctions formed by the insertion of the third metal are at identical temperatures. This law allows the measurement of the thermoelectric emf by introducing a device into the circuit at any point without affecting the net emf, provided that additional junctions introduced are all at the same temperature.
- Law of intermediate temperatures states that if a thermocouple circuit generates an emf e_1 when its two junctions are at temperatures T_1 and T_2, and e_2 when the two junctions are at temperatures T_2 and T_3, then the thermocouple will generate an emf $e_1 + e_2$ when its junction temperatures are maintained at T_1 and T_3.
- An extension of thermocouples is known as a thermopile. A thermopile comprises a number of thermocouples connected in series, wherein the hot junctions are arranged side by side or in star formation.
- The American Society for Testing and Materials has defined the term resistance thermometer as follows: RTD is 'a temperature-measuring device composed of a resistance thermometer element, internal connecting wires, a protective shell with or without the means for mounting a connection head, or connecting wire or other fittings, or both'.
- Semiconductors that are used to measure temperature are called thermistors. When a thermistor is employed for temperature measurement, its resistance decreases with increase in temperature.
- The change in temperature can be measured using pressure thermometers. These thermometers work on the principle of thermal expansion of matter wherein change in temperature is to be measured. Temperature change can be determined using these thermometers, which rely on pressure measurement.

- A bimetallic strip thermometer works on the well-known principle that different metals expand and contract to different degrees, depending on the coefficient of expansion of individual metals.
- Radiation thermometers (pyrometers) measure temperature from the amount of thermal electromagnetic radiation received from a spot on the object of measurement.
- Optical pyrometers measure temperature using the human eye to match the brightness of a hot object to the brightness of a calibrated lamp filament inside the instrument.
- A body that is capable of absorbing all the radiations that fall on it and does not reflect or transmit any radiation is called a black body. A black body is an ideal thermal radiator.
- If E is the radiation impinging on a body (W/m^2), α is the fraction of radiation power absorbed by the body, and e is the total emissivity, then

$$e = \frac{\alpha E}{E} = \frac{\text{Radiation emitted from a body}}{\text{Radiation falling upon the body}}$$

- The Stefan–Boltzmann law states that the total energy radiated by a black body is a function of its absolute temperature only and is proportional to the fourth power of that temperature.
- If E is the total energy radiated per unit area per unit time, then

$$E \propto T_a^4 \text{ or } E = \sigma T_a^4$$

where σ is the Stefan's constant, which is equal to 50×10^{-8} W K^4/m^2 and T_a is the absolute temperature in K.
- At the tip of the fibre optics' free end, a temperature-sensing component is placed. The desired radiation is collected by connecting a measuring system to the other end. The information collected is then processed into a temperature value.
- Every material or matter whose temperature is above absolute zero emits infrared radiations depending on the temperature. Infrared radiations are invisible to the human eye and can be sensed as heat. An infrared thermometer is a non-contact-type sensor that detects the infrared radiation of a heated body.

MULTIPLE-CHOICE QUESTIONS

1. Temperature gives a measure of
   (a) internal energy of the system
   (b) external energy of the system
   (c) total energy of the system
   (d) no relationship with energy

2. The first thermometer was developed by
   (a) Rankine   (c) Kelvin
   (b) Celsius   (d) Galileo Galilei

3. If two bodies that are at different temperatures are so positioned that they are in contact with one another, then
   (a) heat will flow from the colder to the hotter body until they are at the same temperature
   (b) heat will flow from the hotter to the colder body until they are at the same temperature
   (c) heat will not flow
   (d) heat will continue to flow from the hotter to the colder body until you move them apart

4. The measurable property that varies with temperature in a thermocouple is
   (a) expansion   (c) voltage
   (b) thermal radiation   (d) electrical resistance

5. Which of the following temperatures is correct?
   (a) $F = 1.6C + 32$ (c) $F = 1.8C + 273$
   (b) $F = 1.8C + 32$ (d) $F = 1.8C + 212$

6. When there is no heat exchange between two bodies that are at the same temperature, they are said to be in
   (a) thermal equilibrium
   (b) energy equilibrium
   (c) heat equilibrium
   (d) electrical equilibrium

7. Iron–constantan is a
   (a) K-type thermocouple
   (b) J-type thermocouple
   (c) S-type thermocouple
   (d) R-type thermocouple

8. Peltier effect states that when two dissimilar metals are in contact with each other, there exists a
   (a) current difference
   (b) energy difference
   (c) potential difference
   (d) temperature difference

9. Semiconductors used for temperature measurement are called
   (a) thermistors
   (b) thermopiles
   (c) resistance temperature detectors
   (d) pyrometers

10. The metal extensively used in high-accuracy resistance thermometers is
    (a) rhodium   (c) iridium
    (b) nickel   (d) platinum

11. When a thermistor is employed for temperature measurement, its
    (a) resistance increases with the increase in temperature
    (b) resistance decreases with the increase in temperature
    (c) resistance increases with the decrease in temperature
    (d) resistance decreases with the decrease in temperature

12. The Stefan–Boltzmann law states that the total energy radiated by a black body is given by
    (a) $E = \sigma T_a^{5}$ (c) $E = \sigma T_a^{4}$
    (b) $E = \sigma T_a^{3}$ (d) $E^2 = \sigma T_a^{2}$

13. Which of the following statements is true?
    (a) When radiation intensity falls, the wavelength increases.
    (b) When radiation intensity falls, the wavelength decreases.
    (c) When radiation intensity falls, the wavelength remains the same.
    (d) No relationship exists between the wavelength and radiation intensity.

14. Current flows through a circuit spontaneously when two dissimilar metals are joined to form a thermocouple, provided the two junctions formed are maintained at different temperatures. This effect is termed as
    (a) Thomson effect   (c) Rankine effect
    (b) Seebeck effect   (d) Stefan effect

15. To reduce the band width of a wavelength, an optical pyrometer is provided with a
    (a) radiation filter   (c) absorption filter
    (b) blue filter   (d) red filter

REVIEW QUESTIONS

1. Define temperature. How is it different from heat?
2. Compare the different temperature scales.
3. List the factors that influence the response of a temperature-sensing device.
4. Explain Seebeck, Thomson, and Peltier effects.
5. State and explain the different laws of thermocouples.
6. Explain the different thermocouple materials and their designation.
7. Explain thermistors with a neat sketch.
8. With sketches, describe the different forms of thermistors.
9. Write a short note on thermopiles.
10. With the help of a neat diagram, explain the construction and working of an RTD.
11. With a neat sketch, explain a liquid-in-glass thermometer.
12. With a schematic diagram, explain the working of a pressure thermometer.
13. Explain the working principle of a bimetallic strip with neat sketches.
14. What is pyrometry? Briefly explain the theory of pyrometry.
15. With a neat sketch, explain a total radiation pyrometer.
16. Explain the construction and working of an optical pyrometer with the help of a schematic diagram.
17. Mention the advantages and disadvantages of an optical pyrometer.
18. List the different applications of a total radiation pyrometer.
19. Write a note on the measurement of temperature using a fibre-optic pyrometer.
20. With a neat sketch, explain the working of an infrared thermometer. In addition, state its application areas.
21. List the factors that contribute to inaccuracy in temperature measurement using infrared systems.

Answers to Multiple-choice Questions

1. (a)	2. (d)	3. (d)	4. (c)	5. (b)	6. (a)	7. (b)	8. (c)
9. (a)	10. (d)	11. (b)	12. (c)	13. (a)	14. (b)	15. (d)	

Pressure Measurements

After studying this chapter, the reader will be able to

- understand the basics and elucidate the methods of pressure measurement
- discuss manometers
- explain the different types of elastic transducers
- describe the different electrical pressure transducers
- explain the calibration of pressure gauges
- elucidate vacuum measurement systems
- explain high-pressure measurement

16.1 INTRODUCTION

Pressure is an essential component of everyday life of human beings. We talk about atmospheric pressure, blood pressure, gauge pressure, vacuum, etc. Hence, it becomes imperative to know the elementary details about pressure and its measurement. Pressure can be defined in many ways.

Pressure is the force exerted by a medium, usually a fluid, on a unit area. Measuring devices usually register a differential pressure—gauge pressure. Pressure is also defined as the force exerted over a unit area. Force may be exerted by liquids, gases, and solids.

$$\text{Pressure (P)} = \frac{\text{Force (F)}}{\text{Area (A)}}$$

Pressure may be measured in atmospheres, bars, or in terms of the height of a liquid column. Standard atmospheric pressure is usually referred to as 760 mmHg. The standard atmospheric level is always measured at the sea level. It is to be noted that atmospheric pressure decreases with increasing altitude. The units of pressure normally depend on the context in which pressure is measured.

Measurement of pressure becomes an important aspect due to the following reasons:
1. It is a quantity that describes a system.
2. It is invariably a significant process parameter.
3. Many a time, pressure difference is used as a means of measuring the flow rate of a fluid.
4. From the lowest to the highest pressures usually encountered in practice, the level of pressure has a range of nearly 18 orders of magnitude.

16.2 Pressure Measurement Scales

The following four basic scales are employed in pressure measurement:
1. Gauge pressure is measured above the local atmospheric pressure.
2. Total absolute pressure is the total pressure measured from zero pressure as the datum point. When the absolute pressure exceeds the local atmospheric pressure, it may be considered to be the sum of the gauge pressure and the local atmospheric pressure. Total pressure is the sum of atmospheric pressure and gauge pressure.

$$\text{Total absolute pressure} = \text{Atmospheric pressure} + \text{Gauge pressure}$$

3. Differential pressure is the difference in pressure measured between two points.
4. When the pressure to be measured is less than the local atmospheric pressure, it is called vacuum pressure. In other words, when the gauge pressure is negative, it is termed as vacuum. Vacuum is defined by the following relation:

$$\text{Vacuum} = \text{Atmospheric pressure} - \text{Absolute pressure}$$

5. Absolute pressure is measured above total vacuum or zero absolute. Zero absolute represents total lack of pressure.

The relationship between absolute, gauge, and barometric pressures are represented in Fig. 16.1.

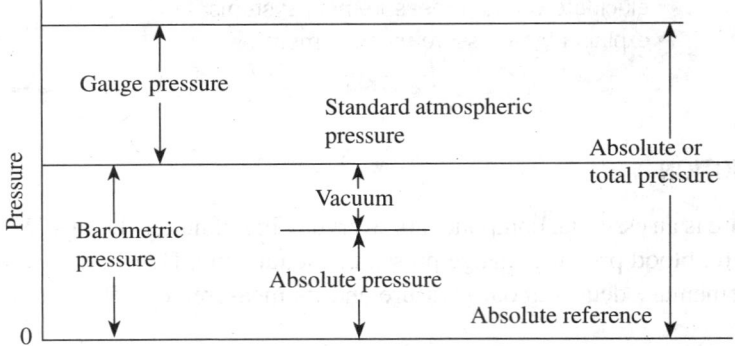

Fig. 16.1 Absolute, gauge, and barometric pressures

The following are the units and conversion factors that are normally used:
1. $1 \text{ Pa} = 1 \text{ N/m}^2$
2. $1 \text{ atm} = 760 \text{ mmHg} = 1.013 \times 10^5 \text{ Pa}$
3. $1 \text{ mmHg} = 1 \text{ Torr}$
4. $1 \text{ Torr} = 1.316 \times 10^{-3} \text{ atm} = 133.3 \text{ Pa}$
5. $1 \text{ bar} = 10^5 \text{ Pa}$

16.3 Methods of Pressure Measurement

Pressure measurements can be grouped into two main categories: *static* and *dynamic* pressures. Static pressure, as the name implies, is the pressure exerted by a fluid when it is in equilibrium or still or static; pressures acting at a point are the same in all directions and do not depend on the direction. For measuring pressures that vary rapidly, the methods that are employed for the measurement of static pressure are not suitable (e.g., pressure in the cylinder of an internal combustion engine). In such cases, pressure transducers are used to convert pressure into signals that are recordable.

16.3.1 Static Pressure Measurement

The static pressure at any point in the fluid is the pressure exerted by the height of the fluid above that point, when the fluid is in static condition. If any attempt is made to restore equilibrium due to the existence of pressure components within a continuous body, the fluid flows from regions of high pressure to those of lower pressure. In such cases, total pressures are direction dependent.

Consider a pipe full of fluid in motion. By attaching a suitable pressure measuring device at the tapping in a pipe wall, the static pressure at the tapping may be determined. We know that in a flowing fluid, several components of pressure exist. Pressure in an air duct can be determined by employing a tube or probe, and the result of such a measurement is dependent on the orientation of the tube or probe.

Figure 16.2 shows the arrangement of pressure probes at two different orientations. These two pressure probes P_1 and P_2 are placed such that their openings receive the impact of the flow. The results of measurements differ from each other. The pressure probe P_1 measures static component of the pressure, whereas the pressure probe P_2 gives stagnation pressure. It can be mentioned here that the static pressure is the pressure sensed while moving along with the stream, and the pressure obtained if the stream is brought to rest entropically may be referred to as the total pressure. The difference between the stagnation pressure and the static pressure is known as the *dynamic* or *velocity pressure*.

It is important to know the different popular devices used for the measurement of static pressure depending on which type of static pressure is under consideration. It is equally important to know the range of measurement of pressure, apart from selecting a proper instrument for measurement. Table 16.1 gives the details of instrumentation and range of pressure to be measured.

Pressure is usually measured by transducing its effects into deflection using the following types of transducers:

1. Gravitational type:
   (a) Liquid columns
   (b) Pistons or loose diaphragms and weights
2. Direct acting elastic type:
   (a) Unsymmetrical loaded tubes
   (b) Symmetrically loaded tubes
   (c) Elastic diaphragms
   (d) Bellows
   (e) Bulk compression

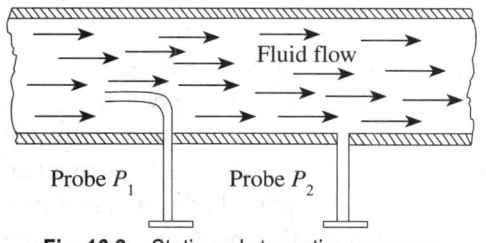

Fig. 16.2 Static and stagnation pressures

Table 16.1 Pressure range and instrumentation

Range of pressure	Instrumentation
Below 1 mmHg	Manometers and low-pressure gauges
Between 1 mmHg and 1000 atm	Bourdon tube, diaphragm gauge, bellow pressure gauge, and dead-weight pressure gauge
Vacuum (760–10^{-9} Torr)	McLeod gauge, thermal conductivity, and ionization gauge
High pressure (1000 atm and above)	Bridgman gauge, electrical resistance-type, Bourdon tube
Variable pressure measurements	Engine indicator, cathode ray oscilloscope

(f) Indirect acting elastic types:
3. Piston with elastic restraining member

16.3.2 Classification of Pressure Measuring Devices

The different instruments/devices used for the measurement of pressure can be classified as follows:
1. Gravitation-type manometers
2. Mechanical displacement-type manometers:
   (a) Ring balance
   (b) Bell type
3. Elastic pressure transducers:
   (a) Bourdon tube pressure gauges
   (b) Diaphragm-type gauges
   (c) Bellow gauges
4. Electrical pressure transducers:
   (a) Resistance-type pressure transducer
   (b) Potentiometer devices
   (c) Inductive-type transducer
   (d) Capacitive-type transducer
   (e) Piezoelectric pressure transducer
   (f) Bridgman gauges
6. Low-pressure measurement gauges:
   (a) McLeod gauges
   (b) Pirani or thermal conductivity gauges
   (c) Ionization gauges
7. Engine indicator (for varying pressure measurements)

16.3.3 Manometers for Pressure Measurement

Manometers have been extensively employed for the measurement of differential pressure. In the following sections, the working of manometers has been discussed. These are sometimes used as primary standards for pressure measurement. It is essential to compensate for the deviations due to gravity by location, compressibility of fluid, and capillary effects to attain the precision required for primary standards. The expansion of the fluid filled in a manometer (due to variation in temperatures) affects their density and, in turn, also the thermal expansion of the read-out scale, affecting the precision of measurement. Better precision in manometers can be accomplished by employing sensors with capacitance or sonar devices instead of visual read-out scales.

Although manometers have several limitations, they are simple and less expensive, which make them popular. They are also used as primary standards for calibration purposes. One of the major disadvantages of manometers is that the filling fluids may vaporize at high vacuum or temperatures. Other limitations include toxicity of mercury, thermal expansion of fluids and read-out scales affecting accuracy of measurement, variations of density, corrosion problems, and evaporation of fluids at low-pressure and high-temperature conditions.

Industrial U Tube Manometer

An improved U tube manometer is used for the measurement of high pressures for industrial purposes, which is given in Fig. 16.3. This type of manometer consists of two limbs, often made of steel, where one limb is of a much larger diameter than the other.

A higher pressure of P_1 to the narrow limb having a cross section of A_1 and a lower pressure of P_2 to the wide limb having a cross section of A_2 are applied. Thus, there exists a differential pressure. It can be observed that the liquid in the wider limb rises and that in the narrow limb falls. Pressure balance can be achieved by using the following equation:

$$P_1 = P_2 + \rho g H = P_2 + \rho g(h + d) \tag{16.1}$$

where H is the total difference in levels in m, h the difference between zero level and the level in the narrow limb in m, and d the difference between zero level and the level in the wider limb in m.

Further, the volume of liquid displaced from the narrow limb to the wider limb must be equal. Thus,

$$A_1 h = A_2 d$$

$$h = \frac{A_2}{A_1} d$$

where A_1 and A_2 are the cross sectional areas (in m^2) of the narrow and wide limbs respectively. Substituting this value of h in equation (16.1), we get the following equation:

$$P_1 = P_2 + \rho g d \left(1 + \frac{A_2}{A_1}\right) \text{ or } P_1 - P_2 = \rho g d \left(1 + \frac{A_2}{A_1}\right)$$

or

$$d = \frac{P_1 - P_2}{\rho g \left(1 + \dfrac{A_2}{A_1}\right)} \tag{16.2}$$

Thus, it can be seen from Eq. 16.2 that the rise of the liquid in the wide limb is proportional to the differential pressure $(P_1 - P_2)$. Industrial manometers comprise a float, which is an integral part of the wide limb, which in turn is connected to a pointer or a pen recorder with the help of a linkage mechanism. Whenever the level of liquid rises due to the differential pressure, the float at the top of the liquid surface records the change in level or, in other words, the differential pressure.

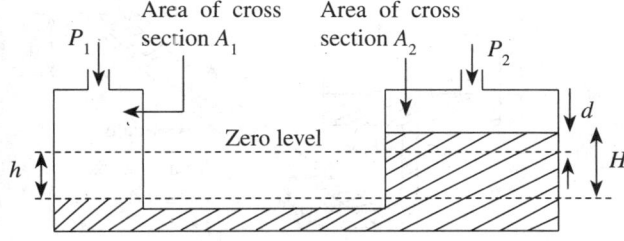

Fig. 16.3   U tube industrial manometer

Cistern Manometer

In these types of manometers, a narrow tube is directly inserted into the wide limb. A differential pressure is applied, as shown in Fig. 16.4. Then

$$P_1 - P_2 = \rho g(h + d)$$

We know that $A_1 d = A_2 h$ and therefore $d = \dfrac{A_2}{A_1} h$

Hence, the following equation is obtained:

$$P_1 - P_2 = \rho g h \left(1 + \dfrac{A_2}{A_1}\right)$$

If the ratio of the areas of narrow and wide tubes, that is, $\dfrac{A_2}{A_1}$, is very small, it can be neglected.

Then, $P_1 - P_2 = \rho g h$.

Hence, it can be concluded that the differential pressure may be measured by measuring the rise in the narrow tube directly (Fig. 16.4).

Inclined Tube Manometer

An inclined tube manometer comprises two limbs. One of the limbs is a narrow glass tube that is inclined at an angle θ to the horizontal. The other limb is a cistern, which is of a wider cross section. Assume that the narrow and wider limbs have cross sectional areas of A_1 and A_2 respectively. A scale is attached to the sloping limb. An inclined tube manometer is shown in Fig. 16.5.

We have the following equation:

$$P_1 - P_2 = \rho g \,(d + x \sin\, \theta)$$

In addition, $A_1 d = A_2 x$; therefore, $d = \dfrac{A_2}{A_1} x$

$$P_1 - P_2 = \rho g x \left(\dfrac{A_2}{A_1} + \sin\, \theta\right)$$

If the ratio of the areas of narrow and wide tube, that is, $\dfrac{A_2}{A_1}$, is very small, it can be neglected.

Hence, $P_1 - P_2 = \rho g x \sin\, \theta$.

The inclined tube manometer is an improvement of the conventional U tube manometer. For the same differential pressure, the inclined tube manometer provides an increased length of scale compared with a simple U tube manometer. The letter h, which is shown in Fig. 16.5, represents the height that would have been registered if an ordinary U tube manometer was employed. We have $x \sin\, \theta = h$ from Fig. 16.5. If the tube is inclined at an angle of 30° to the horizontal, then $\sin\, \theta = \sin 30° = \frac{1}{2}$.

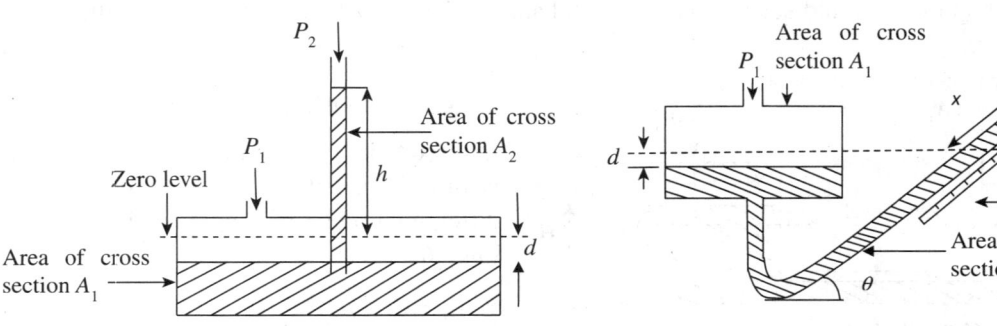

Fig. 16.4 Cistern manometer

Fig. 16.5 Inclined tube manometer

Therefore, $x = 2h$, which means that a scale length of twice the normal value is obtained by inclining the tube at an angle of 30° to the horizontal. Hence, it can be seen that an inclined manometer has the advantage of an increased length of scale.

16.4 RING BALANCE

The ring balance belongs to mechanical displacement-type pressure measuring devices. The ring balance differential manometer, which is also known as a ring balance, is a variation of the U tube manometer. It is composed of an annular ring, which is separated into two parts by a partition. The lower section of the annular ring is also filled with a sealing fluid (either water or mercury). The ring is balanced on a knife edge at its centre so that it is free to rotate. A mass to compensate the difference in pressure is attached to the lower part of the ring. In the ring balance illustrated in Fig. 16.6, P_1 and P_2 represent high and low pressures, respectively. Application of a pressure

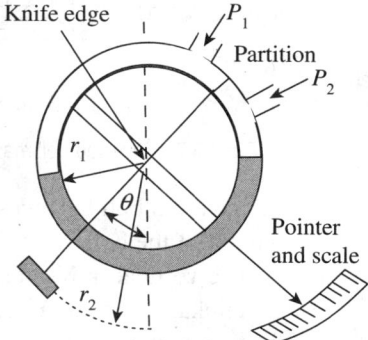

Fig. 16.6 Ring balance

difference across the annular ring causes the displacement of the sealing fluid away from the high-pressure source. This in turn causes a turning moment T_m due to which the annular ring rotates by an angle θ about its centre. The mass attached at the lower part produces an opposing or restoring moment R_m, which balances the turning moment. Thus, the differential pressure can be measured using a pointer and scale arrangement.

The turning moment T_m is given by the following equation:

$$T_m = (P_1 - P_2)Ar_1$$

Here, $P_1 - P_2$ is the differential pressure, A the area of cross section of the annular ring, and r_1 is the mean radius of the annular ring.

The opposing moment, which restores the balance, is given by $R_m = mgr_2 \sin \theta$, where m is the mass attached at the lower part, r_2 is the radius of the point of application of the mass, θ is the angle of rotation acceleration due to gravity, and g is the acceleration due to gravity.

Since the turning moment is balanced by the restoring moment $T_m = R_m$,

$$(P_1 - P_2)Ar_1 = mgr_2 \sin \theta$$

or

$$P_1 - P_2 = \frac{mgr_2}{Ar_1} \sin \theta$$

From the aforementioned equation, it can clearly be seen that the applied differential pressure is proportional to the angle of rotation and the differential pressure can be assessed by determining the angle of rotation.

16.5 INVERTED BELL MANOMETER

An inverted bell manometer is another pressure measuring device that is of the mechanical displacement type.

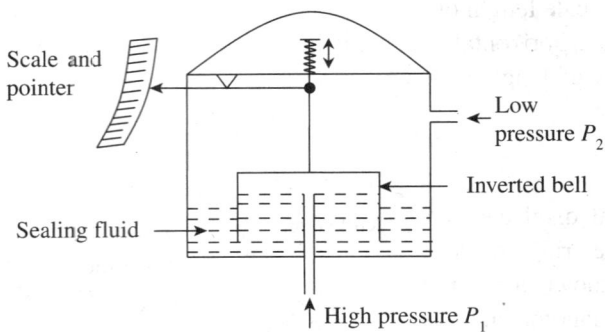

Fig. 16.7 Inverted bell manometer

Figure 16.7 shows the working principle of an inverted bell manometer. In this, as the name suggests, the bell is immersed in the sealing fluid in an upside-down position. The inverted bell moves in the vertical direction due to the differential pressure arising out of the pressure difference between the interior and exterior surfaces of the bell. This type of manometer is capable of measuring absolute, positive, negative, and differential pressures, depending on the pressure on the reference side of the bell. A spring provided on top of the inverted bell balances the vertical motion of the bell due to a pressure difference or by the weight of the bell.

The vertical movement of the bell can be translated into a pointer movement with the help of a linkage system. A variable reluctance pickup can be employed, which converts the vertical motion of the bell to an electrical signal instead of a spring. The inverted bell manometer will measure the absolute pressure if the lower-pressure side is connected to a vacuum line with an appropriate sealing fluid. The displacement of the bell is a linear function of differential pressure.

The upward movement of the bell (F_b) is balanced against the opposing force of the spring (F_s).

When they are at equilibrium, spring force = upward movement, that is,

$$F_s = F_b \text{ or } k \, \Delta y = (P_1 - P_2)A$$

or, $\quad \Delta y = \dfrac{A}{k} (P_1 - P_2)$

where k is the spring constant, Δy is the displacement of the bell, and A is the cross sectional area of the bell.

Further, let l be the length of the pointer and θ the angle of deflection; then $l\theta = \Delta y$.

It follows that $\theta = \dfrac{A}{kl} (P_1 - P_2)$

A double-bell manometer is a variation of this, which is used when high precision is required for differential pressure measurement.

16.6 ELASTIC TRANSDUCERS

Single diaphragms, stacks of diaphragms, and bellows are some of the important elastic transducers used for pressure measurement. Diaphragms are generally used as primary transducers for dynamic pressure measurement. These may be of a flat or corrugated type, as shown in Figs 16.8(a) and (b). Flat diaphragms are used along with electrical secondary transducers for better amplification of small diaphragm deflections. For large deflections, corrugated diaphragms are preferred. Corrugated diaphragms generally find application in

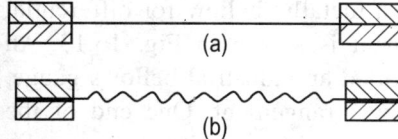

Fig. 16.8 Types of diaphragms (a) Flat diaphragm (b) Corrugated diaphragm

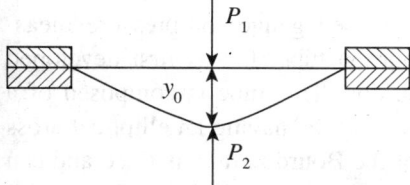

Fig. 16.9 Simple diaphragm

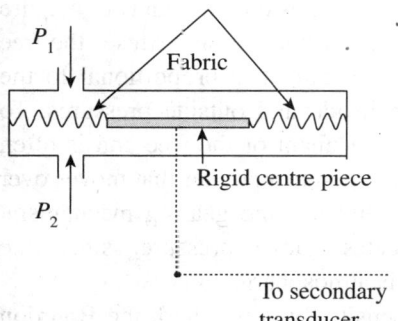

Fig. 16.10 Fabric diaphragm

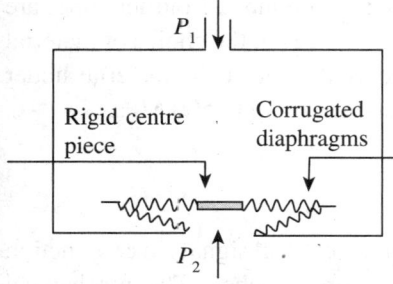

Fig. 16.11 Pressure capsule

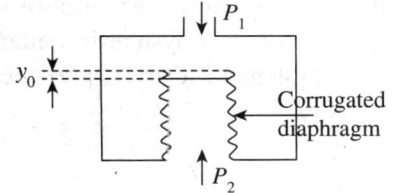

Fig. 16.12 Metallic bellow

static pressure measurement due to their increased size and deflection, which affect the dynamic response.

A single diaphragm in its simplest form is shown in Fig. 16.9. It is a thin, flat, circular plate fixed at the two ends; upon application of pressure, it will deflect as shown in Fig. 16.9, and the resulting differential pressure is given by $P_1 - P_2$. This can be used only for relatively small movements wherein the relationship between pressure and deflection is linear. The deflection attained by flat diaphragms is limited by linearity constraints or stress requirements. However, for practical applications, some modification is required.

Sometimes, a mechanical linkage system or an electrical secondary transducer needs to be connected to the diaphragm at its centre. To enable this, a metal disc or any other rigid material is provided at the centre with diaphragms on either side. The diaphragm may be made up of a variety of materials such as nylon, plastic, leather, silk, or rubberized fabric. This type of transducer, which is used for pressure measurement, is known as the slack diaphragm or fabric diaphragm differential pressure gauge. Construction of a fabric diaphragm is shown in Fig. 16.10.

It comprises a rigid centre piece, which is held on either side by diaphragms made of fabric. A secondary transducer, which may be an electrical or a mechanical linkage system, or a recording pen, is connected at the centre (as shown in Fig. 16.10). The slack diaphragm is used to measure low pressures. Since the centre piece is rigid, there may be a reduction in flexibility of the diaphragm.

A pressure capsule or a metal capsule can be formed by joining two or more diaphragms, as shown in Fig. 16.11. Use of corrugated diaphragms increases linear deflections and reduces stresses. It can be seen from Fig. 16.11 that differential pressure can be created by applying one pressure from inside the capsule and another from the outside. In a metallic capsule, the relationship between deflection and pressure remains linear as long as the movement is not excessive.

Metallic bellows can be employed as pressure-sensing elements. A thin-walled tube is converted into a corrugated diaphragm by using a hydraulic press and is stacked as shown in Fig. 16.12. Due to the differential pressure, there will be a deflection, y_0. Normally, materials such as phosphor bronze, brass, beryllium copper, and stainless steel are used for making bellows. Metallic bellows are often associated with zero shift and hysteresis problems.

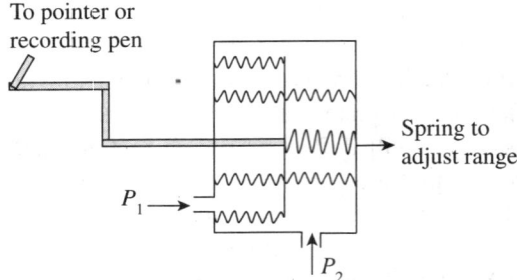

Fig. 16.13 Industrial bellow gauge

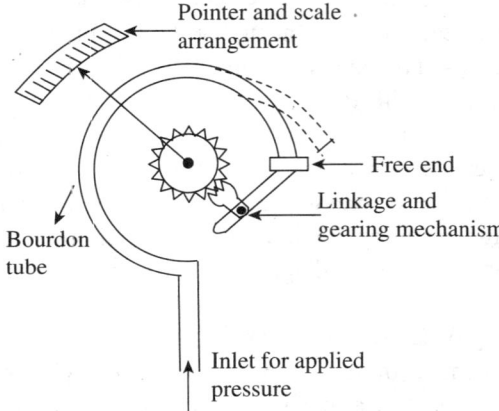

Fig. 16.14 Bourdon tube

Modification of a metallic bellow for differential pressure measurement is shown in Fig. 16.13. An industrial gauge, called an industrial bellows gauge, has a double-bellow arrangement. One end of the double bellow is connected to a pointer or a recorder pen. A high pressure of P_2 and a low pressure of P_1 are applied to create a differential pressure as shown in Fig. 16.13.

The most widely used gauge for pressure measurement is the Bourdon tube. It was first developed in 1849 by E. Bourdon. This tube is composed of a C-shaped hollow metal tube having an elliptical cross section. One end of the Bourdon tube is fixed and can be used as the pressure inlet, as shown in Fig. 16.14. The other end is free and closed. Due to the applied pressure, the tube straightens out and tends to acquire a circular cross section. Thus, pressure causes the free end to move. This movement is proportional to the difference between inside and outside pressures. To measure pressure, movement of the free end is often magnified and transmitted to a pointer that moves over the scale through a linkage and gearing mechanism. The pointer indicates gauge pressure, since the reference pressure is atmospheric.

In case higher sensitivity is required, the Bourdon tube may be formed into a helix containing several turns. Bourdon tubes can also assume helical, twisted, or spiral forms, and the operation of all these gauges is similar to that of C-shaped tubes commonly employed for differential pressure measurement. Bourdon tubes are usually made of phosphor bronze, brass, and beryllium copper. However, the choice of material depends on the range of pressure to be measured and the elastic limit of the material under consideration. Bourdon gauges are employed to measure pressures of up to 500 MPa.

16.7 ELECTRICAL PRESSURE TRANSDUCERS

Electrical pressure transducers translate mechanical output into electrical signals in conjunction with elastic elements such as bellows, diaphragms, and Bourdon tubes. The mechanical displacement is first converted into a change in electrical resistance, which is then converted into an electrical signal, that is, change in either current or voltage. Electric pressure transducers are preferred over mechanical devices because of their quick response, low hysteresis, better linearity properties, and high accuracy in digital measurement systems. Electrical pressure transducers are classified as follows:

1. Resistance-type transducer
2. Potentiometer devices
3. Inductive-type transducer

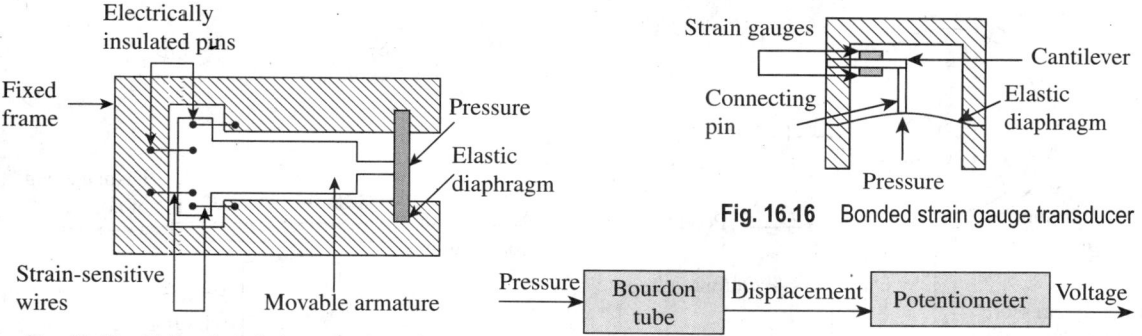

Fig. 16.15 Unbonded strain gauge transducer

Fig. 16.16 Bonded strain gauge transducer

Fig.16.17 Block diagram of a potentiometric pressure transducer

4. Capacitive-type transducer
5. Piezoelectric pressure transducer

The first four are passive electrical transducers and the last one is an active electrical transducer.

16.7.1 Resistance-type Transducer

The basic principle on which a resistance-type pressure transducer works is that a variation in the length of a wire causes a change in its electrical resistance. Figure 16.15 represents an unbonded strain gauge transducer. In between the fixed frame and the movable armature, four strain-sensitive wires are connected. Using electrically insulated pins the wires are located to the frame and movable armature, as shown in Fig. 16.15. The wires that are mounted under initial tension form the active legs of a conventional bridge circuit.

Application of pressure causes a displacement of the armature, which in turn elongates two of the wires and reduces the tension in the other two wires. The applied pressure thus changes the length of the wire due to which the resistance of the wires vary, causing an imbalance in the bridge. Four wires are used to increase the sensitivity of the bridge.

In case of a bonded strain gauge-type pressure transducer, using an appropriate cement, a wire or foil strain gauge is fastened onto a flexible plate, as shown in Fig. 16.16. In order to accomplish temperature compensation, two strain gauge elements are employed in the bridge circuit.

The arrangement shown in the figure consists of a connecting pin, which is connected to a cantilever on which two strain gauges are bonded. The pressure applied on the diaphragm is transmitted to the cantilever through the connecting pin, which causes an imbalance in the bridge circuit.

16.7.2 Potentiometer Devices

Devices such as potentiometers and rheostats with sliding wires, which work on the principle of movable contacts, are also variable-resistance-type pressure transducers (Figs 16.17 and 16.18). The closed end of the Bourdon tube is connected to the potentiometer wiper. A constant voltage is applied to the end terminals of the potentiometer. Application of pressure to the open end of the Bourdon tube results in the deflection of its closed end. Due to this the wiper moves over the potentiometer varying the resistance of the circuit, thus changing the wiper voltage.

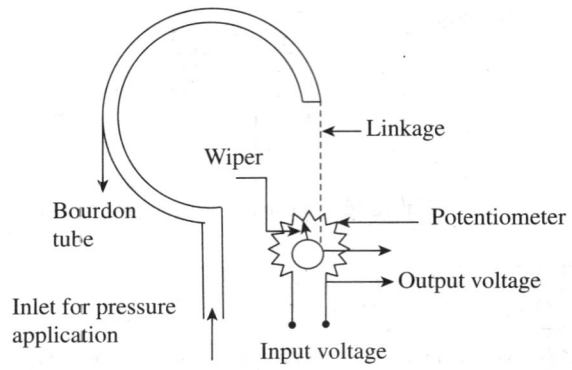

Fig. 16.18 Variable resistance-type transducer

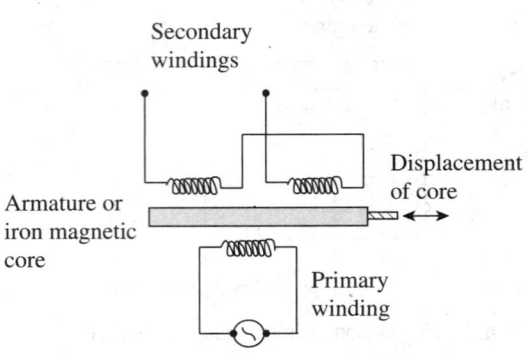

Fig. 16.19 Variable inductance-type transducer

16.7.3 Inductive-type Transducer

The linear variable differential transformer (LVDT) is an inductive type of pressure transducer that works on the mutual inductance principle. It transforms a mechanical displacement into an electrical signal. The magnetic core is connected to an elastic pressure transducer like a Bourdon tube. The Bourdon tube senses the applied pressure and converts it into displacement, which in turn moves the core of the LVDT.

An LVDT comprises one primary and two secondary windings (coils), which are mounted on a common frame, as shown in Fig. 16.19. The three coils are carefully wound on an insulated bobbin. On either side of the primary coil, which is centrally placed, two secondary windings are symmetrically placed. A non-contacting magnetic core moves in the centre of these coils, which are wound on the insulating bobbin. The core, which is made from a uniformly dense cylinder of a nickel–iron alloy, is carefully annealed to enhance and homogenize its magnetic permeability. It is centrally positioned between the secondary windings. When the core is in this position, the induced voltages in the two secondary windings are equal and 180° out of phase, which is taken as the zero position as illustrated in Fig. 16.20. The displacement of core from the zero position due to the applied pressure increases the induced voltage in one of the secondary windings while the voltage in the other decreases. Due to this, the differential voltage, which appears across the two secondary windings, is approximately linear for small core displacements and is hence a measure of applied pressure.

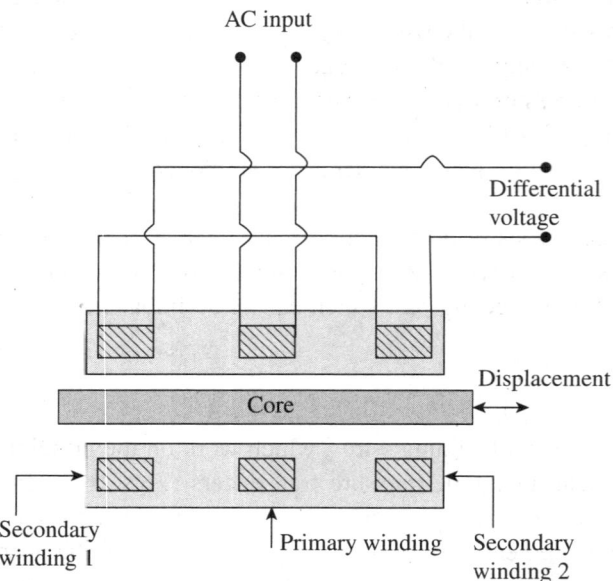

Fig. 16.20 Schematic diagram of an LVDT

16.7.4 Capacitive-type Transducer

The capacitive transducer, shown in Fig. 16.21, works on the principle that when pressure is applied onto the diaphragm, the distance between the two metal plates changes, which in turn changes the capacitance. A variable capacitive transducer comprises a metal diaphragm as the elastic element, which is placed centrally between the two plates. Initially, when the input pressures are the same there is no deflection of the diaphragm and hence capacitance remains constant. Due to the applied pressure, the distance between the fixed plates and the diaphragm varies, and hence the capacitance changes. The resulting change in capacitance can be sensed by a bridge circuit or can be employed to vary the frequency of the oscillator. The capacitive pressure transducer has many advantages such as low hysteresis, superior linearity and repeatability, ability to measure static pressure, and digital output. Capacitive transducers can measure absolute, gauge, or differential pressure.

16.7.5 Piezoelectric-type Transducer

A piezoelectric pressure transducer is an active type of pressure transducer, which works on the principle that when pressure is applied on piezoelectric crystals an electric charge is produced. A block of crystalline material forms the basic element in a piezoelectric crystal, which has the capability to generate an electrical potential due to an applied pressure along a preferred axis. The most widely used materials are quartz and Rochelle salt (potassium sodium tartrate). They possess high mechanical strength, are not expensive, and are easily available. If higher sensitivity is required, materials such as barium titanate and lead zirconate titanate can be used.

A piezoelectric pressure transducer (as shown in Fig. 16.22) comprises a corrugated metal diaphragm on which pressure is applied. Deflection of the diaphragm is transmitted to the piezoelectric crystal through a mechanical link. The piezoelectric crystal is capable of producing the maximum piezoelectric response in one direction and minimum responses in other directions. Thus, the piezoelectric crystal senses the applied pressure and generates a voltage proportional to the applied pressure. The generated voltage can be measured using a calibrated output voltage-measuring instrument, which gives a measure of the applied pressure. This method can be used for high-pressure measurement and can also be employed in systems that require the output of the measured variable in electrical form. It is also used for the measurement of rapidly varying pressures.

Apart from producing an electrical output, the piezoelectric pressure transducer offers other advantages such as smaller size, rugged in construction, and requirement of no external power supply. The limitation is that it cannot be employed for the measurement of static pressure.

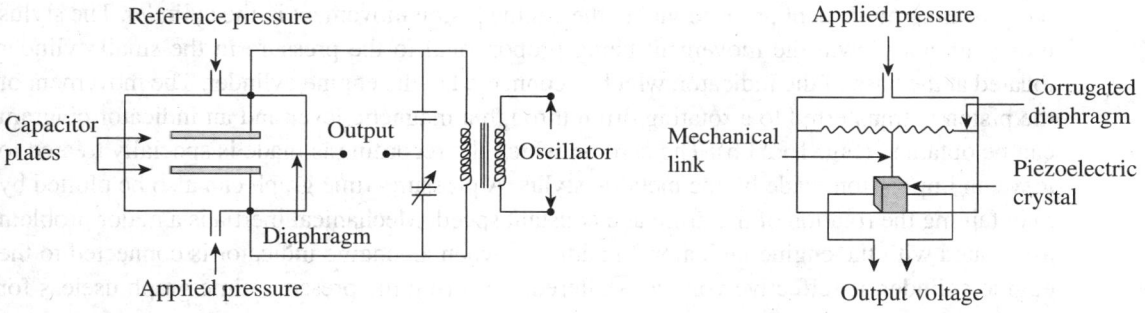

Fig. 16.21   Capacitive-type pressure transducer   **Fig. 16.22**   Piezoelectric pressure transducer

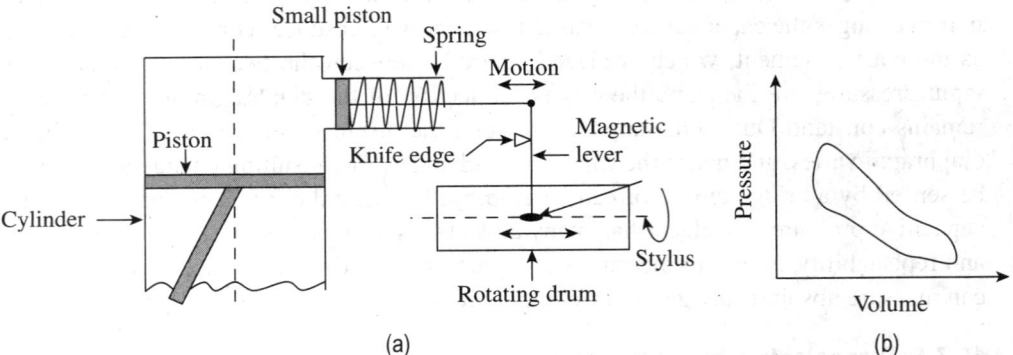

Fig. 16.23 Measurement of cylinder pressure (a) Engine indicator (b) *P–V* diagram

In addition, temperature variations affect the output and a complex circuitry is required for conditioning the signal.

16.7.6 Varying Pressure Measurement

Measurement of variation of pressure is normally simple, especially when the pressure changes slowly. The variation in pressure can be monitored by taking readings periodically. When variation in pressure occurs rapidly, complexity in measurement also increases and static pressure measuring devices will no longer be useful. In fact, when the rate of pressure change increases, it becomes increasingly difficult to take readings, necessitating the use of high-speed recording devices.

Engine Indicator

In order to measure the cylinder pressure in a reciprocating machine, such as an internal combustion engine or an air compressor, it is essential to plot a graph of cylinder pressure versus cylinder volume or time.

An engine indicator illustrated in Fig. 16.23(a) is composed of a small cylinder whose size is known and operates against a spring through which the operating range can be set. An engine indicator records the cylinder pressure corresponding to the piston movement. The drum contains a piece of paper or card on its outer surface on which the simultaneous variation of pressure and cylinder volume is recorded. The rotating drum is designed to provide a reciprocating movement proportional to the engine piston movement in the cylinder. The stylus moves up and down, the movement being proportional to the pressure in the small cylinder located at the base of the indicator, which is connected to the engine cylinder. The movement of the piston is transferred to a rotating drum through a magnetic lever and an indicator diagram can be obtained (Fig. 16.23 b). The card on which the recording is made is specially treated to leave an impression made by the metallic stylus. A pressure–time graph can also be plotted by maintaining the rotation of the drum at a constant speed. Mechanical inertia is a major problem associated with the engine indicator. In addition, when an engine indicator is connected to the engine cylinder, its effective volume is altered, rendering the pressure–time graph useless for small engines.

16.8 DEAD-WEIGHT PRESSURE GAUGE

A dead-weight pressure gauge or piston gauge is a very popular device for measuring static pressure. It works on Archimedes' principle. The air or fluid displaced by the applied weights and the piston exerts a buoyant force, which causes the gauge to indicate the pressure. Dead-weight pressure gauges are normally used to calibrate other pressure measuring devices. A dead-weight tester, shown in Fig. 16.24, is a device used for balancing a fluid pressure with a known weight.

A dead-weight pressure gauge comprises a piston that is inserted into a close-fitting cylinder. The weight of the piston, which is accurately machined, is known. The cross sectional areas of both the piston and the cylinder are known. At the bottom of the dead-weight tester, a chamber with a check valve is provided. The piston is first removed, and the chamber and the cylinder are filled with clean fluid with the plunger in the forwarding position. The plunger is gradually withdrawn, and the entire space is filled with oil. The piston is fixed back into its position, and the gauge to be calibrated is connected to the chamber. The check valve is now opened so that the piston pressure is transmitted to the gauge. Known weights are then placed on the piston. By means of a displacement pump, the plunger can be moved to the forwarding position, thus applying fluid pressure to the other end. Pressure is applied gradually until enough force is attained to lift the piston and the weight combination. When the piston is floating freely within the cylinder, the system is in equilibrium with the system pressure.

Thus, the dead-weight pressure is calculated as follows:

$$P_{dw} = \frac{F_e}{A_e}$$

where F_e is the equivalent force of the piston and weight combination, A_e the equivalent area of the piston and cylinder combination, and P_{dw} the dead-weight pressure.

The reading on the gauge is then compared to the reading thus obtained, and calibration can be carried out if needed. Pressure may be varied by adding several known weights to the piston or by employing different piston cylinder combinations of varying areas. Calibration can be carried out by recording different readings of the gauge in the ascending order of addition of weights, keeping the areas of the piston and cylinder combination the same. Similar exercises can be carried out in the descending order also. Ideally, the readings of the gauge should be the same for both ascending and descending orders; in such a case, the gauge is said to be free from hysteresis. There are some errors associated with a dead-weight tester. One such error is the friction created between the piston and the cylinder wall. The leakage through the clearance between piston and cylinder provides some lubrication, which helps to minimize friction. The rotation of the piston further reduces friction.

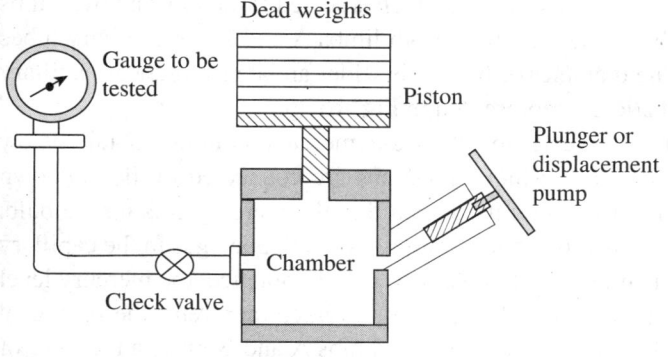

Fig. 16.24 Dead-weight tester

16.9 MEASUREMENT OF VACUUM

Pressures below atmosphere are generally termed as low pressures or vacuum pressures. When the term vacuum is mentioned it means that the gauge pressure is negative. However, atmospheric pressure serves as a reference and absolute pressure is positive. Low pressures are more difficult to measure than medium pressures. Pressures above 1 Torr can easily be measured by the direct measurement method, wherein the force applied causes a displacement. Manometers, diaphragms, bellows, and Bourdon tubes are some examples of the instruments used in direct measurement of pressure. These devices are generally employed to measure a pressure value of about 10 mmHg. For measuring pressures below 1 Torr, indirect or inferential methods are often employed. In these methods, pressure is determined by drawing indirect references to pressure-controlling properties such as volume, thermal conductivity, and ionization of gas. Some of the devices that fall under this category include McLeod gauge, Pirani gauge, and ionization gauge.

16.9.1 McLeod Gauge

McLeod gauge, which was developed in 1874 by Herbert McLeod, is perhaps the most widely used. It is employed as an absolute standard of vacuum measurement for pressures ranging from 10 to 10^{-4} Torr.

A McLeod gauge, which is also known as a compression gauge, is used for vacuum measurement by compressing the low-pressure gas whose pressure is to be measured. The trapped gas gets compressed in a capillary tube. Vacuum is measured by measuring the height of a column of mercury.

McLeod gauge works on Boyle's law, which states that by compressing a known volume of the low-pressure gas to a higher pressure, initial pressure can be calculated by measuring the resulting volume and pressure.

The following fundamental relation represents Boyle's law:

$$P_1 = \frac{P_2 V_2}{V_1}$$

where P_1 and P_2 are the initial and final pressures, respectively, and V_1 and V_2 are the corresponding volumes.

A McLeod gauge is composed of a capillary tube A, which is sealed at the top, and two limbs B and C, which are connected to the vacuum system. Both limbs A and B are capillary tubes and their diameters are the same. The diameter of limb C is wider and hence reduces capillary errors. The McLeod gauge is schematically represented in Fig. 16.25.

Initially, the movable reservoir is lowered to allow the mercury column to fall below the opening level O. In this position, the capillary and limbs are connected to the unknown pressure source. The movable reservoir is then raised such that the mercury fills up the bulb. The mercury level in capillary tube A also rises and compresses the trapped gas in the capillary tube A according to Boyle's law. It is important to note here that, in practice, the mercury level in capillary tube B is raised to the same level as that of limb C, which represents the zero level on the scale. The difference in levels of the two columns in limbs A and B gives a measure of trapped pressure, which can directly be read from the scale.

Let V_1 be the volume of the bulb in capillary A above the level O, P_1 the unknown pressure of the gas in the system connected to B and C, P_2 the pressure of the gas in the limb after compression, and V_2 the volume of the gas in the sealed limb after compression. Then,

$$P_1 V_1 = P_2 V_2$$

where P_1 and P_2 are measured in units of mmHg.

If the cross sectional area of the capillary tube is a and the difference in levels of the two columns in limbs A and B is h, then $V_2 = ah$, where h is the difference between pressures P_1 and P_2, that is, $h = P_2 - P_1$. Therefore, one gets the following equations:

$$P_1 V_1 = P_2 ah$$

$$P_1 V_1 = (h + P_1)ah$$

$$P_1 V_1 = ah^2 + ahP_1$$

$$P_1(V_1 - ah) = ah^2$$

Hence

$$P_1 = \frac{ah^2}{V_1 - ah}$$

$$P_1 = \frac{ah^2}{V_1} \quad (ah \gg V_1)$$

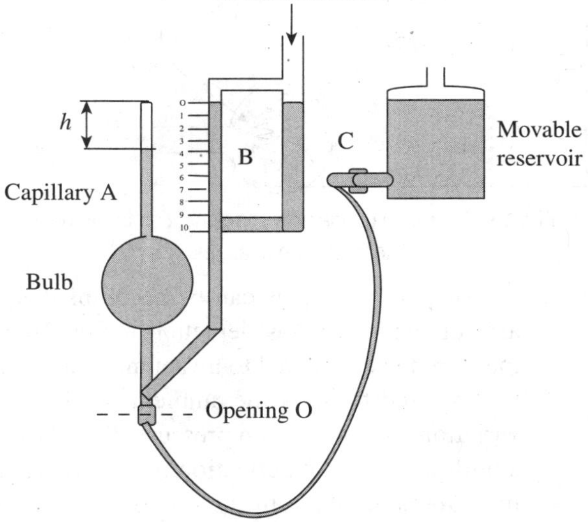

In order to measure low pressures, the value of V_1 is made large compared to that of a. The ratio of V_1 to a is called the compression ratio.

Fig. 16.25 McLeod gauge

If a is made too small, the mercury tends to stick inside the capillary tube; this imposes a restriction on the upper limit of the compression ratio. The compression ratio gets limited due to the excessive weight of mercury if V_1 is very large. McLeod gauges are regularly employed to calibrate other high-vacuum measuring devices. The presence of condensable vapours in the gas whose pressure is to be measured poses a serious limitation as Boyle's law is not followed, which may induce errors.

16.9.2 Pirani Gauge

The principle on which a Pirani gauge (shown in Fig. 16.26) works is thus: when a heated wire is placed in a chamber of gas, thermal conductivity of the gas depends on its pressure. Hence, it follows that energy transfer from the wire to the gas is proportional to the gas pressure. The temperature of the wire can be altered by keeping the heating energy supplied to the wire constant and varying the pressure of the gas, thus providing a method for pressure measurement. On the other hand, a change in the temperature of the wire causes a change in the resistance, providing a second method for the measurement of pressure.

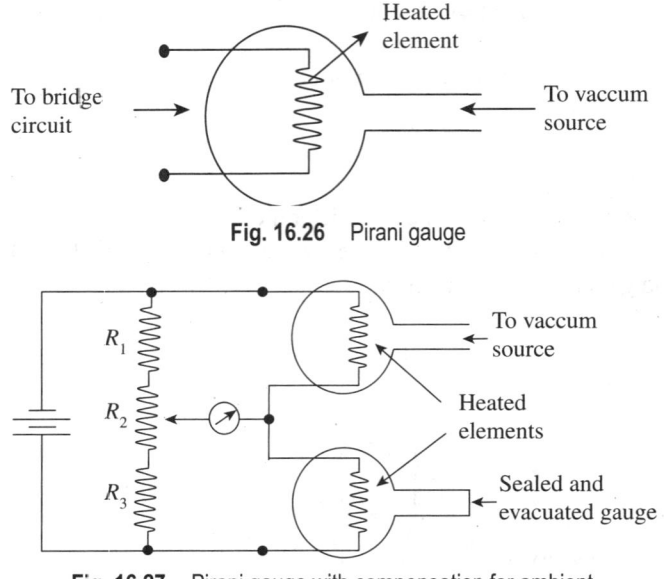

Fig. 16.26 Pirani gauge

Fig. 16.27 Pirani gauge with compensation for ambient temperature changes

Three attributes, namely magnitude of the current, resistivity of the current, and the rate at which heat is dissipated, govern the temperature of the given wire through which an electric current flows. The conductivity of the surrounding media determines the heat dissipation rate. Thermal conductivity reduces due to the reduction in pressure and, consequently, for a given input of electrical energy, the filament attains a higher temperature.

A resistance bridge is employed when the resistance of the wire filament is measured. The bridge is balanced at some reference pressure and the out-of-balance currents are used at all other pressures as a measure of the relative pressures.

Heat loss from the filament due to the variations in ambient temperatures can be compensated. This can be accomplished by connecting the two gauges in series in one arm of the bridge, as depicted in Fig. 16.27. One of the gauges whose pressure is to be measured is connected to a vacuum source and the other is evacuated and sealed. Since both are exposed to the same ambient conditions, the measurement gauge will respond only to variations in the vacuum pressure. By adjusting R_2, the bridge circuit can be balanced to give a null reading. The deflection of the bridge from the null reading, due to the exposure of the measurement gauge to test the pressure environment, will be independent of variations in ambient temperatures.

16.9.3 Ionization Gauge

Ionization gauges are employed for medium- and high-vacuum measurements. These gauges convert neutral gas molecules into positively charged or ionized gas molecules. This gauge is also known as thermionic gauge as electrons are emitted from a heated filament or substance. These emitted electrons are called thermions. The principle of thermionic emission is employed in electron vacuum tubes. When the tungsten filament is heated to a high temperature, electrons acquire sufficient energy and move into the space.

In a hot cathode ionization gauge, electrons emitted from the thermionic cathode can be accelerated in an electric field. These electrons collide with gas molecules and ionize them. The thermionic triode arrangement in an ionization gauge comprises an anode and a cathode encompassed in a glass envelope, which may be connected to the source whose pressure is required to be measured. A schematic representation of the hot ionization gauge is shown in Fig. 16.28.

The anode and grid are at negative and positive potentials respectively, with reference to the filament. Gas molecules collide with the electrons emitted from the heated filament (cathode)

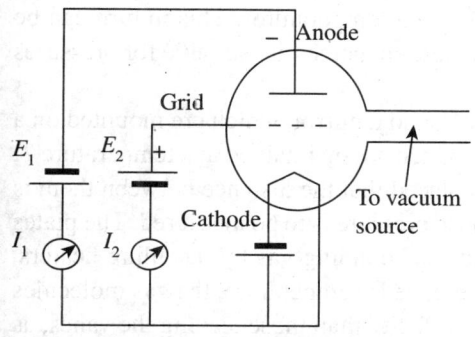

Fig. 16.28 Hot ionization gauge

and become ionized. Positive ions then move towards the negatively charged anode. Thus, an ionization current flows through the circuit, which is a measure of the absolute gas pressure in the gauge. Eventually, the electrons will move towards the positively biased grid, forming an electron current around the grid circuit.

The absolute gas pressure P is given by the following equation:

$$P = KC \frac{I_1}{I_2}$$

where C denotes a probability factor, K is a constant that depends on potentials employed in the gauge and on the electrodes' geometry, and I_1 and I_2 represent ionization current and grid circuit current, respectively.

As long as the potentials and the grid circuit current or electronic current remain unchanged, the absolute gas pressure will be proportional to the ionization current. The filament used in the triode assembly is made of pure or thoriated tungsten or platinum alloy having a coating of barium and strontium oxides. The grid is made of a molybdenum wire, which has a cylindrical cross section and is wound around the filament in the form of a helix. The anode, which is in the form of a cylinder or plate, is made of nickel. The hot ionization gauge has certain disadvantages: production of X-rays, which in turn causes secondary emission; excessive pressure, which causes deterioration of the filament and reduces its life; and careful control of the filament current, as electron bombardment is a function of this current.

16.9.4 Knudsen Gauge

Knudsen gauge, shown in Fig. 16.29, is a device employed to measure very low pressures. Pressure is measured as a function of the net rate of exchange of momentum of molecular

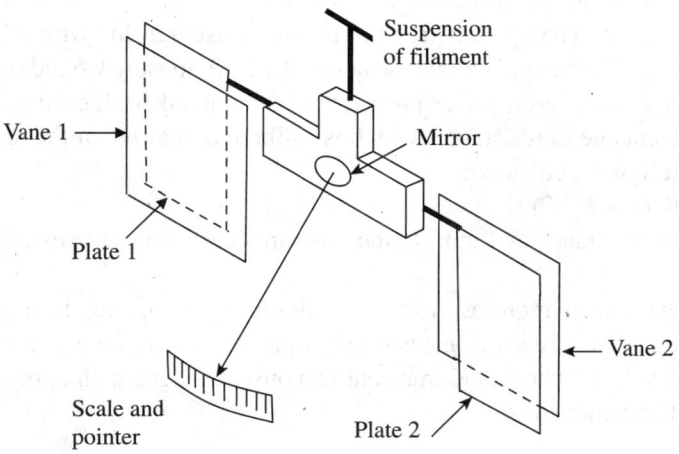

Fig. 16.29 Knudsen gauge

density between vanes and hot plates, which are at different temperatures. This in turn can be correlated to the pressure and temperature of the gas. Knudsen gauges are suitable for pressures ranging between 1 and 10^{-6} Pa.

A Knudsen gauge is composed of two vanes V_1 and V_2 and a mirror, which are mounted on a thin filament suspension. Two heated plates P_1 and P_2, which are maintained at a temperature T, are placed near the vanes. The vanes and plates are so placed that the distance between them is less than the mean free path of the surrounding gas whose pressure is to be measured. The plates are maintained at a temperature higher than that of the surrounding gas by installing heaters. Let the temperature of the vanes be T_g. Due to the difference in temperature, the gas molecules strike the cooler vanes. These molecules have a higher velocity than those leaving the vanes, as per the kinetic theory of gases. The angular displacement of the mirror is a measure of the net momentum F (force) imparted to the vanes due to the difference in velocities. The following equation gives the gas pressure in terms of temperature and measured force:

$$P = 4F \frac{T_g}{T - T_g}$$

This equation can be used for small temperature differences $T - T_g$.

The main advantages of a Knudsen gauge are that it gives absolute pressure and does not need to be calibrated. This gauge is useful for high-precision environments. Unlike the McLeod gauge, a Knudsen gauge does not contain any contaminants such as mercury and oil. It does not need any additional sensors, which are expensive and are required in an ionization gauge. However, it has one disadvantage—the pressure measurement is not accurate, when compared to an ionization gauge.

16.10 HIGH-PRESSURE MEASUREMENT

High-pressure measurement necessitates the use of special devices. The most popular device used is the Bridgman gauge, which is capable of measuring high pressures of around 100,000 atm. The Bridgman gauge works on the principle that the resistance of a fine wire linearly varies with pressure. The applied pressure is sensed using a fine wire of Manganin (84% Cu, 12% Mn, and 4% Ni) having a total resistance of 100 Ω, loosely wound on a coil, and enclosed in a container having appropriate pressure. Conventional bridge circuits are employed for measuring the change in resistance, which is calibrated in terms of the applied pressure and is given by the following equation:

$$R = R_1 (1 + P_r \Delta P)$$

where R_1 is the resistance at 1 atm, P_r the pressure coefficient of resistance, and ΔP the gauge pressure.

A Bridgman gauge requires frequent calibration, as ageing is a problem. A properly calibrated gauge can be used to measure high pressures with an accuracy of 0.1%. Since the Bridgman gauge has a very good transient response, changes with application of pressure are sensed almost instantaneously.

A QUICK OVERVIEW

- Pressure is the force exerted by a medium, usually a fluid, on a unit area. Measuring devices usually register a differential pressure—gauge pressure. Pressure is also defined as the force exerted over a unit area. Force may be exerted by liquids, gases, and solids.

$$\text{Pressure (P)} = \frac{\text{Force (F)}}{\text{Area (A)}}$$

- Gauge pressure is measured above the local atmospheric pressure.
- Total absolute pressure is the total pressure measured from zero pressure as the datum point. When the absolute pressure exceeds the local atmospheric pressure, it may be considered as the sum of the gauge pressure and the local atmospheric pressure. Total pressure is the sum of atmospheric pressure and gauge pressure:
Total absolute pressure = Atmospheric pressure + Gauge pressure
- Differential pressure is the difference in pressure measured between two points.
- When the pressure to be measured is less than the local atmospheric pressure, it is called vacuum pressure. In other words, when the gauge pressure is negative, it is termed vacuum. Vacuum is defined by the following relation:
Vacuum = Atmospheric pressure − Absolute pressure
- Absolute pressure is measured above total vacuum or zero absolute. Zero absolute represents total lack of pressure.
- Pressure measurements can be grouped into two main categories: *static* and *dynamic* pressures. Static pressure, as the name implies, is the pressure exerted by a fluid when it is in equilibrium or still or static; pressures acting at a point are the same in all directions and do not depend on direction.
- For measuring pressures that vary rapidly, the methods that are employed for the measurement of static pressure are not suitable. In such cases, pressure transducers are used to convert pressure into signals that are recordable.

- Single diaphragms, stacks of diaphragms, and bellows are some of the important elastic transducers used for pressure measurement. Diaphragms are generally used as primary transducers for dynamic pressure measurement.
- A dead-weight pressure gauge or piston gauge, which works on the Archimedes' principle, is a very popular device for measuring static pressures. The air or fluid displaced by the applied weights and the piston exerts a buoyant force, which causes the gauge to indicate the pressure. Dead-weight pressure gauges are normally used to calibrate other pressure measuring devices.
- Pressures above 1 Torr can easily be measured by the direct measurement method, wherein the force applied causes displacement. Manometers, diaphragms, bellows, and Bourdon tubes are a few examples of the instruments used in the direct measurement of pressure. These devices are generally employed to measure a pressure value of about 10 mmHg.
- For pressures below 1 Torr, indirect or in-ferential methods are often employed. In these methods, pressure is determined by drawing indirect references to some pressure-controlling properties such as volume, thermal conductivity, and ionization of gas. Some of the devices that fall under this category include McLeod gauge, Pirani gauge, and ionization gauge
- The basic principle on which a resistance-type pressure transducer works is that a variation in the length of a wire causes a change in its electrical resistance.
- Devices such as potentiometers and rheostats with sliding wires, which work on the principle of movable contacts, are also variable resistance-type pressure transducers.
- The linear variable differential transformer (LVDT) is an inductive type of pressure transducer that works on the mutual inductance principle. It transforms a mechanical displacement into an electrical signal.

- The capacitive transducer shown in Fig. 16.21 works on the principle that when pressure is applied onto the diaphragm, the distance between the two metal plates changes, which in turn changes the capacitance.
- A piezoelectric pressure transducer is an active type of pressure transducer, which works on the principle that when pressure is applied on piezoelectric crystals, an electric charge is produced.
- In order to measure the cylinder pressure in a reciprocating machine, such as an internal combustion engine or an air compressor, it is essential to plot a graph of cylinder pressure versus cylinder volume or time.
- A Pirani gauge works on the principle that when a heated wire is placed in a chamber of gas, the thermal conductivity of the gas depends on its pressure. Hence, it follows that the energy transfer from the wire to the gas is proportional to the gas pressure. The temperature of the wire can be altered by keeping the heating energy supplied to the wire constant and varying the pressure of the gas, thus providing a method for pressure measurement.
- A McLeod gauge works on Boyle's law, which states that by compressing a known volume of the low-pressure gas to a higher pressure, initial pressure can be calculated by measuring the resulting volume and pressure. The following fundamental relation represents Boyle's law:

$$P_1 = \frac{P_2 V_2}{V_1}$$

where P_1 and P_2 are the initial and final pressures, respectively, and V_1 and V_2 are the corresponding volumes.

- In a McLeod gauge, to measure low pressures, the value of V_1 is made large compared to a. The ratio of V_1 to a is called the compression ratio. If a is made too small, mercury tends to stick inside the capillary tube; this imposes a restriction on the upper limit of the compression ratio. The compression ratio gets limited due to the excessive weight of mercury if V_1 is very large.

- In a hot ionization gauge, the absolute gas pressure P is determined by the following relation:

$$P = KC\frac{I_1}{I_2}$$

where C denotes a probability factor, K is a constant that depends on potentials employed in the gauge and on the electrodes' geometry, and I_1 and I_2 represent ionization current and grid circuit current, respectively.

- Knudsen gauge is a device employed to measure very low pressures. Pressure is measured as a function of the net rate of exchange of momentum of molecular density between vanes and hot plates, which are at different temperatures. This in turn can be correlated to the pressure and temperature of the gas.
- In a Knudsen gauge, gas pressure is expressed in terms of temperature and measured force:

$$P = 4F\frac{T_g}{T - T_g}$$

This equation can be used for small temperature differences $T - T_g$.

- The main advantages of a Knudsen gauge are that it gives absolute pressure and does not need to be calibrated. This gauge is useful for high-precision environments. A Knudsen gauge does not contain contaminants such as mercury and oil, unlike the McLeod gauge.
- The most popular device that is used is the Bridgman gauge, which is capable of measuring high pressures of around 100,000 atm. The Bridgman gauge works on the principle that the resistance of a fine wire linearly varies with pressure. The change in resistance, which is calibrated in terms of the applied pressure, is given by the following equation:

$$R = R_1(1 + P_r \Delta P)$$

where R_1 is the resistance at 1 atm, P_r the pressure coefficient of resistance, and ΔP the gauge pressure.

MULTIPLE-CHOICE QUESTIONS

1. Gauge pressure is measured
   (a) above the local atmospheric pressure
   (b) below the local atmospheric pressure
   (c) above the zero absolute pressure
   (d) below the zero absolute pressure

2. Total absolute pressure is
   (a) greater than atmospheric pressure + gauge pressure
   (b) equal to atmospheric pressure + gauge pressure
   (c) greater than atmospheric pressure – absolute pressure
   (d) equal to atmospheric pressure – absolute pressure

3. Vacuum is given by
   (a) atmospheric pressure + gauge pressure
   (b) atmospheric pressure + absolute pressure
   (c) atmospheric pressure – gauge pressure
   (d) atmospheric pressure – absolute pressure

4. 1 bar equals to
   (a) 10^6 Pa   (c) 10^4 Pa
   (b) 10^5 Pa   (d) 10^3 Pa

5. A dead-weight pressure gauge works on
   (a) D'Arsonval principle
   (b) Abbe's principle
   (c) Archimedes' principle
   (d) D'alembert's principle

6. A dead-weight pressure gauge is used for
   (a) static pressure measurement
   (b) dynamic pressure measurement
   (c) high-vacuum measurement
   (d) low-volume measurement

7. McLeod gauge works on
   (a) Newton's law   (c) Boyle's law
   (b) Hook's law   (d) Pascal's law

8. With reference to a McLeod gauge, the ratio V_1/a is called
   (a) damping ratio   (c) aspect ratio
   (b) Poisson's ratio   (d) compression ratio

9. In a Pirani gauge, the pressure is related to the
   (a) thermal conductivity of the gas
   (b) volume of the gas
   (c) mass transfer of the gas
   (d) composition of the gas

10. In a Knudsen gauge, gas pressure is expressed in terms of
    (a) temperature
    (b) measured force
    (c) both temperature and measured force
    (d) neither temperature nor measured force

11. Knudsen gauge gives
    (a) gauge pressure
    (b) absolute pressure
    (c) atmospheric and gauge pressures
    (d) atmospheric pressure

12. The cross section of a Bourdon tube is
    (a) circular   (c) rectangular
    (b) elliptical   (d) triangular

13. A McLeod gauge is used to measure
    (a) gauge pressure
    (b) atmospheric pressure
    (c) vacuum pressure
    (d) absolute pressure

14. Which of the following equations represents Boyle's law?
    (a) $P_1 = \dfrac{P_2 V_2}{V_1}$ (c) $P_2 = \dfrac{P_1 V_2}{V_1}$

    (b) $P_1 = \dfrac{P_2 V_1}{V_2}$ (d) $P_2 = \dfrac{V_2 - V_1}{P_1}$

15. The molecules of a gas in a Knudsen's gauge move from hot plates to cooler vanes at high velocity according to
    (a) potential theory of gases
    (b) kinetic theory of gases
    (c) Dalton's theory
    (d) Newton's hypothesis

16. Which of the following devices works on the principle of mutual inductance?
    (a) Potentiometer
    (b) Rheostat
    (c) Piezoelectric crystal
    (d) LVDT

17. When pressure is applied onto the diaphragm, the distance between the two metal plates changes, which in turn changes the
    (a) capacitance   (c) resistance
    (b) inductance   (d) reluctance

18. Which of the following terms is used to denote emitted electrons?

    (a) photons
    (b) electrons
    (c) thermions
    (d) neutrons

REVIEW QUESTIONS

1. Define pressure.
2. Explain why pressure measurement is important.
3. State the four different pressure measurement scales.
4. With a neat diagram, explain the relationship between absolute, gauge, and barometric pressures.
5. With a neat sketch, discuss static pressure measurement.
6. List the different pressure measuring transducers used to transduce the effects of pressure into deflections.
7. Classify pressure measuring devices.
8. With a neat diagram, discuss an industrial U tube manometer. Show that the rise of liquid in the wide limb is proportional to the differential pressure.
9. Explain the following with neat sketches:
    (a) Cistern manometer
    (b) Inclined tube manometer
10. Explain the different types of elastic transducers.
11. Discuss the working of a Bourdon gauge with a neat sketch.
12. Explain the working of a dead-weight tester with a schematic diagram.
13. Write a note on vacuum measurement.
14. Briefly discuss the working of a McLeod gauge with a neat sketch.
15. With a neat diagram, explain the working of a Pirani gauge.
16. Explain in detail how Knudsen gauge is employed for low-pressure measurement.
17. With a neat sketch, explain the working of a piezoelectric transducer for pressure measurement.
18. Explain the following:
    (a) Inductance-type pressure transducers
    (b) Capacitance-type pressure transducers
19. With a neat sketch, explain bonded and unbonded strain gauge pressure transducers.
20. Explain how potentiometer devices can be employed for pressure measurement.
21. Explain the working of an engine indicator with neat sketches.
22. Explain a hot ionization gauge with a neat diagram.
23. Write a short note on high-pressure measurement.

Answers to Multiple-choice Questions

1. (a)	2. (b)	3. (d)	4. (b)	5. (c)	6. (a)	7. (c)	8. (d)	9. (a)
10. (c)	11. (b)	12. (b)	13. (c)	14. (a)	15. (b)	16. (d)	17. (a)	18. (c)

PART III

Nano Impact on Metrology

- **Nanometrology**

Nanometrology

After studying this chapter, the reader will be able to

- acquire basic understanding of the field of nanotechnology
- understand the principal requirements in nanometrology such as morphology, size and shape of particles, crystallographic information, detection of atomic scale defects, topography, and arrangement of atoms
- explain the widely used nanoscale measurement techniques comprising electron microscopy and X-ray diffraction and the instrumentation for the same
- elucidate the typical application areas of these instruments

17.1 INTRODUCTION

Nano in Greek means 'dwarf'. One nanometre is one-billionth of a metre or 10^{-9} m. Comparing an object with a diameter of 1nm with another with a diameter of 1m is like comparing a small pebble with the giant-sized earth. On a lighter note, it is said that a nanometre is the length that a man's beard grows in the time he takes to say 'Hello, how do you do?'.

Nanometrology is the science of measurement at the *nanoscale level*. Figure 17.1 illustrates where a nanoscale stands in relation to a metre and subdivisions of a metre. Nanometrology has a crucial role in the production of nanomaterials and devices with a high degree of accuracy and reliability (nanomanufacturing). It includes length or size measurements (where dimensions are typically given in nanometres and the measurement uncertainty is often less than 1nm) as well as measurement of force, mass, and electrical and other properties.

Nanometrology addresses two main issues: precise measurement of sizes in the nanometre range, and adapting existing or developing new methods to characterize properties as a function of size. A direct consequence of this is the development of methods to characterize sizes based on the evaluation of properties and to compare sizes measured using various methods. Before we move on to the core topics in nanometrology, it is necessary

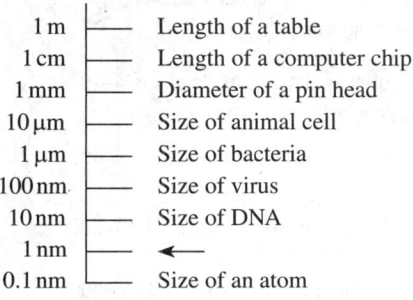

1 m	—	Length of a table
1 cm	—	Length of a computer chip
1 mm	—	Diameter of a pin head
10 μm	—	Size of animal cell
1 μm	—	Size of bacteria
100 nm	—	Size of virus
10 nm	—	Size of DNA
1 nm	— ←	
0.1 nm	—	Size of an atom

Fig. 17.1 Nanoscale in relation to a metre

to provide a formal introduction to nanotechnology. Nanotechnology is a comparatively new field in engineering and we need to understand certain basic concepts before moving on to nanometrology.

17.2 NANOTECHNOLOGY

The properties of the objects that we see around us do not change much with change in size. In the nanoworld, however, change in size affects properties enormously. Richard Feynman, the famous physicist, said, 'Atoms on a small scale behave like nothing similar to those on a large scale. As we go down and fiddle around with atoms down there, we are working with different laws altogether.' For instance, iron loses its magnetic property at nanosize. Gold at 1nm size neither shines nor is chemically neutral. When particles are about 50nm in size, they exhibit properties similar to those of bulk material. As the particle size reduces to about 10–50 nm, properties vary linearly with size, as shown in Fig. 17.2. As the size further reduces, we see unusual new properties, which is basically a result of *quantum effects*. At this size of the particle, electrons are confined in a small volume or box and exhibit energies that depend on the length of the box.

Nanomaterials comprise *nanoparticles* and *nanocrystals*. Small nanoparticles, which possess nanodimension in all the three principal directions, are called *quantum dots*. Table 17.1 illustrates the different types of nanomaterials and their sizes.

17.2.1 Importance of Nanodimension

As has already been pointed out, change in nanosize affects the properties of materials enormously. This is of primary interest in nanoscience, where new discoveries are being made on a regular basis. Some of the interesting phenomena associated with nanodimension are listed here:

1. Due to the contribution of grain boundaries at nanoscale, materials exhibit superior mechanical strength and ductility.

Table 17.1 Types of nanomaterials

Type of nanomaterial	Material	Diameter/ Thickness (nm)
Nanocrystals	Metals, inorganic materials (oxides, nitrides, sulphides, etc.)	1–50
Nanofilms	Layers of quantum dots made of lead selenide, indium arsenide, etc.	1–10
Nanowires	Metals, oxides, nitrides, sulphides, etc.	1–100
Nanotubes	Carbon, metals, inorganic materials	1–1000
Nanosurfaces	Various materials	1–100

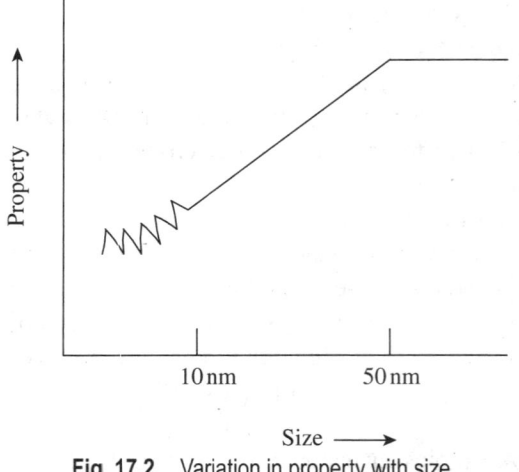

Fig. 17.2 Variation in property with size

2. Thermodynamic properties also undergo a sea change. Thermodynamic phase equilibrium is shifted, due to the contribution of interfaces or interface-related strains, to the free energy of the system. This allows production of new non-equilibrium materials, which exhibit properties that were not known previously.

3. Tribological properties are also affected since the interacting materials are in the nanoscale. These changes facilitate reduced friction and wear in applications of micro-electro-mechanical systems.

4. Magnetic properties undergo remarkable change due to the surface effects of thin layers. This gives a boost to more efficient data storage devices and more sensitive magnetic sensors.

5. If the grain size is smaller than the electron mean free path, the electronic conductivity as well as temperature coefficient decreases.

6. Optical properties also undergo major changes. Band gap changes in nanosized semiconductor particles lead to a blue shift of luminescence. Size-induced control of luminescence and relaxation in oxide nanoparticles leads to changes in the optical properties, which can be used to fabricate interesting optoelectronic devices.

17.2.2 Classification of Nanostructures

Nanostructures are classified in various ways. According to the Royal Society and the Royal Academy of Engineering of the United Kingdom, nanostructures are classified into the following three classes:

Nanoscale in one dimension   Thin film layers and surfaces

Nanoscale in two dimensions   Carbon nanotubes, inorganic nanotubes, nanowires, and biopolymers

Nanoscale in three dimensions   Nanoparticles, fullerenes, dendrimers, and quantum dots

However, most of the recent literature classifies nanostructures as follows:

1. (0-D) Nanoparticles, quantum dots, and nanodots
2. (1-D) Nanowires, nanorods, and nanotubes
3. (2-D) Nanosheets, nanofilms, and nanocoatings
4. (3-D) Bulk and powders

While nanoparticles are crystals of practically zero dimensions made of metals, with sizes below 100 nm, nanowires and nanotubes are single-dimension structures with thickness below 100 nm. Carbon nanotubes are stronger than steel wires, can conduct thousand times more electricity, and can bear weight million times more than their own weight. Two-dimensional (2-D) structures comprising nanosheets, coatings, and thin-film layers are made of nanocrystalline metals or nanocomposites. Nanosheets and coating have grains of diameter in the nanometre scale. A thin film is essentially a 1nm-thick single crystalline film of metal on a single crystal surface. Graphene, the 2-D nanocarbon, is found to exhibit extraordinary properties. Electrons in graphene move differently from other solids and show unusual effects. Graphene is used to make sensors, electrochemical devices, and transistors.

The bulk is a structure in which dimensions of the material are significantly larger than the grain size in all the three principal directions. While a 3-D nanostructured metal is assumed to be a free-standing material, a coating cannot be handled as an item separate from the substrate.

Nanopowders are manufactured by *milling process*. Although the powder is composed of nanosized crystals, it is classified separately from the bulk, since it does not possess a typical 3-D structure. However, in practice, the terms nanopowder, nanoparticle, and nanocrystal are treated as equivalent, and they refer to nanosized particles that are single crystals.

17.2.3 Applications

Nanotechnology has wide-ranging applications in material science, medicine and drug delivery, sensorics, biotechnology, optics, tissue engineering, and cancer therapy, among others. This section describes some of the applications that are of significant interest to mechanical engineers.

Nanosensors Nanosensors are devices that use biological, chemical, or mechanical sensory points to detect and convey information about the nanoregime under investigation. Nanosensors can be used to detect harmful gases and changes in electrical or magnetic forces, or as biosensors to explore the mysteries of living cells. Gas sensors are used in industries for process control, monitoring environmental pollution, detection of fire, etc. The most commonly used gas sensors are metal-oxide-based gas sensors, which can detect gases such as hydrogen, ammonia, nitric oxide, and hydrocarbons.

Water purification Bioactive nanoparticles provide effective solution for water purification. Nanosilver particles form silver ions when dispersed in water. We know well that, compared to silver, nanosilver has a larger surface area, which can increase its contact with microorganisms. Silver ions prevent build-up of bacteria and algae in water. They also have many other useful properties; for example, they are non-toxic, non-stimulating, and non-allergic and are easily dispersed in water. They can be used in water reservoirs, water tanks, and community water systems.

Lighting Nanotechnology-driven *solid-state light-emitting diodes* and *organic light-emitting diodes* are becoming popular. They can provide cheaper, energy-efficient, and environment-friendly options for lighting. In the near future, we hope to see flexible and miniature devices that can provide extremely bright light.

Nanocomputers A nanocomputer has parts that are only a few nanometres in size. *Y-junction carbon nanotubes* have successfully been used as transistors in fabricating a nanocomputer. At present, transistors used in processors are approaching 45 nm in size. Nano-integrated circuits of nanocomputers are fabricated by *nanolithography*. Once manufacturing begins on a large scale, nanocomputers are expected to offer immense potential for storing vast quantities of data and retrieving stored data with negligible power consumption.

Nanotechnology-based garments Nanocrystals repel water droplets and minute solid particles, providing the scope for producing waterproof and dust-proof clothing. Textile scientists have been successful in producing stain- and wrinkle-proof textile material by coating the fabric with a thin layer of extremely *hydrophobic* (water-resistant) silicon nanofilaments. The unique spiky structure of the filament creates a 100% waterproof coating.

17.3 IMPORTANCE OF NANOMETROLOGY

Nanometrology is one of the most exciting and challenging areas for mechanical engineers.

It has really caught the imagination of scientists the world over, and it cannot progress at a fast pace unless we find ways and means to make accurate measurements up to 0.1 nm. In many cases, measurement techniques developed for conventional materials cannot be applied to nanostructures. The following are some of the reasons why nanometrology is a priority field today:

1. Special protocols for nanostructures and nanomaterials must be developed to avoid severe mistakes in evaluating results.
2. New phenomena at the nanoscale level require an understanding of and the ability to measure the physics of very small objects. The arrangement of atoms or particles in nanostructures has new and sometimes even exotic forms.
3. Measurement *standards* have to be developed to match technology advances and support the increasing applications of nanostructures.
4. Nanotechnology is interdisciplinary in nature and comprises varied fields such as biology, chemistry, and materials technology with underpinned knowledge of physics. Therefore, measurement techniques are also quite complex.

Table 17.2 illustrates the measurement parameters/properties and the popular measurement techniques used for measuring the same.

Table 17.2   Nanometrology techniques

Parameter/Property	Measurement technique
Morphology: size and shape of particles; crystallographic information: detection of atomic scale defects	Transmission electron microscopy
Topography: surface features; morphology: shape and size of the particles; composition: elements and compounds the sample is composed of; crystallographic information: arrangement of atoms	Scanning electron microscopy
Three-dimensional surface topology: size, shape, roughness, defects, electronic structures	Scanning tunnelling microscopy
Topology, roughness, and elasticity of surface, grain size, frictional characteristics, specific molecular interactions, and magnetic features on surface	Atomic force microscopy or scanning force microscopy
Crystallographic information: type of crystal structure, film thickness, interface roughness, and surface topology	X-ray diffraction (XRD)

17.4 INTRODUCTION TO MICROSCOPY

The first recorded use of microscopes can be traced back to the early 1600s, when the Dutch scientist Anton van Leeuwenhoek developed tiny glass lenses. He demonstrated his invention, which could be used to observe blood cells, bacteria, and structures within the cells of animal tissue. However, the instrumentation part was primitive by today's standards. A simple one-lens device had to be positioned very accurately, making observation very tiring in practice. Soon it became a standard practice to have a *compound microscope*, containing at least two lenses: an *objective* (placed close to the *object* to be magnified) and an *eyepiece* (placed fairly close to the *eye*). By increasing its dimensions or employing a larger number of lenses, the magnification M of a compound microscope can be increased substantially. The spatial resolution of a compound microscope is limited by the diameter (aperture) of the lens, just as in the case of diffraction at the pupil of the eye or at a circular hole in an opaque screen. With a large-aperture lens, a resolution limit of just over half the wavelength of light can be achieved, as first deduced by Abbé in 1873. For light in the middle of the visible spectrum ($\lambda = 0.5\,\mu m$), this means the best possible object resolution that can be achieved is about 0.3 μm.

One possibility for improving resolution is to decrease the wavelength λ of the incident light. Greater improvement in resolution comes from using ultraviolet (UV) radiation, with wavelengths in the range of 100–300 nm. The light source can be a gas-discharge lamp, and the final image is viewed on a phosphor screen that converts the UV radiation to visible light. Since ordinary glass absorbs UV light strongly, the focusing lenses must be made from a material such as quartz (transparent down to 190 nm) or lithium fluoride (transparent down to about 100 nm).

X-rays offer the possibility of even better spatial resolution, since they are electromagnetic waves with a wavelength shorter than those of UV light. X-ray microscopes more commonly use *soft X-rays*, with wavelengths in the range of 1–10 nm. However, laboratory X-ray sources are relatively weak (XRD patterns are often recorded over many minutes or hours). This situation prevented the practical realization of an X-ray microscope until the development of an intense radiation source, the *synchrotron*, in which electrons circulate at high speed in vacuum within a *storage ring*. Guided around a circular path by strong electromagnets, their centripetal acceleration results in the emission of *bremsstrahlung X-rays*. However, synchrotron X-ray sources are large and expensive.

Early in the 20th century, physicists discovered that material particles such as electrons possess a wavelike character. Inspired by Einstein's photon description of electromagnetic radiation, the French quantum physicist Louis de Broglie proposed that their wavelength is given by

$$\lambda = h/p = h/(mv)$$

where $h \,(= 6.626 \times 10^{-34}\,\text{J s})$ is the Planck constant; p, m, and v represent the momentum, mass, and speed of the electron, respectively. This discovery earned Louis de Broglie the Nobel Prize in Physics in the year 1929.

When electrons are emitted into vacuum from a heated filament and accelerated through a potential difference of 50 V, the speed of electrons shoot up to 4.2×10^6 m/s with a wavelength of 0.17 nm. Since this wavelength is comparable to atomic dimensions, such 'slow' electrons are strongly diffracted from the regular array of atoms at the surface of a crystal. Raising the accelerating potential to 50 kV, the wavelength shrinks to about 5 pm (0.005 nm), and such higher-energy electrons can penetrate distances of several microns into a solid. If the solid is crystalline, the electrons are diffracted by atomic planes inside the material, as in the case of X-rays. It is, therefore, possible to form a transmission electron diffraction pattern from electrons that have passed through a thin specimen. If these transmitted electrons are focused, their very short wavelength will allow the specimen to be imaged with a spatial resolution much better than the light-optical microscope. In a *transmission electron microscope* (TEM), electrons penetrate a thin specimen and are then imaged by appropriate lenses, quite similar to an optical microscope. One limitation of a TEM is that, unless the specimen is made very thin, electrons are strongly scattered within the specimen, or even absorbed rather than being transmitted. This constraint provided the motivation to develop electron microscopes that are capable of examining relatively thick (so-called bulk) specimens.

In a *scanning electron microscope* (SEM), primary electrons are focused into a small-diameter electron probe that is scanned across the specimen. An electrostatic or magnetic field is applied at right angles to the beam in order to change its direction of travel. By scanning simultaneously in two perpendicular directions, a square or rectangular area of specimen (known as a raster) can be covered and an image of this area can be formed by collecting secondary electrons from each point on the specimen. A modern SEM provides an image resolution typically between 1 and 10 nm. The images have a relatively large depth of focus: specimen features that are displaced from the plane of focus appear almost sharply in focus.

It is possible to employ the fine-probe/scanning technique with a thin sample and record, instead of secondary electrons, the electrons that emerge (in a particular direction) from the opposite side of the specimen. The result is a *scanning-transmission electron microscope* (STEM). The first STEM was developed by von Ardenne in 1938 by adding scanning coils to a TEM, and today many TEMs are equipped with scanning attachments, making them dual-mode (TEM/STEM) instruments. The raster method of image formation is also employed in a scanning-probe microscope like the *scanning tunnelling microscope* (STM). In an STM, a sharply pointed tip (or probe) is mechanically scanned in close proximity (up to 1 nm) to the surface of a specimen, and a small potential difference (≈ 1 V) is applied. Provided the tip and specimen are electrically conducting, electrons move between the tip and the specimen by the process of quantum-mechanical tunnelling. A motorized mechanism is provided in the instrument to give precise motion to the probe. To perform scanning microscopy, the tip is raster-scanned across the surface of the specimen in the X- and Y-directions. A negative-feedback mechanism ensures that the gap between the tip and the sample will remain constant. The tip will move in the z-direction in exact synchronization with the undulations of the surface (the specimen topography). This z-motion is represented by variations in the z-piezo voltage, which in turn can be used to modulate the beam in a cathode-ray tube (CRT) display device (as in an SEM) or stored in the computer memory as a topographical image.

The following sections explain the principle of working and applications of some of the important microscopes used in nanometrology.

17.4.1 Transmission Electron Microscope

A TEM comprises three lenses: an objective lens, an intermediate lens, and a projector lens. The microscope is built in such a way that it allows easy switching from the high-magnification imaging mode to the selected-area diffraction mode. The optical system of a TEM is shown in Fig. 17.3. Movable selection apertures are placed in the following manner: one in the image plane of the objective lens and a second one close to the back focal plane. While the former is useful for selecting a small area ($<1\,\mu m^2$) of the specimen while viewing the image, the latter enables the user to select either a single beam or a number of image-forming diffracted beams.

Figure 17.3(a) illustrates two modes of usage: the high-resolution, high-magnification *imaging mode* and the *diffraction mode*. In the imaging mode, the electron beam produced by an electron source is collimated by the condenser lens system (not shown in Fig. 17.3) and scattered by the specimen. An image is formed in the image plane of the objective lens. The aperture provided near the objective lens enables the selection of one small area of interest of the specimen. The image is then magnified by the intermediate lens. Since the intermediate lens is focused on the image plane of the objective lens, an intermediate image is formed in the image

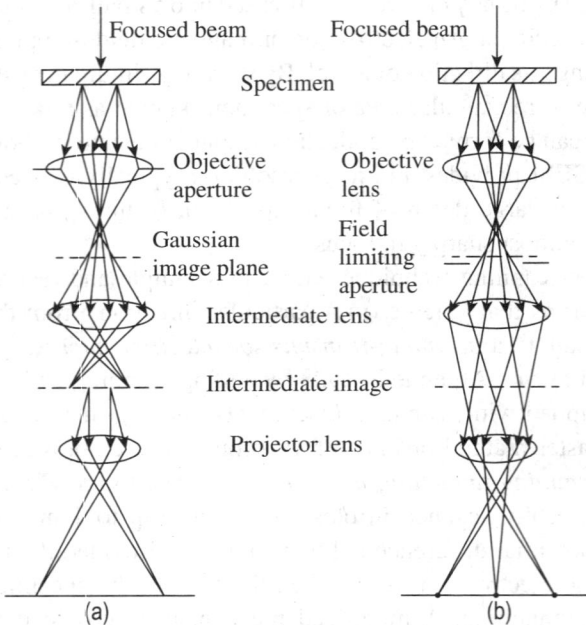

Fig. 17.3 Optical system of a TEM (a) Bright field image (b) Diffraction pattern

plane of the intermediate lens. This image is the object for the projector lens that forms a final image on a fluorescent screen or on the entrance plane of a recording device for a permanent recording of the image, which can be used for further analysis.

In the diffraction mode shown in Fig. 17.3(b), the focal length of the intermediate lens is made larger, such that the back focal plane of the objective lens coincides with the object plane of the projector lens. A magnified representation of the diffraction pattern is then produced on the fluorescent screen. In the process, the selected area is not changed since only the strength of the intermediate lens is modified. The diffraction pattern is thus representative of the selected area. However, under high-resolution conditions, the field of view in the image is much smaller than the selected area in the diffraction mode.

Electron Gun

An electron gun produces a beam of electrons whose kinetic energy is high enough to enable them to pass through thin areas of a TEM specimen. The gun consists of an electron source, also known as the cathode because it is at a high negative potential, and an electron-accelerating chamber. While there are several types of electron sources operating on different physical principles, Fig. 17.4 illustrates a common form of an electron gun, called a *thermionic electron gun*. The electron source is a V-shaped filament made of tungsten wire, spot-welded to straight-wire leads that are mounted in a ceramic or glass socket. This allows easy assembly and disassembly of the unit. A direct current heats the filament to about 2700 K, at which temperature tungsten emits electrons into the surrounding vacuum by a process known as *thermionic emission*. The thermionic-electron gun consists of a tungsten filament F, a Wehnelt electrode W, a ceramic high-voltage insulator C, and an o-ring seal O to the lower part of the TEM column. An autobias resistor R_b, located inside the high-voltage generator, is used to generate a potential difference between W and F, thereby controlling the electron-emission current I_e. Arrows denote the direction of *electron* flow that gives rise to the emission current.

The dependence of electron-beam current on the filament heating current is shown in Fig. 17.5. As the current is increased from zero, the filament temperature eventually becomes high enough to give some emission current. As the filament temperature is further increased, the beam current becomes saturated and hence approximately *independent* of the filament temperature. The filament heating current (which is adjustable by the TEM operator) should never be set higher than the value required for current saturation. Higher values result in a slight increase in the beam current and a *decrease* in source lifetime, due to evaporation of tungsten. The change

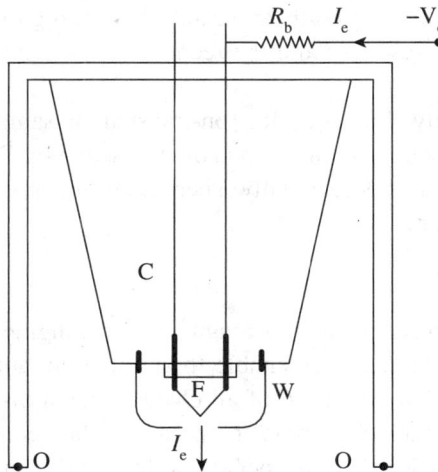

Fig. 17.4 Thermionic electron gun

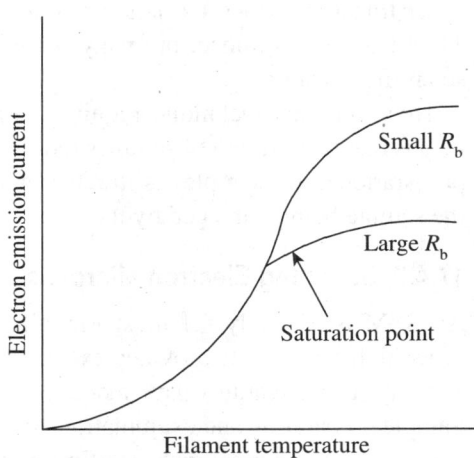

Fig. 17.5 Relationship between emission current and cathode temperature

in I_e can be monitored from an emission-current meter or by observing the brightness of the TEM screen, allowing the filament current to be set appropriately. If the beam current needs to be changed, this is done using a *bias-control* knob that selects a different value of R_b, as indicated in Fig. 17.5.

Clean and vibration-free vacuum systems are essential to provide stability to the specimen and avoid its contamination by a carbon film resulting from the cracking of organic molecules present in the residual gas. Anti-contamination devices such as metal blades surrounding the specimen, which are cooled to liquid nitrogen temperature, are available in most microscopes. Images are formed on a fluorescent screen and can be viewed by the observer. Exposing the image on a photographic medium, a permanent recording can be obtained. Degassing the photographic material prior to its use is strongly recommended. In recent microscopes, electronic viewing and recording methods are increasingly being used. Development of the charge-coupled device (CCD) camera has led to a revolution across the entire field of electron microscopy, and it has particular value for quantitative TEM applications. CCD cameras offer high sensitivity, wide dynamic range, and overall usefulness for the user. The fixed location of a CCD camera enables geometric distortions of the imaging system to be compensated accurately, which is advantageous for extracting quantitative phase information during off-axis electron holography.

Applications of TEM

TEMs enable the study of individual atomic columns in most inorganic materials, making it possible to determine the atomic-scale microstructure of lattice defects and other in-homogeneities. Structural features of interest include planar faults such as grain boundaries, interfaces, and crystallographic shear planes; linear faults such as dislocations and nanowires, as well as point defects; nanosized particles; and local surface morphology. Additional information can be extracted from high-resolution studies, including unique insights into the controlling influence of structural discontinuities on a range of physical and chemical processes such as phase transformations, oxidation reactions, epitaxial growth, and catalysis. The high-resolution TEM has had an impact on many scientific disciplines and the technique has generated vast scientific literature.

However, this technique requires extremely thin and electron-transparent samples. This means that preparation of samples requires special attention and is time consuming. During the preparation of the sample, its structure may change occasionally. There is also the possibility of the sample being damaged by the electron beam.

17.4.2 Scanning Electron Microscope

An SEM is arguably the most versatile microscope, with a magnification ranging from 5× to as high as 10^6×. It provides excellent resolution, is amenable to automation, and is user-friendly. These features have made it the most widely used of all electron beam instruments. Sample preparation and examination are also relatively simple compared to other techniques. A wide range of nanomaterials, starting from powders to films, pellets, wafers, carbon nanotubes, and even wet samples, can be examined using an SEM. It is also possible to correlate the observations made at nanoscale to those at macroscale and draw reliable conclusions.

The use of a field-emission gun in an SEM makes it possible to image individual heavy atoms in the transmission mode by collecting scattered electrons with a sensitive detector. When an electron beam strikes a bulk specimen, a variety of electrons, photons, phonons, and other signals are generated (Fig. 17.6). Three types of electrons are emitted from the electron-entrance surface of the specimen: secondary electrons with energies <50 eV, Auger electrons produced by the decay of the excited atoms, and backscattered electrons that have energies close to those of the incident electrons. All these signals can be used to form images or diffraction patterns of the specimen or can be analysed to provide spectroscopic information. De-excitation of atoms that are excited by the primary electrons also produces continuous and characteristic X-rays as well as visible light. These signals can be utilized to provide qualitative or quantitative information on the elements or phases present in the regions of interest. All these signals are the product of strong electron–specimen interactions, which depend on the energy of the incident electrons and the nature of the specimen.

Figure 17.7 illustrates the components of an SEM. A tungsten filament is used as the source of electrons. Since the maximum accelerating voltage for the filament is lower than that for TEM, the electron gun is smaller. The beam size is also quite small, of the order of 10 nm, which necessitates the use of two or three lenses to condense the beam to this size. The final lens that forms this very small beam is named the *objective lens* and its performance largely determines the spatial resolution of the instrument.

The electron beam of an SEM is scanned horizontally across the specimen in two mutually perpendicular (X and Y) directions. The X-scan is relatively fast and is generated by a sawtooth wave generator. This generator supplies scanning current to two coils, connected in series and located on either side of the optic axis, just above

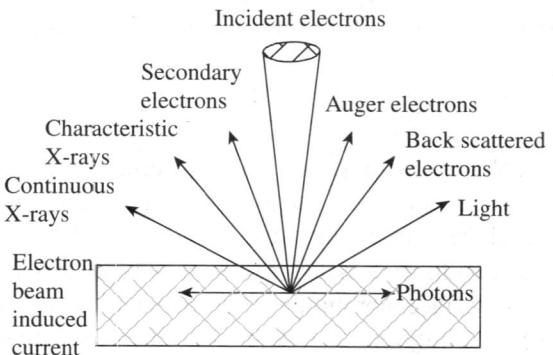

Fig. 17.6 Signals generated in an SEM

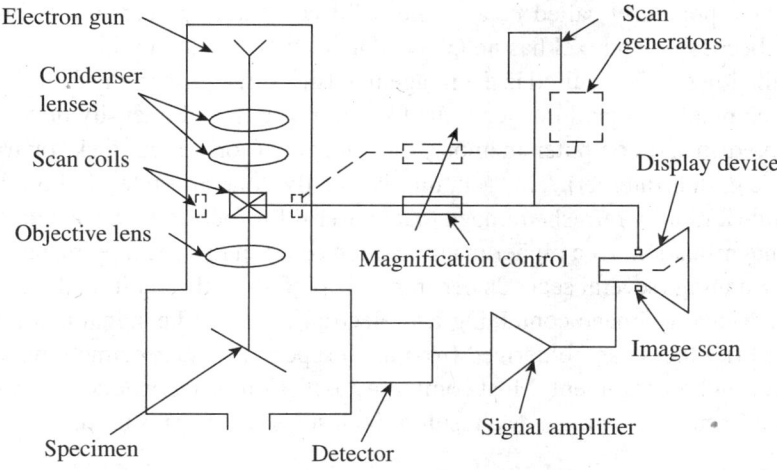

Fig. 17.7 Components of an SEM

the objective lens. The coils generate a magnetic field in the Y-direction, creating a force on an electron (traveling in the Z-direction) that deflects it in the X-direction. The Y-scan is much slower and is generated by a second sawtooth wave generator. The entire procedure is known as *raster scanning* and causes the beam to sequentially cover a rectangular area on the specimen, as shown in Fig. 17.8.

The beam traces a straight line path from A to A_1, which is from left to right, during its X-deflection signal. However, while taking the reverse path, the beam is deflected by a small amount in the Y-direction, and it takes a diagonal path from A_1 to B. A second line scan takes the probe to point B_1, at which point it flies back to C; the process is repeated until n lines have been scanned and the beam arrives at point Z_1. This entire sequence constitutes a single frame of the raster scan. From point Z_1, the probe quickly returns to A, as a result of the rapid flyback of both the line and the frame generators, and the next frame is executed. This process may continuously run for many frames, as happens in a raster scan terminal.

The outputs of the two scan generators can be used to generate the display on a CRT. The electron beam in the CRT scans exactly in synchronization with the beam in the SEM, so for every point on the specimen (within the raster-scanned area) there is an *equivalent* point on the display screen, displayed at the same instant of time. In order to introduce *contrast* into the image, a voltage signal must be applied to the electron gun of the CRT, to vary the brightness of the scanning spot. This voltage is derived from a detector that responds to some change in the specimen induced by the SEM incident probe.

In recent times, the CRT display devices have become obsolete. The scan signals are generated digitally, by computer-controlled circuitry. The image is divided into a total of $m \times n$ picture

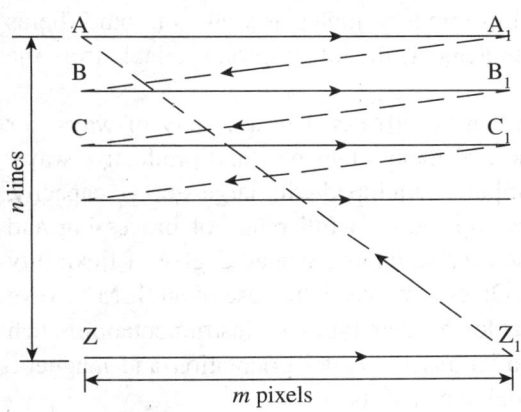

Fig. 17.8 Scan pattern in an SEM

elements, popularly called *pixels*. The SEM computer can capture an image up to the pixel level, because each pixel has an (x, y) address, which is stored in the memory. The additional information that is required is the image intensity value (also in the form of a digitized number) for each pixel. A digital image, in the form of position and intensity information, can therefore be stored in the computer memory, on a magnetic or optical disk, or transmitted over data lines (e.g., the Internet). The scanning is usually done at a rate of about 60 frames/second to generate a rapidly refreshed image that is useful for focusing the specimen or for viewing it at low magnification. At a higher magnification or when making a permanent record of an image, slow scanning (several seconds per frame) is preferred; the additional recording time results in a higher-quality image containing less electronic noise. The signal that modulates (alters) the image brightness can be derived from any property of the specimen that changes in response to electron bombardment. Most commonly, emission of *secondary electrons* (atomic electrons ejected from the specimen as a result of inelastic scattering) is used.

SEM Specimen Preparation

As long as the test specimen is made of a conducting material, no special preparation is required before the microscopic examination. On the other hand, specimens of insulating materials do not provide a path to ground the specimen current I_s and may undergo electrostatic charging when exposed to an electron probe. This problem is addressed by coating the surface of the specimen with a thin film of metal (gold or chromium) or conducting carbon. This is done in vacuum, using the evaporation or sublimation techniques. Films having a thickness of 10–20 nm conduct sufficiently to prevent electrostatic charging of most specimens. However, the external contours of a very thin film closely follow those of the specimen, providing the possibility of a faithful topographical image.

Applications of SEM

An important feature of an SEM is its large depth of field, which is responsible, in part, for the 3-D appearance of the specimen image. The greater depth of field of the SEM provides much more information about the specimen. Most SEM micrographs, in fact, have been produced with magnifications below 8000 diameters (8000×). At these magnifications, the SEM operates well within its resolution capabilities. In addition, the SEM is also capable of examining objects at very low magnification. This feature is useful in forensic studies as well as in other fields such as archaeology because the SEM image complements the information available from the light microscope.

Once it is in a digital form, an SEM image can be processed in a variety of ways, for example, nonlinear amplification, differentiation, and many other new and productive ways. The availability of powerful and inexpensive computers equipped with large storage capacity, high-resolution displays, and software packages capable of a full range of processing and quantitative functions on digital images gives the user an unprecedented degree of flexibility and convenience in using the output of the SEM. Other advances in the use of an SEM involve contrast mechanisms that are not readily available in other types of instrumentation, such as electron channelling contrast produced by variations in crystal orientation and magnetic contrast from magnetic domains in uniaxial and cubic materials.

For a metallurgist, an SEM provides the capability to determine the crystal structure

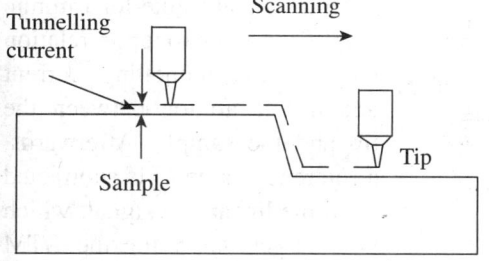

Fig. 17.9 Principle of an STM

and grain orientation of crystals on the surface of prepared specimens. This capability makes use of the diffraction pattern of the backscattered electrons emerging from the specimen surface and is known as electron backscattering diffraction. These patterns are then analysed with a computer-assisted indexing method. Automated indexing of patterns and computer-automated crystal lattice orientation mapping allow this technique to identify phases and show misorientation across grain boundaries.

17.4.3 Scanning Tunnelling Microscope

The STM was invented by Binnig, Rohrer, and their co-workers at IBM Research Laboratory, Zurich, Switzerland, in the early 1980s. The 1986 Nobel Prize in Physics was awarded to Binnig and Rohrer for their design of the STM. An STM provides 3-D atomic-scale images of the surface of a sample. It has a stylus with an extremely sharp tip. The stylus scans the surface of the sample from a fixed distance. It is a powerful tool for viewing surfaces at the atomic level. An STM works on the principle of *quantum tunnelling*. When an atomically sharpened tip under a small voltage is brought close to the surface of a sample, so that the separation is of the order of a nanometre, there is a small change in current in the circuit. This effect is called the quantum tunnelling effect. The induced current is referred to as the *tunnelling current*. This current increases as the gap between the tip and the sample decreases. The change in the tunnelling current can be calibrated with respect to the change in gap. In other words, if we scan the tip over the sample surface while keeping the tunnelling current constant, the tip movement depicts the surface topography, because the separation between the tip apex and the sample surface is always constant. Figure 17.9 illustrates the working principle of an STM. The resolution obtained in an STM is so high that individual atoms can be resolved when the tip apex is atomically sharp.

An STM requires a very sharp stylus tip and an extremely clean sample surface. It uses a sharp metal wire, usually made of tungsten or Pt–Ir alloy as the probe. The tip is prepared either by a mechanical cutter in case of Pt–Ir tip or by electromechanical etching for tungsten tip. The recent advances have made it possible to have an in situ tip growth with application of high voltage while the tip is being faced towards the sample. Thermal field treatments, a nanopillar growth on a tip by pulling it from a heated sample with a special purpose machine (SPM), and so on, have been proposed. In addition, attaching a carbon nanotube to the tip apex has also attracted much interest.

Figure 17.10 shows the components of an STM system. A tip is attached at a corner of a scanner, consisting of three rectangular rods of piezo ceramics [Pb (Zr, Ti) O3 (PZT)] that are crossing perpendicularly. The PZT rod can be elongated by increasing the voltage applied between two electrodes on its opposite longitudinal faces. For example, the rod elongates 1–2 nm per 1 V. To scan the tip faster, either a compact and tube-type piezo scanner or a shear piezo scanner is used.

A tunnelling current less than the order of a nanoampere in magnitude is detected by a current amplifier with a conversion ratio of $10^{7-9} VA^{-1}$. The output of the current amplifier is

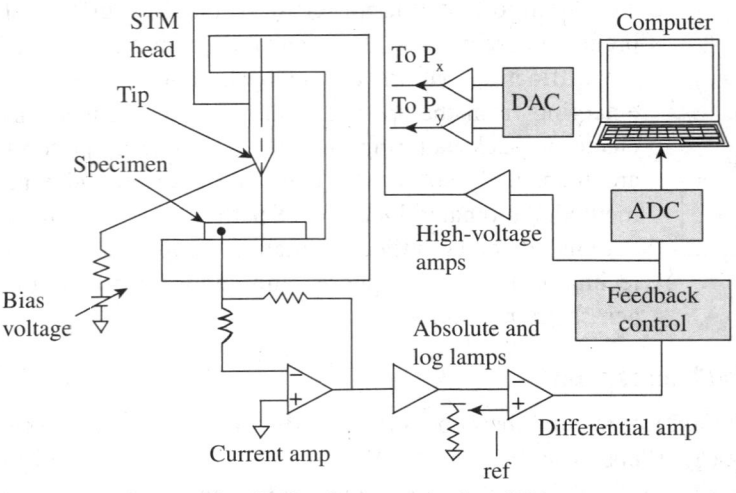

Fig. 17.10 Components of an STM

fed into an absolute-logarithmic amplifier to linearize the relation between the tunnelling current and the separation between the tip and the sample. Afterwards, a reference value I_{re} is subtracted from the linearized signal, which is a target value for the STM feedback operation to keep the current constant. Then, the signal is input to the feedback control. A suitable set of gain and time constants is selected to maintain a constant current. Finally, the output from the feedback control is amplified with a high-voltage amplifier having an output range higher than ~100 V, which is applied to the z-piezo. When the tunnelling current exceeds the target value, the feedback control retracts the tip, and conversely, when the tunnelling current decreases from that value, the control brings the tip closer to the sample. To observe an STM image, X–Y piezos are scanned by changing voltages applied to them in saw-like waveforms that are generated by a computer with digital-to-analog converters (DACs). The signal output from the feedback control is fed into an analog-to-digital converter (ADC) installed in the computer. The STM image, processed from the 3-D data of X–Y–Z voltages applied to the scanner, is displayed on a computer monitor and stored in the computer memory.

An STM requires a vibration-free environment. The instrument is provided with airlegs or a mechanical or gas spring system, and a big steel platform. The tip is suspended from a piezo-tripod. The three piezo legs control the tip motion within a fraction of an angstrom. This set-up permits the STM to be used under high vacuum and also at low temperatures. The entire set-up needs an environmental control system, including an ultra-high vacuum chamber and pumps to keep the tip and the sample clean, a clean gas purging system, a liquid cell with an electrochemical control, and temperature controls for high- and low-temperature observations.

Applications of STM

An STM is a novel surface imaging microscope with atomic resolution. At present, the applications of an STM are numerous. It is used as a powerful high-resolution surface microscope in materials science for samples with electrical conductivity, and its applications will continue to expand in various other fields as well. An STM is a powerful tool for viewing surfaces at the atomic level. It is versatile as it can be used in ultra-high vacuum, in air and various other liquid or gas ambients, and at temperatures ranging from near 0 K to 1000 °C.

Surface Topography Measurement

A microscope called the *topographiner* had been developed by a team of scientists, led by R. Young, in the USA in the early 1970s. Young applied a high voltage to a sharpened metal tip and scanned it over the sample surface. Although they succeeded in obtaining surface topography on a nanometre scale, they could not achieve atomic resolution due to the shortcomings of the vibration isolation part of the instrument. On the other hand, Binnig and Rohrer successfully developed a stable vibration isolation stage, which made it possible to use the tunnelling mechanism to achieve the desired results.

Today, an STM is the best choice for plotting surface topography of nanomaterials. As long as the structure of the specimen remains stable during scanning and the specimen is a conductor of electricity, the STM provides a high-resolution image of surface topography. When beginning the STM scanning, it is required to obtain the tunnelling current. This is achieved by bringing the tip closer to the sample (the tip and the sample being separated by a few millimetres) using a coarse positioning system. Several types of coarse positioning systems, consisting of piezo ceramics as a main drive, have been developed. To observe an STM image, $X–Y$ piezos are scanned by changing voltages applied to them in saw-like waveforms that are generated by a computer with DACs. As already pointed out, the signal output from the feedback control is fed into an ADC installed in the computer. The STM image, processed from the 3-D data of $x–y–z$ voltages applied to the scanner, is displayed on a computer monitor.

17.4.4 Atomic Force Microscope

Although STM was considered a fundamental advancement for scientific research, it had limited applications, because it worked only on electrically conductive samples. This limitation led the inventors to think about a new instrument that would be able to image insulating samples. In 1986, Binnig, Quate, and Gerber showed how improvization could be done by replacing the wire of a tunnelling probe from an STM with a lever prepared by carefully gluing a tiny diamond onto the end of a spring made from a thin strip of gold. This was the cantilever of the first atomic force microscope (AFM). The movement of the cantilever was monitored by measuring the tunnelling current between the gold spring and a wire suspended above it. This set-up was highly sensitive to the movement of the probe as it scanned the sample, again moved by piezoelectric elements. This created new excitement in nanometrology.

Atomic force microscopy allows the researcher to see and measure surface structure with unprecedented resolution and accuracy. One can even get images of the arrangement of individual atoms in a sample or see the structure of individual molecules. An AFM is rather different from other microscopes, because it does not form an image by focusing light or electrons onto a surface, like an optical or electron microscope. An AFM physically 'feels' the sample's surface with a sharp probe, building up a map of the height of the sample's surface. By scanning a probe over the sample surface, it builds a map of the height or topography of the surface as it goes along. This is very different from an imaging microscope, which measures a 2-D projection of a sample's surface. It has been given the name AFM since it operates by measuring attractive or repulsive forces between the tip and the sample in constant height or constant force mode. Most practical applications deal with samples of (sub)micrometre dimensions in the $X–Y$ plane and of nanorange in the Z-axis. Since its invention in the 1980s, AFMs have come to be used in all fields of science, such as chemistry, biology, physics, materials science, nanotechnology, astronomy, and medicine.

The basic component of an AFM is the piezoelectric transducer. The piezoelectric transducer moves the tip over the sample surface, a force transducer senses the force

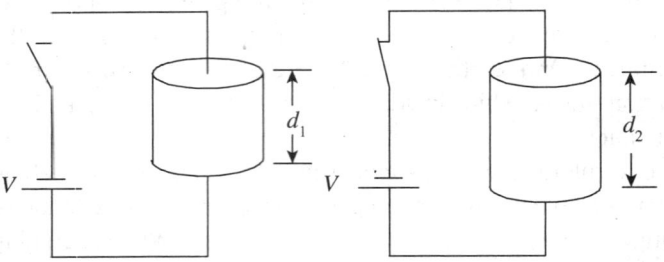

Fig. 17.11 Piezoelectric material

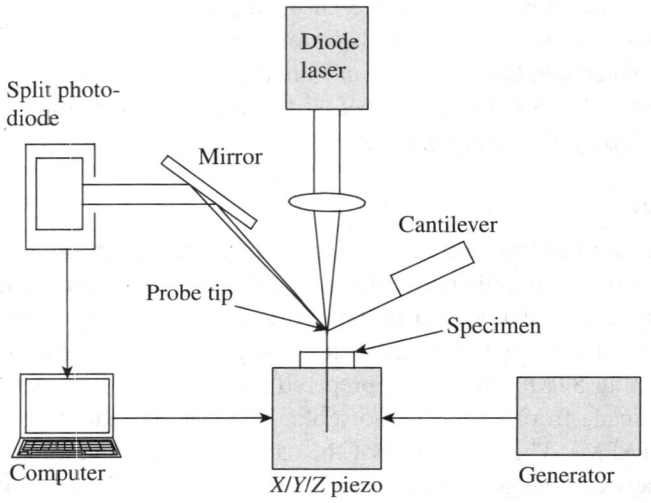

Fig. 17.12 Laser deflection contact AFM

between the tip and the surface, and the feedback control feeds the signal from the force transducer back into the piezoelectric, in order to maintain a fixed force between the tip and the sample. Piezoelectric materials are electromechanical transducers that convert electrical potential into mechanical motion. When a potential is applied across two opposite sides of the piezoelectric device, as shown in Fig. 17.11, it changes geometry. The magnitude of the dimensional change is of the order of 0.1 nm per applied voltage of 1 V. Thus, the ability to control such tiny movement makes piezoelectric materials the key for making measurements in an AFM.

The component of a laser deflection-type instrument is shown in Fig. 17.12. The necessary parts are the X, Y, and Z-piezo that are separately actuated by the X/Y drive and Z-control with extreme precision. The sample is mounted on the XYZ piezo, close to a sharp tip under the inclined cantilever with its mount. The diode laser light is focused at the end of the cantilever, reflected via a mirror to a split diode that provides the feedback signal (topologic information) for maintaining the force by Z-piezo response. The force between an AFM probe and a surface is measured with a force transducer. The force transducer must have a force resolution of 1 nN or less so that the probe is not broken while scanning. The control electronics take the signal from the force transducers and use it to drive the piezoelectrics, so as to maintain the probe–sample distance and thus the interaction force at a set level. Thus, if the probe registers an increase in force (for instance, while scanning, the tip encounters a particle on the surface), the feedback control causes the piezoelectrics to move the probe away from the surface. Conversely, if the force transducer registers a decrease in force, the probe is moved towards the surface. Data sampling is made at discrete steps by means of an ADC. A computer reconstructs the 3-D topological image or projections from the data matrix. Imaging software adds colour, height contrast, and illumination from variable directions.

The electronic control unit is assembled in a separate cabinet in the instrument. The primary function of the electronic control unit in an AFM is to

• generate scanning signals for the X–Y piezoelectrics;
• take an input signal from the force sensor and then generate the control signal for the Z piezo;

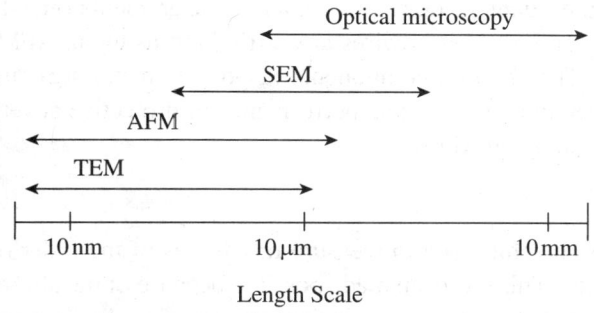

Fig. 17.13 Comparison of length scales of various microscopes

- control output signals for X–Y–Z stepper motors;
- generate signals for oscillating the probe and measuring phase or amplitude when an oscillating mode is used for scanning; and
- collect signals for display by the computer.

One of the major advantages of an AFM is its ability to magnify in the X, Y, and Z axes. Figure 17.13 shows a comparison between several types of microscopes. It can be seen that the maximum length of the specimen is limited to slightly more than 100 µm. This is because an AFM requires scanning the probe mechanically over a surface, and scanning such large areas will consume a lot of time. However, the length-scale of an AFM overlaps nicely with a conventional optical microscope. Since AFM is also quite compact in construction, it can be integrated with an optical microscope. This combination will cover a length-scale from nanometres to millimetres. Table 17.3 gives a comparison between AFM, SEM, and TEM.

Table 17.3 Comparison between AFM, SEM, and TEM

	AFM	SEM	TEM
Sample preparation	Almost nil	Required	Required
Resolution (nm)	0.1	5	0.1
Relative cost	Low	Medium	High
Sample environment	Any clean environment	Vacuum or gas	Vacuum
Depth of field	Poor	Good	Poor
Sample type	Conductive or insulating	Conductive	Conductive
Maximum sample dimension	Unlimited (in theory, but limited in practice)	30 mm	1 mm
Measurement	Three-dimensional	Two-dimensional	Two-dimensional

Applications of AFM

The AFM has been developed into a novel technique for obtaining high-resolution images of both conductors and insulators. As already pointed out, an AFM is compact and hardly requires any specimen preparation. Therefore, it is perceived to be very user-friendly and quite popular among users. One of the most popular applications is in the field of surface science. It can be employed to get an image's atomic resolution as well as to measure electrical, magnetic, and mechanical properties of nanomaterials. It is possible to capture images of fine structure of metals and absorbed impurities. An AFM can also be used to study structures of organic and inorganic insulators. It can reveal topographic, tribological, roughness, and adhesion/fouling characterization of a wide variety of nanomaterials.

AFMs can be used to measure an extremely wide range of nanoparticles, including different metal nanoparticles, metal oxide particles, many types of composite metal/organic particles, synthetic polymer particles, nanorods, and quantum dots. Mechanical measurements on 1-D nanostructures such as carbon nanotubes and metal nanowires are carried out using an AFM. It has many applications in metallurgy. The sintered components in powder metallurgy are subjected to roughness measurement. Even the topology and performance of protective covers or coatings can most reliably be determined by an AFM.

Force Measurement Using AFM

An AFM is not only a microscope. It can be employed to measure tiny forces of the order of tens of piconewtons to tens of nanonewtons. This has been made possible because of the ability of AFMs to interact with the sample physically. One can give the analogy of a blind person visualizing samples by touching them. With the sense of touch comes the ability to push, pull, deform, and manipulate objects. Thus, the physical contact with the specimen gives the AFM extraordinary sensitivity, which aids in measurement of atomic forces. The atomic forces are created in the process of interaction between molecules in covalent and electrostatic bonds, multiple hydrogen bonding, etc. Even biological processes such as inter-cellular binding and adhesion create tiny forces that can be measured by an AFM.

The AFM tip is engaged with the specimen surface, and a 'force versus distance' experiment is performed. The deflection of the lever is recorded over a known distance, and the subsequent electrical signal generated by the photodetector is recorded. The X, Y channels of the AFM scanners are frozen, the feedback loop is suspended, and the tip and the sample are pushed together (or, in some cases, pulled apart) by ramping the Z-piezo channel of the scanner. The resulting photodiode signal is plotted. By calibrating this voltage signal with force, employing any one of several mathematical approaches recommended by research studies, it is possible to determine atomic forces.

17.5 X-RAY DIFFRACTION SYSTEM

An X-ray diffraction (XRD) system is an ideal method for examining samples of metals, polymers, ceramics, semiconductors, thin films, and coatings. It can also be employed for forensic and archaeological analysis. A 2-D diffraction pattern provides abundant information on the atomic arrangement, microstructure, and defects of a solid or liquid material.

17.5.1 Principles of XRD

X-rays are electromagnetic radiations with wavelengths in the range of 0.01–100 Å. When a monochromatic X-ray beam hits a sample, in addition to absorption and other phenomena, it generates scattered X-rays with the same wavelength as the incident beam. This type of scattering is also known as elastic scattering or coherent scattering. The X-rays scattered from a sample are not evenly distributed in space, but are a function of the electron distribution in the sample. The intensities and spatial distributions of the scattered X-rays form a specific diffraction pattern that is uniquely determined by the structure of the sample. The Bragg law, named after the nobel laureate (1914), Lawrence Bragg, brings out the relationship between the diffraction pattern and the material structure.

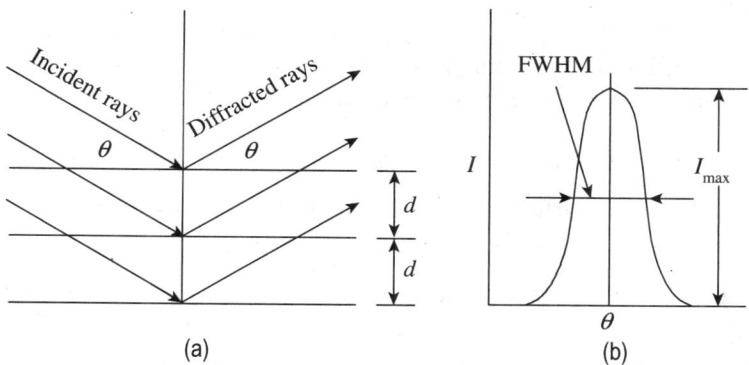

Fig. 17.14   Bragg law (a) Incident rays and refracted rays
(b) Diffraction peak at Bragg angle

17.5.2 Bragg Law

The Bragg law describes the relationship between the diffraction pattern and the material structure. If the incident X-rays hit the crystal planes with an incident angle θ and reflection angle θ as shown in Fig. 17.14(a), the diffraction peak is observed when the Bragg condition is satisfied, that is, $n\lambda = 2d \sin \theta$.

Here, λ is the wavelength, d is the distance between adjacent crystal planes, θ is the *Bragg angle* at which one observes a diffraction peak, and n is an integer number, called the order of reflection. This means that the Bragg condition with the same d-spacing and Bragg angle can be satisfied by various X-ray wavelengths. A typical diffraction peak is a broadened peak displayed by the curved line in Fig. 17.14(b). The peak broadening can be due to many effects, including imperfect crystal conditions such as strain and mosaic structure. The curved line gives a peak profile, which is the diffracted intensity distribution in the vicinity of the Bragg angle. The highest point on the curve gives the maximum intensity of the peak, I_{max}. The width of a peak is typically measured by its full width at half maximum (FWHM). The total diffracted energy of a diffracted beam for a peak can be measured by the area under the curve, which is referred to as integrated intensity.

XRD can provide information on the atomic arrangement in materials with long-range order, short-range order, or no order at all, such as gases, liquids, and amorphous solids. Typical diffraction patterns in solids, liquids, and gases are illustrated in Fig. 17.15. The diffraction pattern from crystals has many sharp peaks corresponding to various crystal planes, based on the Bragg law. Both amorphous solid and liquid materials do not have the long-range order

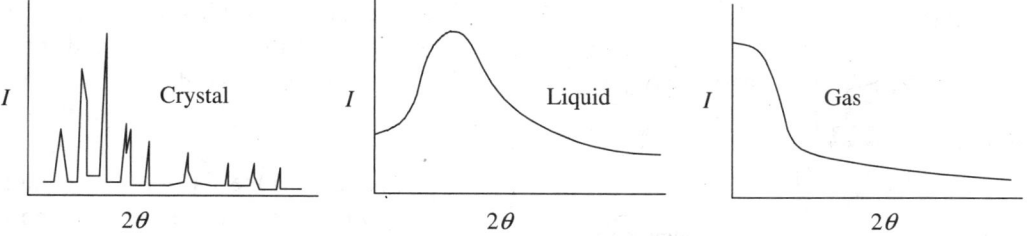

Fig. 17.15   Diffraction patterns in solids, liquids, and gases

that a crystal does, but the atomic distance has a narrow distribution due to the atoms being tightly packed. The integrated diffraction profiles can be analysed with existing algorithms and methods. Profiling and matching with existing templates in the database enable the identification of variations in atomic structures due to various defects.

17.5.3 Two-dimensional XRD System

A typical 2-D XRD system, referred to as XRD^2, comprises five basic components, as shown in Fig. 17.16.

X-ray source X-rays are produced with the required radiation energy, focal spot size, and intensity.

X-ray optics It conditions the primary X-ray beam to the required wavelength, beam focus size, beam profile, and divergence.

Goniometer and sample stage Its function is to establish and manoeuvre the geometric relationship between primary beam, sample, and detector.

Sample alignment and monitor This component assists users with positioning the sample at the centre of the instrument and monitors the sample state and position.

Area detector It intercepts and records the scattering X-rays from a sample, and saves and displays the diffraction pattern into a 2-D frame.

A variety of X-ray sources, from sealed X-ray tube and rotating anode generator to synchrotron radiation, can be used in XRD. The sealed tube generator and rotating anode generator produce X-ray radiation with the same physical principle. Electrons are emitted from the cathode and are accelerated by high voltages between the cathode and the anode. The anode is made of the selected metal, so it is also called a metal target. When the electron beam hits the target, X-rays are produced and radiate in all directions. Intensity of the X-ray beam depends on X-ray optics, focal spot brightness, and focal spot profile. Cooling water circulation is provided to the X-ray generator to avoid meltdown of the anode. Depending on the cooling efficiency, only limited power can be applied to an X-ray generator. The total amount of X-rays generated is proportional to the total power load on the anode.

The function of X-ray optics in XRD is to condition the X-ray beam into a spectrum of desired purity, intensity, and cross section. The space between the focal spot of the X-ray tube and the sample is referred to as the primary beam path. X-rays travelling through this beam path are scattered by the air with two adverse effects. One is the attenuation of the primary beam intensity. The more harmful effect is that the scattered X-rays travel in

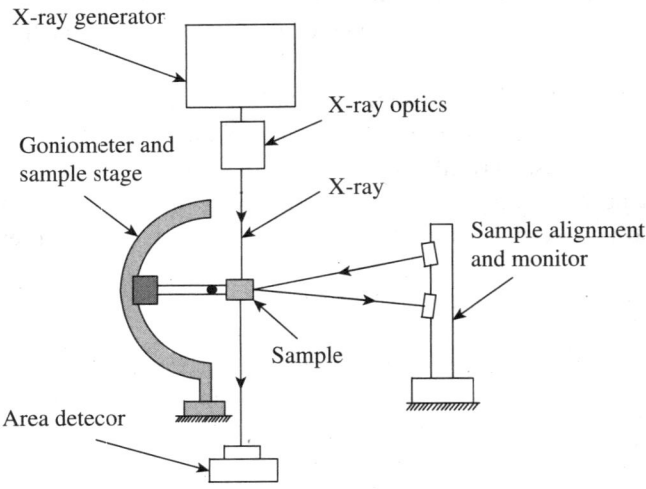

Fig. 17.16 Two-dimensional X-ray diffraction system

all directions and some reach the detector. This air scatter introduces a background noise over the diffraction pattern. As a consequence, weak diffraction patterns may be buried under the background. Therefore, the open incident beam path should be kept as small as possible. To reduce air attenuation and air scatter of the incident beam, a helium-purged beam path or a vacuum beam path is sometimes used in an XRD system.

The function of the goniometer and sample positioning stages is to establish and control the geometric relationship between the incident beam, sample, and detector. The goniometer is also the supporting base to many components such as X-ray sources, X-ray optics, sample environment stages, sample-aligning microscopes, and so on. In a 2-D XRD system, the goniometer should facilitate at least three rotation axes in order to cover all the possible orientations of a sample in the diffractometer.

Sample alignment systems are required to position the sample at the centre of the instrument and to monitor the sample state and position before and during data collection. Either optical microscopes or video microscopes can be used for sample alignment and visualization. The optical microscope allows the user to directly observe the sample as a magnified image, with a crosshair to determine the sample position. Video images can be captured more conveniently with the X-ray safety enclosure.

17.5.4 Applications of XRD System

XRD is the ideal, non-destructive, analytical method for examining samples of many types, such as metals, polymers, ceramics, semiconductors, thin films, and coatings. It is increasingly being employed for drug discovery and processing, forensic analysis, archaeological analysis, and many emerging applications. In the long history of powder XRD, data collection and analysis have been based mainly on 1-D diffraction profiles measured with scanning point detectors or linear position-sensitive detectors. Therefore, almost all X-ray powder diffraction applications, such as phase identification, texture, residual stress, crystallite size, and per cent crystallinity, are developed in accord with the diffraction profiles collected by conventional diffractometers. A 2-D diffraction pattern contains abundant information about the atomic arrangement, microstructure, and defects of a solid or liquid material.

A QUICK OVERVIEW

- Nanotechnology has wide-ranging applications in material science, medicine and drug delivery, sensorics, biotechnology, optics, tissue engineering, cancer therapy, etc.
- Nanometrology has a crucial role in the production of nanomaterials and devices with a high degree of accuracy and reliability. It includes length or size measurements (where dimensions are typically given in nanometres and the measurement uncertainty is often less than 1 nm) as well as measurement of force, mass, and electrical and other properties.
- A transmission electron microscope (TEM) comprises three lenses: an objective lens, an intermediate lens, and a projector lens. The microscope is built in such a way that it allows easy switching from the high-magnification imaging mode to the selected-area diffraction mode. The final image is formed on a fluorescent screen or on the entrance plane of a recording device for a permanent recording of the image, which can be used for further analysis.

- The scanning electron microscope (SEM) is arguably the most versatile microscope with magnification ranging from 5× to as high as 10^6×. The use of a field-emission gun in an SEM makes it possible to image individual heavy atoms in the transmission mode by collecting scattered electrons with a sensitive detector. A tungsten filament is used as the source of electrons and the beam size is also quite small, of the order of 10 nm. The electron beam of an SEM is scanned horizontally across the specimen in two mutually perpendicular (X and Y) directions. The outputs of the two scan generators can either be used to generate the display on a cathode-ray tube or be stored in the computer memory in digital format.
- A scanning tunnelling microscope (STM) provides three-dimensional (3-D) atomic-scale images of the surface of a sample. The STM works on the principle of quantum tunnelling. When an atomically sharpened tip under a small voltage is brought close to the surface of a sample, so that the separation is of the order of a nanometre, there is a small change in current in the circuit. The induced current is referred to as the tunnelling current. This current increases as the gap between the tip and the sample decreases. The change in the tunnelling current can be calibrated with respect to the change in gap. In other words, if we scan the tip over the sample surface while keeping the tunnelling current constant, the tip movement depicts the surface topography, because the separation between the tip apex and the sample surface is always constant.
- An atomic force microscope (AFM) is different from other microscopes because it does not form an image by focusing light or electrons onto a surface, like an optical or electron microscope. An AFM physically 'feels' the sample's surface with a sharp probe, building a map of the height of the sample's surface. By scanning a probe over the sample surface, it builds a map of the height or topography of the surface as it goes along. It has been given the name AFM since it operates by measuring attractive or repulsive forces between the tip and the sample in constant height or constant force mode. The basic component of an AFM is the piezoelectric transducer. The control electronics take the signal from the force transducers and use it to drive the piezoelectrics, so as to maintain the probe–sample distance and thus the interaction force at a set level. A computer reconstructs the 3-D topological image or projections from the data matrix.
- X-ray diffraction is an ideal method for examining samples of metals, polymers, ceramics, semi-conductors, thin films, and coatings. When a monochromatic X-ray beam hits a sample, in addition to absorption and other phenomena, it generates scattered X-rays with the same wavelength as the incident beam. The intensities and spatial distributions of the scattered X-rays form a specific diffraction pattern that depends on the structure of the sample.

MULTIPLE-CHOICE QUESTIONS

1. Nanoparticles, which possess nanodimension in all the three principal directions, are called
   (a) quantum crystals
   (b) quantum dots
   (c) nanotubes
   (d) nanowires
2. The 2-D nanocarbon is called
   (a) graphene
   (c) nanoplate
   (b) graphite
   (d) none of these
3. Which of the following is used as transistors in fabricating a nanocomputer?
   (a) Nanomers
   (b) Y-junction carbon nanotubes
   (c) Graphene

(d) Quantum crystals

4. The resolution of an optical microscope can be improved by
   (a) decreasing the wavelength of incident light
   (b) increasing the wavelength of incident light
   (c) increasing the intensity of incident light
   (d) decreasing the intensity of incident light

5. Which microscope allows the user to switch between image mode and diffraction mode?
   (a) SEM   (c) AFM
   (b) STEM   (d) TEM

6. When an electron beam strikes the specimen in an SEM, _____ are emitted from the electron-entrance surface of the specimen.
   (a) two types of electrons
   (b) three types of electrons
   (c) four types of electrons
   (d) no electrons

7. The beam size in an SEM is of the order of
   (a) 1 nm   (c) 10 nm
   (b) 10 nm   (d) 1 μm

8. _____ type of scanning is employed in SEM.
   (a) Raster scanning
   (b) Linear scanning
   (c) Interlaced scanning
   (d) Non-interlaced scanning

9. Which of the following preparations is required before testing to facilitate the use of specimens with insulating materials in an SEM?
   (a) Glazing the surface
   (b) Electroplating the surface
   (c) Polishing the surface
   (d) Coating the surface with a thin film of metal

10. Which of the following microscopes provides 3-D atomic-scale images of the surface of a sample?
    (a) Optical high-resolution microscope
    (b) TEM
    (c) SEM
    (d) STM

11. An STM works on the principle of
    (a) quantum mechanics
    (b) quantum tunnelling
    (c) XRD
    (d) laser microscopy

12. Which of the following microscopes is employed to measure tiny forces of the order of piconewtons?
    (a) TEM   (c) AFM
    (b) STEM   (d) XRD

13. _____ is the best choice for plotting surface topography of nanomaterials.
    (a) Perthometer   (c) STM
    (b) Talyserf   (d) AFM

14. Which of the following microscopes has the best resolution (up to atomic level)?
    (a) AFM   (c) SEM
    (b) STM   (d) Topographiner

15. The Bragg law brings out the relationship between
    (a) the diffraction pattern and the material structure
    (b) the atomic forces and material structure
    (c) wavelength of incident light and microscope resolution
    (d) none of these

16. In an XRD system, _____ is used to control the geometric relationship between the incident beam, sample, and detector.
    (a) X-ray optics   (c) goniometer
    (b) carbon nanotubes   (d) area detector

17. In an XRD system, the total diffracted energy of a diffracted beam for a peak can be measured by the area under the curve and is referred to as
    (a) integrated intensity
    (b) differential intensity
    (c) diffracted intensity
    (d) thermal intensity

18. In which of the following nanomaterials, does XRD pattern have sharp peaks and valleys?
    (a) Crystals   (c) Gases
    (b) Liquids   (d) Amorphous solids

19. Among the following microscopes, which is the most expensive?
    (a) AFM   (c) TEM
    (b) SEM   (d) Optical microscope

20. Which of the following is not a 2-D nano-structure?
    (a) Nanosheet   (c) Nanorod
    (b) Nanofilm   (d) Nanocoating

REVIEW QUESTIONS

1. Define nanometrology. What is the scope of nanometrology?
2. Explain how nanosize affects the properties of materials.
3. How are nanomaterials classified? List any five major applications of nanotechnology.
4. Discuss how the technique of microscopy has evolved from the micron level to the nano level.
5. Differentiate between image mode and diffraction mode in the usage of a TEM.
6. Explain the construction and working principle of thermionic electron gun used in TEM. Discuss the relationship between emission current and cathode temperature.
7. With the help of a sketch, explain the working principle of an SEM. What is the preferred scan pattern in an SEM?
8. Explain the working principle of an STM.
9. What is the requirement of the stylus tip in an STM? How does it come into play in the measurement of surface topography?
10. What is the role of a piezoelectric transducer in an AFM?
11. Explain the working principle of a laser deflection contact AFM.
12. What is the approach for measuring force using an AFM?
13. What is Bragg law? What is its significance in measurement using XRD?
14. Explain the 2-D XRD system.
15. In the form of a table, highlight the advantages and applications of all the nanometrology instruments explained in this chapter.

Answers to Multiple-choice Questions

1. (b)	2. (a)	3. (b)	4. (a)	5. (d)	6. (b)	7. (b)	8. (a)	9. (d)
10. (d)	11. (b)	12. (d)	13. (c)	14. (a)	15. (a)	16. (c)	17. (a)	
18. (a)	19. (c)	20. (d)						

APPENDICES

Universal Measuring Machine

A universal measuring machine (UMM) is used to measure the geometric features of components. The measurement can be performed in both absolute and comparative modes, thereby providing a convenient means to carry out inspection of length and diameters of both plain and threaded work, tapers, and pitch of screw threads to a high degree of accuracy. Figure A.1 illustrates the constructional features of a UMM. It consists of a fixed head and a movable head, also called the measuring carriage. The part to be inspected is mounted on the clamping device. The spindle centres, which are fitted with sleeves of hardened steel, ensure accurate centring of various tools such as the locating microscope, feller microscope, goniometer microscope, spotting tool, and locating indicator.

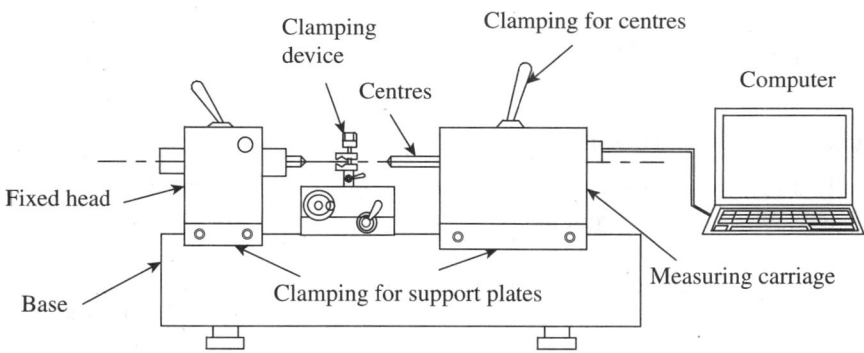

Fig. A.1 Universal measuring machine

The spindle has an electronic test indicator rather than a probe. The indicator can be accurately controlled and moved across a part, either along a linear axis or radially around the spindle. It continuously records the profile and geometry of the component being inspected, which proves advantageous for the UMM while profiling radii, contours, and holes. A component can also be tested optically by a short-focus microscope carried in a special tool holder. The UMMs, which are presently available in the market, are interfaced with a computer, which enables easier recording of test results and subsequent analysis. A user-friendly software interface enables the user to select the testing procedure, resolution, and type of test probe.

Presently, UMMs are being used as special-purpose machines in metrology laboratories. These are valuable devices for comparing master gauges and length standards. However, since UMMs have largely been replaced by coordinate measuring machines, they have limited usage in companies. Whenever the need to use a UMM arises, manufacturers opt to subcontract the measurement to a laboratory that has the facility.

Flow
Measurement

Measurement of fluid flow is a phenomenon that is studied in great detail in courses on fluid mechanics. Hence, we are treating this topic as an appendix for the book. Here we will try to present the basics of some of the popular equipment used in the measurement of flow.

B.1 ORIFICE METER

An orifice meter or an orifice plate, as illustrated in Fig. B.1, is a device that has been designed to measure the volumetric flow and mass flow rates of the fluid flowing within a pipe. Ironically, this instrument is actually a disk with a hole (an orifice) usually at its centre or sometimes placed eccentrically.

Such a disk is simply introduced in pipe flanges facing the flow of any incoming fluid within a pipe.

Mechanism of Fluid Flow Through Orifices

The fluid flowing through the pipe first experiences a convergence effect to enter through the orifice. However, the maximum convergence of the fluid does not occur exactly at the orifice

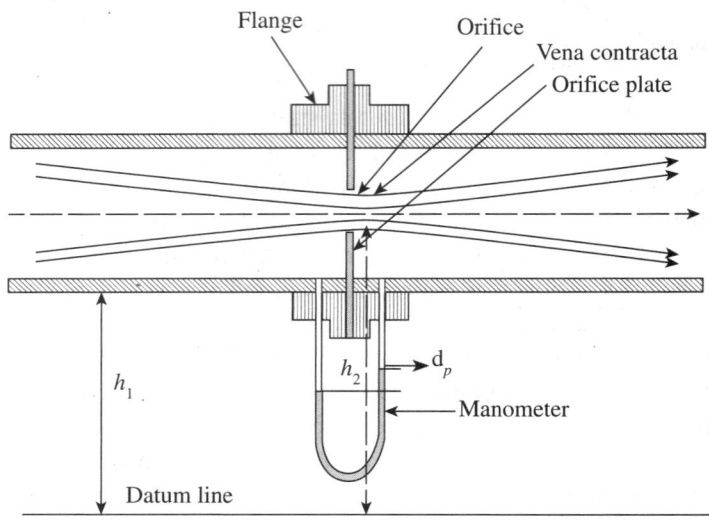

Fig. B.1 Orifice meter

point, but a little downstream from it. This position is termed *vena contracta*, where the area of cross section is minimum, the fluid velocity is maximum, and the pressure is minimum.

Beyond the vena contracta point, the fluid flow again diverges steadily to reach the equilibrium.

The velocity (v) of the fluid is determined by measuring the pressure difference between the flow at the vena contracta position and the normal flow. A manometer is connected to this set-up to measure the pressure difference. The obtained pressure values are used to derive the velocity of the flowing fluid using *Bernoulli's equation*.

$$\frac{v_1^2}{2} + gh_1 + \frac{P_1}{\rho_1} = \frac{v_2^2}{2} + gh_2 + \frac{P_2}{\rho_2}$$

Here,

v_1 and v_2 refer to the velocity of the fluid at the normal flow and vena contracta positions, respectively;

g is the acceleration due to gravity;

h_1 and h_2 refer to the vertical distance of the normal flow and vena contracta positions, from the datum line, respectively;

P_1 and P_2 refer to the pressure values at the normal flow and vena contracta positions, respectively; and

ρ_1 and ρ_2 are the fluid densities at the normal flow and vena contracta position, respectively.

Case 1. In incompressible fluids, $\rho_1 = \rho_2 = \rho$.
Assuming the pipe is horizontal, $gh_1 = gh_2$
The equation could be simplified as follows:

$$\left[\frac{v_2^2 - v_1^2}{2} = \frac{P_1 - P_2}{\rho}\right]$$

This equation can be further simplified by substituting $v_1 = \frac{Q}{A_1}; v_2 = \frac{Q}{A_2}$, where Q is the volumetric flow rate.

Since A_1 and A_2 cannot be directly equated to the areas of pipe and vena contracta respectively, a coefficient of discharge (C_D) is introduced into the equation.

$$Q = C_D \frac{A_2}{\sqrt{1 - \left(\frac{A_2}{A_1}\right)^2}} \sqrt{\frac{2(P_1 - P_2)}{\rho}}$$

Further, $Q = K\sqrt{\Delta p}$

Here,

$$K = C_D \frac{A_2(1.414)}{\sqrt{\rho\left(1 - \left(\frac{A_2}{A_1}\right)^2\right)}}$$

is a constant with respect to the application of the orifice meter.

Case 2.   In compressible fluids, density is a variable component through various stages of flow. Hence $\rho_1 \neq \rho_2$

In addition, in the case of compressible fluid, mass flow rate is more important than volumetric flow and the following equation is derived:

$$Q_{mass} = Q = \varepsilon\, C_D \frac{A_2}{\sqrt{1 - \left(\dfrac{A_2}{A_1}\right)^2}} \sqrt{2\left(P_1 - P_2\right)}$$

Here, ε is called the expansion ratio.

B.2 VENTURI METER

The venturi meter uses the same principle as an orifice meter. However, unlike an orifice meter, which is just a plate, a venturi meter is a complete device, which can be fitted between the pipe lines.

It consists of a short length of pipe steadily narrowing down to a lesser area of cross section called a *throat* and gradually expanding to the original pipe diameter. This is shown in Fig. B.2. As in the case of an orifice meter, the fluid flow converges through the constricted section of the pipe and gradually expands to reach equilibrium. The difference in pressure measured using a manometer is used to determine the velocity of a flowing fluid. Here, the fluid convergence towards the area of constriction is termed venturi effect.

Applying Bernoulli's equation, we obtain the following equation:

$$Q = C_D \frac{A_1 A_2}{\sqrt{A_1^2 - A_2^2}} \sqrt{2g\Delta h}$$

Here, C_D is the coefficient of discharge; A_1 and A_2 are the cross-sectional areas of the inlet and constricted sections, respectively; and Δh is the venturi head.

On substituting $C = \dfrac{A_1 A_2}{\sqrt{A_1^2 - A_2^2}} \sqrt{2g}$, the aforementioned equation becomes $Q = C_D C \sqrt{\Delta h}$

C is termed the venturi constant

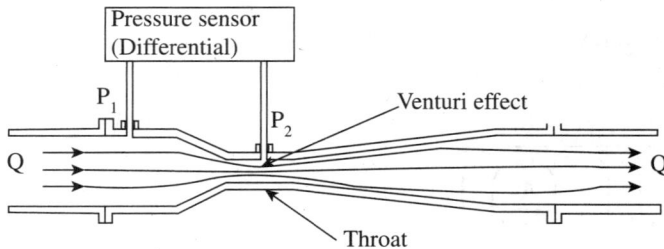

Fig. B.2   Venturi meter

The advantage of a venturi meter is that it offers minimum head loss when compared to the orifice meter. However, the practical disadvantage of a venturi meter is that the differential pressure is reduced during the flow, leading to a less efficient output value. Moreover, the construction and assembly of a venturi meter is much more expensive than an orifice meter. These devices are chosen on the basis of the nature of the flow set-up and the end requirement of the engineer.

B.3 PITOT TUBE

The pitot tube, illustrated in Fig. B.3, is a device that is used to measure the local velocity and total pressure of a fluid, as against the velocity measured across the tube in case of an orifice plate and a venturi meter. It finds extensive application in aircrafts.

As illustrated in Fig. B.3, a pitot tube consists of an 'L' shaped structure held upfront against the fluid flow. It consists of a static probe and an impact probe. The impact probe should face a direction that is against the fluid flow. While the static probe measures the static pressure in the system, the impact probe is meant to measure the total pressure of the system.

When the fluid flows against the tip of the impact probe, it is brought to rest. This has an effect (rise) on the pressure (P_2) corresponding to P_1 at the static probe.

Applying Bernoulli's Theorem for an incompressible fluid, we get the following equation:

$$V = \sqrt{\frac{2\left(P_2 - P_1\right)}{\rho}}$$

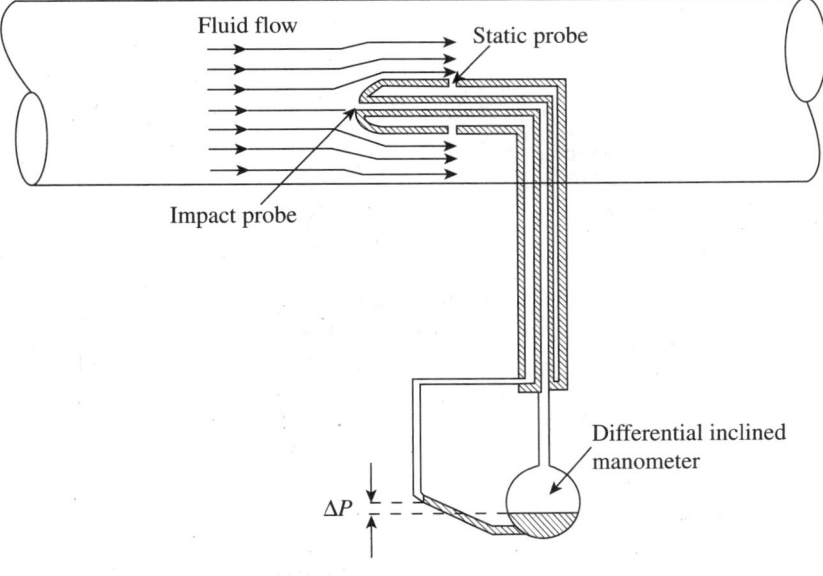

Fig. B.3   Pitot tube

B.4 ROTAMETER

The rotameters is a class of variable area meters that is designed to study the flow rate of a fluid within a closed tube.

It uses the natural principles of buoyancy, drag, and acceleration due to gravity to measure the fluid flow. A typical rotameter illustrated in Fig. B.4 consists of a tapered glass tube with a liquid and a float. The float is made to balance by the action and reaction of buoyant, drag, and gravity forces.

When this set-up is introduced into the pipeline, by virtue of fluid flow, the system should experience a change in the pressure interval (ΔP) and also a displacement in the float.

According to the drag equation, ΔP varies with the square of flow rate of the fluid. To counter this action and maintain a constant pressure interval, the area of the meter is varied and hence, the tapered design of the rotameter.

The float moves upwards till an equilibrium is reached. The scale on the glass measuring the displacement of the float is directly proportional to the fluid flow rate.

$$Q = K(A_t - A_f)$$

There are also rotameters where the flow rate values are directly calibrated on the surface of the glass, thus enabling direct measurement.

The advantages of a rotameter are that it can directly measure flow rate values and its construction is also relatively simple and inexpensive. However, it cannot be used when the fluid is opaque because the float cannot be seen from outside. In addition, when there is an increased velocity of fluid flow, there is a possibility of breakage of the glass tube.

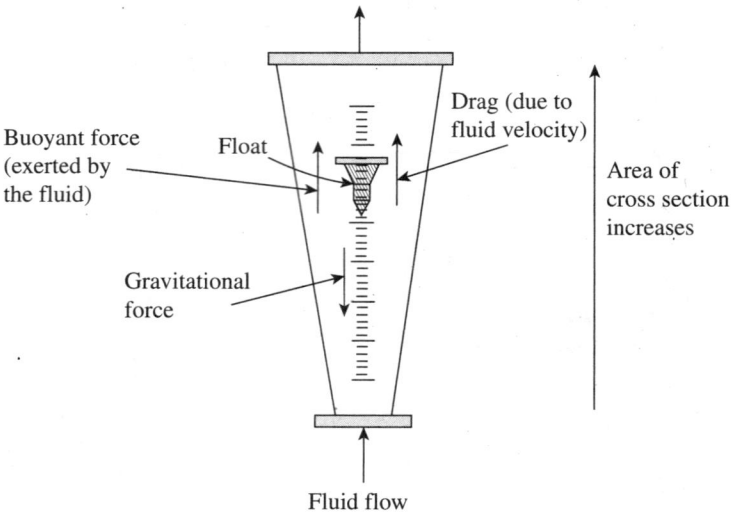

Fig. B.4 Rotameter

APPENDIX C

Experiments in Metrology and Measurements

Metrology and inspection process are as important as any manufacturing process. These enable manufacturing industries to produce zero-defect products, improve productivity, and enhance the image of the company. The quality control engineer should have a sound understanding of the proper selection, measurement, and maintenance of various instruments and devices at his/her disposal. A clear understanding of practical metrology is essential for design engineers, production engineers, as well as maintenance engineers. Understandably, a laboratory course on engineering metrology and mechanical measurements is an indispensable part of the mechanical engineering curriculum in universities around the world.

In this book, we have discussed in detail the various measurement techniques and instrumentation available for evaluating the accuracy and precision of manufactured components. However, only a laboratory course on the subject can provide students with the required knowledge about the proper use and application of various techniques. The experiments described here are an attempt in this direction and are of help to both instructors and students. A total of 20 experiments have been explained, covering all the major measurement techniques in metrology and mechanical measurements. The experiments discussed here were selected after a perusal of the curricula of the major universities in India.

Each experiment has been outlined in several sections, which comprise the aim of the experiment, apparatus used, theoretical background, experimental procedure, relevant formulae, tabular columns, and standard graphs. The major experiments in metrology include the use of comparators such as dial gauges, sine bars, clinometers, and autocollimators. Students will learn the proper use of instruments such as micrometers, bevel protractors, gear tooth callipers, and gear tooth micrometers. Optical instruments provide a high-resolution, non-contact mode for inspection. Accordingly, devices such as optical flat, tool maker's microscope, and profile projector have also been discussed. The remaining experiments in metrology pertain to measurement of the effective diameter of a screw and the measurement of surface roughness value using Handysurf. In addition to the aforementioned instruments and devices, students will also learn to use slip gauges, angle gauges, and feeler gauges.

Mechanical measurement techniques such as determination of force, pressure, and temperature are frequently used in the industry, more so in process industries such as power plants, cement industries, steel industries, and sugar industries. Experiments in mechanical measurements deal with the major techniques involved in measuring forces in metal cutting and calibration of devices such as thermocouples, load cells, linear variable differential transformers (LVDTs), strain gauges, and pressure gauges. After performing the experiments following the step-by-step procedure given in each experiment, students are expected to separately document their inference and the final results obtained. Students are also advised to analyse the experimental results meticulously and draw pertinent inferences, with the help of the course instructor. We understand that the experiments compiled in this appendix are not exhaustive, and there is scope for adding more to the list. However, one has to draw a line somewhere, keeping in mind the course duration and capability of engineering colleges to spare resources for the establishment of a metrology laboratory. Therefore, we have made an attempt to judiciously select experiments that are of utmost importance in the manufacturing industry.

Contents

C.1 CALIBRATION OF MICROMETERS

C.1.1 Aim

To calibrate the given micrometer

C.1.2 Apparatus

The following are the apparatuses used in this experiment:
1. Micrometer
2. Slip gauges

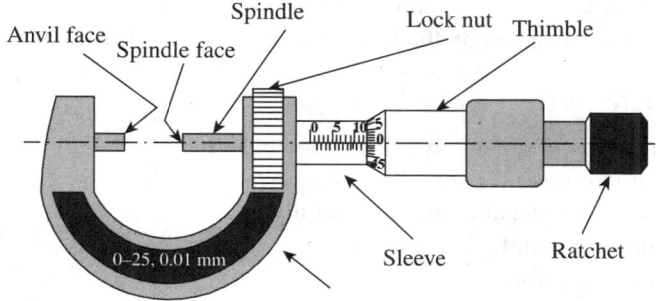

Fig. C.1   Micrometer

C.1.3 Theory

A micrometer, as illustrated in Fig. C.1, is one of the most common and most popular form of measuring instruments. It is used to measure the linear dimensions of small objects very accurately.

Principle

It utilizes a screw to provide a linear scale conversion of the circular motion of the screw head into the linear motion of the screw.

Terminology

Pitch   When a screw works in a fixed nut, the screw tip moves through a definite distance, called the pitch, for one complete rotation. It is equal to the distance between the two consecutive threads of a screw.

Least count   It can be defined as the distance moved by the screw when it is rotated through one division on the head scale. This is the smallest distance that can be measured accurately with the instrument. It is therefore the ratio of the rotation of the pitch of the screw to the number of head scale divisions.

Components

The major components of a micrometer are as follows:
1. Frame
2. Anvil and spindle
3. Thimble and barrel

4. Ratchet driver
5. Adjusting nut

Errors

The possible sources of errors, which may result in incorrect functioning of the instrument, are as follows:
1. Lack of flatness of the anvils
2. Lack of parallelism of anvils at some or all parts of the scale
3. Inaccurate setting of the zero reading
4. Inaccurate reading due to inaccurate zero position
5. Inaccurate readings shown by the fractional divisions on the thimble

C.1.4 Procedure

The experimental procedure involves the following steps:
1. Check the micrometer for smooth running over its range.
2. Clean the anvils of the micrometer carefully.
3. Find out the least count.
4. Find out the zero error.
5. Choose standard slip gauges.
6. Note down the dimensions of the chosen slip gauges.
7. Clean the chosen slip gauge.
8. Measure the slip gauge dimension over the micrometer.
9. Note down the measured reading in a tabular form as 'Indicated reading'.
10. Repeat steps 4–8 for at least 20 trials so that the entire range of the micrometer is covered.
11. Calculate the error and the percentage error.

Formulae

The formulae used for the calculation of some parameters are described here.

Least count, LC = Pitch of the screw/No. of head scale divisions = _____ mm

Pitch, P = Distance moved by the spindle/One revolution of the thimble = _____ mm

Indicated or micrometer reading = (Pitch scale + Coinciding head scale reading) × LC

Error = Indicated reading – Actual reading

Percentage error = (Error/Actual reading) × 100

C.1.5 Tabular Column

The experimental values can be documented in the format given in Table C.1.

Table C.1 Observation table for the calibration of a micrometer

Trial no.	Actual reading	Indicated reading			Error	% Error
		PSR (mm)	CHSD	Micrometer reading (mm)		

C.1.6 Graphs

In this experiment, the following graphs need to be drawn:

1. A graph of % error versus actual reading is drawn. The model graph is illustrated in Fig. C.2.
2. A graph of indicated reading versus actual reading is drawn, as illustrated in Fig. C.3.

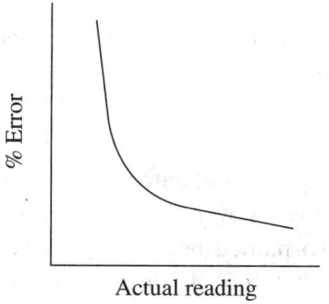

Fig. C.2   Model graph of % error versus actual reading

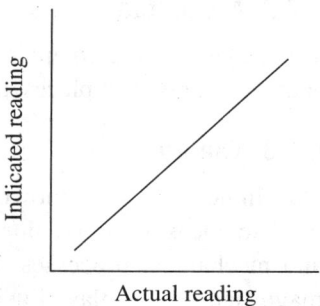

Fig. C.3   Model graph of indicated reading versus actual reading

C.2 USE OF DIAL GAUGES AS MECHANICAL COMPARATORS

C.2.1 Aim

To compare the given workpiece with a standard specimen using a mechanical dial indicator

C.2.2 Apparatus

The apparatuses used in this experiment include dial indicators with stands, the standard specimen, brass test pieces, a V-block, vernier callipers, etc.

C.2.3 Theory

Dial indicators, also known as dial gauges and probe indicators, are instruments that are used to measure small linear distances accurately; they are frequently used in industrial and mechanical processes. Dial indicators are so named because measurement results are magnified and displayed using a dial. A typical sketch of the instrument is depicted in Fig. C.4.

A special type of the dial indicator is the dial test indicator (DTI), which is primarily used in machine set-ups. A DTI measures displacement at an angle of a lever or plunger perpendicular to the axis of the indicator. A regular dial indicator measures linear displacement along that axis.

Dial indicators may be used to check the variations in tolerance during the inspection of a machined part and to measure the deflection of a beam or ring under laboratory conditions, and also in many other situations where a small measurement needs to be registered or indicated. Dial indicators typically measure ranges from 0.25 to 300 mm (0.015–12.0 in), with graduations of 0.001–0.01 mm (metric) or 0.00005–0.001 in (imperial).

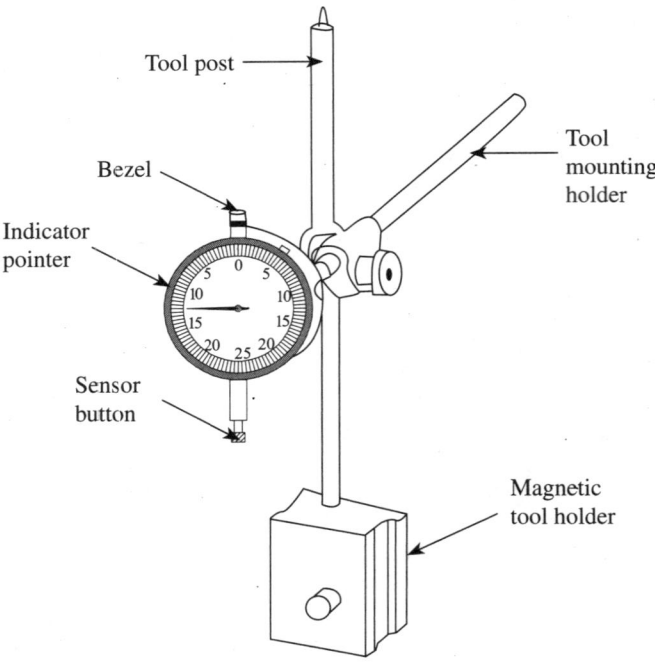

Fig. C.4 Dial gauge

Comparators

A comparator works on relative measurements, that is, it gives only the dimensional difference in relation to a basic dimension. A comparator compares the unknown dimensions of a part with some standard or master setting, which represents the basic size; dimensional variations from the master setting have to be amplified and measured. The advantage of a comparator is that the operator does not require much skill to use it.

Applications

Dial indicators are used in the following cases:

1. To check for runout when fitting a

new disc to an automotive disc brake (runout can ruin the disc rapidly if it exceeds the specified tolerance, typically 0.05 mm or less)
2. In a quality environment, to check for consistency and accuracy in the manufacturing process
3. On the workshop floor, to set up or calibrate a machine initially, prior to a production run
4. By tool makers (mould makers) in the process of manufacturing precision tools
5. In metal engineering workshops, where a typical application is the centring of a lathe's workpiece in a four-jaw chuck; a DTI is used to indicate the runout (misalignment between the axis of rotational symmetry of the workpiece and the axis of rotation of the spindle) of the workpiece, with the ultimate aim of reducing it to a suitably small range using small chuck jaw adjustments.
6. In areas other than manufacturing where accurate measurements need to be recorded (e.g., physics)

C.2.4 Procedure

The steps involved in the experimental procedure are as follows:
1. Note down the least count of the dial gauge.
2. Note down the least count of the vernier calliper.
3. Measure the thickness and diameter of the standard specimen.
4. Set the dial indicator to the standard specimen and adjust it to read zero.
5. Note down the number of sample pieces (N).
6. Check all the test pieces, note down the variations, and mention if it is 'Accepted' (or) 'Rejected' (or) 'Sent to rework'. While checking the diameter, the sample pieces are placed on the V-block.
7. Draw the 'control chart'.

Thickness and Diameter Measurement

The thickness and diameter obtained through step 3 of the procedure are documented in Tables C.2 and C.3, respectively.

Table C.2 Thickness measurement

Sample no.	Dial gauge reading		Actual reading (mm)	Remarks
	Divisions	mm		

Table C.3 Diameter measurement

Sample no.	Dial gauge reading		Actual reading (mm)	Remarks
	Divisions	mm		

C.3 STUDY OF SINE BARS

C.3.1 Aim

To set the sine bar to the given angle using slip gauges

C.3.2 Apparatus

The apparatuses used in this experiment include the following:
1. Sine bar
2. Set of slip gauges
3. Surface plate

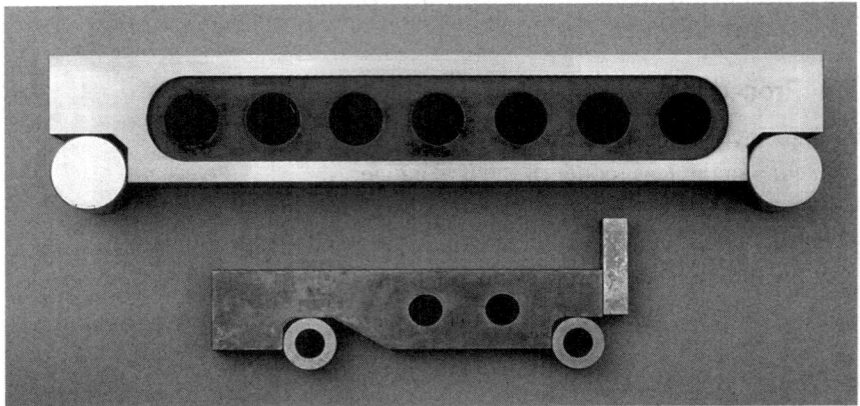

Fig. C.5 Sine bar

C.3.3 Theory

Figure C.5 shows a typical sine bar. Angles may be set and measured using a sine bar. Although simple in design and easy to use, a sine bar can provide accurate results, especially when small angles are being tested, or it can be used along with other measuring instruments such as slip gauges or dial gauges.

A sine bar consists of a hardened, precision ground body, with two precision ground cylinders fixed at the ends. The distance between the centres of the cylinders is controlled precisely, and the top of the bar is parallel to a line through the centres of the two rollers. The dimension between the two rollers is chosen to be a whole number (for ease of calculations) and forms the hypotenuse of a triangle when in use. The image shows a 10 in and a 100 mm sine bar. Sine bars are also available in 200 and 300 mm lengths.

Limitations

The following are the limitations of a sine bar:
1. The accuracy in measurement is limited by the measurement of the distance between centres of the two precision rollers. This fundamental limitation alone includes the use of the sine bars as a primary standard of angle.
2. A sine bar becomes impractical for use as the angle exceeds 45° because of the following reasons:

(a) A sine bar is physically difficult to handle in that position.
(b) The body of the sine bar obstructs the gauge block stack, even if relieved.
(c) A slight error in the sine bar causes a large angular error.
(d) Long gauge blocks are not as accurate as shorter gauge blocks.
(e) Temperature variation becomes a critical issue.
(f) A difference in deformation occurs at the point of contact of the roller with the support surface and the gauge blocks. Due to higher angles, the weight load is shifted more towards the fulcrum roller.

Sine Principle

Figure C.6 illustrates the fundamentals of the sine formula, from which, we can arrive at the following equation:

$$\text{Sin } \theta = h/L$$

Here, h is the required slip gauge combination in mm and L is the spacing between the rollers in mm.

Wringing Phenomenon

When two clean and perfectly flat surfaces are made to slide together under pressure, they adhere firmly. This phenomenon, as illustrated in Fig. C.7, is called 'wringing' of slip gauges. The phenomenon of wringing occurs due to molecular adhesion between a liquid film and the mating surface. Wringing ensures that any moisture, oil, dust, or similar foreign matters are pushed away from the gauging surfaces instead of being trapped between the gauges. Therefore, the first two surfaces should be made to slide transversely, as shown in Fig. C.7, and then twisted together to form a wrung pair.

C.3.4 Procedure

The steps to be adopted in carrying out the experimental procedure are as follows:
1. Clean the surface plate.
2. Place the set-up on the surface plate.
3. Note down the length of the sine bar.
4. Note down the dimensions of the slip gauges available.
5. For the given angle, calculate the height of the slip gauge required.
6. Select a minimum number of slip gauges to give an accuracy of at least up to the third decimal of h.
7. Bring all the selected slip gauges together, so that the combination gives the height h.

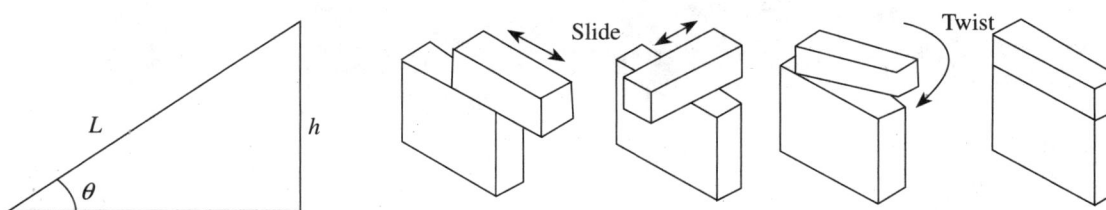

Fig. C.6   Illustration of the sine formula   **Fig. C.7**   Wringing of slip gauges

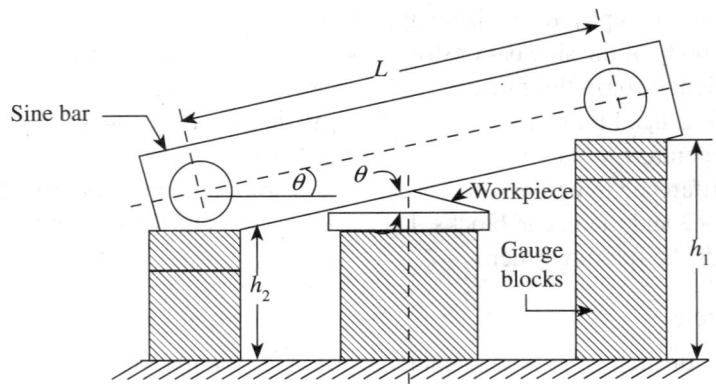

Fig. C.8 Setting up of a sine bar for the given angle

8. Place the sine bar with the contact roller on top of the combination of slip gauges on the surface plate, as shown in Fig. C.8.
9. Measure the built-up angle to ascertain accuracy, if angle/gap gauges are available.

C.3.5 Applications of Sine Bar

A sine bar is also used to locate any work to a given angle and to check unknown angles.

C.3.6 Observation

An observation of the following parameters should be made.

$$\text{Sin } \theta = [(h_1 - h_2)/L],$$

where $(h_1 - h_2)$ is the height built up to correct amounts with precision gauge blocks.

$h = L \sin \theta = $ _____

Length of the sine bar $L = $ _____

C.3.7 Tabular Column

Each trial along with the angle, height measurement, and combination measurement is tabulated in the format given in Table C.4.

Table C.4 Documentation of results of measurement of the given angles

Trial no.		Angle (degree)	Height h (mm)	Combination			
				Left	(mm)	Right	(mm)

C.4 STUDY OF BEVEL PROTRACTORS

C.4.1 Aim

To determine the angle of a given specimen

C.4.2 Apparatus

The following items are used for measurement:
1. Bevel protractor
2. Specimen

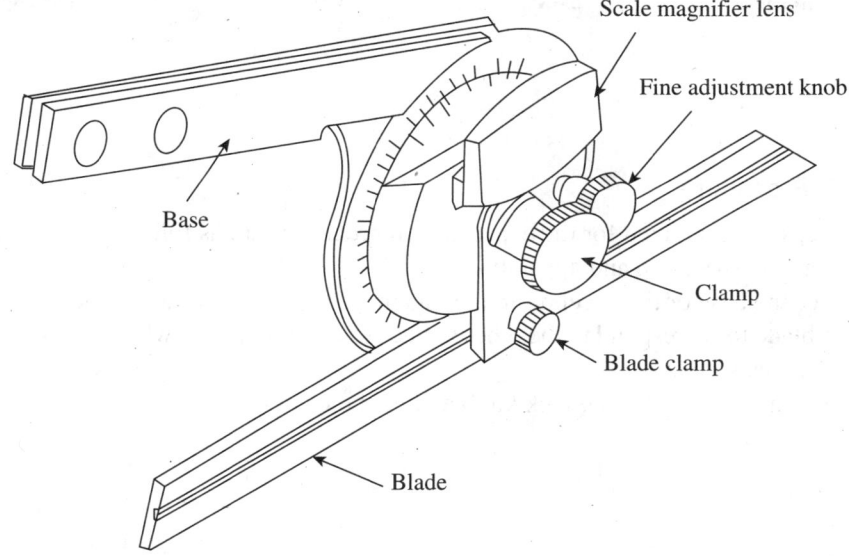

Fig. C.9 Bevel protractor

C.4.3 Theory

A bevel protractor, as shown in Fig. C.9, is used for measuring angles and laying out the angles within an accuracy of 5'. The protractor dial is slotted to hold a blade, which can be rotated with the dial to the required angle and also adjusted independently to any desired length. The blade can be located in any position. Universal bevel protractors are classified into types A, B, C, and D. In types A and B, the vernier is graduated to read up to 5' of arc. The difference between types A and B is that, type A is provided with fine adjustment whereas type B is not. In the case of type C, the scale is graduated to read in degrees. Type D is also graduated in degrees, but is not provided with either vernier or fine adjustment.

C.4.4 Observation

The acute angles are computed for each trial using the main scale reading (MSR), and the corresponding vernier scale reading (VSR) is obtained. These values can be documented in model Tables C.5 and C.6; an acute angle can be computed using the following expression:

Q = MSR + VSR.

Table C.5 Documentation of results of acute angle measurement

Trial no.	MSR	VSR	Acute angle
1			
2			

Similarly, the expression for an obtuse angle is $Q = 180 - (MSR + VSR)$.

Table C.6 Documentation of results of obtuse angle measurement

Trial no.	MSR	VSR	Obtuse angle
1			
2			

C.4.5 Procedure

The steps to be adopted for the experimental procedure are as follows:
1. Clean the specimen and apparatus.
2. Place the base of the bevel protractor against a reference surface of the workpiece and rotate the blade to correspond to the other surface of the workpiece, which gives the included angle being measured.
3. Lock the blade using the lock knob and read the angle.

C.5 GEAR TOOTH MICROMETER

C.5.1 Aim

To determine tooth thickness of the given spur gear

C.5.2 Apparatus

The following are the apparatuses used for the required experiment:
1. Gear tooth micrometer
2. Spur gear

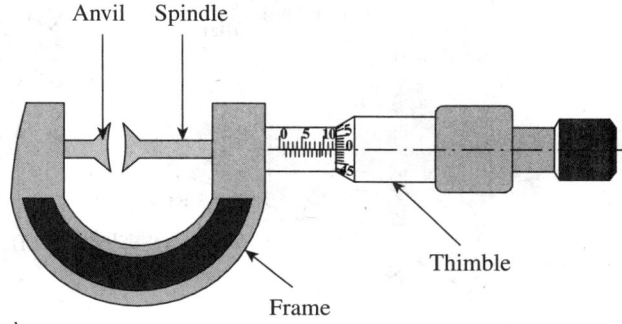

Fig. C.10   Gear tooth micrometer

C.5.3 Theory

A gear tooth micrometer is illustrated in Fig. C.10. It is used to measure the variation in the base tangent length W (a straight line connecting the point of contact of two unlike tooth flanks by two parallel measuring surfaces).

A tooth span micrometer differs from a conventional external micrometer in that its anvil and spindle are fitted with replaceable disk-type cups. Micrometers are available with the following measuring capacities: 0–25, 25–50, 50–75, and 75–100 mm.

C.5.4 Spur Gear Terminology

It is essential to learn the basic definitions of terms used in gear terminology. These terms are represented in Fig. C.11.

Base circle It is the circle from which the involute form is generated. It is fixed and unalterable.

Pitch circle It is an imaginary circle on the gear tooth, which, by pure rolling action, would produce the same motion as the toothed gear.

Pressure angle It is the angle between the line of action and the common tangent to the pitch circle.

Circular pitch It is the distance measured along the circumference of the pitch circle from a point on one tooth to a corresponding point on the next tooth.

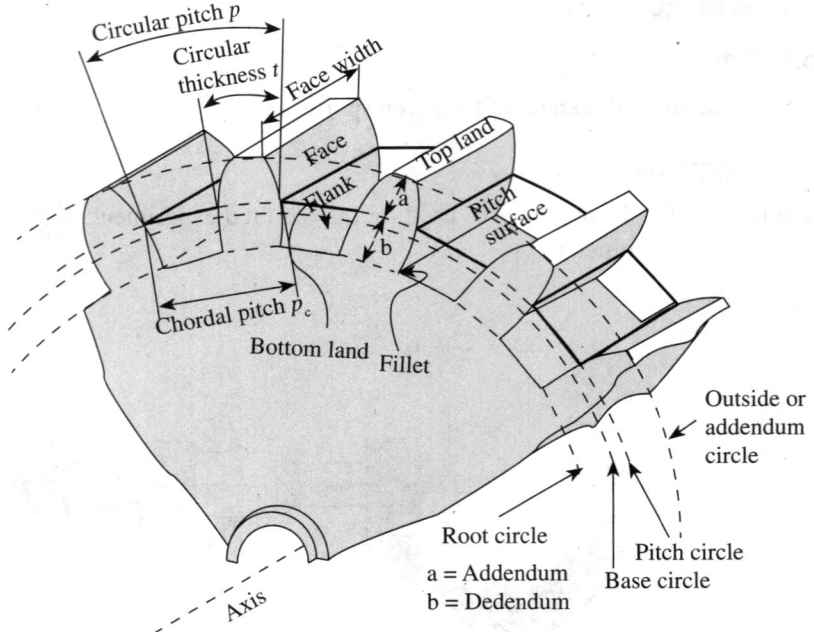

Fig. C.11   Gear terminology

Diametrical pitch   It is the number of teeth of the gear per millimetre of the pitch circle diameter.

Module   It is the length of the pitch circle diameter per tooth and is denoted by m: $m = (1/dp)$.

Addendum   It is the radical distance from the pitch circle to the lip of the tooth. Its nominal value is 1 module.

Clearance   It is the radial distance from the tip of a tooth to the bottom of a mating tooth space.

Blank diameter   The diameter of the gear blank is equal to the pitch circle diameter plus twice the addenda.

Tooth thickness   It is the actual distance measured along the pitch circle from its intercept, with one flank to its intercept on the blank.

Profile   It is the portion of the tooth flank between the specified form circle and the outside circle.

Face   It is the part of the tooth surface that is lying above the pitch circle surface.

Flank   It is the part of the tooth that is lying below the pitch surface.

C.5.5 Base Tangent Method

In this method, the span of a convenient number of teeth is measured with the help of a tangent comparator. This uses a single vernier calliper and therefore has the following advantages over gear tooth vernier, which uses two vernier scales:

1. Measurement does not depend on two vernier readings, each being a function of the other.
2. Measurement is not made with an edge of the measuring jaw but with the face.

Consider a straight generator (edge) ABC being rolled back and forth along a base circle, as shown in Fig. C.12. Its ends sweep out opposed involutes A_2AA_1 and C_2CC_1, respectively. Thus, the measurements made across these opposed involutes by span gauging will be constant,

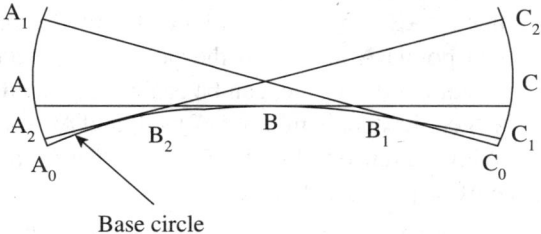

Fig. C.12   Generation of a pair of opposed involutes by a common generator

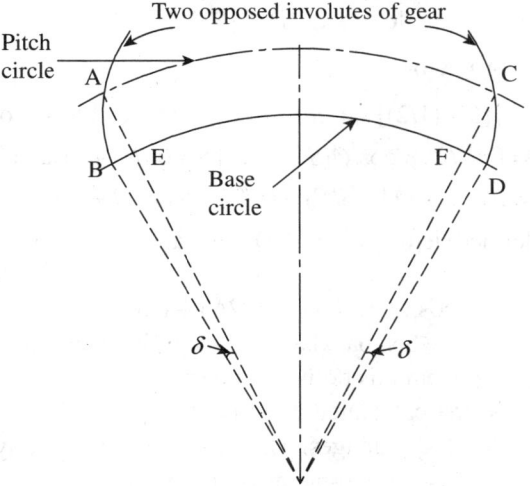

Fig. C.13   Representation of the length of an arc

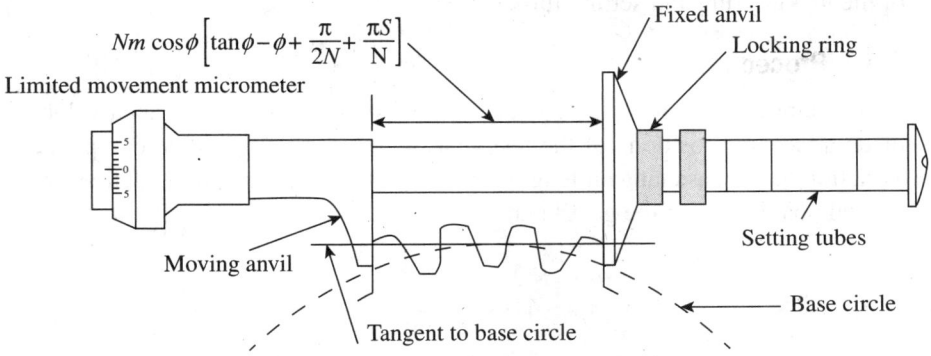

Fig. C.14   David Brown base tangent comparator

that is, $AC = A_1C_1 = A_2C_2 = A_0C_0$, equal to the arc length of the base circle between the origins of the involutes.

Further, the position of the measuring faces is not important as long as they are parallel and on an opposed pair of the true involutes. As the tooth form is most likely to conform to a true involute at the pitch point of the gear, it is always preferable to choose a number of teeth such that the measurement is made approximately at the pitch circle of the gear.

The value of the distance between two opposed involutes or the dimension over parallel faces is equal to the distance around the base circle between the points where the corresponding tooth flanks cut it, that is, ABC in Fig. C.12. It can be divided mathematically as follows:.

The angle between points A and C on the pitch circle, where the flanks of the opposed involute teeth of the gear cut this circle, can be easily calculated.

Let us say that the gear has got N number of teeth and AC on the pitch circle corresponds to S number of teeth. With reference to Fig. C.13, we get the following:

Distance $AC = [S - (1/2)]$ pitches

Angle subtended by $AC = [S - (1/2)] \times (2\pi/N)$ radians

Angles of arcs BE and FD = Involute function of pressure angle $= \delta = \tan \emptyset - \emptyset$

Therefore, angle of arc $BD = [S - (1/2)] \times (2\pi/N) + 2(\tan \emptyset - \emptyset)$

BD = Angle of arc $BD \times R_b$

$\qquad = \{[S - (1/2)] \times (2\pi/N) + 2(\tan \emptyset - \emptyset)\} \times R_p \cos \emptyset \qquad$ (since $R_b = R_p \cos \emptyset$)

$\qquad = (mN/2) \times \cos \emptyset [S - (1/2)] \times (2\pi/N) + 2(\tan \emptyset - \emptyset) \qquad$ (since $R_p = mN/2$)

$\qquad = Nm \cos \emptyset [(\pi S/N) - (\pi/2N) + \tan \emptyset - \emptyset]$

As already defined, length of arc BD = distance between two opposed involutes, and is thus as follows:

$BD = Nm \cos \emptyset [(\pi S/N) - (\pi/2N) + \tan \emptyset - \emptyset]$

It may be noted that when backlash allowance is specified normal to the tooth flanks, it must simply be subtracted from this derived value.

This distance is first calculated and then set in the David Brown tangent comparator (Fig. C.14) with the help of slip gauges. The instrument essentially consists of a fixed anvil and a movable anvil. There is a micrometer on the moving anvil side, and this has a very limited movement on either side of the setting. Set the fixed anvil at the desired position with the help of the locking ring and setting tubes.

C.5.6 Procedure

A base tangent length variable is determined by computing the mean value of measurement made at several positions of the test gear, with the nominal value computed by the formula. Variation of the base tangent length measured at different positions defines the accuracy of the mutual positioning of the gear teeth.

C.5.7 Observation

Observation Table C.7 can be used to document the observations of the experiment.

Table C.7 Measurement of an addendum diameter

Trial no.	Addendum diameter (mm)			
	MSR	CVSR	Total reading	Mean

Least count of a micrometer = _____

Module, m = Addendum mean/$(z + 2)$

Chordal addendum, $d = Zm/2 \times [1 + 2/z - \cos(90/z)]$

Theoretical base tangent length, $W_t = Zm \cos \emptyset [\tan \emptyset - \emptyset - (\pi/2z) + (\pi s/Z)]$, where Z is the number of teeth on the gear and s is the span; $\emptyset$ should be in radians.

C.5.8 Finding Actual Width of Gear Tooth

Actual width can be documented in Table C.8.

Table C.8 Documentation of actual width

Trial no.	Width (mm)			
	PSR	CHSD	Total reading	Mean (W_a)

Actual reading, W_a = _____

Theoretical reading, W_t = _____

% Error = $[(W_t - W_a)/W_t] \times 100$

Measurement of gear tooth thickness (W) can lead to the following errors:

1. Circular pitch error over span of teeth
2. Pitch variation
3. Base pitch variation
4. Eccentricity
5. Error due to inefficient handling
6. Error due to feel of the user

C.6 VERNIER GEAR TOOTH CALLIPER

C.6.1 Aim

To measure the given gear tooth thickness using a vernier gear

C.6.2 Apparatus

The apparatuses required for the measurement are as follows:
1. Specimen spur gear
2. Lens
3. Gear tooth calliper

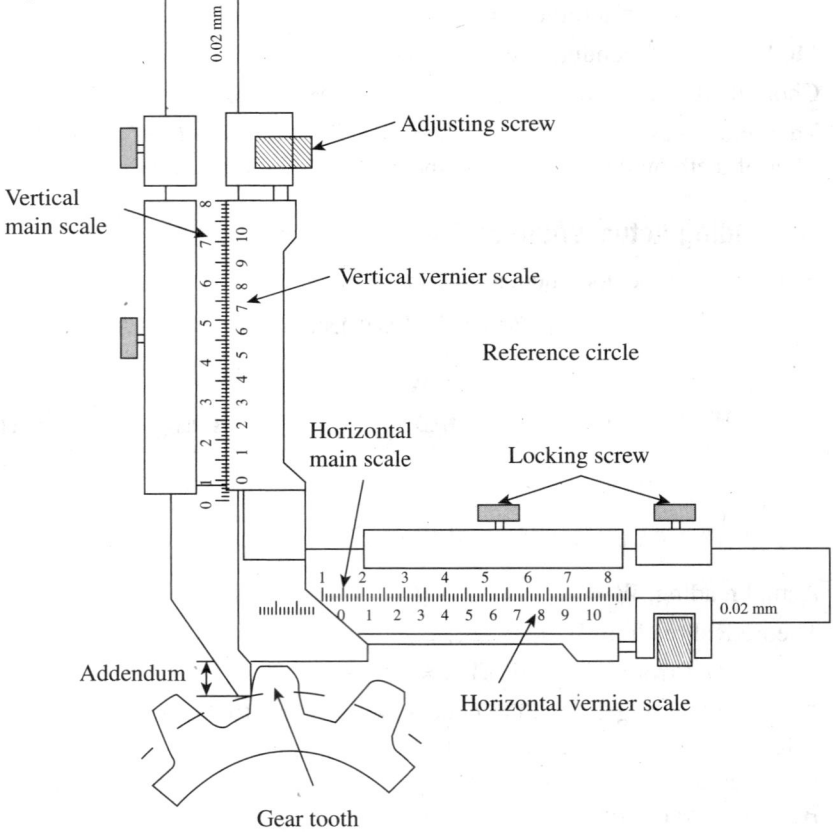

Fig. C.15 Vernier gear tooth calliper

C.6.3 Theory

A vernier gear tooth calliper is designed to measure the chordal thickness of the gear tooth. Two sides carry verniers and are movable on mutually perpendicular beams of frame. The vertical side is coupled to a tongue and the horizontal side to a movable jaw, as shown in Fig. C.15. Gear tooth callipers are available in three sizes for testing gears over a wide range of modules. The main difference between a gear tooth micrometer and a vernier calliper is that in the former case, two errors are encountered: error on the vertical scale and that on the horizontal scale.

Errors can arise due to the following reasons:
1. Circular pitch error over span of teeth
2. Pitch variation
3. Base pitch variation
4. Eccentricity
5. Inefficient handling
6. Feel of the user

C.6.4 Procedure

Experiments can be performed in the following way:
1. Calculate chordal thickness W and chordal addendum d using the following formula:
   $$W = N \times m \sin (90/N)$$
   $$d = N \times m/2 \ [1 + 2/n - \cos 90/N]$$
   where N is the number of teeth and m is the module.
2. Determine the error in the chord with a gear tooth vernier calliper.
3. Determine the error in the instrument in the vertical and horizontal vernier jaws coinciding with mainframe.
4. Note down the error.
5. Adjust the vertical vernier jaw for the chordal addendum d and place the tongue over the space of the gear tooth (i.e., d + error value).
6. Measure the gear tooth thickness w in a manner similar to the measurement by a conventional calliper.

C.6.5 Observation

An observation of the following parameters should be made:
1. Chord addendum (d) = _____
2. Calculated gear tooth thickness (W) = _____
3. Measured gear tooth thickness (W_m) = _____
4. % Error = $[(W_t - W_a)/W_t] \times 100$ = _____

C.7 DETERMINATION OF EFFECTIVE DIAMETER OF SCREW THREADS BY TWO- AND THREE-WIRE METHODS

C.7.1 Aim

To find out the effective diameter of the screw thread by the two-wire/three-wire method

C.7.2 Apparatus

The apparatuses required in the experiment are as follows:
1. Two micrometers, with ranges 0.25 and 25–50 mm
2. Standard wires
3. Straight edge
4. Specimen screw

C.7.3 Theory

The effective diameter of a screw thread can be measured using a floating carriage diameter-measuring machine (Fig. C.16).

Definition   Screw thread is the helical ridge produced by forming a continuous helical groove of uniform section on the external or internal surface of a cylinder or a cone.

Screw terminology   It is essential to learn the definitions of some of the terms related to screw threads. These terms are represented in Fig. C.17.

Types

The following are the types of screw threads:

External thread   A thread formed on the outside of a cylinder is called an external thread.

Internal thread   A thread formed on the inside of a cylinder is called an internal thread.

Single-start thread   A single-start thread is a thread in which the lead is equal to the pitch.

Fig. C.16   Floating carriage diameter-measuring machine for the measurement of the effective diameter of a screw thread

Multiple-start thread   A multi-start thread is a thread in which the lead is an integral multiple of the pitch.

Pitch   The pitch of a thread having uniform spacing is the distance measured parallel to the axis, between corresponding points on adjacent thread forms in the same axial plane and on the same side of the axis. The basic pitch is equal to the lead divided by the number of thread starts.

Lead   Lead is the axial distance moved by the threaded part, when it is given one complete revolution about its axis with respect to a fixed mating surface.

Threads per inch   It is the reciprocal of the pitch in inches.

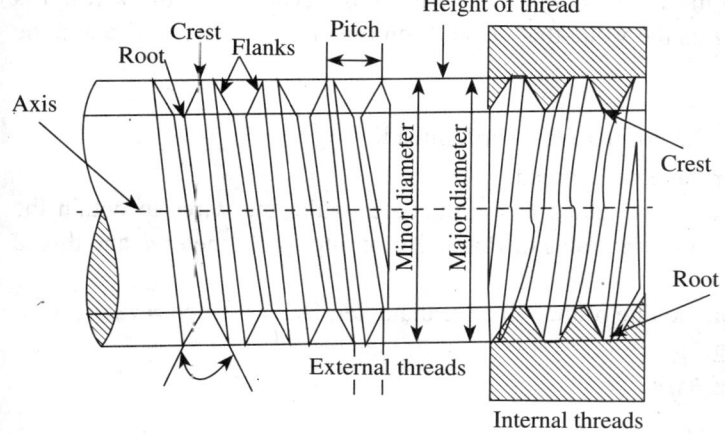

Fig. C.17 Screw thread terminology

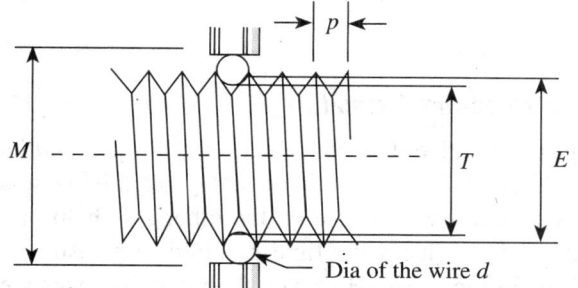

Fig. C.18 Two-wire method of measuring the effective diameter

Measurement of Pitch Diameter Using Two-wire Method

The two-wire method of measuring the effective diameter makes use of two similar wires. The main advantage of this method is that it can be used for measuring pitch diameter of a tapered as well as a straight thread. However, a disadvantage of this method is that it is not accurate. This is because alignment is not possible with two wires. While measuring, the micrometer should be allowed to align itself in order to avoid any sine errors that are likely to be introduced if the thread on centres and the micrometer are misaligned. Thus, two-wire methods are carried out on a floating carriage diameter-measuring machine, in which the alignment is inherent.

The effective diameter of a screw thread may be ascertained by placing two wires of identical diameter between the flanks of the thread, as shown in Figs C.18 and C.19, and measuring the distance over the outside of these wires. The effective diameter E is then given by the following relation:

$$E = T + P$$

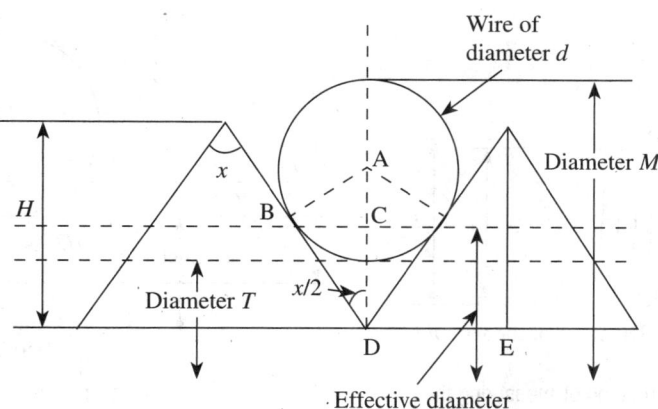

Fig. C.19 Measurement of the mean diameter over wires

Here, T is the dimension under the wires ($= M - 2d$), M is the dimension over the wires, d is the diameter of each wire, and P is the value that depends on the diameter of the wire and the pitch of the thread.

Notes:

$P = (0.9605p - 1.1657d)$ for the Whitworth thread.

$P = (0.866p - d)$ for the metric thread.

P is a constant value that has to be added to the diameter under the wires to obtain the effective diameter. The expression for the value of P in terms of pitch p, diameter d, and thread angle x can be derived as follows:

In Fig. C.19, since BC lies on the line of the effective diameter, BC = ½ pitch = ½p.

$$OP = d \, \text{cosec} \, (x/2)/2$$
$$PA = d \, [\text{cosec} \, (x/2) - 1]/2$$
$$PQ = QC \cot (x/2) = p/4 \cot (x/2)$$
$$AQ = PQ - AP = [p/4 \cot (x/2)] - d[\text{cosec} \, (x/2) - 1]/2$$

AQ is half the value of P.

Therefore, $P = 2AQ$

$$= (p/2) \cot (x/2) - d \, [\text{cosec} \, (x/2) - 1]$$

Measurement of Pitch Diameter Using Three-wire Method

The three-wire method of measuring the effective diameter makes use of three similar wires. The arrangement is shown in Figs C.20 and C.21. The main advantage of this method is that it gives an accurate value. Two wires on one side and one on the other side help in aligning the micrometer square to the thread. It also reduces the time required to measure the pitch diameters of a number of threads of nearly same size. However, the main disadvantage of this method is that it cannot be used for measuring a tapered thread.

From Fig. C.21,

$$AD = AB \, \text{cosec} \, (x/2) = r \, \text{cosec} \, (x/2)$$
$$H = DE \cot (x/2) = (p/2) \cot (x/2)$$

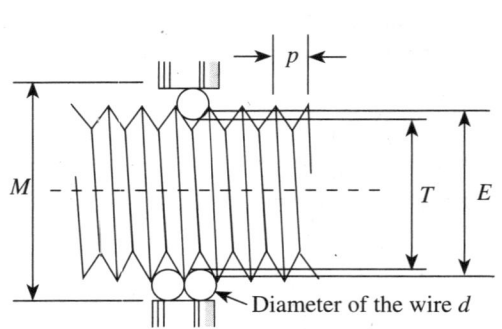

Fig. C.20 Three-wire method of measuring the effective diameter

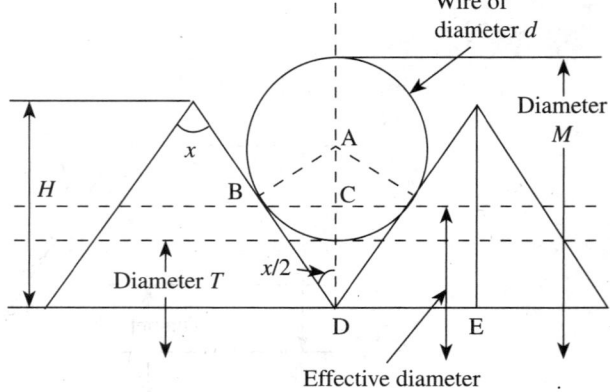

Fig. C.21 Measurement of the mean diameter over wires

$CD = (1/2)H = (p/4) \cot (x/2)$

$h = AD - CD$

$= r \operatorname{cosec} (x/2) - (p/4) \cot (x/2)$

Distance over the wires $= M = E + 2h + 2r$

$= E + 2 [r \operatorname{cosec} (x/2) - (p/4) \cot (x/2)] + 2r$

$= E + 2r [(1 + \operatorname{cosec} (x/2)] - (p/4) \cot (x/2)$

$= E + d [(1 + \operatorname{cosec} (x/2)] - (p/4) \cot (x/2) \text{ (since } 2r = d)$

In these relations, M is the distance over wires, E is the effective diameter, r is the radius of the wires, d is the diameter of the wires, h is the height of the centre of the wire or the rod from the effective diameter, and x is the angle of the thread.

Note:

In case of the Whitworth thread: $M = D + 3.1657d - 1.6p$.

In case of the metric thread: $M = D + 3d - 1.5155p$.

Best-size Wire

This wire is of such a diameter that it makes contact with the flanks of the thread on the effective diameter or the pitch line. Actually, the effective diameter can be measured with a wire of any diameter that makes contact with the true flank of the thread, but the values so obtained will differ from those obtained using best-size wires if there is any error in angle or form of thread. It is advised to take a wire within a range of $\pm (1/5)$ of flank length.

It can be shown that diameter of the best-size wire $= d = \{p/[2 \operatorname{cosec} (x/2)]\}$.

C.7.4 Procedure

The experimental procedure involves the following steps:

1. Note down the zero error of both micrometers and tabulate in the format shown in Table C.9.
2. Note down the range and least count of both micrometers and document in Table C.9.
3. Note the thread angle $(x°)$.
4. Measure the length of 10 threads using a steel rule.
5. Calculate the thread pitch (p).
6. Calculate the best size wire (D).
7. After measuring all the standard wires available, choose two wires randomly.
8. Place the chosen two wires over the specimen screw thread and measure the mean diameter (M) over the wires.
9. Calculate the correction factor (P), whose value depends upon the diameter of the wire and the pitch of the thread.
10. Calculate the dimension under the wire (T).
11. Calculate the effective or pitch diameter of the thread (E).

The experimental results thus obtained from steps 1–11 are documented in Tables C.10 and C.11 for two- and three-wire methods, respectively.

C.7.5 Observation—Two-wire Method

An observation of the following parameters should be made:

1. Thread pitch $= p =$ [Length of the thread/Number of threads] $=$ _____

2. Wire size = d_b = D = [P/2 sec (x/2)] = _____
3. Correction factor = P = p/2 cot (x/2) − d[cosec (x/2) − 1] = _____
4. Dimensions under the wire = T = [M − 2d] = _____
5. Effective diameter or pitch diameter = E = [T + P] = _____
6. Length of the thread = _____
7. Number of threads = _____

Table C.9 Details of a micrometer

		Micrometer 1	Micrometer 2
Range			
Least count			
Zero error			

Table C.10 Documentation of results of the measurement of the effective diameter using the two-wire method

Trial no.	Wire diameter d (mm)	Mean diameter M (mm)	Correction factor P (mm)	Dimension under wires T (mm)	Effective diameter E (mm)

C.7.6 Observation—Three-wire Method

An observation of the following parameters should be made:
1. Thread pitch = p = [Length of the thread/Number of threads] = _____
2. Wire size = d_b = D = [P/2 sec (x/2)] = _____
3. Correction factor = P = p/2 cot (x/2) − d[cosec (x/2) − 1] = _____
4. Dimensions under the wire = T = [M − 2d] = _____
5. Effective diameter or pitch diameter = E = [T + P] = _____
6. Length of the thread = _____
7. Number of threads = _____

Table C.11 Documentation of results of the measurement of effective diameter using the three-wire method

Trial no.	Wire diameter d (mm)	Mean diameter M (mm)	Correction factor P (mm)	Dimension under wires T (mm)	Effective diameter E (mm)

C.8 MEASUREMENT OF STRAIGHTNESS USING AUTOCOLLIMATORS

C.8.1 Aim

To measure straightness using an autocollimator

C.8.2 Apparatus

The apparatuses used include a surface plate and an autocollimator.

C.8.3 Theory

The following sections discuss straightness and its measurement using an autocollimator.

Straightness

A line is said to be straight over a given length if the deviation of various points on the line from two mutually perpendicular reference planes remains within stipulated limits. The reference planes are so chosen that their intersection is parallel to the straight line lying between the two specific end points. The tolerance on the straightness of a line is defined as the maximum deviation of the spread of points on either side of the reference line, as shown in Fig. C.22.

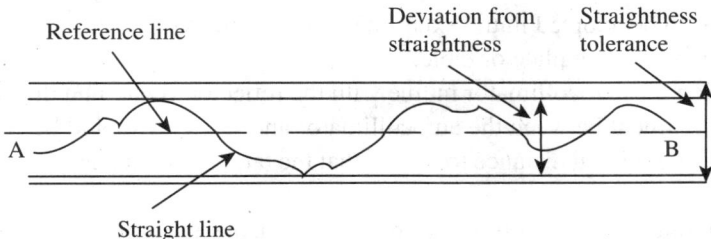

Fig. C.22 Graphical representation of the definition of straightness

Autocollimator

An autocollimator (shown in Fig. C.23) is a special form of telescope, which is used to measure small angles with a high degree of resolution. It is used for various applications such as precision alignment, verification of angle standards, detection of angular movement, and so on. It projects a beam of collimated light onto a reflector, which is deflected by a small angle in the vertical plane. The light reflected back is magnified and focused on either an eyepiece or a photo detector. The deflection between the beam and the reflected beam is a measure of the angular tilt of the reflector. Figure C.24 illustrates the working principle of an autocollimator.

The reticle is an illuminated target with a cross-hair pattern, which is positioned in the focal plane of an objective lens. A plane mirror perpendicular to the optical axis serves the purpose of reflecting an image

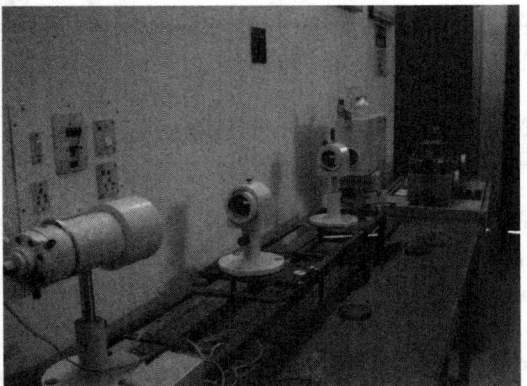

Fig. C.23 Autocollimator for the measurement of straightness

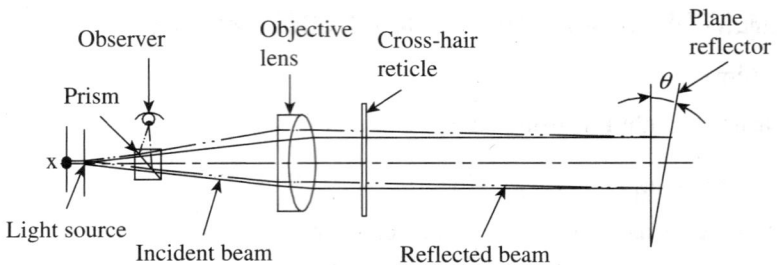

Fig. C.24 Working principle of an autocollimator

of the pattern back to the observation point. A viewing system is required to observe the relative position of the image of the cross-wires. In most of the autocollimators, this is performed by means of a simple eyepiece. If rotation of the plane reflector by an angle θ results in the displacement of the image by an amount d, then $d = 2f\theta$, where f is the focal length of the objective lens. The results of measurement are documented in Table C.12.

C.8.4 Procedure

The steps to be adopted in the experiment are as follows:
1. Clean the surface plate or table.
2. Position the autocollimator in line with the reflector. Switch on the lamp in the autocollimator; alignment between the autocollimator and reflector should be checked at both extremes of the operational distance to ensure that the target graticule is contained within the eyepiece field.
3. Fix a guide strip to control the horizontal displacement of the reflector and minimize the movement of the target graticule.
4. Mark off the positions along the surface plate equal to the pitch positions on the reflector base (usually, 100 mm). Column 1 in Table C.12 should indicate this position.
5. Take the reading at the initial position and tabulate it (column 2).
6. Move the reflector to the next position and again tabulate the reading.
7. Continue this method until the final outward position is recorded. To improve the accuracy and ensure that no errors have been introduced, readings should also be taken on the inward run. If this exercise is followed, the average of the two readings has to be shown in column 2 of the table.
8. The remainder of the table should be filled by adopting the following steps:
   (a) Column 3 is the difference between the tilt observed at the position at which the reading is taken and that at the original position.
   (b) The angular position in column 3 is converted into a linear measure (1" = 0.5 µm) in column 4. Insert a zero at the top of the column to represent the datum.
   (c) Column 5 is the cumulative algebraic sum of the displacements. Calculate the mean displacement, which is the amount by which the displacement must be adjusted to relate them to the zero datum. *The value of mean displacement is referred to as an error of straightness.*
   (d) Plot the values of column 5 versus those of column 1.

Table C.12 Documentation of results of straightness measurement

Position of reflector	Autocollimator reading (seconds)	Difference from first reading	Rise (+) or fall (−) from first reading (μm)	Cumulative rise or fall (μm)
1			0	0
2				
3				
–				
–				

Conversion of Angular Deviation to Linear Measure

$\tan \theta = h/\text{radius}$

If $\theta = 1''$ of arc,

$h = \tan (1'') \times \text{radius}$

$h = 4.848 \times 10^{-6}$ m

$h = 5\,\mu\text{m/m}$ approximately

C.9 MEASUREMENT OF FLATNESS USING CLINOMETERS

C.9.1 Aim

To measure the flatness of a surface plate using a clinometer

C.9.2 Apparatus

The apparatuses used are a surface plate and a clinometer.

C.9.3 Theory

The following sections discuss the flatness of a surface plate and its measurement using a clinometer.

Flatness

Machine tool tables, which hold workpieces during machining, should have a high degree of flatness. Many metrological devices such as sine bars invariably need a surface plate that should be perfectly flat. Flatness error may be defined as the minimum separation of a pair of parallel planes that will just contain all points on the surface. Figure C.25 illustrates the measure of flatness error a.

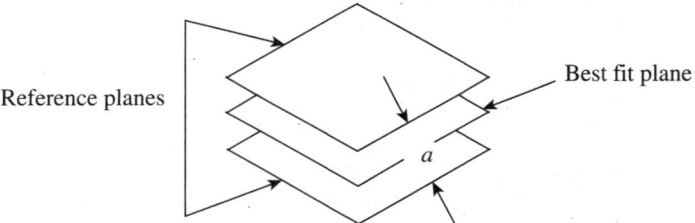

Fig. C.25 Graphical representation of definition of flatness

It is possible, using a simple geometrical approach, to fit a best-fit plane for the macro-surface topography. Flatness is the deviation of the surface from the best-fit plane. According to IS: 2063-1962, a surface is deemed to be flat within a range of measurements when the variation of the perpendicular distance of its points from a geometrical plane (this plane should be exterior to the surface to be tested) parallel to the general trajectory of the plane to be tested remains below a given value.

Clinometer

A clinometer is a special case of a spirit level. While spirit levels are restricted to relatively small angles, clinometers can be used for much larger angles. It comprises a level mounted on a frame so that the frame may be turned to any desired angle to a horizontal reference. Clinometers are used to determine straightness and flatness of surfaces. They are also used for setting an inclinable table on a jig boring machine and for angular jobs on surface grinding machines. They provide superior accuracy compared to ordinary spirit levels. Figure C.26 shows the picture of a clinometer.

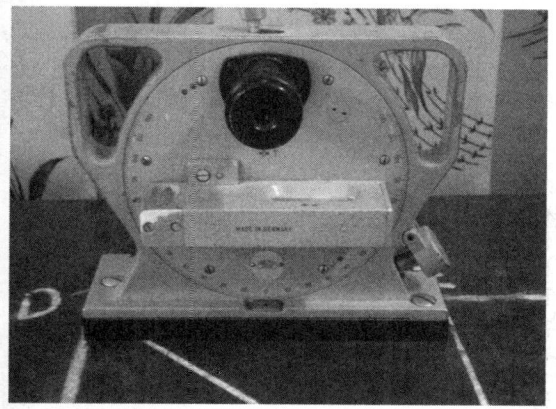

Fig. C.26 Clinometer

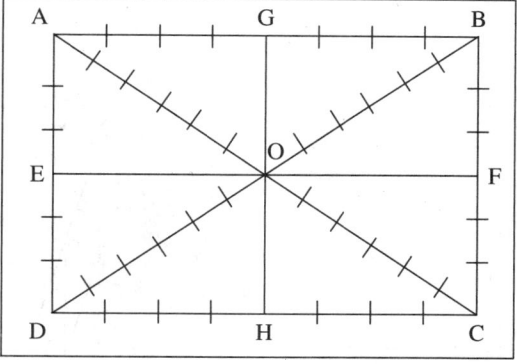

Fig. C.27 Grid of straight lines for flatness measurement

C.9.4 Procedure

Assuming that a clinometer is used for measuring angular deviations, a grid of straight lines, as shown in Fig. C.27, is formulated. Care is taken to ensure that maximum area of the flat table or the surface plate being tested is covered by the grid. While lines AB, DC, AD, and BC are drawn parallel to the edges of the flat surface, two diagonal lines DB and AC are drawn, which will intersect at the centre point O. Markings are made on each line at distances corresponding to the base length of the clinometer.

1. The straightness test is carried out on all the lines and the readings are tabulated. For line AB, for example, the clinometer base is positioned at successive marked steps and readings are noted down. A graphical plot is sketched using the first position as the reference and angular readings of the clinometer as ordinate values. The corresponding readings in microns are calculated, and a plot of cumulative readings in microns for successive positions of the clinometer base is drawn. As an example, the probable plot for line AB is shown in Fig. C.28.
2. We know that a plane is defined as a two-dimensional entity passing through a minimum of

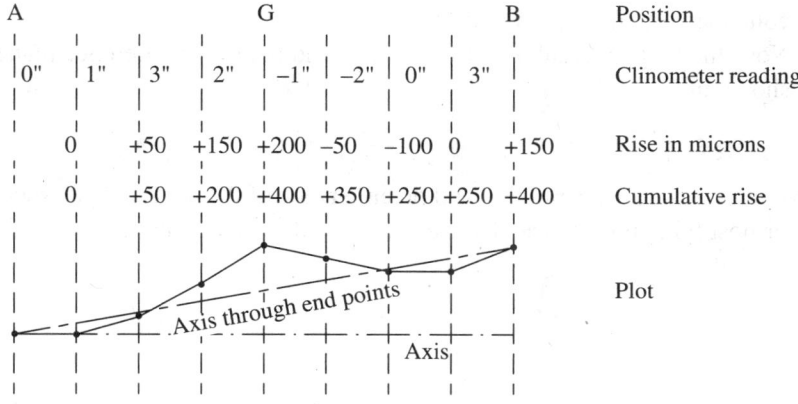

Fig. C.28 Plot of actual readings drawn from measured values

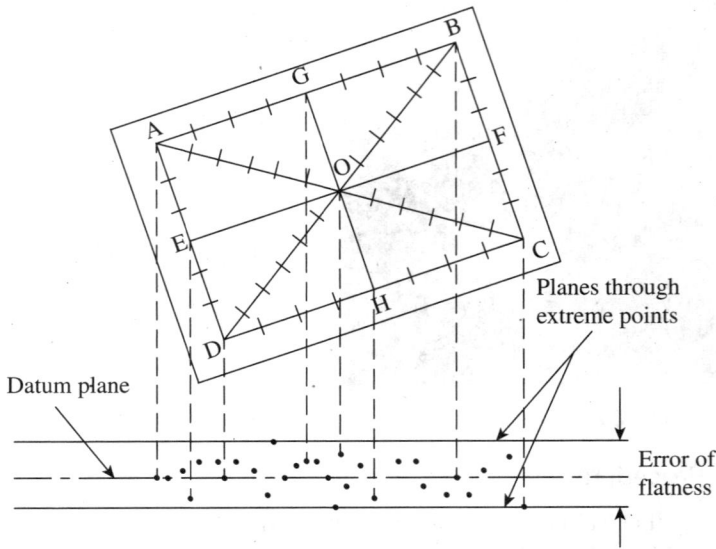

Fig. C.29   Plot of height of all the points with respect to the reference plane

three points not lying in the same straight line. Accordingly, a plane passing through points A, B, and D is assumed to be an arbitrary plane, relative to which the heights of all other points are determined. Therefore, the ends of lines AB, AD, and BD are corrected to zero, and heights of points A, B, and D are forced to zero.

3. The height of the centre 'O' is determined relative to the arbitrary plane ABD. Since O is also the mid-point of line AC, all the points on AC can be fixed relative to the arbitrary plane ABD. Assume A = 0 and reassign the value of O on AC to the value of O on BD. This will readjust all the values on AC in relation to the arbitrary plane ABD.

4. After the previous step, point C is fixed relative to plane ABD, and points B and D are set to zero; all intermediate points on BC and DC are also adjusted accordingly.

5. The same procedure applies to lines EF and GH. The mid-points of these lines should also coincide with the known mid-point value of 'O'.

6. Now, the heights of all the points, above and below the reference plane ABD, are plotted as shown in Fig. C.29.

C.9.5 Result

Two lines are drawn parallel to and on either side of the datum plane, such that they enclose the outermost lying points. The distance between these two outer lines is the flatness error.

C.10 TOOL MAKER'S MICROSCOPE

C.10.1 Aim

To measure the major diameter, minor diameter, and pitch of the given screw thread using a tool marker's microscope

C.10.2 Apparatus

The apparatuses required for the experiment are as follows:
1. Tool maker's microscope
2. Steel scale
3. Specimen screw

C.10.3 Theory

A large tool maker's microscope, shown in Fig. C.30, essentially consists of a cast base, the main lighting unit, the upright with carrying arm, and the sighting microscope. The rigid cast base rests on three foot screws by means of which the equipment can be levelled with reference to the build-in box level. The base carries the coordinate measuring table, consisting of two measuring slides, one along the X and the other along the Y direction, and a rotary circular table provided with a glass plate. The slides run on precision balls in hardened guideways warranting reliable travel. Two micrometer screws, each of them measuring a specified range given by the equipment manufacturer, permit the measuring table to be displaced in the directions X and Y. The range of movement of the carriage can be widened up to 150 mm in the X direction and up to 50 mm in the Y direction with the use of gauge blocks.

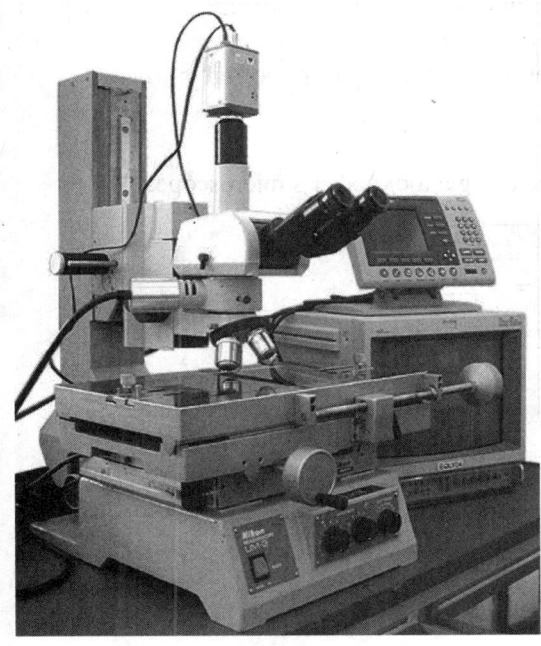

Fig. C.30   Tool maker's microscope

Construction

The general view of a tool maker's microscope, giving its design, optical systems, and measuring techniques, is as shown in Fig. C.31.

It consists of an optical head having various interchangeable eyepieces, which moves along the guideways of a vertical column. The optical head can be clamped in any position by a screw. The working table on which the part to be inspected is placed, is secured on a heavy hollow base. The table has a compound slide, by means of which the measured part can have longitudinal and lateral movements. The slight movement given to these screws can be easily determined using scales and verniers.

Cross-lines are engraved on the ground glass screen, which can be rotated through 360°, and measurements are made using these cross-lines. The optical eyepiece/head tube is adjusted in height for focusing purposes, until a sharp image can be seen on

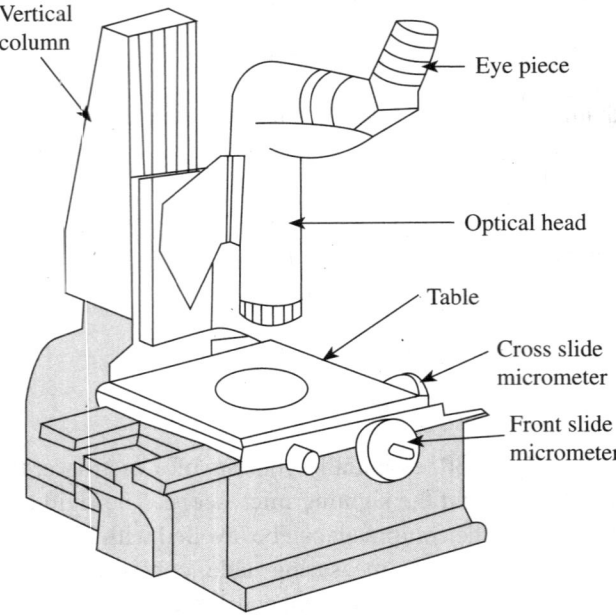

Fig. C.31 Line sketch of a tool maker's microscope

the ground glass screen. The magnification obtained depends on the objectives of the microscope, but instruments are commonly supplied with facilities to give a magnification of 10–100×. In order to adapt the apparatus to deal with a variety of works, various attachments may be fitted to the work table, such as clamp attachment and light arrangement. Linear movement of the table is controlled by micrometer screws having an accuracy of 0.0025 mm.

Applications of Tool Maker's Microscope

A tool maker's microscope can be used for the following purposes:

1. Measurement of the parts of complex forms, for example, profile of external threads, tools, gauges, and templates
2. Determination of relative position of various points on a workpiece
3. Measurement of angle
4. Length measurement in Cartesian and polar coordinates
5. Angle measurements of tools, threading tools, punches, gauges, templates, etc.
6. Thread measurements, including measurements of profile of major and minor diameters, height of lead, thread angle, profile position with respect to the thread axis, and the shape of thread (rounding, flattering, and straightness of flanks)
7. Comparison between centres and drawn patterns, and drawing of projected profiles

C.10.4 Observations

The following observations are to be made while using a tool maker's microscope:
1. Least count of the horizontal travel scale = _____
2. Least count of the vertical travel scale = _____
3. Difference = (Initial reading − Final reading) = _____
4. Type of screw thread = _____
5. Total length of the test piece = _____
6. Magnification used = _____
7. Maximum horizontal travel of the microscope = _____
8. Maximum vertical travel of the microscope = _____

C.10.5 Tabular Column

The experimental results are documented in Table C.13.

Table C.13 Documentation of experimental results

Parameter	Trial no.	Initial reading (IR)	Final reading (FR)	Difference (IR – FR)	Average (mm)
Outer diameter (D)	1.				
	2.				
Inner diameter (d)	1.				
	2.				
Pitch (p)	1.				
	2.				

C.10.6 Procedure

The steps to be adopted for the experimental procedure are as follows:
1. Clamp the given screw thread on the table keeping the axis approximately horizontal.
2. Move the microscope in the X and Y directions to the vicinity of the screw thread profile.
3. Adjust the optical head for focusing the screw thread profile at the cross-wires.

Measurement of Major or Outer Diameter

The major or outer diameter can be measured in the following way:
1. Coincide the horizontal cross-wire to the crest (outermost point) of the screw thread profile. Note down the reading (IR).
2. Move the optical head along the Y-axis and coincide the horizontal cross-wire with the outermost point on the other side of the profile. Note down the reading (FR).
3. The difference between initial reading and final reading gives the major diameter.

Measurement of Inner or Minor Diameter

The inner or minor diameter can be measured in the following way:
1. Coincide the horizontal cross-wire with the root of the screw thread profile. Note down the reading (IR).
2. Move the optical head along the Y-axis and coincide the horizontal cross with the root of the screw thread on the other side.
3. The difference between final reading and initial reading gives the inner or minor diameter.

Measurement of Pitch

The steps involved in pitch measurement are as follows:
1. Coincide the vertical cross-wire approximately with the line of symmetry passing through the screw thread, apex, or notch. Note down the reading (IR).
2. Move the optical head along the X-axis and coincide the cross-wire with the adjacent screw thread. Note down the reading (FR).
3. The difference between initial reading and final reading gives the pitch of the screw thread. The experimental results are documented in Table C.13.

C.11 MEASUREMENT OF GEAR ELEMENTS USING PROFILE PROJECTOR

C.11.1 Aim

To measure the elements of a given gear using a profile projector

C.11.2 Apparatus Required

The following are the apparatuses used for the measurement:
1. Profile projector
2. Gear (specimen)

C.11.3 Parameters to be Measured

The parameters shown in Fig. C.32, which are measured using a profile projector are as follows:

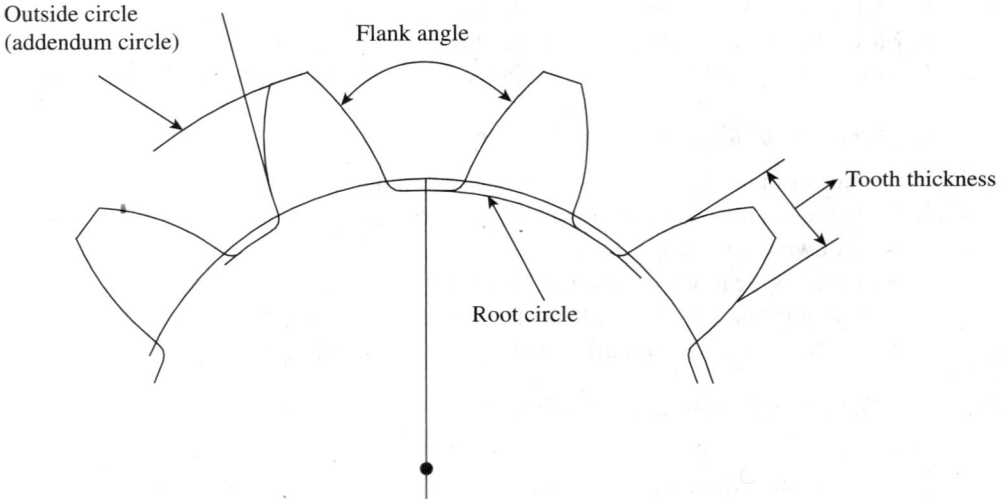

Fig. C.32 Parameters of a gear

1. Addendum circle diameter
2. Root circle diameter
3. Tooth thickness
4. Flank angle

C.11.4 Theory

Figure C.33 shows a profile projector (also called an optical projector). It is a versatile comparator, which is widely used for the purpose of inspection. It is especially used in tool room applications. It projects a two-dimensional magnified image of the workpiece onto a viewing screen to facilitate measurement. The device comprises three main elements: the projector consisting of a light source and a set of lenses housed inside an enclosure, a work table to hold the workpiece in place, and a transparent screen with or without a chart gauge for comparison or measurement of parts.

Figure C.34 illustrates the various parts of a profile projector. The workpiece to be inspected is mounted on a table such that it is in line with the light beam coming from the light source. The table may be either stationary or movable. In most projectors, the table can be moved in two mutually perpendicular directions in the horizontal plane. The movement is effected by operating a knob attached with a double vernier micrometer, which can provide a positional accuracy up to 5 μm or better. The light beam originating from the lamp is condensed by means of a condenser and falls on the workpiece. The image of the workpiece is carried by the light beam, which passes through a projection lens. The projection lens magnifies the image, which falls on a highly polished mirror kept at an angle. The reflected light beam carrying the image of the workpiece now falls on a transparent screen. Selection of high-quality optical elements and lamp, and mounting them at the right location will ensure a clear and sharp image, which in turn ensures accuracy of the measurement.

C.11.5 Procedure

The experimental procedure involves the following steps:
1. Clean the instrument and its accessories using a fine cotton cloth.
2. Place the gear specimen on the transparent platform provided for this purpose.
3. Switch on the power and adjust the degree of magnification in the profile projector (the magnification depends upon the distance between the focal plane of the lens and the screen).
4. Make the horizontal cross-wire coincide with the uppermost visible point of the gear (on the addendum circle) and note down the initial reading on the corresponding micrometer. Now, operate the micrometer and move the image downwards until the horizontal cross-wire coincides with the lowermost point of the gear; note down the reading on the micrometer. The difference in the two readings gives the value of the addendum circle diameter of the gear. Take at least three readings and record the values in Table C.14. The average gives a more accurate value of the addendum circle diameter.

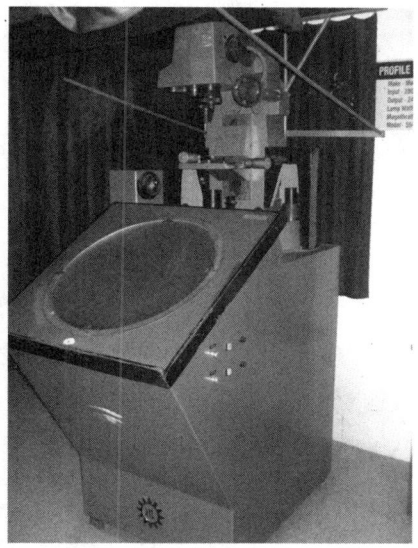

Fig. C.33 Profile projector

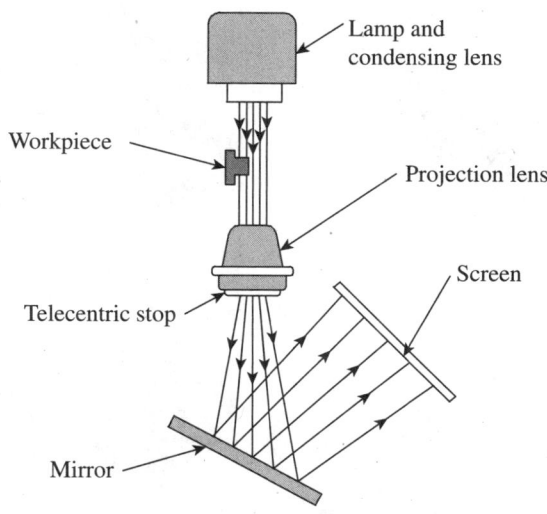

Fig. C.34 Parts of a profile projector

5. Repeat the procedure in step 4, in order to measure the root diameter and tooth thickness. Tabulate the results.
6. Now, use the circular scale provided with the instrument to measure the flank angle and document the results in Table C.14.

Table C.14   Documentation of results of the experiment

Parameter	Value			Average
	Trial 1	Trial 2	Trial 3	
Addendum circle diameter				
Dedendum circle diameter				
Tooth thickness				
Flank angle				

C.12 USE OF OPTICAL FLATS

C.12.1 Aim

To check the height difference between the two slip gauges

C.12.2 Apparatus

The apparatuses used for this experiment are as follows:
1. Optical flat
2. Monochromatic light
3. Slip gauge

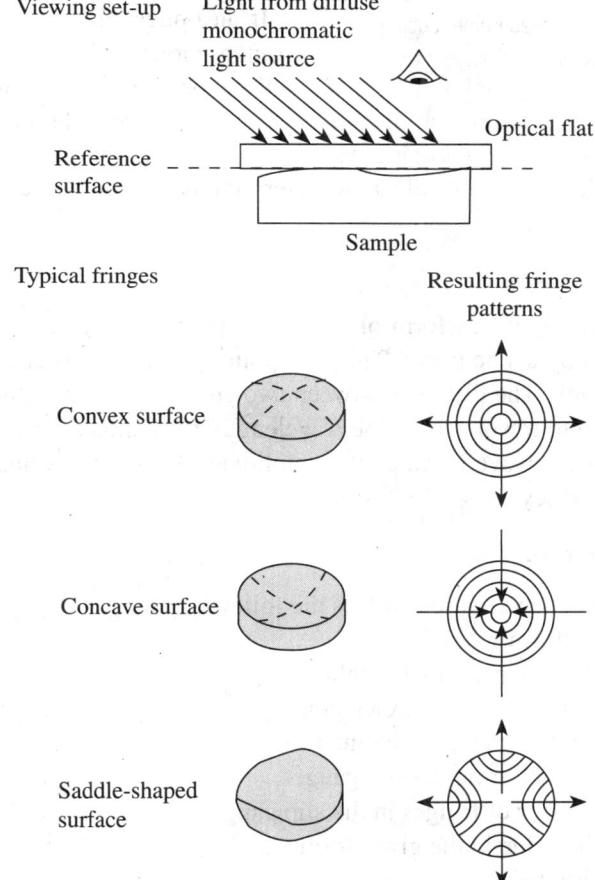

Fig. C.35   Working principle of an optical flat and different fringe patterns

C.12.3 Theory

Optical flats are cylindrical pieces 25–300 mm in diameter, with a thickness of about one-sixth of the diameter. They are made of transparent materials such as quartz, glass, and sapphire; optical flats made up of quartz are more commonly used because of their hardness, low coefficient of expansion, resistance to corrosion, and much longer useful life. One or both

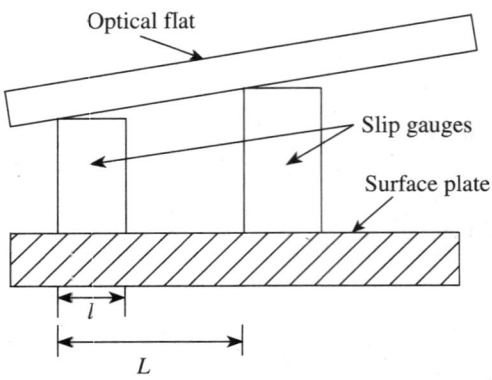

Fig. C.36   Arrangement for measurement using an optical flat

surfaces of optical flats may be highly polished. For measuring flatness, in addition to an optical flat, a monochromatic light source emitting light of a single wavelength is also required. The yellow-orange light radiated by helium gas is most satisfactory for use in an optical flat. Optical flats are used to test the flatness of lapped surfaces such as surfaces of gauge blocks, gauges, and micrometer anvils. When an optical flat is placed on the surface of a workpiece, it will not form an intimate contact but will be at a slight inclination to the surface, forming an air wedge between the surfaces.

If an optical flat is now illuminated by light rays from a monochromatic source, interference fringes will be observed, as shown in Fig. C.35. These are produced by the interference of light rays reflected, from the bottom face of the optical flat and the top face of the workpiece being tested, through the layers of air. An arrangement for measurement using an optical flat to determine height difference between two slip gauges is shown in Fig. C.36.

Monochromatic Light

Monochromatic light is a form of energy propagated by electromagnetic waves, which may be represented by a sine curve. The high point of the wave is called a crest and the low point is called a trough. The distance between two crests or two troughs is called the wavelength λ. The time taken to travel across one wavelength λ is called the time period (T). The maximum disturbance of the wave is called the amplitude (A), and the velocity of transmission (λ/T) is called the frequency.

C.12.4 Procedure

The experimental procedure involves the following steps:
1. Clean the platform.
2. Switch on the monochromatic light.
3. Select two slip gauges and clean them.
4. Place the slip gauges and clean them.
5. Place the optical flat on the slip gauges.
6. Count the number of fringes in the slip gauges.
7. Substitute the value in the given formula.
8. Compare with the actual value.

C.12.5 Observation

$$H = \lambda LN/2l$$

where $\lambda = 0.0002794$ mm, l is the width of the slip gauge in mm, L is the distance between the slip gauges in mm, and N is the number of fringes.

C.13 MEASUREMENT OF SURFACE ROUGHNESS VALUES USING HANDYSURF

C.13.1 Aim

To measure the surface roughness parameters R_z, R_a, and R_{max} using a Handysurf

C.13.2 Apparatus

To carry out the measurement, a Handysurf and a specimen are required.

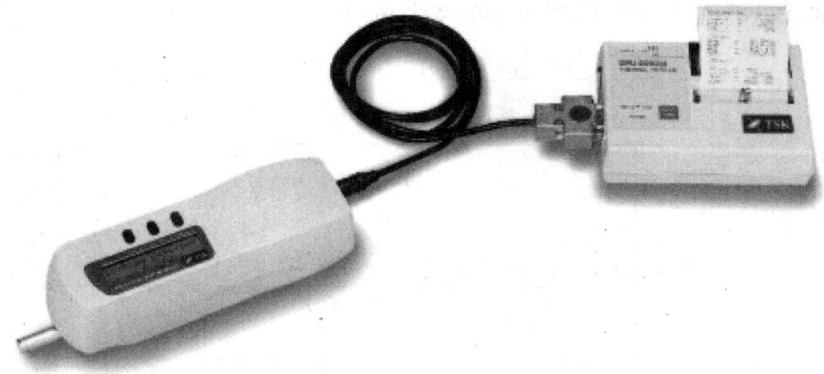

Fig. C.37 Handysurf

C.13.3 Theory

A Handysurf is required to assign a numerical value to the degree of surface roughness. This will enable the analyst to assess whether the surface quality meets the functional requirements of a component. A typical Handysurf device is shown in Fig. C.37. Various methodologies are employed to arrive at a representative parameter of surface roughness. Some of these are 10-point height average (R_z), centre-line average value (R_a), and R_{max}.

Ten-point Height Average Value

The 10-point height average (R_z) value is also referred to as the *peak-to-valley height*. We are basically looking at the average height encompassing a number of successive peaks and valleys of the asperities. As can be seen in Fig. C.38, a line AA parallel to the general lay of the trace is drawn. The heights of five consecutive peaks and valleys from line AA are noted down.

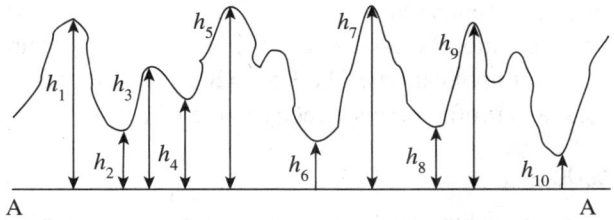

Fig. C.38 Graphical representation of peaks and valleys

The average peak to valley height R_z is given by the following expression:

$$R_z = \frac{(h_1 + h_3 + h_5 + h_7 + h_9) - (h_2 + h_4 + h_6 + h_8 + h_{10})}{5} \times \frac{1000}{\text{Vertical magnification}} \, \mu m$$

Centre-line Average Value

The centre-line average (R_a) value is the prevalent standard for measuring surface roughness. It is defined as the average height from a mean line of all ordinates of the surface, regardless of sign. Referring to Fig. C.39, it can be shown that

$$R_a = \frac{A_1 + A_2 + \dots + A_N}{L} = \Sigma A/L$$

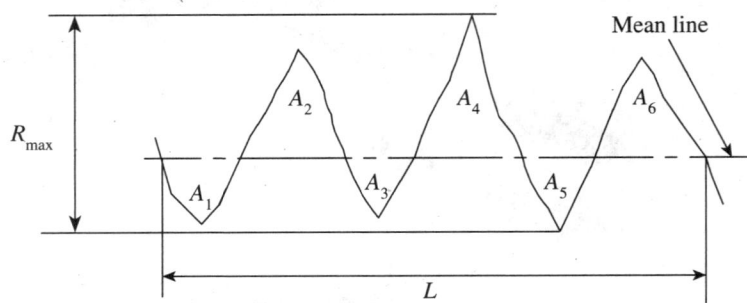

Fig. C.39   Graphical representation of R_a and R_{max} values

R_{max} Value

It is the maximum separation between a peak and a valley, as illustrated in Fig. C.39.

C.13.4 Handysurf

A Handysurf is a small, portable surface-measuring system. It is easy to operate, is capable of instant data processing, and has adequate memory to store data. Perhaps, there is no other instrument that can measure, evaluate, and document surface roughness in an easier way than the Handysurf.

It can be used in the receiving inspection, production, or final inspection. It makes quality assurance incredibly flexible because one can easily carry this compact, hand-held surface tester to the place where the part to be measured is kept. The small and light Handysurf measures not only flat and horizontal surfaces, but also vertical and overhead surfaces. In addition, the traversing unit can be separated from the display unit and used with optional holding devices for more flexible operation. A range of easily exchangeable pickup systems are provided by the supplier for different applications. The RS 232C serial interface is used to transfer measured data, parameters, and profile points directly to the PC.

C.13.5 Procedure

The experimental procedure can be carried out in the following way:
1. Select the measuring range, evaluation length, and cut-off value, depending on the surface to be measured.

2. Choose the pickup system that is suited for the application.
3. Carry out a freehand measurement of the specimen surface.
4. Transfer the data to a computer (if available); the Handysurf is also capable of storing a minimum of 10 records.
5. Generate the required surface roughness values and record the results in the format shown in Table C.15.

C.13.6 Observations

An observation of some parameters should be made to estimate surface roughness using a Handysurf.

Traversing Unit

Traversing length: _____ mm
Tracing speed: _____ mm/ s
Evaluation length: _____ mm

Pick-up System

Measuring range: _____ μm
Resolution: _____ μm
Probing force: _____ mN
Material of stylus: _____
Radius of stylus: _____ μm

C.13.7 Tabular Column

The experimental values can be documented in the format shown in Table C.15.

Table C.15   Documentation of surface roughness values

Specimen number	R_z	R_a	R_{max}
1.			
2.			

C.14 LATHE TOOL DYNAMOMETER

C.14.1 Aim

To determine the cutting forces in turning using a lathe tool dynamometer

C.14.2 Apparatus

The instrument required for measurement is a lathe tool dynamometer.

C.14.3 Theory

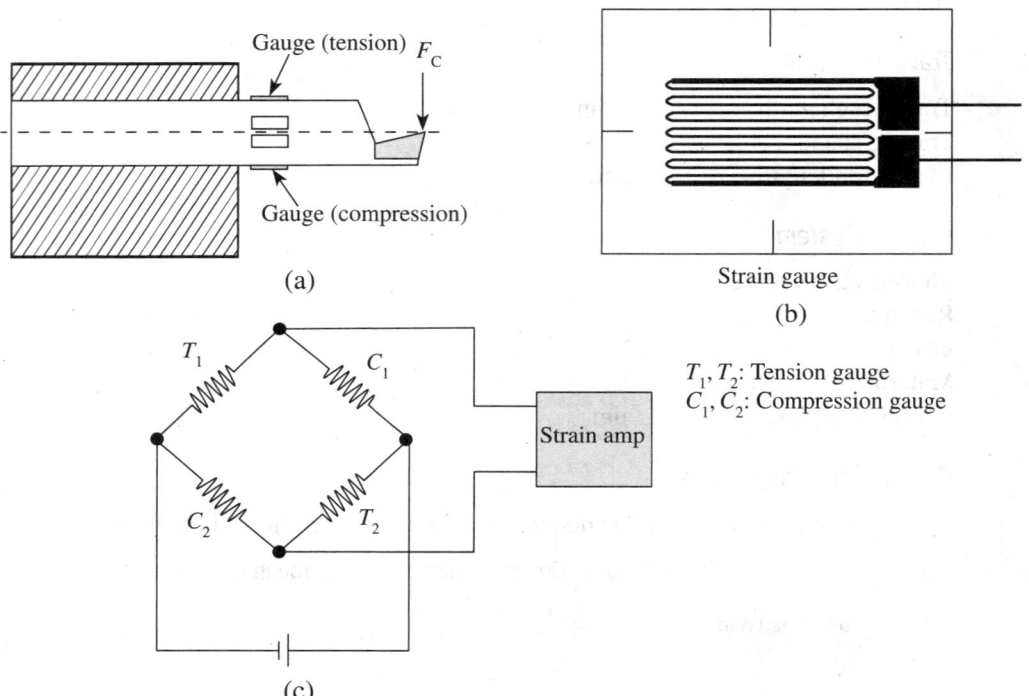

Fig. C.40 Arrangement of a strain gauge for cutting force measurement (a) Basic elastic dynamometer (b) Wire-type gauge (c) Wheatstone bridge and amplifier

A common type of dynamometer uses strain gauges to sense elastic strains caused by cutting forces. Figure C.40(a) shows a basic elastic beam-type dynamometer with gauges bonded to its surface. Figure C.40(b) shows an example of a wire-type gauge, and Fig. C.40(c) shows a Wheatstone bridge and amplifier system usually used to measure strain changes in the gauges. The main cutting force (F) will cause the beam to bend, so that the gauges on the top surface will be placed in tension, on the bottom surface will be placed in compression, and on the side surfaces (at the neutral axis) will experience no strain. Likewise, a feed force will strain the side-face gauges but not those at the top or bottom. The arrangement shown in Fig. C.40 is not sensitive to force along the axis of the beam, as this causes equal strain changes in all gauges. The fractional resistance change of a strain gauge ($\Delta R/R$) is related to its fractional length change or direct strain (MIL) by its gauge factor K; $K = (\Delta R/R)/(\Delta L/L)$ for wire strain gauges. The value of K typically ranges from 1.75 to 3.5. Strains due to load may be detected with a bridge circuit. The upper limit of strain is around 2×10^{-3}, determined by the elastic limit of the beam. A disadvantage of the simple cantilever dynamometer is that the strain of the gauges basically depends on the moment applied to the section at which they are positioned. They therefore depend on the distance of the gauges from where the load is applied, as well as on the size of the load. Octagonal ring and parallel beam are better-designed instruments that are less sensitive to where the load is applied. Supporting the load on well-separated thin sections results in the sum of the strains in the gauges being unchanged when the point of application of the load is changed, even though the strains are redistributed between the sections. It is possible to connect the strain gauges in a bridge circuit so that the output is not sensitive to where the force is applied.

To investigate the performance of cutting tools during metal cutting, measurement of the cutting forces is essential; this helps in the analysis of metal cutting with respect to the following factors:

1. Effects of speed, feed, and depth of cut on the action of the cutting tool
2. Effects of mechanical properties of work material on cutting forces
3. Values of forces exerted on machine components including jig and fixture
4. Uses and effect of these forces on geometrical accuracy of the workpiece

C.14.4 Procedure

The major objective of this experiment is to determine the influence of cutting parameters— cutting speed, feed rate, and depth of cut—on cutting forces. In turning, there are predominantly three force components: tangential cutting force, feed force, and radial force. Tangential cutting force (F_x) acts tangentially to the workpiece surface at the point of contact of tool and workpiece. The feed force (F_y) is the reaction force, directly opposite to the feed direction. The radial force component (F_z) acts radially and is directed towards the axis of the workpiece.

First, keep the feed and depth of cut constant, vary the speed, and take readings of forces in the x, y, and z directions. Next, keep the speed and depth of cut constant, vary the feed value, and take readings of forces along x, y, and z directions. Finally, keep the feed and speed constant, vary the depth of cut, and take readings of forces in the x, y, and z directions. Document the respective results in Tables C.16, C.17, and C.18, respectively.

C.14.5 Observation

The experimental values for the following variables can be documented in tabular formats:

1. Feed and depth of cut are kept constant; speed is varied.

Table C.16   Documentation of experimental results with variation in speed

Trial no.	Feed (mm)	Speed (rpm)	Depth of cut	D_o (mm)	$(F_x) \times 9.81$ (N)	$(F_y) \times 9.81$ (N)	$(F_z) \times 9.81$ (N)	$(F_R) \times 9.81$ (N)

$$F_R^2 = (F_x^2 + F_y^2 + F_z^2).$$

2. Speed and depth of cut are kept constant; feed is varied.

Table C.17 Documentation of experimental results with variation in feed

Trial no.	Feed (mm)	Speed (rpm)	Depth of cut	D_o (mm)	$(F_x) \times 9.81$ (N)	$(F_y) \times 9.81$ (N)	$(F_z) \times 9.81$ (N)	$(F_R) \times 9.81$ (N)

3. Speed and feed are kept constant; depth of cut is varied.

Table C.18 Documentation of experimental results with variation in depth of cut

Trial no.	Feed (mm)	Speed in (rpm)	Depth of cut	D_o (mm)	$(F_x) \times 9.81$ (N)	$(F_y) \times 9.81$ (N)	$(F_z) \times 9.81$ (N)	$(F_R) \times 9.81$ (N)

C.14.6 Graph

Draw the following graphs:
1. Resultant force versus speed (rpm)
2. Resultant force versus feed (mm/rev)
3. Resultant force versus depth of cut (mm), along the y and x axes, respectively.

C.15 DRILL TOOL DYNAMOMETER

C.15.1 Aim

To determine torque and thrust force using a drill tool dynamometer and to determine specific power consumption

C.15.2 Apparatus

For the experiment, a drill tool dynamometer and a CI specimen are required.

C.15.3 Theory

Figure C.41 shows a typical sketch of a drill tool dynamometer. Due to torque and thrust, forces will develop and act on the several components of the drilling machine as follows:

Foundation

Both T and P_X will be transmitted to the base and the foundation from the job and through the bed, clamps, and the foundation bolts.

Spindle

As this salient component will be subjected to both torque and thrust, it is designed accordingly; the motor is selected based on the maximum torque and spindle speed.

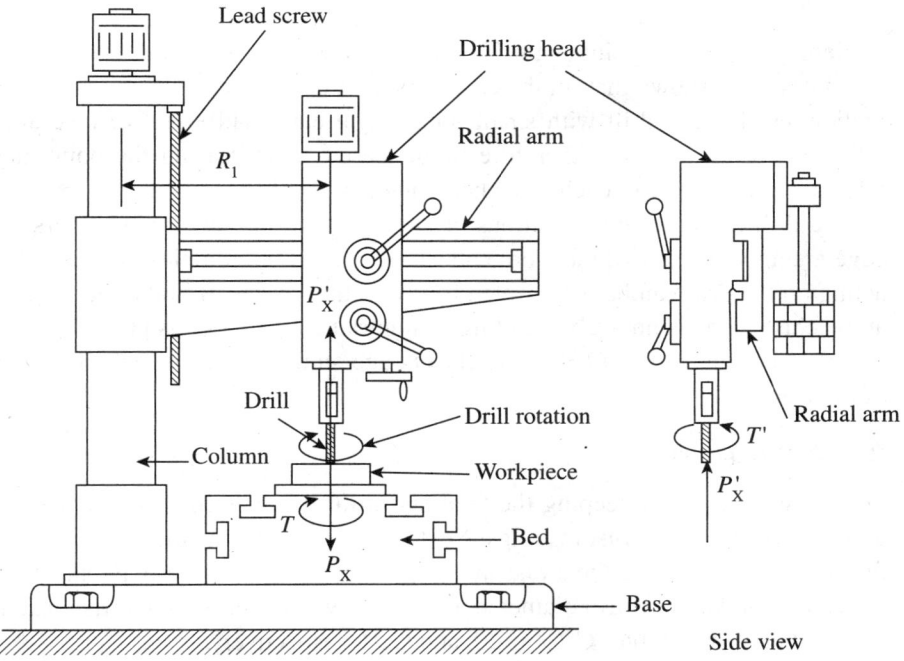

Fig. C.41 Drill tool dynamometer

Radial Arm

This cantilever beam is subjected to a large bending moment, $P_X \times R_i$, depending upon the magnitude of P_X and the distance R_i of the drilling head from the column axis. This arm will also be subjected to another twist in the other vertical plane due to P_X, depending upon its distance from the mid-plane of the radial arm. In addition, the arm will bear the weight of the drilling head and its balancing weight, which will also induce bending moment.

Column

This main structural part will have two axial forces: weight and P_X' acting vertically downwards and upwards, respectively. The force P_X' will also induce a large bending moment, equal to $P_X \times R_i$ in the column.

Drilling

The most common hole-making operation is drilling and it is usually performed with the help of a twist drill. Figure C.41 shows a drilling operation. If the total advancement to the drill per revolution (the feed rate) is f, then the share of each cutting edge is $f/2$ because each lip is getting the uncut layer, the top surface of which has been finished by the other lip 180° ahead (during a 180° revolution, the vertical displacement of the drill is $f/2$); the uncut thickness t_1 and the width of cut w are given as follows:

$$t_1 = (f/2)\, r_{in}\, \beta$$
$$w = (D/2)/r_{in}\, \beta$$

Here, β is the half-point angle. The rake angle can be found out from the section view, but it can easily be shown that, in the case of a twist drill, it depends on the radial location of the sectioning plane. A drill with small point angle has small effective rake angle. The cutting efficiency of the drill can therefore be increased by increasing the point angle because the effective rake angle approaches the helix angle of the drill.

In a drilling operation, variations of cutting speed and other parameters along the cutting edge are appreciable and the whole phenomenon is very complex. The effect of all the forces acting on the drill can be represented by a resisting torque M and a thrust force F. The action at the chisel edge is not truly a cutting action; rather, the chisel is pushed into the material like a wedge. However, the effect of the chisel edge on the torque is negligible, as it is on the axis of rotation.

C.15.4 Procedure

On the drill machine, keeping the feed and drill diameter constant, vary the speed and note down the torque and thrust readings. Next, keep the speed and drill diameter constant, vary the feed rate, and note down the thrust and torque readings in each case. Finally, keep the speed and feed constant, vary the drill diameter, and note down the thrust and torque readings. The results are documented in Tables C.19, C.20, and C.21, respectively.

C.15.5 Observation

First, feed and drill diameter are kept constant and speed is varied.

Table C.19   Documentation of experimental results with variation in speed

Speed variation (rpm)	Feed (mm/rev)	Drill diameter (mm)	Torque (kg m)	Thrust (kgf)	Angular velocity (rad/sec)	Cutting speed (m/min)

Power consumed (W)	Power in revolution of drill bit (kW)	MRR (mm³/min)	Specific power consumption

1. Angular velocity = $2\pi N/60$ = _____.
2. Cutting speed = $\pi DN/1000$ = _____.
3. Power consumption = (Thrust force × Cutting speed) (W).
   Hence, power consumption = (Thrust force in kgf × Cutting speed in m/m × 0.17) (W)
4. Power per revolution of drill bit = (Power consumption/Rpm) (kW)
5. Material removal rate = $[(\pi d^2/4)$ × feed rate] mm³/min
6. Specific power consumption = (Power required/Rev)/(MRR). MRR is the material removal rate.

   In the next step, speed and drill diameter are kept constant, and feed is varied.

Table C.20   Documentation of experimental results with variation in feed

Speed variation (rpm)	Feed (mm/rev)	Drill diameter (mm)	Torque (kg m)	Thrust (kgf)	Angular velocity (rad/sec)	Cutting speed (m/min)	Power consumed (W)

Power per revolution of drill bit (kW)	MRR (mm³/min)	Specific power consumption

In the final step, speed and feed are kept constant, and drill diameter is varied.

Table C.21 Documentation of experimental results with variation in drill diameter

Speed variation (rpm)	Feed (mm/rev)	Drill diameter (mm)	Torque (kg m)	Thrust (kgf)	Angular velocity (rad/sec)	Cutting speed (m/min)	Power consumed (W)

Power per revolution of drill bit (kW)	MRR (mm³/min)	Specific power consumption

C.15.6 Graph

In this experiment, the following graphs need to be drawn:

1. Torque in kg m versus speed in rpm
2. Torque in kg m versus feed in mm/rev, drawn along the y and x axes, respectively.

C.16 CALIBRATION OF PRESSURE GAUGES

C.16.1 Aim

To calibrate the given pressure gauge

C.16.2 Theory

A pressure gauge, especially a Bourdon's gauge, is calibrated by means of a dead weight tester. Figure C.42 gives the constructional details of the dead weight pressure tester. The essential components of such a tester are reservoir 'R', cylinder 'C', barrel 'B', and passage up to 'C' to hold clean dry oil. A spindle 'S' with a highly finished surface and a precise cross-sectional area slides vertically in the barrel 'B' through a close-fitting, highly-polished bearing 'b' and carries the table 'T' at its upper end. Its lower end rests on 'HP' piston. Screws 'Ls1' and 'Ls2' lock the passage of oil when required. The tester is mounted on a stand 'ST'. It is provided with special precise weights marked in terms of pressure. A double-acting piston 'DP' can be moved forward and backward by rotating the handle 'H'. The double-acting piston works both ways and does not allow the oil to leak. The gauge under calibration can be connected at 'D' such that the connection is leak tight. Keeping 'Ls1' and 'Ls2' open, the handle is rotated in such a way that the oil is just in level with the gauge connecting point 'D', and 'Ls2' is now locked. The gauge 'G' is mounted carefully. 'H' is rotated until the table 'T' is at a raised position in line with the upper edge of the colour band. 'Ls1' is then locked. 'Ls2' is opened, and the handle H is rotated in such a way that the gauge needle just moves and reads some minimum pressure characteristics of the tester because of the weight of the unloaded table acting through the piston 'LP'. A weight is placed on the table, which increases the pressure on the oil in the tester. The gauge pressure reading should correspond to the amount scribed on the weight if it is operating correctly. If not, the dial is rotated so that the needle points to the correct pressure. Another weight is added and another gauge reading is noted, and so on. A combination of weights can be used with a thinner piston 'HP', with a multiplying factor given by the manufacturer, say 20. Such a tester can give pressure values accurate up to +0.05% of the pressure being measured.

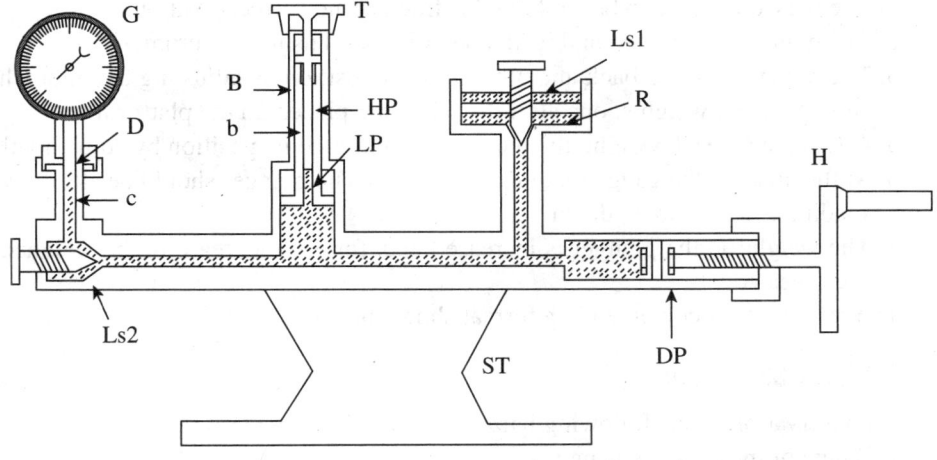

Fig. C.42 Constructional set-up

The range of pressure is typically 0.5–10,000 kg/cm², using dual spindles. The pressure range that a dead weight tester can measure is limited by the area of cross section of the spindle S and the number of weights that can be placed safely on the table. In order to increase the range, another spindle of a smaller cross section can be provided, increasing the pressure range for the same weight placed on the table by a factor equal to the ratio of the cross-sectional area of the two spindles. The weights are typically for 0.05, 0.1, 0.5, 1, 2, 5, 9, and 10 kg/cm².With a thinner spindle and a factor of 20, these weights can become 1, 2, 10, 20, 40, 100, 130, and 200 kg/cm², respectively. The dead weight tester is depicted in Fig. C.43.

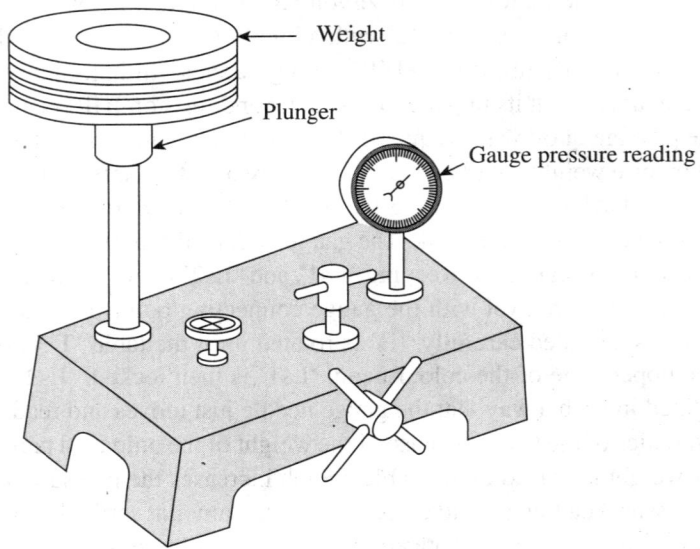

Fig. C.43   Dead weight tester

C.16.3 Procedure

The experimental procedure involves the following steps:
1. The pressure gauge to be calibrated is fitted at the correct position.
2. The plunger is weighed and its diameter is found using a micrometer.
3. The plunger is fitted back and brought to its position by adjusting the pilot wheel.
4. Now different weights, one after the other, are placed on the platform.
5. After placing each weight, the plunger is brought to its position by adjusting the pilot wheel.
6. At this instant, the gauge reading is recorded. The plunger should be steady while the gauge reading is being noted down.
7. The weight on the plunger is increased and, for each increase in the weights, steps 4 and 5 are repeated.

The results are documented in a format shown in Table C.22.

C.16.4 Observation

An observation of the following parameters should be made:
1. Range of the pressure gauge: _____
2. Weight of the plunger: _____

3. Diameter of the plunger (D): _____

4. Cross-sectional area of the plunger: $A = (\pi D^2/4) =$ _____ m^2

5. Actual pressure $(P_a) = [(\text{Load in g})/(C/S \text{ area of plunger})] =$ _____ g/mm^2 _____ $kgf/mm^2 =$ _____ N/m^2

6. Percentage error $= [(P_a - P_g)/P_a] \times 100$

7. Gauge pressure $(P_g) =$ _____ $kg/cm^2 =$ _____ N/m^2

Table C.22 Documentation of experimental results for pressure measurement

Weight of the plunger (g)	Weight added (g)	Total weight of the plunger (g)	Actual pressure (kgf/cm²)	Gauge pressure (increasing weights)	Gauge pressure (decreasing weights)	Average gauge pressure (kg/cm²)	% Error

C.16.5 Graph

Draw the following graphs:

1. Pressure on gauge (y axis) versus applied weights (x axis)
2. Percentage error (y axis) versus average gauge pressure (x axis)

C.17 CALIBRATION OF LOAD CELLS

C.17.1 Aim

To calibrate the given load cell

C.17.2 Apparatus

The apparatuses required for this experiment include a load cell, a digital load indicator, and dead weights.

C.17.3 Theory

A load cell, as shown in Fig. C.44, is an electronic device (transducer) that is used to convert force into an electrical signal. This conversion is indirect and happens in two stages. Through a mechanical arrangement, the force being sensed deforms a strain gauge. The strain gauge converts the deformation (strain) to electrical signals. Usually, a load cell consists of four strain gauges in a Wheatstone bridge configuration. However, load cells of one or two strain gauges are also available. The electrical signal output is typically in the order of a few millivolts and requires amplification by an instrumentation amplifier before it can be used. The output of the transducer is plugged into an algorithm to calculate the force applied to the transducer.

Although strain-gauge load cells are most common, there are other types of load cells as well. In industrial applications, the hydraulic (or hydrostatic) type is probably the second most common load cell, which is used to eliminate some of the problems associated with strain-gauge load cell devices. As an example, a hydraulic load cell is immune to transient voltages (lightning) and so may be a more effective device in outdoor environments.

Other types include piezo-electric load cells (useful for dynamic measurement of force) and vibrating wire load cells, which are useful in geomechanical applications due to low amount of drift.

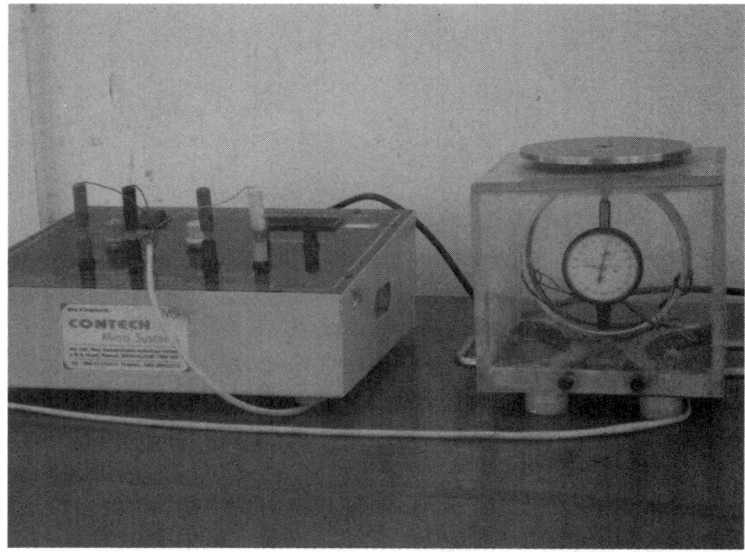

Fig. C.44 Typical load cell

Every load cell is affected by 'ringing' when subjected to abrupt load changes. This stems from the spring-like behaviour of load cells. In order to measure the loads, they have to deform. As such, a load cell of finite stiffness must have spring-like behaviour, exhibiting vibrations at its natural frequency. An oscillating data pattern can be the result of ringing. Ringing can be suppressed in a limited fashion by passive means. Alternatively, a control system can use an actuator to actively damp out the ringing of a load cell. This method offers better performance even when there is significant increase in complexity.

Strain-gauge load cells convert the load acting on them into electrical signals. The gauges themselves are bonded onto a beam or a structural member that deforms when weight is applied. In most cases, four strain gauges are used to obtain maximum sensitivity and temperature compensation. Two of the gauges are usually in tension and two in compression, and they are wired with compensation adjustments. When weight is applied, the strain changes the electrical resistance of the gauges in proportion to the load. The arrangement of a strain gauge is as shown in Fig. C.45. Other load cells are fading into obscurity, as strain-gauge load cells continue to increase their accuracy and lower their unit costs.

Digital Load Indicator

Digital load indicators are highly stable amplifiers compatible with the load cells to read force (measured in Newton) directly, as shown in Fig. C.46. Linear potentiometers with fine adjustments for the zero setting help operate the instrument smoothly.

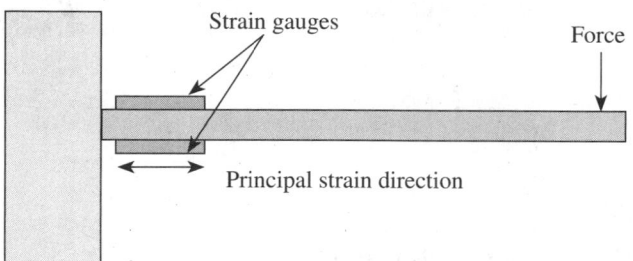

Fig. C.45   Strain-gauge load cell

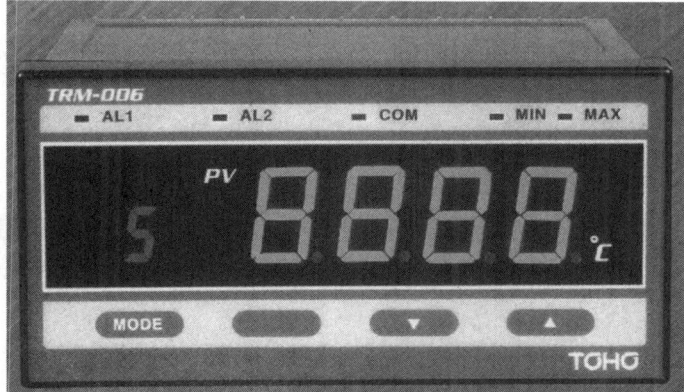

Fig. C.46   Digital load indicator

C.17.4 Procedure

The steps involved in the experimental procedure are as follows:
1. Connect the load cell to a digital indicator inserting the corresponding colour codes.
2. Connect the digital indicator to the mains and switch it on.
3. Adjust the zero knob of the indicator to 'zero'.
4. Apply weights of up to 8 kg.
5. Apply the calibration knob of the indicator to read 78.48 N (i.e., 8 kg × 9.81 N).
6. Remove the weights from the load cell.
7. Set the zero knob to zero and repeat the calibration.
8. Now the instrument is ready for taking measurements.
9. Place the weights one by one and take down the indicator reading.
10. Calculate the correction error and percentage error.
11. Document the results in Table C.23.

C.17.5 Observation

An observation of the following parameters are to be made:
1. N_i = Indicated reading
2. N_a = Actual reading

Table C.23 Documentation of experimental results

| Trial no. | Load | | Indicator reading | Correction | Error | % Error |
	W (kg)	N_a (N)	N_i (N)	$N_a - N_i$	$N_i - N_a$	$[(N_i - N_a)/N_i] \times 100$

C.17.6 Graph

The following graphs have to be drawn:
1. % Error versus N_i
2. N_i versus N_a
3. Correction versus N_i

C.18 CALIBRATION OF LVDT

C.18.1 Aim

To calibrate the given LVDT

C.18.2 Apparatus

For performing this experiment, an LVDT of ±10 mm capacity and a digital displacement indicator are required.

C.18.3 Theory

An LVDT is a type of electrical transformer that is used for measuring linear displacement. Figure C.47 shows a picture of a standard LVDT. The transformer has three solenoidal coils placed end-to-end around a tube. The central coil is the primary and the two outer coils are the secondary coils. A cylindrical ferromagnetic core, attached to the object whose position is to be measured, slides along the axis of the tube.

An alternating current is driven through the primary coil, causing a voltage to be induced in each secondary coil proportional to its mutual inductance with the primary coil. The frequency is usually in the range of 1–10 kHz.

As the core moves, these mutual inductances change, causing the voltages induced in the secondary coils to change. The coils are connected in reverse series, so that the output voltage is the difference (hence 'differential') between the two secondary voltages. When the core is in its central position, equidistant between the two secondary coils, equal but opposite voltages are induced in these two coils, and hence the output voltage is zero.

When the core is displaced in one direction, voltage increases in one coil and decreases in the other, causing the output voltage to increase from zero to a maximum. This voltage is in phase with the primary voltage. When the core moves in the other direction, the output voltage also increases from zero to a maximum, but its phase is opposite to that of the primary. The magnitude of the output voltage is proportional to the distance moved by the core (up to its limit of travel), which is why the device is described as 'linear'. The phase of the voltage indicates the direction of displacement.

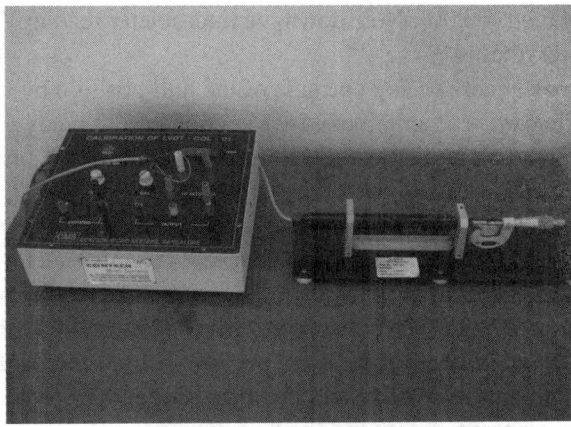

Fig. C.47   Linear variable differential transformer

Since the sliding core does not touch the inside of the tube, it can move without friction, making the LVDT (Fig. C.48) a highly reliable device. The absence of any sliding or rotating contacts allows the LVDT to be completely sealed against the environment.

LVDTs are commonly used for position feedback in servomechanisms, and for automated measurement in machine tools and many other industrial and scientific applications.

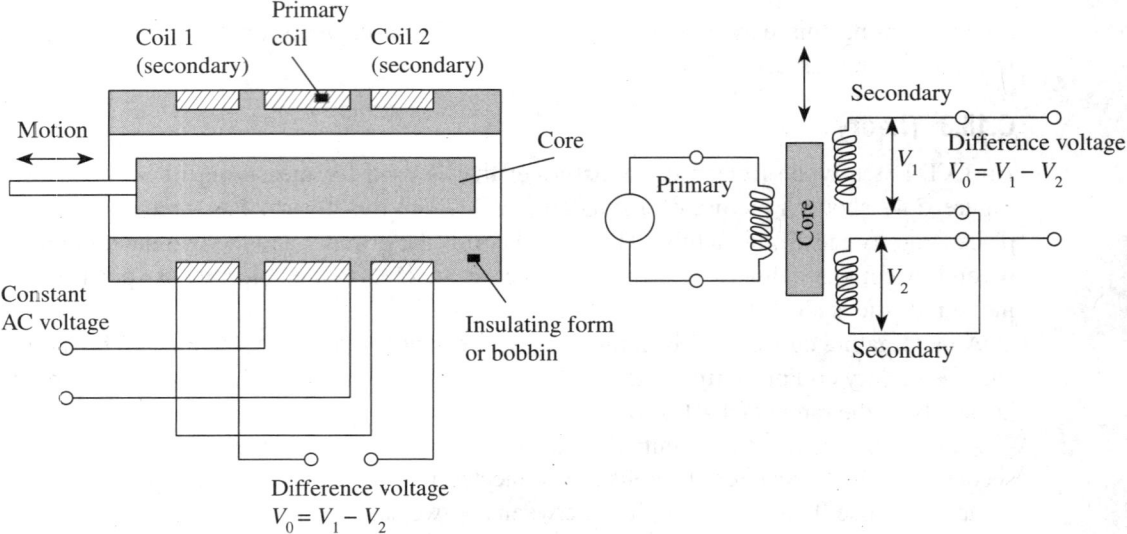

Fig. C.48 Schematic diagram of an LVDT (a) Sectional view of an LVDT (b) Circuit diagram of an LVDT

C.18.4 Procedure

The steps to be adopted for the experiment are as follows:

1. Connect the instrument to a 230 V power supply and switch 'ON' the instrument.
2. Adjust the displacement indicator to read zero.
3. Connect the LVDT sensor to the displacement indicator through the given cable.
4. Rotate the micrometer knob in clockwise or anticlockwise direction to bring the LVDT core to null position of the sensor where there is no induced emf. At this position, the indicator will read '00.00'. Note down the micrometer reading. This is the initial reading of the micrometer, that is, S.
5. Now move the core to any one side of the null position by slowly rotating the screw gauge knob in clockwise or anticlockwise direction until it reads ±10 mm in the micrometer. Note down the reading and adjust the calibration knob of the instrument to read '10.00 mm' in the indicator.
6. Repeat the previous operation once again and then calibrate the given LVDT.
7. Bring the LVDT to null position and rotate the micrometer in clockwise or anticlockwise direction by 1 or 2 mm; note down the readings of the micrometer as well as the displacement indicator. Note down the micrometer reading as the final reading, that is, S_f.
8. Repeat the experiment for different positions of the core and simultaneously note down the readings of the screw gauge and the displacement indicator at every step.
9. Calculate the error, if any, and the % error.
10. Document the results in Table C.24.

C.18.5 Observation

The experimental results are documented in a tabular format.

Table C.24   Documentation of experimental results

Trial no.	Core position		LVDT indicator reading S (mm)	Micrometer reading			Correction $S_m - S$	Error	% Error $[(S - S_m)/S] \times 100$
	Towards left of none	Towards right of none		S_i	S_f	S_m			

Note:
Error = $(S - S_m)$. S_i, S_f, and S_m are measured in millimetres.

C.18.6 Graph

In this experiment, the following graphs are required to be drawn:
1. S_m versus S_a
2. Correction versus S
3. % Error versus S, along the y and x axes, respectively.

C.19 DETERMINATION OF MODULUS OF ELASTICITY OF MILD STEEL SPECIMEN USING STRAIN GAUGES

C.19.1 Aim

To determine the modulus of elasticity or elastic constant of a cantilever beam subjected to a concentrated end load by using strain gauges

C.19.2 Apparatus

The apparatuses required for the measurement include a cantilever beam with a concentrated end load arrangement, strain gauges, and a strain indicator.

C.19.3 Theory

Strain is the amount of deformation of a body due to an applied force. More specifically, strain (e) is defined as the fractional change in length, as shown in Fig. C.49.

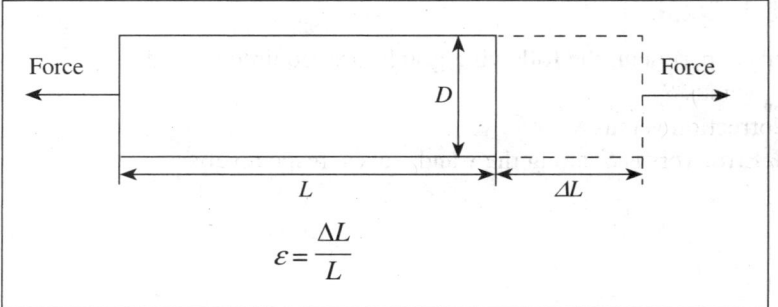

Fig. C.49 Illustration of strain

Strain can be positive (tensile) or negative (compressive). Although dimensionless, it is sometimes expressed in units such as in/in or mm/mm. In practice, the magnitude of measured strain is very small. Therefore, strain is often expressed as microstrain (me), which is $e \times 10^{-6}$.

When a bar is strained with a uniaxial force, as in Fig. C.49, a phenomenon known as Poisson strain causes the girth of the bar, D, to contract in the transverse or perpendicular direction. The magnitude of this transverse contraction is a material property indicated by its Poisson's ratio. The Poisson's ratio (n) of a material is defined as the negative ratio of the strain in the transverse direction (perpendicular to the force) to the strain in the axial direction (parallel to the force), or $n = e_T/e$. The Poisson's ratio for steel, for example, ranges from 0.25 to 0.3.

Strain Gauge

While there are several methods of measuring strain, the most common is measurement using a strain gauge, a device whose electrical resistance varies in proportion to the amount of strain in the device. The most widely used gauge is the bonded metallic strain gauge. A metallic strain gauge consists of a very fine wire or, more commonly, a metallic foil arranged in a grid pattern. The grid pattern maximizes the amount of metallic wire or foil subject to strain in the parallel direction (Fig. C.50). The cross-sectional area of the grid is minimized to reduce the effect of shear strain and Poisson strain. The grid is bonded to a thin backing,

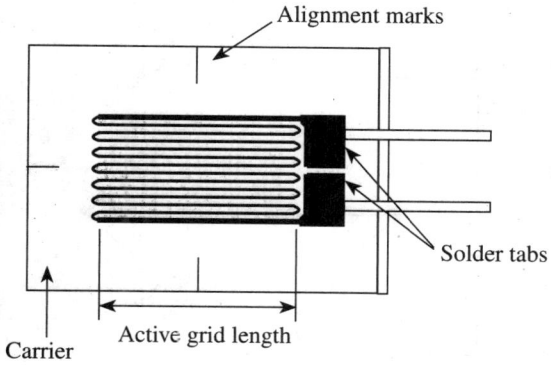

Fig. C.50 Bonded metallic strain gauge

called the carrier, which is directly attached to the test specimen. Therefore, the strain experienced by the test specimen is directly transferred to the strain gauge, which responds with a linear change in electrical resistance. Strain gauges that are commercially available have nominal resistance values from 30 to 3000 Ω, with those having resistance of 120, 350, and 1000 Ω being the most commonly available strain gauges.

It is very important that the strain gauge is properly mounted onto the test specimen so that the strain is accurately transferred from the test specimen, through the adhesive and strain gauge backing, to the foil itself.

A fundamental parameter of the strain gauge is its sensitivity to strain, expressed quantitatively as the gauge factor (GF). Gauge factor is defined as the ratio of fractional change in electrical resistance to the fractional change in length (strain):

$$GF = \frac{\Delta R/R}{\Delta L/L} = \frac{\Delta R/R}{\varepsilon}$$

The value of the gauge factor for metallic strain gauges is typically around 2.

Strain Measurement

A typical set-up for strain measurement is shown in Fig. C.51.

In practice, strain measurements rarely involve quantities larger than a few millistrain ($e \times 10^{-3}$). Therefore, measurement of strain requires accurate measurement of very small changes in resistance. For example, suppose a test specimen undergoes a strain of 500 me. A strain gauge with a gauge factor of 2 will exhibit a change in electrical resistance of only 2 (500×10^{-6}) = 0.1%. For a 120 Ω gauge, this is a change of only 0.12 Ω.

To measure such small changes in resistance, strain gauges are almost always used in a bridge configuration with a voltage excitation source. The general Wheatstone bridge, illustrated in Fig. C.52, consists of four resistive arms, with an excitation voltage, V_{EX}, applied across the bridge.

The output voltage of the bridge, V_O, is given as

$$V_O = \left[\frac{R_3}{R_3 + R_4} - \frac{R_2}{R_1 + R_2} \right] \cdot V_{EX}$$

Gauge Factor

Gauge factor (GF) or strain factor of a strain gauge is the ratio of the relative change in electrical resistance to the relative change in length (which is the mechanical strain e). The resistance is also dependent on temperature. The higher the gauge factor, the greater the electrical output for indication or recording purpose. The value of the gauge factor may vary from 1.7 to 4.0.

The gauge factor GF is defined as follows:

Fig. C.51 Typical set-up for strain measurement

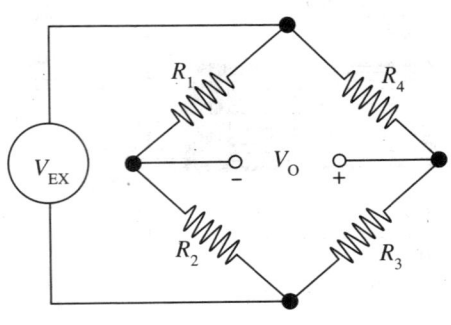

Fig. C.52 Wheatstone bridge

$$GF = \frac{\Delta R / R_{\mathrm{G}}}{\varepsilon}$$

Here, ΔR is the change in resistance caused by strain, R_{G} is the resistance of the undeformed gauge, and ε is the strain.

C.19.4 Observation

An observation of the following parameters should be made:

Distance between gauge centres to the point of application of load l: _____ mm

Width of the beam b: _____ mm

Thickness of the beam t: _____ mm

Table C.25 Documentation of experimental results

Trial no.	Load applied W (N)	Strain indicator reading ε	Measured strain $\varepsilon_m = [\varepsilon \times 10^5]$	Bending stress $\sigma = 6W_l/bt^2$	Modulus of elasticity $E = \sigma\varepsilon_m$ (N/mm^2)

W = Load applied = _____

ε = Bending stress = $6Wl/bt^2$

ε_m = Measured strain = $(\varepsilon \times 10^5)$ = _____

E = Young's modulus = σ/ε_m = _____

The results are recorded in Table C.25.

C.19.5 Graph

A graph of ε_m (x axis) versus σ (y axis) is plotted. Modulus of elasticity (E) is the slope of the line.

C.20 CALIBRATION OF THERMOCOUPLES

C.20.1 Aim

To calibrate the given thermocouple using a resistance thermometer

C.20.2 Apparatus

The calibration experiment requires a thermocouple, a heating coil to heat the water in the water bath, a resistance thermometer, and a digital indicator to indicate the temperature of both the thermocouple and the resistance thermometer.

C.20.3 Theory

A thermocouple is a junction between two different metals that produces a voltage related to a temperature difference between the two ends. A simple schematic diagram of a thermocouple is illustrated in Fig. C.53. Thermocouples are a widely used type of temperature sensor and can also be used to convert heat into electrical power. They are inexpensive and interchangeable, are supplied fitted with standard connectors, and can measure a wide range of temperatures. The main limitation is their accuracy: system errors of less than 1 K can be difficult to achieve.

Any junction of dissimilar metals will produce an electric potential related to the temperature difference between the metals. Thermocouples used for practical measurement of temperature consist of junctions of specific alloys, which have a predictable and repeatable relationship between temperature and voltage. Different alloys are used for different temperature ranges. Properties such as resistance to corrosion may also be important when choosing a type of thermocouple. Where the measurement point is far from the measuring instrument, the intermediate connection can be made by extension wires, which are less costly than the materials used to make the sensor. Thermocouples are usually standardized against a reference temperature of 0 °C; practical instruments use electronic methods of cold-junction compensation to adjust for varying temperature at the instrument terminals. Electronic instruments can also compensate for the varying characteristics of the thermocouple, and thereby improve the precision and accuracy of measurements.

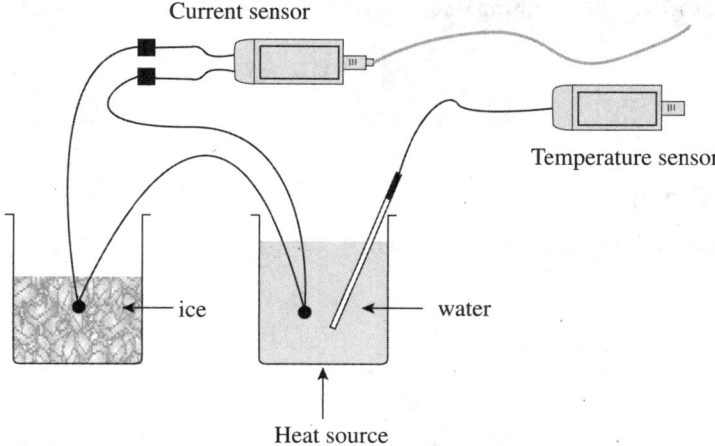

Fig. C.53   Schematic of a thermocouple

Thermocouples are widely used in science and industrial applications, including temperature measurement for kilns, gas turbine exhaust, diesel engines, and other industrial processes.

C.20.4 Procedure

The experimental procedure involves the following steps:

1. Turn the type selector to the desired position according to the given thermocouple probe.
2. Connect the resistance temperature detector (RTD) probe to the RTD display.
3. Connect the given thermocouple to the thermocouple temperature display.
4. Place the thermocouple cold junction and the RTD probe in a beaker containing water at room temperature.
5. Connect the power supply to the temperature indicator.
6. Record the room temperature from the RTD temperature indicator.
7. Adjust the zero-setting knob of the thermocouple temperature indicator until the display shows the room temperature.
8. Connect the power supply to the heating coil and heat the water in the water bath.
9. Set the temperature of the thermocouple to the temperature of the RTD indicator when the water is boiling, using the calibration knob.
10. Repeat this procedure at least three to four times until the RTD display and the thermocouple display match.
11. Now calibrate the thermocouple with reference to the RTD.
12. Simultaneously record the readings of the RTD and the thermocouple temperature indicator at regular intervals.
13. The experimental results are documented in Table C.26.

C.20.5 Observation

The experimental results are documented in a tabular format.

Table C.26   Documentation of experimental results

Trial no.	Temp. of water by RTD (t_a) (°C)	Temp. of water by thermocouple (t_m) (°C)	Correction $(t_a) - (t_m)$	Error $[t_m] - [t_a]$	% Error $[(t_m) - (t_a)]/(t_m) \times 100$

C.20.6 Graph

The following graphs should be drawn:

1. t_m versus t_a
2. Correction versus t_m
3. % Error (y axis) versus t_m (x axis)

VIVA QUESTIONS

1. What is a sine bar? Describe its use in angle measurement.
2. State the limitations of a sine bar. Give reasons.
3. List the different instruments used for angle measurements.
4. Briefly explain the sine principle.
5. What is a comparator? State its advantages and disadvantages.
6. Give the complete classification of comparators.
7. Explain the following and state when they are used:
   (a) Interchangeable manufacture
   (b) Selective assembly
8. Define the following:
   (a) Least count
   (b) Tolerance
   (c) Unilateral tolerance
   (d) Bilateral tolerance
   (e) Allowance
   (f) Fits
   (g) Limits
9. Define accuracy and precision, and differentiate between them.
10. Briefly explain the different elements of a screw thread.
11. Define a screw thread.
12. Classify errors.
13. Define calibration and explain why instruments need calibration.
14. What are constant errors?
15. Define the effective diameter of a screw thread.
16. What do you mean by the best-size wire?
17. Briefly explain the need for correction factor in order to determine the effective diameter.
18. State the advantages and disadvantages of the two-wire method and explain how the disadvantages can be overcome.
19. What are the limitations of a bevel protractor?
20. Out of the two methods used to measure chordal thickness or chordal width, which one is more accurate and why?
21. Briefly explain why chordal width of a gear should be measured.
22. Define metal cutting.
23. Define speed, feed, and depth of cut.
24. List out the different properties of cutting forces.
25. What do you mean by orthographic and oblique cutting?
26. Differentiate between a cutting tool and machine tool with suitable examples.
27. Define stress, strain, and Young's modulus.
28. With the help of a block diagram, explain the relationship between the following:
    (a) Gauge pressure
    (b) Atmospheric pressure
    (c) Absolute pressure

(d) Vacuum pressure
29. Define gauge factor and its importance.
30. Explain Thompson–Pelton effect and Seebeck effect.
31. What is a thermocouple?
32. Explain the working principle of a strain gauge.
33. Explain the principle of dynamometry.
34. What is a proving ring? Explain how it is used to calibrate a load cell.
35. List the different materials used for thermocouples.
36. Explain the importance of bonding a strain gauge.
37. List the different strain gauge materials.
38. What do you mean by hysteresis loss? Explain the term.

Control Chart Factors

Sample size, n	$d_2 = R/\sigma$	$d_3 = \sigma_R/\sigma$
2	1.128	0.893
3	1.693	0.888
4	2.059	0.880
5	2.326	0.864
6	2.534	0.848
7	2.704	0.833
8	2.847	0.820
9	2.970	0.808
10	3.078	0.797
11	3.173	0.787
12	3.258	0.779
13	3.336	0.770
14	3.407	0.763
15	3.472	0.756
16	3.532	0.750
17	3.588	0.744
18	3.640	0.739
19	3.689	0.734
20	3.735	0.729
21	3.778	0.724
22	3.819	0.720
23	3.858	0.716
24	3.895	0.712
25	3.931	0.708

References

Amelinckx, S., D. Van Dyck, J. Van Landuyt, and G. Van Tendeloo, *Electron Microscopy—Principles and Fundamentals*, VCH (A Wiley company), 1997.

ASTME (American Society of Tool and Manufacturing Engineers), Handbook of Industrial Metrology, Eastern Economy Edition, Prentice Hall of India Private Limited, New Delhi, 1988.

Beckwith, T.G. and N. Lewis Buck, *Mechanical Measurements*, Oxford & IBH Publishing Co., New Delhi, 1969.

Bob B. He, *Two-Dimensional X-Ray Diffraction*, John Wiley & Sons, Hoboken, 2009.

Bozzola J. John and Lonnie D. Russell, *Electron Microscopy Principles and Techniques for Biologists*, 2nd edition, Jones and Bartlett Publishers, Boston, 1999.

Brue G., *Six Sigma for Managers*, Tata McGraw Hill Publishing Co. Ltd., New Delhi, 2005.

Busch, Ted, Roger Harlow, and Richard Thompson, *Fundamentals of Dimensional Metrology*, 3rd edition, Delmar Publishers Inc., New York, 1998.

Chapman, W. *Workshop Technology*, Part II, 4th edition, reprint 2003 e-book, Elsevier Butterworth-Heinemann.

Collet, C.V. and A.D. Hope, *Engineering Measurements*, The English Language Book Society, Pitman Publishing Limited, reprint 1980, New Delhi.

Crompton, T.R., *Characterisation of Polymers*, Vol. 2, Smithers, 2010.

Doebelin, E.O., *Measurement Systems Application and Design*, 5th edition, McGraw-Hill Publishing Company, New Delhi, 2004.

Eaton, Peter and Paul West, *Atomic Force Microscopy*, Oxford University Press, Oxford, 2010.

García, Ricardo, *Amplitude Modulation Atomic Force Microscopy*, Wiley-VCH e-book, Germany, 2010.

Gayler, J.F.W. and C.R. Shotbolt, *Metrology for Engineers*, 5th edition, English Language Weinheim Book Society/Cassell, Cassell Publishers Ltd, London, 1990.

Goldstein, J. I., Dale E. N., Patrick E., David C. J., Charles E. L., Eric L., Linda S. and Joseph R. M., *Scanning Electron Microscopy and X-ray Microanalysis*, 3rd edition, Kluwer Academic/Plenum Publishers, New York, 2003.

Gupta, I.C., *A Textbook of Engineering Metrology*, 7th edition, Dhanpat Rai and Co. Pvt Ltd, New Delhi, 2012.

Holman, J.P., *Experimental Methods for Engineers*, McGraw-Hill, 2007.

ISA Bulletin, 'ISO System of Limits and Fits—Part I', Published in ISA Bulletin 25(1940).

Jain, R.K., *Engineering Metrology*, Khanna Publishers, Delhi, 2009.

Lal, H., *Organisational Excellence Through Total Quality Management—A Practical Approach*, 1st edition, New Age International Publishers, New Delhi, 2008.

Krishnaswamy, K. and S. Vijayachitra, *Industrial Instrumentation*, 1st edition, New Age International Publishers, 2005.

Levin, R.I. and D.S. Rubin, *Statistics for Management*, 7th edition, PHI, New Delhi, 1997.

Lindford, A., *Flow Measurement*, 2nd edition, E & FN Spon, London, 1961.

Menon, Shashi, *Liquid Pipeline Hydraulics*, Marcel Dekker, New York, 2004.

Michler, G.H., *Electron Microscopy of Polymers*, Springer-Verlag Berlin Heidelberg, 2008.

NPTEL (National Programme on Technology Enhanced Learning), Lectures on Metrology, Version 2 EE IIT, Kharagpur.

Pennycook, S.J. and Peter D. Nellist (editors), *Scanning Transmission Electron Microscopy—Imaging and Analysis*, Springer, UK, 2011.

Rao, C.N.R., *Nanoworld—An Introduction to Nanoscience and Technology*, Published by the Jawaharlal Centre for Advanced Scientific Research, Bangalore, 2010.

Sawhney, A.K. and Puneet Sawney, *A Course in Mechanical Measurements and Instrumentation*, 9th edition, Dhanpat Rai and Co. Pvt Ltd, New Delhi, 1998.

Sirohi, R.S. and H.C. Radha Krishna, *Mechanical Measurements*, 3rd edition, New Age International Publishers, reprint 2010.

Venkateshan, S.P., 'E-courses—sub modules 2.8 and 4.4', Indian Institute of Technology, Madras, ecourses.vtu.ac.in/nptel/courses/IIT.pdf, last accessed on 2 July 2012.

Waseda. Y., Eiichiro M., and Kozo S, *X-Ray Diffraction Crystallography—Introduction*, Examples and Solved Problems, Springer, 2011.

Yao, N. and Zhong L.W. (editors), *Handbook Of Microscopy for Nanotechnology*, Kluwer Academic Publishers, Boston, 2005.

'Accurate Sensor Technologies, Measurement of Temperature in Extreme Condition', *Pyrometer Handbook*, 2010, www.accuratesensors.com/hb/finalhandbook.pdf, last accessed on 2 July 2013.

'Basic Concepts of Measurements', www.pioneer.netserv.chula.ac.th/~tarporn/2141375/HandOut/M Method.pdf, last accessed on 2 July 2012.

'Telemetry Tutorial-L-3 Communications', www.lcom.com/tw/tutorial/telemetry_tutorial.pdf, last accessed on 2 July 2012.

Index

About the Authors

N.V. Raghavendra is Professor and Head, Department of Mechanical Engineering, the National Institute of Engineering (NIE), Mysore. He has more than 25 years of teaching and research experience. A PhD holder from the Indian Institute of Science (IISc), his doctoral research area was on 'Acquisition of Technological Capability and Economic Performance in Clusters', done with an objective to understand the major factors that influence acquisition of technological capability (especially in smaller firms) and also recommend policy measures for their growth and sustenance. Dr Raghavendra was a member of the Implementation Committee of the prestigious VTU–Bosch Rexroth Centre of Automation Technology, Mysore. He has also served as the Special Officer of the VTU–Bosch Rexroth Centre, Mysore from September 2007 to December 2008 and Director of the NIE–Eicher Centre for Automobile Technology, NIE, from January 2011 to January 2013.

L. Krishnamurthy is Professor, Department of Mechanical Engineering, the National Institute of Engineering, Mysore. He has more than 25 years of teaching cum research experience. Prof. Krishnamurthy holds a doctoral degree from Kuvempu University, Karnataka. His research area was on 'Machinability Studies on Metal Matrix Hybrid Composites'. He co-investigates the research project titled 'Characterization of Composite Materials and Application of Nanomaterials for Sustainable Energy' sanctioned by Nano Mission, Department of Science and Technology, Government of India.

Related Titles

Machine Drawing
9780198070771

Basudeb Bhattacharyya,
Associate Professor, Department of Applied Mechanics and Aerospace Engineering, Bengal Engineering and Science University, Shibpur

Machine Drawing has been designed as a textbook for undergraduate students of mechanical engineering. While taking a holistic view of the subject, the book aims to help students learn the art of drafting precise and accurate drawings of machine parts.

Key Features
- Contains neatly drawn and well-labelled illustrations to ensure absolute visualization of the concepts taught
- Provides numerous solved examples in each chapter with step-by-step solutions
- Includes plenty of multiple-choice questions (with answers), review questions, and drawing exercises

Engineering Thermodynamics (Revised first edition)
9780198078876

Parthasarthi Chattopadhyay,
Senior Faculty, Department of Mechanical Engineering, Techno India College of Technology, Kolkata

This revised edition of *Engineering Thermodynamics* is designed as a textbook for undergraduate students of mechanical engineering. It provides in-depth coverage of the fundamental principles of thermodynamics. While providing mathematical representations, it lays emphasis on the physical aspects of the subject.

Key Features
- Contains more than 250 numerical problems and above 450 multiple-choice questions for practice
- Provides important notes and interesting facts throughout the text

Heat Transfer 2e
9780198079972

P.S. Ghoshdastidar,
Professor, Department of Mechanical Engineering, Indian Institute of Technology, Kanpur

The second edition of *Heat Transfer*, including a new chapter, is a comprehensive textbook aimed at undergraduate students of mechanical, chemical, metallurgical, and aerospace engineering. The book provides an exhaustive coverage of two- and three-dimensional heat conduction, forced and free convection, boiling and radiation heat transfer, heat exchangers, mass transfer, and computer methods in heat transfer. The main emphasis is on fundamental concepts and their applications in complex problems.

Key Features
- Explains key concepts with easy-to-understand mathematics
- Provides detailed solution methodologies

New to the Edition
- Introduces a new chapter on solidification and melting
- Includes new sections such as heat generation in a solid sphere, flow regimes in free convection over a vertical plate, and formation of vapour bubbles

Other Related Titles

9780195696554   Bhattacharya: *Engineering Mechanics*
9780195673913   Ghosal: *Robotics*
9780195687811   Appuu Kuttan: *Introduction to Mechatronics*
9780198062325   Uicker, Pennock, Shigley: *Theory of Machines and Mechanisms (SI Edition), International Version, 3e*
9780198086116   Shaw: *Metal Cutting Principles, 2e*
9780198062240   Vela Murali: *Engineering Mechanics*